IEE TELECOMMUNICATIONS SERIES 38

Series Editors: Professor Charles J. Hughes
Professor David Parsons
Professor Gerry White

SATELLITE COMMUNICATION SYSTEMS

3rd Edition

Other volumes in this series:

SATELLITE COMMUNICATION SYSTEMS

3rd Edition

Edited by

B. G. Evans

The Institution of Electrical Engineers

Published by: The Institution of Electrical Engineers, London,
United Kingdom

© 1999: The Institution of Electrical Engineers

The Institution of Electrical Engineers,
Michael Faraday House,
Six Hills Way, Stevenage,
Herts. SG1 2AY, United Kingdom

British Library Cataloguing in Publication Data

A CIP catalogue record for this book
is available from the British Library

ISBN 0 85296 899 X

Printed in England by Short Run Press Ltd., Exeter

Contents

Preface and acknowledgments

For the last 15 years we have hosted vacation schools organised by the IEE on satellite communications at the University of Surrey. Over this time, Surrey University has become synonomous with satellites, both from an education and training viewpoint, as well as a centre for research and latterly for the commercial production and sale of small satellites. The vacation courses have produced generations of trained engineers who are now the backbone of the satellite industry in the UK and abroad. Many ex-students have returned as lecturers (and are authors of chapters in this book) and some have now risen to become the captains of the satellite industry itself.

This is the third edition of the book which is based on the material presented at this *de-facto* industry-standard training course. It is my privilege to once again edit the book and oversee the updating process. It is different from the many other texts in the same field that have come and gone over the last ten years. The material has been designed to enable those with a basic engineering or mathematically based education to enter, and to specialise in, the field of satellite communications with the minimum of effort. The approach has been to concentrate on the design and planning of systems and thus the reader will not find a highly theoretical approach to the subject. Basic equations are quoted rather than derived, but their use in the planning and design procedures is explained in detail. The authors of the chapters have between them a wealth of experience as practitioners in the satellite field, and we aim to capture this and to pass it on to the next generation in a digestable manner.

We have also aimed at a broad approach to the subject and it is noticeable that the third edition is larger than its predecessors. We have added new material on the history and background to the subject and on the business aspects of satellite communications. I was concerned that there was overlap in some of these chapters, but I came to the conclusion that each author had a different and valuable perspective on the scene which was complementary and worth retaining. Many of the chapters have been updated to reflect the changes that have occurred in the ITU and its structure and recommendations in the last few years, as well as the general privatisation of satellite organisations. We acknowledge with gratitude the ITU in particular, and other sources of material used in the text. We have attempted to give credits in the text where due, but we apologise in advance to any that have been omitted.

The advance of the satellite business in areas such as mobile and personal communication systems, multimedia systems, military business and small satellites, navigation and positioning are all reflected by new chapters. I feel that we have produced a book that truly covers the subject better than any other in the market and I hope that you will agree.

As general editor of the book, I would like to acknowledge the contributions of all of the individual authors, lecturers and excellent administrators from the IEE at the successive vacation schools. Almost 700 students at the schools themselves have made valuable contributions and suggestions and I would especially like to thank them. I would however single out Tim Tozer, Paul Thompson, Barry Claydon and Dave O'Connor for their contribution on the organising committee and in particular Tim for his contributions as tutor at the many schools.

Many thanks also to Jonathan Simpson and Fiona MacDonald at the IEE Publications Department for their help and pressure, without which this book would have not appeared.

B.G. Evans
Guildford, October 1998

Contributors

Dr D Betram
Satellite Communications Centre
Defence Research Agency
DRA Defford
Worcester
WR8 9DU

Mr N Cartwright
Tithe House
Stutton Lane
Tattingstone
IP9 2NZ

**Mr I E Casewell and
Mr A Bachelor**
Racal Research Ltd
Worton Drive
Worton Grange Industrial Estate
Reading
Berkshire
RG2 0SB

Dr B Claydon
ERA Technology
Cleeve Road
Leatherhead
Surrey
KT22 7SA

Mr G Drury
NDS Limited
Gamma House
Enterprise Road
Chilworth
Hants
SO16 7NS

Professor B G Evans
Centre for Communications Systems
    Research
University of Surrey
Guildford
Surrey
GU2 5XH

Mr L Ghedia
ICO Global Communications Inc
1 Caroline Street
London
W6 9BN

Mr R Heron
Delta Communication Ltd
Business and Innovation Centre
Angel Way, Listerhills
Bradford
West Yorkshire
BD7 1BX

Mr T G Jeans
Centre for Communications Systems
    Research
University of Surrey
Guildford
Surrey
GU2 5XH

Mr M A Kent
BT
PP1.01 Broadway House
3 High Street
Bromley
BR1 1LF

Mr J Miller
Cable and Wireless plc
Gretton House
28–30 Kirby Street
London
EC1N 8RN

Mr D R O'Connor
Matra Marconi Space UK Ltd.
Gunnels Wood Road
Stevenage
Herts.
SG1 2AS

Mr J J Pocha and Mr P Harris
Matra Marconi Space UK Ltd
COB Post Code 205
Gunnels Wood Road
Stevenage
SG1 2AS

Mr P Shellswell
BBC Research and Development
    Department
Kingswood Warren
Tadworth
Surrey
KT20 6NP

Dr P Sweeney
Department of Electronic and
    Electrical
Engineering
University of Surrey
Guildford
Surrey
GU2 5XH

Professor M N Sweeting
Surrey Satellite Technology Ltd
University of Surrey
Guildford
Surrey
GU2 5XH

**Drs P T and J D Thompson and
    Mr A G Reed**
ERA Technology Ltd
Cleeve Road
Leatherhead
Surrey
KT22 7SA

Mr T C Tozer
Department of Electronics
University of York
Heslington
York
YO10 5DD

Chapter 1

Introduction

B.G. Evans

This initial Chapter provides an overview of the components of a satellite system and the major parameters for consideration in its design. It also attempts to give a brief review of the current status and position of satellite communications.

1.1 Satellite systems

Although we shall deal with the communication aspects, satellite systems are in fact used for many different services as defined by the ITU and given in Chapter 4. Those specifically addressed in this book are:

- fixed satellite service (FSS);
- broadcast satellite service (BSS);
- mobile satellite service (MSS);

although communications clearly remains a major part of other satellite services as well.

FSS includes all of the current radiocommunication services operated via the major operators such as INTELSAT, EUTELSAT, PANAMSAT etc. (see Chapter 2), and operates essentially to fixed earth stations. BSS covers the area of direct broadcasting satellites (DBS), which are addressed in Chapter 17. This consists of much smaller earth stations on domestic premises together with fixed earth stations providing the uplink feeder to the satellite. MSS currently operates in the maritime mobile service (MMS), aeronautical mobile service (AMS), land mobile service (LMS) via INMARSAT, plus a number of regional operators e.g. AMSC, OPTUS etc. These services consist of earth terminals located on the mobiles as well as fixed base stations for connection back into major terrestrial networks. A number of global mobile satellite personal communication systems (GMPCS), e.g. Iridium, Globalstar and ICO, will start to operate in the 1998–2000 era (see Chapters 17, 18).

With these systems in mind we shall look at the design of networks consisting of satellites and earth stations together with their connection to users, which may involve the use of existing terrestrial-network tails. It is important to realise that the design of such networks is based upon the provision of a service, be this voice, data, facsimile, video etc, and the quality of the service as defined by the ITU-R and ITU-T (see Chapter 4) is the major requirement to be met by the design. A point-to-point satellite link may be all that the network consists of (e.g. a private business link), but the satellite link could be just one part of a major network consisting of many other links. In the latter case the satellite portion cannot be considered in isolation from the rest of the network. Hence, the design of satellite systems is complex and involves many variables which all need to be traded off in order to reach an optimum economic engineering design which meets the service requirements. The quality of service (QoS) and its availability are the key design aims for users in the coverage area of the satellite.

In this book we shall explore the interplay between these variables and the engineering, organisational and management constraints which dictate their choice.

1.2 Radio regulations and frequency bands

As satellite systems employ the transmission and reception of radio waves we have a potential interference situation where users in similar frequency bands could interfere with each other, to the common degradation of their system quality. Hence the international regulation of their transmission is crucial to satisfactory performance. The mechanisms for achieving this are discussed in Chapter 4; it is sufficient to note here that there are international agreements (via the ITU) for spectrum allocation for different services (IFRB), e.g. Figure 1.1 shows the allocations for satellite services. Some of the bands are shared with terrestrial systems and nearly all are shared among different satellite operators and hence coordination is necessary to avoid excessive interference. The ITU provides radio regulations which outline in detail the methods to be employed in order to avoid this excessive interference between users (see Chapter 4).

These are broken down into:

(i)   Satellite internetwork coordination.
(ii)   Earth station coordination with fixed terrestrial links.
(iii) Interference between orbits.

Such procedures need to be completed prior to authorisation for transmission being given.

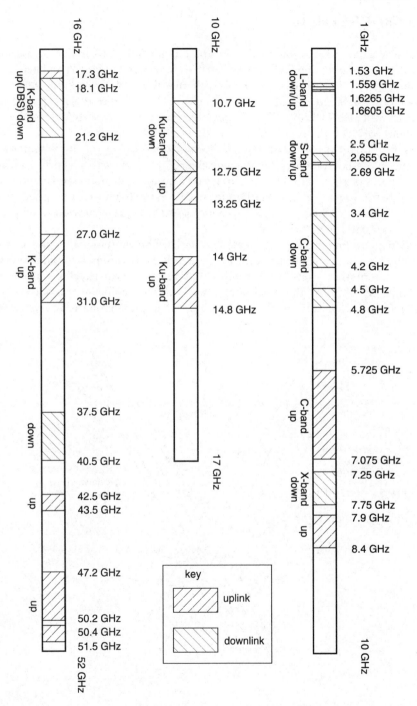

Figure 1.1   Frequency allocations for satellite communication services

1.3 Satellite orbits

There is a large range of satellite orbits, but not all of them are of use for satellite communications. Those most used are shown in Figure 1.2, and here the 24-hour geostationary (circular) orbit with an altitude of 35 786 km is the most commonly used for fixed communications. All of the major operators, INTELSAT, EUTELSAT, INMARSAT etc., have used this orbit, as, having a 24-hour period, it imposes minimal tracking constraints on the earth stations. It is, in fact, only the perturbations of the orbit caused by the gravitational forces of the stars and planets and the nonsphericity of the earth which require tracking, and then only for the larger earth-station antennas (see Chapters 5 and 12). However, for earth stations located at extreme latitudes, the elevation angles become very small and this causes propagation problems associated with the longer paths in the troposphere.

In the extreme, at the polar caps, the geostationary (circular) orbit is not visible. Thus, for systems that require coverage of these regions, we need to investigate alternative orbits. It transpires that highly eccentric, elliptical orbits inclined by 63.4° to the equatorial plane exhibit significant apsidal dwells around their apogee. Thus the satellites appear to remain quasistationary and

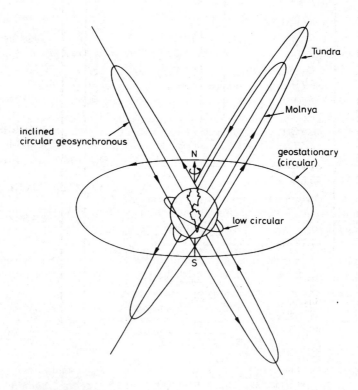

Figure 1.2   Satellite orbits

are useable for periods of eight to 12 hours for such coverage regions. A full 24-hour service can be maintained by two or three such satellites with suitable phased orbits and hand-over facilities between them. Owing to the motion of the satellite around the apogee with respect to an observer on earth a sizeable Doppler shift is associated with these orbits and the radio receivers must be designed to cope with this. Two such orbits are shown in Figure 1.2, the first being the Molnya orbit, first used by the Soviet Union in the 1960s for television transmission to its remote areas. This has an apogee of around 40 000 km and a perigee of around 1000 km. Similar characteristics are exhibited by the second, the Tundra orbit, which has an apogee of 46 300 km and a perigee of 25 250 km. Both orbits are candidates for mobile (in particular land mobile) satellite systems and for complementing the geostationary orbit in achieving better worldwide coverage. As an example of the different views of the earth from these orbits we compare in Figures 1.3 and 1.4 a view of the earth from a Molnya and geostationary satellite placed at 3.5°W. This clearly demonstrates the improved coverage and elevation for the Molnya system. The advantage in terms of reduced link margin for the Molnya can be used to offset other parameters in the satellite link design (see Chapter 11). Such highly elliptic orbits (HEO) are now being considered for digital audio broadcasting (DAB) services (see Chapter 17) as well as mid-earth orbit (MEO) applications of HEOs for mobile applications (see Chapter 20).

Finally, we should mention the low-earth circular orbits (at altitudes of 500–1500 km) shown in Figure 1.2. These have in the past been used for earth resources, data-relay and navigation satellites as well as low-cost store-and-forward communications systems (see Chapter 24).

In recent years the use of LEOs at around 1000 km altitudes has increased owing to their ability to provide global coverage for mobile and personal communication users. Two examples are the Iridium system of 66 (six planes of 11 satellites each) satellites and Globalstar system of 48 (eight planes with six satellites each) satellites. The coverage from these systems is shown in Figure 1.5. The lower altitude improves the power budget especially for omnidirectional handheld terminals where transmitted power is severely constrained. The advantages are obtained at the expense of considerable complexity in handover between the multiple spot beams (16 for Globalstar and 48 for Iridium) and to the gateway earth stations (GES) linking with the terrestrial network which can have up to five separate tracking antennas and terrestrial interconnection networks. (*Note:* Iridium uses intersatellite links (ISLs) to avoid terrestrial GES interconnections.)

The LEO orbits are not the only ones to allow good global coverage, and the GPS series of navigation satellites has used inclined circular orbits at 24 000 km altitude for a number of years (see Chapter 21). Between LEO and GEO, satellites are said to be in mid-earth orbit and ICO, an offshoot of INMARSAT, plans to operate ten (five satellites in two planes) satellites in inclined circular orbit (hence ICO) in MEO at 10 350 km altitude to provide mobile communications to handhelds. The coverage of ICO is shown in Figure 1.6 and it can be

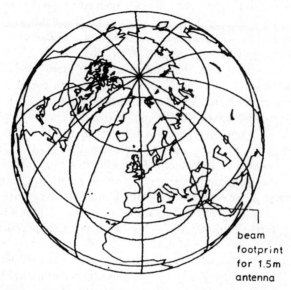

beam
footprint
for 1.5m
antenna

Figure 1.3   *View of the earth from the apogee of a 12-hour Molnya orbit centred at*
            3.5 °W

Figure 1.4   *View of the earth from a geostationary position at 3.5 °W with a similar size*
            footprint to that of Figure 1.3, centred on the UK

seen that a much larger number of spot beams (163 for ICO) is needed to
achieve the same power budgets for the higher altitudes (see Chapter 20).
However, the number of GESs needed is much lower, at around 12, than for
LEO constellations. In all of the above orbits the choice of the number of satel-

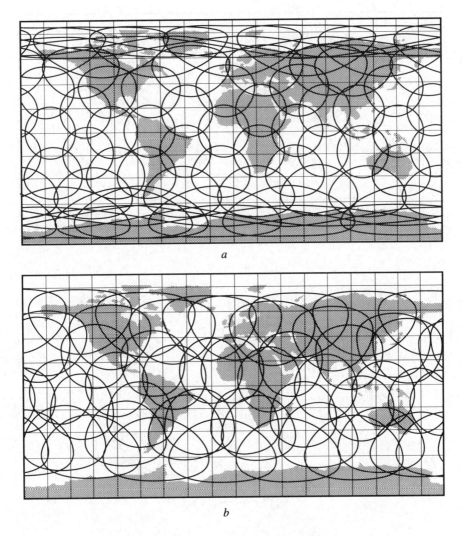

a

b

Figure 1.5   Coverage of (a) Iridium and (b) Globalstar LEO constellations

lites, planes and phasing within the planes needs to be performed for optimum coverage of the service area and efficient system design. LEO and MEO constellations have also been proposed for the next generation of multimedia service satellites at Ka- and V-bands via such systems as Skybridge, EuroSkyWay, West, Astrolink, Spaceway and Teledesic etc (see Chapter 25).

One of the major problems with the extensive use of the geostationary orbit has been the congestion (see Figure 1.7) which is caused by operators seeking to use the more preferential parking positions for both international and domestic use (particularly over the Americas and the oceans). Clearly, satellites must be

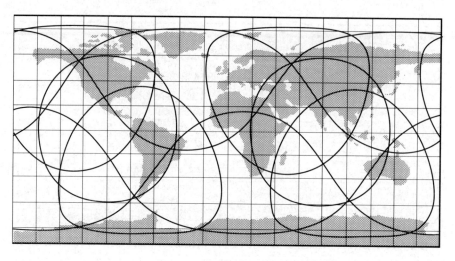

Figure 1.6   Coverage of ICO, MEO constellations

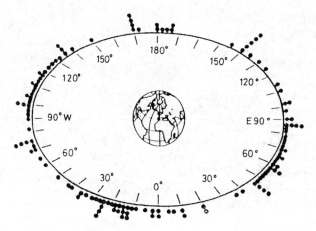

Figure 1.7   Congestion in the geostationary orbit

restricted in their proximity in the orbit (currently 2°), and this has implications for internetwork coordination and on antenna performance (see Chapter 12). The philosophy adopted in allocating orbit positions on an equitable basis is discussed in Chapter 4. Such allocations must maximise the efficiency of the use of the orbit and frequency bands as both are valuable resources.

One major effect on communications in using the geostationary orbit is the coverage, which has already been mentioned, but note from Figure 1.8 that a global coverage beam is produced by an antenna on the spacecraft with 17.4° beam width, and that this precludes coverage of polar regions.

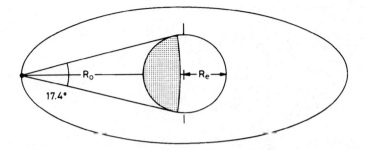

Figure 1.8   Geostationary earth coverage

R_e = earth radius = 6378 km

R_o = satellite altitude = 35 786 km

The other effect of importance is the propagation time delay on an earth-satellite-earth link which has limits given by:

$$\text{min} = 2R_o/c = 238 \text{ ms}$$

$$\text{max} = \frac{2(R_o + R_e)\cos\left(\dfrac{17.4°}{2}\right)}{c} = 284 \text{ ms}$$

Such delays have severe implications for the use of some services such as voice circuits which are required, therefore, to use echo cancellors to produce good quality on a single hop. However, the GEO orbit is perfectly acceptable for television and some interactive computer/data services. Double-hop working (linking a domestic to an international satellite) is not normally used for speech owing to the unacceptable degradations which result.

It is important to note that the echoes are generated within the terrestrial network on transition from two- to four-wire PSTN working. Full four-wire working will therefore not generate such echoes. Even if echoes are generated they can be controlled successfully by the use of echo cancellors, even in GEO systems, but this does imply careful implementation on the part of the terrestrial operators.

Owing to the dynamic orbital perturbations of the satellite, the delays will vary in time and this is an important issue for consideration in the synchronisation and control of time-division multiple-access (TDMA) systems (see Chapter 8).

For more details of orbit dynamics the reader is referred to Chapter 13.

1.4 The basic satellite system

The basic satellite system consists of a space segment and a ground segment, as shown in Figure 1.9. The space segment consists of the satellites plus control, or

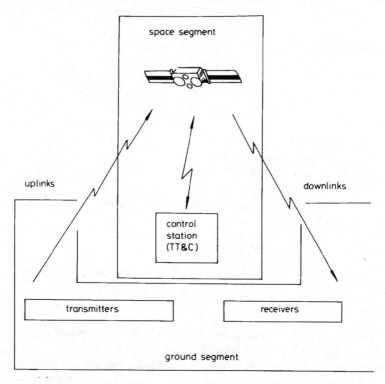

Figure 1.9   Satellite-communication-system architecture

telemetry, tracking and command (TT&C) stations, to maintain the satellites in orbit. For a GEO operational system to be considered secure the operational satellite is backed up by an in-orbit spare satellite as well as, in some cases, a ground spare, ready for launch in case of malfunction of either of the orbiting satellites. The TT&C station is necessary to keep the satellites operating in space. It provides constant checking of the satellite subsystems' status, monitors outputs, provides ranging data, acts as a testing facility and updates the satellite configuration via the telemetry links. It generally performs all the housekeeping routines needed to maintain the satellites as operational repeaters. The TT&C station is usually duplicated for security reasons. For satellite constellations (LEO or MEO) the control of a satellite in orbit transfers from GES to GES as the satellites precess in their orbits around the earth. Thus, the TT&C control is a distributed function rather than a centralised one as in the GEO case.

The satellites themselves consist of two major components:

(i)   Communications payload.
(ii)   Spacecraft bus.

The communications payload (see Chapter 14) consists of the satellite antennas plus the repeater itself. The latter provides for low-noise reception via an RF front end, frequency conversion between the up- and down-link frequencies and a final power amplifier to boost the signal prior to transmission on the downlink. Two different types of payload are shown in Figure 1.10. All of the existing payloads are of the transparent type shown in Figure 1.10a and consist of only RF amplification and frequency conversion. Some future payloads will be regenerative or processing in nature and demodulate the signals to baseband, regenerate digitally and remodulate (and recode) for downward transmission. This is an important innovation as it will enable the up- and downlink designs to be separated and much more efficient systems should thus result. There are, of course, additional problems of reliability and radiation-hardened baseband equipment to be considered (see Chapter 12). The bandwidth handled by the satellite is usually broken down (demultiplexed) into traffic-manageable segments (40–80 MHz for FSS and 5–10 MHz for MSS) each of which is handled by separate repeaters (called transponders), which are connected by a switching matrix to the various onboard antennas.

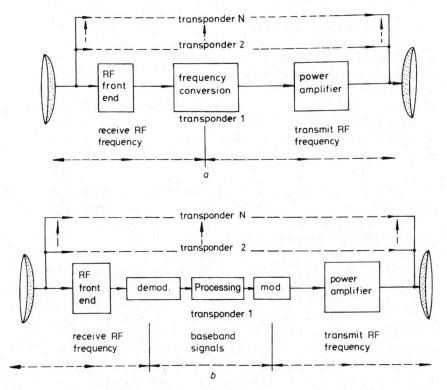

Figure 1.10   Communications payloads

    a Conventional transparent (nonregenerative) satellite
    b Processing (regenerative) satellite

The spacecraft bus contains the housekeeping systems to support the payload and consists of:

- spacecraft structure;
- electrical-power subsystem;
- propulsion subsystem;
- altitude-control subsystem;
- thermal-control subsystem;
- TT&C subsystem.

For a particular satellite service and choice of orbit each of the above can be designed to support the payload (or payloads) and this process is discussed in Chapter 13. The mass, size and volume constraints are also very much determined by the available launchers.

The ground segment of the satellite system consists of all of the communicating earth stations which access the operational satellite. As shown in Figure 1.11, these earth stations consist of:

- antenna (plus tracking subsystem);
- feed system (polarisers, duplexers, orthomode junctions etc.);
- high-power amplifiers (HPAs);
- low-noise amplifiers (LNAs);
- up convertors/down convertors (between microwave to IF);
- ground communications equipment (GCE) (modems, coders, multiplex etc.);
- control and monitoring equipment (CME);
- power supplies.

The larger stations involved in the INTELSAT global network have full provision of these subsystems, but the smaller business and mobile stations are of much smaller scale and have much reduced provision. The latter point is discussed in detail in Chapters 12 and 22.

A network may consist of a few to hundreds of earth stations, and these all have to access the satellite in an equitable manner. This is usually accomplished by either frequency-division multiple access (FDMA), time-division multiple access (TDMA), code-division multiple access (CDMA) or a random-access scheme (RA). Sometimes a hybrid combination of these is used. All of these schemes are discussed in Chapter 8 as well as the optimum choice of access for a particular service provision and network.

Finally, the satellite system (consisting of the earth station to satellite to earth station link) must be interfaced to the user, either directly or via a network e.g. the PSTN, ISDN or PLMN. Standards in the interconnection of earth stations to users are an important feature of the design and are discussed in Chapter 16.

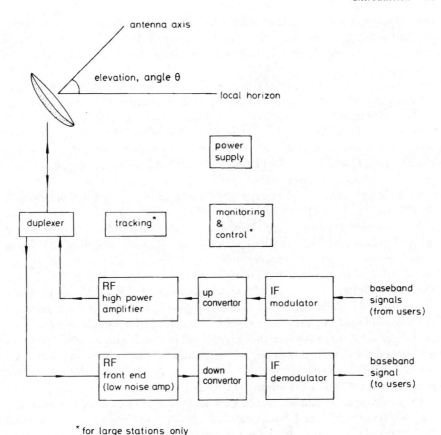

Figure 1.11   Earth-station architecture

1.5 Satellite communications in transition

From the early days of satellite communication systems (1964 onwards for FSS and INTELSAT and 1976 onwards for MSS and INMARSAT) up until the late 1980s/early 1990s, the major operators were intergovernmental organisations (IGOs) with specific missions. The INTELSAT treaty was formed as an offshot of the UN treaty and had the assistance of developing countries as an aim. INMARSAT was a similar IGO with specific responsibility for safety at sea and search and rescue operations. The only other satellite organisations in these first twenty years or so were either regional systems e.g. EUTELSAT for Europe or ARABSAT, set up to provide educational, cultural and television links between the countries of the Arab league, which were also IGOs, or standalone domestic systems. Examples of the latter include Indonesia and its PALAPA series of satellites, ANIK satellites in Canada or the OPTUS system in Australia. Such systems were national in their structure and organisation and government controlled or influenced. Much of the technological developments e.g. single channel per carrier (SCPC) and demand assignment (DA), TDMA, shaped and multibeam antennas, frequency reuse using geographic and orthogonal polarisation separations, digital signal processing such as DSI, speech and video compression etc. were results of R&D supported by these IGOs. Such R&D was centrally funded from the member country contributions. The members subscribed according to their usage of the systems and became shareholders on such a *pro rata* basis. Thus, the development and finance of the space system was centrally planned as well as the resulting standards of the earth stations to use the systems. Hence, we saw INTELSAT and INMARSAT standard A, B, C etc. earth terminals (see Chapter 15).

Communications over the major oceans had been a major feature of both INTELSAT and INMARSAT, which initially concentrated entirely on the maritime sector. It was competition in this area that initiated a change in the old order. This occurred in the 1980s with the introduction of digital fibre-optic cables across the Atlantic with capacities much greater than those of the satellites which had hitherto ruled supreme at these rates. Although satellites fought back using digital-circuit multiplication equipment (DCME) the cables would always provide greater capacity and hence lower circuit costs on a point-to-point link. Thus the balance of traffic was transferred from satellite to cable with the former being an important back up and still remaining on some world routes the cheaper option. Diversification in telecommunications was an established principle and thus satellites would never be completely replaced, but on these point-to-point routes cables now had the edge.

In the FSS band, satellites could now look to exploit their real advantages e.g. their broadcast nature. Small stations (VSATs) were beginning to appear on the scene and satellites could be designed with particular spot-beam coverage to allow interconnection of large networks (100s of terminals) of such smaller stations. In addition, television transmission either direct to home small dishes or to cable head collectors was becoming big business and dominating systems such as EUTELSAT. INTELSAT could not respond to these new markets, largely in the developed world, because of its mission, highly bureaucratic control and committee structure and because its satellites had been designed for different missions and were thus expensive for the new markets. Inevitably this promoted an opportunity for competition, and private systems such as PANAMSAT and ORION appeared to address the new FSS markets.

In the MSS arena similar moves were afoot with new regional systems such as AMSC in America and OPTUS in Australia challenging INMARSAT's dominance. Also emerging were new private organisations to tackle the global handheld telephone market which, although successful in urban areas via cellular, was still dogged with poor coverage outside urban areas and with a plethora of standards which mitigated against true global roaming. Much of the rural world still did not have access to a telephone and this represented an enormous market. Thus major system proposals (Iridium, Globalstar) using constellations of satellites and costing $2–10 billion appeared. INMARSAT recognised the challenge but could not compete as an IGO and thus spun off a private subsidiary called ICO Global Series in which it had a minority shareholding. In 1998 both INTELSAT and INMARSAT announced plans to make the transition from IGOs to private operating companies by the year 2000. Hence, the transition for satellite communications had taken place, and this merely mirrored what was happening in the broader field of telecommunications—liberalisation and deregulation—the onset of a new era of competition and private companies and organisations. We must not forget, however, the important role of the IGOs, without which satellite communication would not be in the position in which it is today.

Perhaps we thus stand at the crossroads in the development of satellite communications which is represented by the current small-dish/mobile era. This era is characterised by new markets of:

(i)   Direct digital broadcasting of television.
(ii) Small-dish business systems.
(iii) Mobile, portable multimedia systems.

The first of these, DBS, has been a long time coming, but 1989 saw the first operational analogue system and 1998 the first digital systems. It is still a speculative market with the major technological innovations being higher power satellites and very small, low-cost integrated domestic earth stations. The real challenge lies in the other two markets, and especially the mobile systems for which the broadcast nature of satellites is ideal.

Business (VSAT) systems have been very successful in the USA and some of the developing world (China and East Europe) but less so in Europe where regulating regimes have persisted and space segment has remained expensive. The maritime mobile and latterly land mobile areas have been very successful, with the aeronautical area less so. Again, expensive space segment as a result of lower power, wider beam and nonprocessing satellites has been the reason. The key to opening these new markets lies in the development of:

● onboard processing;
● multibeam coverage deployable antennas;

and when developed, we will be in the next era of satellite communications, the intelligent era.

Thus, our market-driven crossroads also leads us to a technology crossroads, from dumb to intelligent satellites. This involves the placement of processors on board, regenerative transponders and the use of a baseband switch to interconnect between the beams of a multibeam coverage antenna as shown in Figure 1.12. Besides the functions of baseband signal processing (regeneration) and traffic or message switching, the presence of onboard processing power opens up the possibility of overall systems resource control. Thus, the satellite is designed to meet the needs of the user rather than the user fitting in with the satellite.

Users with a whole range of traffic requirements and capacities can then intercommunicate efficiently via the satellite, which sorts out the relative bit rates,

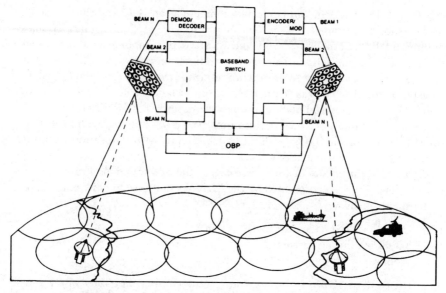

Figure 1.12   Onboard processing (intelligent) satellite using multiple-beam antennas and onboard switching

switches, reformats the information and assembles it into suitable formats for transmission and reception by simple and cheap earth stations. Such a satellite, as a switching node in the sky, is likely to alter the currently accepted hierarchical communication-network structures and allow direct access to a range of levels simultaneously.

The intelligent era is not likely to arrive until around the year 2003 when we should also be well on our way to personal multimedia communication systems, with satellites playing an important role in a synergy with terrestrial systems to produce a truly integrated network infrastructure.

1.6 Towards the future

As we enter the second millennium satellite communications has already established direct broadcasting of television, small-dish VSAT systems, maritime, aeronautical and land mobile systems and is about to embark on global hand-held satellite telephone systems. So what of the future?

Surprisingly technology has not progressed as rapidly as the organisational changes which have beset the field. Satellites have become more powerful, have larger numbers of beams and have some digital signal processing on board but have not yet (1998) reached the full onboard processing (OBP) stages as illustrated in Figure 1.12. The Iridium satellites are arguably the most advanced with some OBP and with ISLs. These will be the model for future satellites, both

FSS and MSS, and this represents a major change in the satellite payload, enabling more complex routing onboard right down to the level of service switching — the switchboard in the sky.

In the FSS area there are plans for satellite systems designed for multimedia Ka-band communications using ATM switching onboard the satellite. Systems will use LEO, MEO and GEO orbits and some will use ISLs and IOLs in order to optimise the service delivery. ATM systems will be designed for interactive multimedia at BERs of 10^{-10} and availabilities in excess of 95%. This will be a challenge both to the satellite payload designers as well as the communication engineers. As the Internet becomes more pervasive, Internet protocol (IP) over satellite will feature more and a complete multicasting IP satellite network (Teledesic's constellation of 288 satellites in LEO is based upon such a concept) will become a reality by 2005.

The digital-processing revolution has seen compression of broadcast video down to around 2–6 Mbit/s and hence a tremendous saving in satellite bandwidth from the old analogue television of 27 MHz to the new digital of around 4–6 MHz. This heralds the era of 100s of digital television channels from the sky. Standards such as MPEG 2 and 4 for the source coding and DVB-S for the transmission have played important roles in the new digital satellite revolution. Once in digital format the TV satellites can provide for interactive multimedia to the home and open up a vast range of new services. Digital audio broadcasting is another area in which satellites will benefit and we see in 1998 the first DAB systems (Worldspace) coming into existence to deliver high quality radio channels globally. Again, once in the digital domain such systems can also be used for interactive multimedia services.

In the mobile/personal satellite area the first-generation GMPCS systems using constellations for global coverage and super GEOs for regional coverage are limited to voice and low-rate data. The second-generation systems in 2003–2006 will be part of the IMT-2000 standard set and provide multimedia services in the range 64 kbits–2 Mbits/s to mobiles and portables. Beyond this it is likely that mobile, portable and fixed will merge into giant hybrid constellations providing a full multimedia service — set across the old boundaries in a seamless manner and by 2010 we should have reached the fully integrated systems that have so long been the dream — or will there be unseen disruptions?

Chapter 2

Historical overview of satellite communications

P.T. Thompson and J.D. Thompson

2.1 The visionaries

The concept of satellite communications is normally accredited to an Englishman, Arthur C. Clarke, because of a famous paper[1] published in the British *Wireless World*. However, Dr Clarke produced at least two documents prior to this in which elements of the idea of satellite communications were presented. He published a letter[2] in the 'Letters to the Editor' column of *Wireless World* on 'Peacetime uses for V2' in February 1945 in which he postulates an 'artificial satellite' in a 24-hour orbit and even goes on to suggest the use of three such satellites at 120 degree spacing. Modestly, he finishes the article with 'I'm afraid this isn't going to be of the slightest use to our post war planners, but I think is the *ultimate* solution to the problem.'

Following that short letter he wrote a more extensive paper on 25 May 1945 entitled 'The space station: its radio applications' which he circulated to several key council members of the British Interplanetary Society (whose motto is aptly *From imagination to reality*) who gave it their whole-hearted support. The top copy is now in the archives of the Smithsonian Institute in Washington DC and a facsimile is published in 'How the world was one'[3] along with a copy of his later and now famous *Wireless World* paper of October 1945. In this paper Arthur C. Clarke proposed that three communications stations be placed in synchronous 24-hour orbit which could form a global communications system and make worldwide communications possible. An interesting but seldom recognised characteristic of his idea is that he envisioned the use of staffed space stations in synchronous orbit. The communications equipment would be installed, operated and maintained by the space-station crew. To date, we have seen hundreds of unstaffed communications satellites launched into geosynchronous orbit for commercial and military use, with the newest spacecraft lasting ten or more years without maintenance, but as yet permanent staffed space stations are rare and none are in geostationary orbit.

Arthur C. Clarke was not the only person thinking about communications satellites. In 1946, the US Army's *Project Rand* pointed out in a classified study the potential commercial use of synchronous communications- satellites. Unfortunately, this report remained secret for so long that it had little impact.

In 1954, John Pierce[4] of Bell Laboratories considered the communications satellite problem independently of Clarke. To Pierce there seemed little reason at that time to replace overland cables or terrestrial microwave relays with satellites. Satellites, the electronic equivalent of high microwave relay towers, seemed best suited for spanning the oceans, which so far could only be done with expensive undersea cables of limited capacity or via high-frequency waves bounced off the ionosphere. There were two general possibilities for such satellites: passive reflectors that would bounce the radio waves between ground antennas, and active repeaters which would amplify the received signal and retransmit it to the ground. Either kind of satellite could be placed in medium-altitude orbits, requiring a constellation of many satellites and steerable antennas on the ground, or one satellite in synchronous orbit, where it would appear to remain stationary at one location.

At the time it seemed to Pierce that nothing practical could be done to facilitate satellite communications, although the communications equipment was not the problem. The invention of the transistor, the solar cell and the travelling-wave tube amplifier in the 1940s and 1950s allowed relatively compact highly reliable repeaters to be built. The difficulty was the rockets. It was not until the development of suitable launch vehicles that these concepts could be realised. Launch vehicles of the power required became available in the mid 1960s as a by-product of the military development of the intercontinental ballistic missile.

Several private companies in the United States, including RCA and Lockheed in the early 1950s, investigated the possibility of communications satellites before the government became interested. The Hughes Aircraft Company spent company money from 1959 to 1961 to demonstrate the feasibility of a design for synchronous satellites before convincing NASA and the Defence Department to fund the rest of the project. However, by far the greatest activity was in AT&T's Bell Telephone Laboratories.

In the UK two major studies were undertaken, both of which showed significant innovation and application of advanced satellite technologies. The first was 'A study of satellite systems for civil communications', RAE, Farnborough, report 26 (March 1961). The second was undertaken by G.K.C. Pardoe of the British Satellite Development Company entitled 'World communications satellite system' and published at the 13th congress of the International Astronautical Federation (IAF) in September 1962.[5] However, the pace of the evolution and developments in this field, the cautiousness of the UK government and the problems which beset the UK/European launch vehicles resulted in a UK preference for international cooperation with the USA.

2.1.1 The start of space activities

The V-2 was the catalyst for the artificial unmanned satellite. In the days of early rocketry researchers concentrated on manned flight. There was also scepticism regarding radio waves penetrating the upper atmosphere. More important, in those years of radio valves, there simply was no adequate electronic technology. For these reasons, radiocommunications studies are conspicuously absent from the early astronautical literature. Beginning in 1945, given the large-scale rocket technology, wartime advances in electronics and the development of transistors, proposals for satellites started to be published.[6,7] In 1955 Radio Moscow had announced a prospective satellite launch, but this and subsequent similar announcements were not considered seriously. The USA had long taken its technological superiority for granted; besides, satellite and space-station proposals were at that time quite common, and most of them were considered to be pie-in-the-sky fantasies. However, on 29 July 1955 the USA announced that it would 'launch small earth-circling satellites' as part of its contribution to the International Geophysical Year (Project Vanguard).[8]

On 4 October 1957, the entire world received a shock: the Soviets launched SPUTNIK 1.[9] The Americans suddenly lost their complacency over their presumed technological superiority. Another shock for them came when the first complete Vanguard exploded in a ball of flame less than a second after lift off, dropping its four pound test satellite. It was not until the 31 January 1958 that America launched its first satellite, the Explorer 1, but the USSR countered this by launching a 7000 pound 'flying laboratory', and thus the space race had begun.

It was to be the space race that accelerated the development of the technology and the will to launch satellites into space. However, it was only when the race settled down that worthy application satellites began to be developed and launched, riding on the spin-offs from the race.

2.2 The pioneers

Arthur C. Clarke's book 'How the world was one' was 'Dedicated to the real fathers of the communications satellite, John Pierce and Harold Rosen, by the godfather'. It is reasonable to assume that Arthur was the visionary and others, especially these two, were pioneers. Both had the backing of significant commercial organisations behind them (AT&T and Hughes Aircraft Company, respectively) and therefore had the resources to push the pioneering efforts required at that time.

2.3 The early days

The early days of satellite communications comprised many experiments and fact-finding missions which eventually led to the active transponder-based satellites in common use today. The background of these early days is too extensive to detail here. Table 2.1 gives some basic information but the reader is referred to the material in the references at the end of this chapter to gain further insight.

2.4 International activities

2.4.1 History and development of INTELSAT

In 1962 the United States, Canada and the United Kingdom held discussions on forming an international satellite organisation based upon the concept established in the US Congress Communications Satellite Act of 1962.[10] Later discussions, in 1963, were expanded to include most of the European countries. Thus, the UK played a key role in initiating this organisation. Furthermore, along with other European countries, the UK made it clear that its interests were not only in using the satellite system, but also in having ownership rights which carried with them an active participation in all aspects of the system.

Serious negotiations were begun in Rome in February 1964 at which participants included representatives from western Europe, the USA and Canada. Shortly thereafter, Australia and Japan were included, recognising that together

Table 2.1   Early communications satellites

Satellite	Country	Launch date	Notes
SCORE	USA	18/12/1958	active for 13 days
ECHO A–10	USA	13/5/1960	failure
ECHO 1	USA	12/8/1960	100 ft balloon
COURIER 1A	USA	18/8/1960	launch failure
COURIER 1B	USA	4/10/1960	lost command capability after 17 days
ECHO (AVT–1)	USA	14/1/1962	failure, balloon ruptured
TELSTAR 1	USA	10/7/1962	ok until 21/2/1963
ECHO (AVT–2)	USA	18/7/1962	
RELAY 1	USA	12/12/1962	1 transponder, ok until February 1965
SYNCOM 1	USA	14/2/1963	communications lost at orbit injection
TELSTAR 2	USA	7/5/1963	ok until May 1965
SYNCOM 2	USA	26/7/1963	used until 1966, turned off April 1969
RELAY 2	USA	21/1/1964	used until September 1965
ECHO 2	USA	25/1/1964	used with USSR, decayed June 1969
MOLNIYA 1–F1	USSR	19/2/1964	failure
SYNCOM 3	USA	19/8/1964	used until 1966, turned off April 1969
INTELSAT 1 (Early Bird)	USA	6/4/1965	used until Jan 1969, temp use Jun–Aug 1969
MOLNIYA 1–1	USSR	23/4/1965	decayed 16/8/1979

these countries accounted for some 85 per cent of the world's international telephone traffic.

In the remarkably brief period of just over six months two agreements entered into force and the new international entity, INTELSAT, was created. These interim arrangements on how the organisation was to operate and disputes settled were opened for signature on 20 August 1964 and came into force for the 11 founder members.

With regard to the definite arrangements, two interrelated agreements were opened for signature on 20 August 1971. About one and a half years later on 12 February 1973 these two new agreements entered into force, having received the necessary ratifications. One is an agreement between governments party to the agreements (parties) and the other is between signatories which are telecommunications entities (public or private) designated by a party and as such are signatories of the operating agreement. In the transition to the definitive arrangements the most radical changes in the organisation were in its structure. This was mainly undertaken as a result of significant pressure from the UK which led the rest of Europe on this matter.[11] The interim arrangements had revealed difficulties for the Europeans in the use of COMSAT* as a manager of the system while also having the role of the major shareholder. This was overcome by the gradual phasing out of the COMSAT management role and the introduction of an executive organ which took on a permanent status in the Washington DC Headquarters to manage the organisation in accordance with the Board of Governors' wishes. This transition took some six years to implement because of the need to properly handle the ongoing contracts etc. The role of INTELSAT undertook major changes in the 1990s as detailed later.

2.4.2 *History and development of EUTELSAT*

In 1964, the European Conference on Satellite Communications (CETS), which was originally created to coordinate a European position in the INTELSAT negotiations, began to focus attention on a possible European satellite programme. The objective of this work was to give Europe, and in particular its industry, technical capability in this area based on an experimental satellite programme. 1966 saw the formation and first meeting of the European Space Conference (ESC), designed to harmonise the work of the different European bodies dealing with space activities. The European satellite programme under study was originally conceived for the provision of Eurovision television programmes for the European Broadcasting Union (EBU) as well as some telephony in Europe and the Mediterranean basin. The economic viability of such a project was later called into question and the emphasis, in 1969, was redirected towards having a system primarily dedicated to European telephony requirements with some capacity available for EBU television.

---

*The Communication Satellite Corporation (COMSAT) was established by the US Government as an independent body to manage the international satellite communications access within the US.

In August 1970, a European telecommunications satellite working group (SET) was established to collaborate with ESRO/ELDO in carrying out studies in this area. The outcome of the initial work was the publication in July 1971 of a study on a European satellite system. One of the conclusions was that the space segment of the system should be owned, operated and managed by telecommunications authorities, acting through a new organisation to be formed, for which the name of EUTELSAT had been suggested. With the above decisions in mind a two-phase approach was envisaged for the telecommunications programme comprising an experimental and technological phase (1972–1976) which would culminate in the launch of an experimental satellite in 1976, and a further phase (1976–1980) which would include earth-segment implementation, final space-segment development and the procurement of operational satellites. At the December 1971 ESRO/ELDO Council meeting this was approved and work on the experimental satellite, to be known as the Orbital Test Satellite (OTS), commenced at the end of 1972. The operational phase satellites were to be known as the European Regional Communications Satellite System (ECS).

In March 1977, a conference was held to prepare for the establishment of an interim organisation to manage the space segment, called Interim EUTELSAT. As a result an agreement was opened on 13 May 1977 for signature and entered into force on 30 June 1977.

Under the ECS arrangement, the European Space Agency (ESA) would secure the provision, launch and maintenance in orbit of the ECS satellites, and provide replacement satellites, with a view to having a continuity of the initial space segment over ten years.

OTS–2 was successfully launched on 11 May 1978 (after an unsuccessful attempt in September 1977) and was to be utilised initially for three years. However, on the completion of three years in orbit, the satellite was still working properly, and special financial arrangements were concluded with ESA to keep it there. To finance such payments, it was decided that 80 per cent of EUTELSAT's contributions to ESA regarding these costs should be provided by user signatories. In the event the satellite was used until the end of 1983 and the above funding requirement was not only met but slightly exceeded.

In 1979 Interim EUTELSAT began to consider arrangements for the definitive organisation. After much debate the definitive agreements entered into force on 20 July 1985.

ECS flight 1 was successfully launched on 16 June 1983 and followed by four more in the period up to July 1988. Unfortunately, flight 3 suffered a launch failure in 1985.

The period 1983–1984 witnessed considerable activity in the field of future programmes. The initial system was unable to adequately provide sufficient capacity beyond 1990 with the resources available under the EUTELSAT-ESA arrangement. It was concluded that this situation could be addressed by procuring enhanced satellites which could be available for operation by the end of 1989. In the event, the first of the Eutelsat-II satellite was launched in January

1991 with more to follow shortly. By 1998 EUTELSAT had added several new generations of satellite to its fleet, some specifically designed for television services (HOT BIRDS).

2.4.3 History and development of INMARSAT

In 1966 the International Maritime Organisation (IMO), based in London, undertook studies on the possibilities of satisfying the communication needs of the maritime mobile service (MMS) by the use of satellite communications and the need to provide radio frequencies for this purpose. In 1967 the ITU World Administrative Radio Conference (WARC) for the MMS adopted a recommendation relating to the utilisation of space communication techniques in the MMS outlining further work to be conducted prior to the 1971 WARC.

Following the allocation of frequencies to the maritime mobile satellite service by WARC 1971, it was decided that the IMO should play an active role in the early organisation and introduction of an international maritime satellite system in full cooperation with the telecommunication authorities of its member governments. In March 1973 the IMO decided to recommend that an international conference of governments be convened in early 1975 to decide on the principle of setting up an international system and, if it accepted that principle, to conclude agreements to give effect to its decision. It transpired that three such conferences would be needed before the INMARSAT convention was adopted. After protracted discussions most matters were resolved by the third session of the conference which took place in London from 1–3 September 1976. The conference decided to establish a preparatory committee to carry out activities between the closing of the conference and the coming into force of the instruments establishing INMARSAT. Twenty-two countries participated in this, which provided the background for the INMARSAT organisation following its establishment on 16 July 1979. They agreed that the first session of the INMARSAT council should take place immediately after entry into force of the convention from 16 to 27 July 1979 and the first meeting of the INMARSAT assembly should take place after 3 September 1979.

The UK government was a prime mover in all aspects of establishing INMARSAT. It funded the early European-based MARECS system to the tune of 39 per cent of the total, made way for the establishment of the INMARSAT headquarters in London and played a vital role with the IMO in getting an equitable share arrangement.

In December 1997 INMARSAT celebrated the installation of its 100 000th terminal, indicating the significant extent of its operations.

Figure 2.1 indicates the magnitude and growth of the revenues earned by the three major satellite consortia. Growth is still ongoing with no evidence of diminishing interest in satellite services as yet.

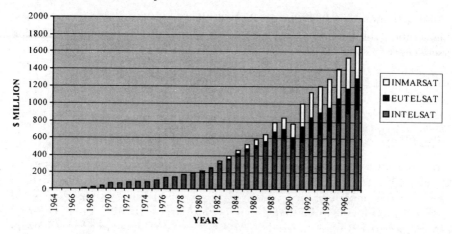

Figure 2.1   Revenue of the three major satellite consortia

2.5 Television satellite broadcasting

It has been recognised for many years that for television, with its wide coverage areas which span both national and international boundaries, satellites provide an ideal transmission medium. From the very start all communication satellites were capable of carrying television transmissions. Conventional communications satellites are generally only capable of operating with large receive antennas on the ground owing to power limitations in the satellite. Special high-powered direct broadcasting satellites (DBS) that can transmit directly to small individual receivers were conceived in the early days of satellite communications[12] but were not economically or technically viable until the mid 1980s.

It was probably the satellite transmission of the 1964 Tokyo Olympic television coverage that, as 400 million viewers watched, alerted the governments and operators to the potential of television coverage using satellites. This gave the impression that satellite-based television transmission was a very lucrative market with enormous potential for exploitation. Consequently, as early as 1965, satellite broadcasting direct to the individual receiver was proposed with an estimated cost of about £49 per household to convert existing televisions.[13]

In 1966 there were calls for direct broadcasting of television to Europe, for which the costs were estimated to be about one fifth to one tenth that of the conventional land relay networks,[14] and ESRO proposed a satellite-based European DBS. In the UK the electronics industry was keen to see DBS develop and it was considered feasible by the British Space Development Company that satellites would be feeding four UK national television channels within five years. This proved to be too ambitious and the UK had to wait until 1989 for its own DBS services, although television distribution services were initiated in 1988 via the ASTRA-based Sky services operating in the fixed satellite service bands.

By 1972 concerns had been expressed over the possibility of DBS transmissions infringing national boundaries. The USSR applied to the UN to 'protect the sovereignty of states against any outside interference and prevent the turning of direct-television broadcasting into a source of international conflicts and aggravation of relations between states.'[15] This factor was one that led to the convening of an ITU WARC on broadcasting services (known as WARC-BS or WARC-77): From this conference came the allocation of frequency bands dedicated to DBS and separate from telecommunication frequencies. These frequencies were to allow national DBS services to be developed and to exploit all possible technical approaches to stop infringement of boundaries and transmission into other states. Each country was allocated a number of satellite channels and an associated national coverage area for DBS services within an overall plan. This plan came into effect in January 1979 and although it essentially provides for a worldwide allocation of resources for DBS use it has, to date, had a poor take-up. There are about 20 satellites currently in orbit although the plan allows for in excess of 300!

It is worthy of note that Japan, Russia and Europe were the first to exploit television transmission to the home or cable feeder heads.

In early 1981 BT, a major satellite-system operator in the UK, began actively studying the concept of establishing a UK-owned satellite which would provide UK DBS television as well as international telecommunications facilities. The government was keen on free market forces and permitted United Satellites Ltd. (UNISAT) to use the UK allocation, but imposed a constraint that the BBC must operate two of the DBS television channels. In 1984 the BBC announced it had difficulties in leasing the whole satellite system and ITV was invited to participate. In addition, owing to the prevailing situation, the government allowed private companies into the project. A consortium to programme UNISAT's three channels was established under joint ownership of the BBC, with a 50 per cent stake, ITV network companies, which had 30 per cent, and a group of twenty-one companies with 20 per cent, which became known as the 21 Club. However, by 1985 the BBC had estimated that providing programming for UNISAT was going to prove expensive, and that it might have to resort to advertising to raise the necessary finance. This approach was not liked by ITV and the public, and their concerns led to the BBC withdrawing from the programme. Without the BBC's support as the major user of the DBS payload the project collapsed.[16]

Following this the Independent Broadcasting Authority (IBA) licensed, under a franchise agreement, a private company called British Satellite Broadcasting (BSB) to provide three channels of direct-broadcast television. In the meantime BT leased several transponders on INTELSAT and EUTELSAT, as well as beginning to use ASTRA.[17,18]

In 1988 a major competitor, Sky, backed by Rupert Murdoch, arose. This enterprise had the Prime Minister's blessing; at the annual Press Association luncheon she said 'I think the opportunity of more channels can enable us to have some very upmarket television'.[19] The government, being of a view that

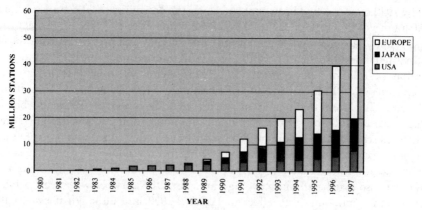

Figure 2.2   Installed direct to the home terminals

supported market forces and competition, was unable to give BSB any preferential treatment. BSB, as an enterprising concern, began to lose confidence of getting government support but the money already invested by the participants meant they would have to continue to try and achieve a marketable product or make a major loss. Sky announced it was to be operational by February 1989 while BSB did not broadcast until several months later, thus having lost the initiative and the audience. From then on BSB went down hill, until it was eventually merged with Sky as BSkyB in late 1989. Sky operates with the ASTRA satellites provided by Société Européenne de Satellites (SES) of Luxembourg in the telecommunications (FSS) frequency bands and has a wide European coverage area.

In competition EUTELSAT carries similar services for a very wide range of countries including Turkey, Italy, Croatia, Hungary, Portugal, Greece etc. and has major plans for enhancing its service offering.

As indicated in Figure 2.2, activities on DBS in the UK and Europe were ahead of other parts of the world. Digital direct-to-the-home (DTH) services are developing rapidly and this will result in a more rapid penetration of satellite-based television.

2.6 Technological considerations

The development of satellites used for civil communications in the GSO has been paced by the evolution of adequate payload-bearing launch vehicles. Figure 2.3 indicates the growth of such capability up to 1994 and Figure 2.4 shows the number of communications payloads launched per year up to 1997.

The development of the satellites themselves can be best presented in a few charts indicating graphically the evolution in terms of mass, power, lifetime and effort to build (Figures 2.5 to 2.7).

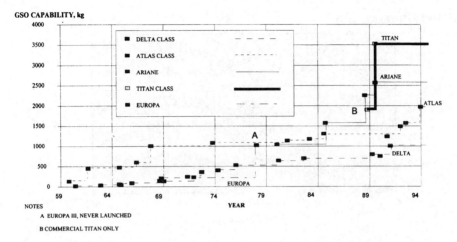

Figure 2.3   Evolution of GSO payload capacity for commercial launch vehicles

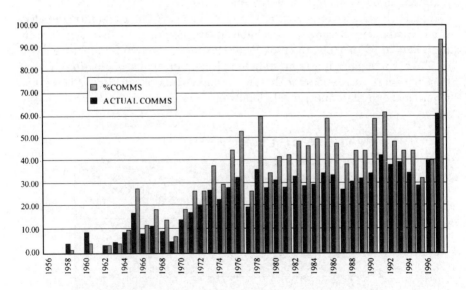

Figure 2.4   Payload launch history

%   COMMS indicates the percentage of all payloads launched into orbit which are communications payloads

ACTUAL COMMS indicates the number of communications payloads launched per year

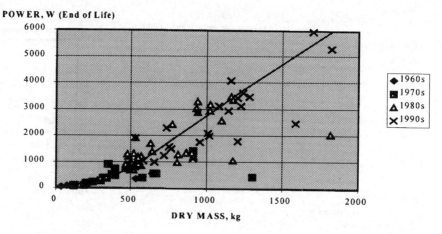

Figure 2.5   Evolution of the dry-mass/electrical power capabilities of spacecraft

Examination of data on the mass of the satellites and the effort required to build them leads to Figure 2.6 which emphasises the fact that large satellite programmes absorb a lot of working hours. It is interesting that, despite the somewhat exotic materials used in spacecraft, a reasonable estimate of the cost of a programme can be derived from a simple conversion of the salary costs required (along with a representative overhead).

As launch vehicles improved in terms of mass-carrying capability and electronic components became more reliable, it was possible to increase significantly the potential lifetime of a satellite. Figure 2.7 shows the trends in lifetime based on 98 known designs; it is interesting to reflect on the value of a lifetime of more than ten years. It is apparent that one of the virtues of satellite systems is the requirement to plan and provide follow-on systems which use improved tech-

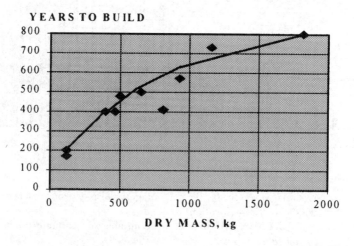

Figure 2.6   Effort required to build a satellite (based on ten known satellite programmes)

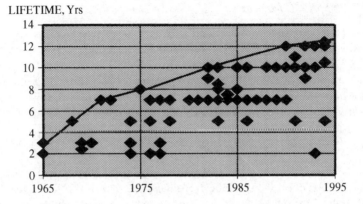

Figure 2.7   Growth of satellite lifetimes (based on 98 communication satellites since 1965)

nology and result in higher-capacity systems or improved quality of service. Such improvements were paced by the technology and not by the marketplace. With 20-year lifetimes now being feasible (especially with the advent of in-orbit use of ion propulsion), it is the market which will in future pace the lifetime. Currently, system planners are grappling with the conundrum of geriatric satellites against competition from new and improved entrants into orbit.

Other major technological developments include:

- improved filters of novel design which permit improved multiplexing;
- solid-state power amplifiers that can have improved performance at moderate power levels;
- adoption of heat-pipe technologies for thermal management;
- improved solar arrays and glass cover slides;
- autonomous onboard controllers which assist in fault control and routine manoeuvres;
- adoption of liquid propellant satellite apogee kick motors along with more complex launch-vehicle mission profiles.

2.7 Overall developments

The overall development of satellite communications can be segmented as follows.

2.7.1 1945–1960 Imagination

This period was characterised by the imaginative writings of a very few far-sighted science-oriented people who could foresee with some vision the potential of using spaceborne communications repeaters.

2.7.2 1960–1970 Innovation

This decade saw the establishment of many key space organisations such as NASA, ELDO, ESRO, CNES and satellite organisations such as COMSAT and INTELSAT. What appears to have taken all by surprise was the pace of the evolution and developments. Furthermore, many governments found that they did not have the machinery to cope with events happening at this pace and consequently some significant opportunities were missed.

2.7.3 1970–1980 Commercialisation

With the early success of INTELSAT using Early Bird,[20] it became evident that the results of previous financial studies were realistic, at least on an international scale. Many clamoured for a piece of the action, which was until then strongly biased towards the USA; Britain and Europe were no exception to this clamouring.

SKYNET 2, the second-generation UK military satellite system, was a UK-manufactured satellite with only assistance being provided by the USA. Thus, UK and European satellite companies became commercial and developed their own product lines. In addition, UK manufacturers started making significant elements of INTELSAT satellites. The UK also had gained experience with launch vehicles, especially its Black Arrow launcher programme. Ironically, in 1971 the Black Arrow launched a UK satellite, after the Black Arrow programme itself had been cancelled!

The European OTS put European manufactures in a viable position to be commercial in satellite supply. The success of OTS spurred on the efforts underway to establish a European regional system and EUTELSAT was born, albeit with many more gestation pains than for INTELSAT.

2.7.4 1980–1990s Liberalisation

The operation of telecommunication systems has, in most countries, been strictly regulated by government. Indeed, one of the basic rules of the ITU stipulates that, although administrations should adhere to the ITU regulations, it is the sovereign right of each country to determine its own telecommunications policy and its implementation. The main drivers for such sovereignty are, of course, of a political and national security nature. However, with politics in many developed countries turning towards market-led approaches and free-market trading, any barriers imposed by national telecommunications regulation become an anathema. Consequently in the 1980s we saw these barriers being slowly removed in a carefully planned and controlled manner. This situation is reflected into the satellite communications arm of telecommunications. The US, followed closely by the UK, were prime movers in deregulation.[21] In addition the European Commission (EC) also initiated efforts to ensure proper market operation while trying to harmonise European activity.[22] Other countries have followed in the efforts to deregulate.

In the UK, access to the space segment operated by the satellite consortia had proved to be difficult for potential nonsignatory users, as highlighted in the evidence given to the House of Lords Select Committee in its report on science and technology, covering the United Kingdom space policy 1987.[23]

In February 1988 the UK government announced that it intended to issue licences for specialised satellite services which were for operating one-way point-to-multipoint systems within the UK. The UK government announced at the same time its intention to issue class licences for receive-only terminals, not just for TVRO as was previously the case. Work also progressed on reviewing access to the satellite consortia space segment.[24] Thus, the start of the liberalisation had begun and was due to develop much more in the 1990s.

2.7.5 *1990s Privatisation and private ventures*

The late 1980s and the 1990s are best characterised in terms of the advent of major private ventures into a predominantly regulated scene. This is true in the traditional fixed satellite services as well as the broadcasting satellite arena, where private operators have given significant competition to the more traditional broadcasters.

Such activity has resulted in structural change to the major satellite consortia, INTELSAT, INMARSAT and EUTELSAT. In 1995 INMARSAT created a private company to implement handheld satellite telephone communications known as ICO Global Communications. The formation of this company resulted from the INMARSAT-led Project 21 initiatives launched in 1991.

In 1998 INTELSAT decided to create a new company (INC) with the transfer of assets comprising five satellites to the privatised company. Discussions were also well in hand on restructuring the remainder of INTELSAT.

EUTELSAT is somewhat behind the other satellite consortia in restructuring. There is a perceived need for EUTELSAT to be more flexible in order to be competitive with organisations such as SES/ASTRA and a treaty amendment is expected to be tabled in 1999.

Two fields of satellite communications have provoked unprecedented levels of private investment and promise a stimulating future for the satellite communications industry at all levels. Constellations of satellites forming the GMPCS are discussed in detail in Chapter 20 and the new multimedia Ka-band satellites in Chapter 25.

2.8 The future

The Teal Group in its first 'Worldwide satellite market forecast: 1998–2007' publication estimates that about 1200 satellites will be put into commercial operation representing a market value of about $58 billion. These figures exclude military, earth observation and scientific missions.

Another estimate by Pioneer Consulting in a report 'Satellite data networks: the Internet's next frontier' indicates investment in broadband satellite systems between 1998 and 2010 of the order of $76 billion with a potential 36 million subscribers globally.

If only a fraction of such predictions become reality it is clear that satellite systems have a promising future with significant scope for engineers as well as the system operators.

2.9 Conclusions

The development of satellite communications has been an exciting one with many challenges. Technical problems had to be solved, the science of operation of equipment in outer space understood and ways found to meet the unending demand for services. The visionaries, pioneers and all those involved in this almost unparalleled evolution are to be congratulated. Historically it is evident that the developments that have led to the world as we know it today have been driven very significantly by the role of enhanced, large-volume global communications of all kinds (a subject which would warrant further material in its own right).

Considering the British Interplanetary Society motto *From imagination to reality*, it is the authors' opinion that this is most apt regarding satellite communications and should not to be forgotten for the future, which looks as equally exciting, if not more so, as the past!

2.10 References

1 CLARKE, A.C.: 'Extra-terrestrial relays', *Wireless World*, October 1945, pp. 305–308
2 CLARKE, A.C.: 'Peacetime uses for V2', *Wireless World*, February 1945, p. 58
3 CLARKE, A.C.: 'How the world was one' (Victor Gollancz Ltd., 1992), pp. 267–279
4 PIERCE, J.R.: 'Orbital radio relays', *Jet Propulsion*, April 1955
5 Slightly modified versions of these papers are published in GATLAND, 'Telecommunications satellites' under the titles of 'A design study for an equatorial circular-orbit, COMSAT system (GPO/RAE)' and 'A design study for an equatorial circular-orbit, COMSAT system (BSDC)' as chapters 9 & 10
6 op.cit. CLARKE, 'Extra-terrestrial relays'
7 PIERCE, J.R.: 'The beginnings of satellite communications' (San Francisco Press Inc., 1968)
8 ANDERSON, F.W. Jr.: 'Orders of magnitude: a history of NACA and NASA 1915–1976' (NASA, Scientific & Technical Information Office, Washington DC, 1976) p. 11
9 BAKER, D.: 'The rocket' (New Cavendish Books, London, 1978), p. 17
10 Proposal of President Kennedy of 7 February 1962, introduced by Senators Magnuson and Kerr
11 CROOME, A.: *Daily Telegraph*, 5 September 1970
12 *Flight International*, 3 November 1966, p. 772
13 *The Times*, 6 April 1965
14 *Flight International*, 3 November 1966, p. 772

15 *Flight International*, 24 August 1972, p. 291
16 HOWELL, W.J. Jr.: 'World broadcasting in the age of the satellite' (Ablex, Norwood NJ, 1986) p. 260
17 HUDSON, H.: 'Communication satellites' (The Free Press, 1990)
18 CHIPPENDALE, P. and FRANKS, S.: 'Dished! The rise and fall of British satellite broadcasting' (Simon & Schuster, London, 1991), p. 14
19 op. cit. CHIPPENDALE, p. 38
20 *Daily Telegraph*, 2 April 1966, (cited at having made £720 k in its first six months of operations)
21 HILLS, J.: 'Deregulating telecoms' (Frances Pinter (Publishers), London, 1986) p. 157
22 LIPPENS DE CERF, P.: 'International satellite telecommunications and EEC law'. *AIAA* paper IISL-87-83, AIAA space conference 1987
23 House of Lords select committee report on science and technology, covering the United Kingdom space policy, 17 December 1987, (HMSO, London), *HL paper 41-I*
24 Consultative document, Office of Telecommunications (London, July 1993) paragraph 1.2

Chapter 3

The satellite communications business

N. Cartwright

3.1 Introduction

What is the satellite communications business today? As already mentioned in Chapter 2, the satellite communications business is still the only commercially viable space-based activity. A surge in interest and activity in satellite-based applications is taking place in the late 1990s, driven by the increasing capability of digital technology to broaden the range of applications through complex circuit miniaturisation and high-volume production costs — a familiar process throughout all sectors of electronics. Much of the recent activity is in the private, commercial sector with advanced, innovative, corporate and consumer digital services such as compressed digital television broadcasting (DVB), satellite newsgathering, business television, wideband multimedia applications, VSATs, asymmetric interactive services for the home and global mobile personal communications by satellite (GMPCS). To highlight the availability of consumer technology in a related area, global positioning system (GPS) satellite navigation receivers employing low-noise L-band front ends and spread-spectrum processing technology, once expensive equipment for professional users, are now very low-cost units, facilitated by developments in surface mount technology and ASICs, and the amortisation of their development costs over large production volumes. One can now buy GPS receivers and advanced digital direct-to-home television (DTH-TV) systems, with Ku-band LNB noise figures unheard of ten years ago, in many high street stores.

The industry is now well represented on stock exchanges around the world with a market capitalisation of tens of billions of US dollars invested in manufacturing, operating and media companies. Revenues from the global satellite communications market, including the launcher, space and ground-segment manufacturing and services industries, are similarly large, exceeding US$50 thousand million in 1997 and growing. Not many years ago few people would have predicted such rapid and diverse growth in the business from its beginnings in the early 1960s. Transponders now number several thousand, installed on over 200 GEO civilian communications satellites (including those of the former USSR incorporated into a commercially revitalised INTERSPUTNIK). So many, if not all, of us will be affected by this increasingly competitive, consumer-oriented business in the future. The original international, regional and national satellite organisations still play a major role but most of these organisations are now changing to meet the technological challenges of the 21st century presented by the rapidly increasing influence of the information society.

This Chapter aims to give a flavour of the key activities and decision-making processes involved in some of the new ventures. In such a brief contribution it will not be possible to cover all aspects from conception through launch to service operation, so we will focus on some of the commercial aspects for one particular new service, multimedia. The issues we will touch upon will be the ones which are felt to be the most important; therefore this text to some extent gives a personal view based on experience in the commercial aspects of the business.

We will not address the very large elements of the business that include scientific, remote-sensing, weather and military satellite systems. These form a very large proportion of the hardware now orbiting the earth. With the exception of remote-sensing satellite systems, few of these applications have been developed and operated with truly commercial considerations in mind and they are often largely public-funded activities — communications remains by far the most profitable space application to date.

Multimedia may offer a mix of digital data in the form of text, audio and visual information, where the visual images range from monochrome graphics to full-motion colour, and the audio content from toll-quality voice to high-quality stereo music standard. The service may be broadcast, narrowcast or interactive. Multimedia information clearly has highly variable bandwidth requirements dependent upon its application. Interactivity may typically be very asymmetrical, with wide bandwidth in the forward direction, from the server to the client, and narrow bandwidth in the return direction, from the client to the server. Such asymmetry may be required by the individual residential or SOHO (small office, home office) user to access material from the worldwide web (WWW). On the other hand, a symmetrical service with equally wide bandwidths in both directions may also be required by some SOHO users for videoconferencing or multimedia material preparation where high-quality professional audio-visual work is to be regularly transferred in both directions.

Before describing the business-planning activities for such a new multimedia service, a few facts and figures about satellite systems today may be appropriate, starting with the formation of the principle international satellite communications organisations which have largely driven the market until quite recently.

3.2 Satellite organisations

The satellite communications business has evolved greatly over the last 20–30 years from the early days when there were just INTELSAT (formed in 1964), INTERSPUTNIK (established in 1971), EUTELSAT (founded as Interim EUTELSAT in 1977), ARABSAT (formed in 1976) and INMARSAT (founded in 1978) providing international and regional fixed, point-to-point, satellite services and international maritime mobile satellite services for their numerous members. Two domestic national systems also evolved in the 1970s: the ANIK satellite networks serving the vast landmass of Canada and the PALAPA networks serving the thousands of islands in the geographically extensive Indonesian archipelago with the world's fourth largest population.

The INTELSAT organisation developed an Agreement and Operating Agreement that some subsequent organisations have broadly adopted. The organisation's agreements or conventions are signed by the governments of each member state, which then appoint a national telecommunications operator, i.e. the monopoly PTT, as sole signatory of the organisation operating agreement. Some of these PTTs, such as BT in the UK, may now be wholly privatised companies; others are still wholly state owned and controlled, and some are in the process of transition from the public to private sector.

The international satellite organisations have experienced difficulties in recent years with competition from private satellite operators. Competition is likely to become tougher as more private operators come onto the scene. The structure of the organisations, where the principal signatories are all competing with each other to provide similar services, tends to inhibit the organisation's ability to respond to the changing competitive situation. This situation has been recognised by INTELSAT, INMARSAT and EUTELSAT and various measures are in hand to try to meet the challenge of increasing competition.

3.2.1 INTELSAT

INTELSAT was the first global communications satellite organisation and was formed as soon as it became clear that there was both a future for international public telecommunications by satellite and a viable satellite and launcher construction industry. As of June 1998 INTELSAT's membership comprises 143 nations. INTELSAT provides voice and data services for public-switched telecommunications and private-line networks for business applications on a global basis. It also provides video services for broadcast contribution and distribution, satellite newsgathering, special event services and DTH-TV.

The organisation has a staff of around 650 people involved in the daily management of its operations and is governed by an Assembly of Parties which meets every two years. This is a meeting of representatives from governments that are signatories to the INTELSAT agreement and deals with policies, long-term objectives, the principles and scope of INTELSAT activities and considers resolutions passed to it by the Meeting of Signatories or the Board of Governors. Each party has a single vote and the decision-making bodies are all composed of delegates from the member nations.

The Meeting of Signatories takes place annually and comprises all those signatories that are bound by the INTELSAT Operating Agreement. The signatories consider all high-level matters related to the technical, operational and financial aspects of the INTELSAT system. The signatories can set the capital-investment ceiling, decide the minimum investment share for representation on the Board of Governors, approve earth stations, allot satellite capacity and establish nondiscriminatory charges. Each signatory has a single vote at the Meeting of Signatories.

The Board of Governors meets quarterly and deals with general matters related to the running of the INTELSAT space segment, such as its design, development, construction, operation and maintenance. The Board is a management board consisting of a maximum of 20 delegates who may be representatives of signatories whose investment share exceeds a certain level set by the Meeting of Signatories, groupings of signatory representatives whose combined investment share exceeds this level or up to five groups each of at least five signatories from the same ITU region, irrespective of investment share.

Each investment shareholder contributes to INTELSAT and receives capital repayments and compensation for the use of capital in proportion to its investment share. Each shareholder's investment share is calculated annually based on its percentage of use of INTELSAT capacity, subject to a minimum investment share of 0.05% for each member country. Currently a 0.05% share represents a financial investment of approximately US$1 018 650. INTELSAT's revenue from its telecommunications operations in 1997 was more than US$960 million. The operating surplus has been such that signatories have received an average return on their capital employed of approximately 18% in the period 1993 – 1996.

At the end of 1997 the US signatory, COMSAT, had the largest signatory investment share, at just under 18%, followed by the UK signatory, BT plc, at about 5.7%. The UK entity also supports a number of investing entities, such as Cable & Wireless plc, Mercury Communications Ltd and Hong Kong Telecom which, combined with the UK signatory investment, produce a total UK investment share of 7.3%. Among the remaining 141 member nations Italy, Norway, Germany, India, Argentina, China and France have the next largest investment shares of approximately 5.17%, 4.61%, 3.45%, 3.35%, 3.32%, 3.07% and 2.92%, respectively.

INTELSAT's principal response to growing competition has been to develop plans to restructure the organisation. An amendment to signatory operating agreements has been approved permitting multiple signatories, recognising the growth in operators following telecommunications liberalisation policies throughout its members. In March 1998 the INTELSAT Assembly of Parties also approved the formation of a commercial spin-off company, provisionally named 'New Skies Satellites N.V.', to be based in the Netherlands. It is proposed that five operational satellites be transferred to the new company and one under construction. This is INTELSAT K, the INTELSAT K-TV satellite being procured for deployment over Asia at 95 °E. All INTELSAT's Ka-band filings at 95 °E and 319.5 °E will also be transferred. It is likely that the new company will concentrate on the provision of regional consumer and corporate video, multimedia and new services, including DTH-TV.

INTELSAT's total assets in 1997, prior to the creation of the independent spin-off company, amounted to approximately US$3.5 billion, most of which is related to spacecraft investments. After the planned transfer the organisation's assets will include a reduced global constellation of 19 satellites comprising a mix of INTELSAT V/V-A, VI, VII/VII-A and VIII spacecraft. Five new spacecraft in the INTELSAT VIII/VIII-A series have been launched, the last, INTELSAT 805, in June 1998. Four INTELSAT IX satellites will be launched by 2001.

INTELSAT has also filed details of Ka-band networks with the ITU at ten orbital locations to enable it to offer VSAT, videoconferencing, television and satellite newsgathering, interactive DTH services, and interactive multimedia services. Some of these network filings are now planned to be transferred to the spin-off company.

For further information see the INTELSAT worldwide web site at www.intelsat.int.

3.2.2 EUTELSAT

EUTELSAT membership stands at 47 countries, including several former members of the Soviet Union. The governments of all member countries, called parties in the organisation, ratify the EUTELSAT convention. The parties meet once a year in an Assembly of Parties to take decisions on long-term objectives.

As of October 1998, EUTELSAT operates a constellation of 13 satellites in the European region: two EUTELSAT-Is, four EUTELSAT-IIs, HOT BIRD^{TM}s 1, 2, 3, 4, 5, W2 and TDF-2. Six satellites are planned for launch and future operations, including EUTELSAT-W1, W3 and W4, RESSAT (a backup for a possible W-series satellite failure), SESAT (for Far Eastern, Russian and African coverage) and EUROPESAT-1B. HOT BIRD^{TM} 5 and W2 were launched in 1998. The current EUTELSAT satellites, not including the recently acquired TDF-2 craft, now support a total of around 200 transponders with bandwidths varying from 33 to 72 MHz.

The parties nominate signatories to the Operating Agreement who will be responsible for investment in the organisation. As with INTELSAT and INMARSAT, the signatories are historically the existing national monopolistic public telecommunications operators, although with privatisation gathering pace in Europe this situation is now changing.

The Board of Signatories is the main decision-making body and meets at least four times a year to discuss commercial and financial strategies, design and deployment of satellites and planning of new services.

In 1997 EUTELSAT generated revenues of 376 million ecu and provided a return on total capital employed of over 14.2% (which is paid to the signatories as compensation for the use of their capital). At the end of 1997 the UK signatory, BT plc, had the largest signatory investment share, at about 22.6%, followed by France Telecom, Telecom Italia S.p.A, Royal PTT Nederland N.V. (KPN), Deutsche Telekom AG and Belgacom with 15.37%, 12.74%, 8.85%, 8.09% and 6.54%, respectively.

In a similar manner to INTELSAT, the EUTELSAT Assembly of Parties has recently been studying structural changes which will enable it to meet the challenges of the future. The objective is to complete such changes by 2001. In one move the Assembly has approved an amendment to the convention permitting multiple signatories per member country, thereby recognising the growth in telecommunications liberalisation within Europe and the number of competing operators and broadcasters. Multiple signatories allows these new operators to invest in the organisation and to receive the benefits and rewards.

A deeper restructuring process, approved in May 1998, involves the formation of a French limited company acting under the regulation of an intergovernmental organisation thereby ensuring that the existing regional service obligations of EUTELSAT are preserved.

The story of EUTELSAT's history is interesting because the organisation has been subject to change from market pressures since its inception as an entity providing telecommunications services to European PTTs to the present-day situation, where the provision of analogue and digital television and multimedia service distribution direct to consumers through the lease of transponders to broadcasters represents a major part of the business. The early beginnings of European communications satellites can be traced to the 1960s, but in the early 1970s the European Space Research Organisation, ESRO, developed the European Communication Satellite system (ECS) using the relatively new Ku-band technology. ESRO later became the European Space Agency, ESA, and launched the first successful European orbiting test satellite, OTS-2, in 1978.

Ku-band technology was chosen instead of the more conventional and mature C-band technology for several reasons:

- to enable European industry to benefit from participation in the relatively new Ku-band payload technology;
- to avoid interference with C-band INTELSAT and terrestrial networks;
- to exploit the wider bandwidths available at the higher frequencies;
- to enable the deployment of smaller earth stations by virtue of the smaller beam footprints possible, and hence higher power flux densities at the earth's surface, provided by the then current spacecraft antenna-design and fabrication capability and power-amplifier technology.

Interim EUTELSAT was formed in 1977 and managed the OTS test programme. The subsequent ECS system using the first ECS-1 and -2 satellites procured in the early 1980s was developed by ESA and operated by Interim EUTELSAT until the final ratification and formalisation of the EUTELSAT Convention in 1985. ECS was designed to provide regional 120 Mbit/s TDMA public-switched telephony facilities between the 20 plus European PTT members of CEPT and television and radio distribution services between EUTELSAT members for the EBU. The satellites were named EUTELSAT I-F1 and F2 when EUTELSAT became operational. They have been replaced and augmented over the years by newer and more powerful designs creating the present-day EUTELSAT constellation of 13 satellites between 36° and 7° east (as of October 1998).

The expected primary use of the EUTELSAT networks for regional public telecommunications services to members appears to have attracted insufficient traffic to totally justify the investment for public-switched trunk telephony. This is partly because of the cost of the TDMA earth stations and partly because the falling costs of international optical-fibre terrestrial networks made the use of satellite communications over the relatively small distances between European population centres uneconomic compared to the distances bridged by transoceanic networks. However, the availability of satellite networks has recently become more critical with the urgent need for good telecommunications services to Eastern Europe.

To increase the flexibility and utilisation of the system, EUTELSAT also developed a Ku-band small-dish business system using the SMS (satellite multiservices) payload operating in the 12.5–12.75 GHz FSS band. SMS is capable of facilitating point-to-point low-speed (9.6 kbit/s–2 Mbit/s) digital open or closed networks using the European wide-coverage beam linked to the SMS transponders, SCPC/FDMA technology and customer premises or Teleport-based 2.4–5.5 m diameter earth-station antenna designs. Open networks permit interconnection to public networks, whereas closed networks only permit connection to specific private networks. The SMS facility is now extensively used for VSAT operations. EUTELSAT has also now begun to operate a demand-assignment multiple-access (DAMA) service on an interim basis for on-demand pay-per-use/per minute digital voice and facsimile services, using technology developed by ESA.

It is reported that 17 transponders on five satellites out of the total of EUTELSAT's 160 transponders are currently used for all forms of public and private telecommunications, including the SMS, trunk telephony and leased digital capacity (for further information see the EUTELSAT annual report at its worldwide web site www.eutelsat.org). In addition, several EUTELSAT II F4M transponders are used to provide television and radio programme distribution and exchange capacity between broadcaster members of the European Broadcasting Union (EBU).

Entertainment television broadcasting to television receive-only (TVRO) terminals was tested and piloted in the early 1980s and the EUTELSAT satellites have become increasingly used for television transmission and broadcasting since that time. Initially, television transmissions were carried out in the fixed satellite service 10.7–11.7 GHz (FSS) forward-link band and were intended for feeding CATV head ends and SMATV installations in hotels, apartment blocks etc. However, as in the US TVRO market, improvements in low-noise receiver technology and falling equipment costs led to the eventual development of a substantial DTH-TV market and the opportunity arose to lease transponders to broadcasters offering such services.

EUTELSAT's continuing response to the increasing competition in the provision of television transmission services, and particularly analogue (PAL/ D2-MAC) and digital (MPEG-2) entertainment DTH-TV in Europe, has therefore been to reorient its satellite deployment strategy, exploiting the hot-bird concept of colocating a number of satellites at a single orbital slot, i.e. 13 °E, and to utilise the whole of the Ku-band from 10.7 to 12.75 GHz. The hot-bird location competes for European viewers with the SES ASTRA constellation at 19.2 °E. A feature of the EUTELSAT entertainment television channels is that a large proportion of them are free-to-air and can be received by anyone with a suitable antenna and analogue or digital receiver, unlike the greater proportion of encrypted pay channels carried by the SES ASTRA constellation that require a decoder and subscription. The EUTELSAT HOT BIRD constellation is growing and competing with the ASTRA system, but the task is made more difficult because of EUTELSAT's ownership by multiple European signatories, all with equal rights to transponder capacity, which limits the formation of a unified group of transponder allocations in one orbital location with a common cultural composition, unlike the ASTRA allocations. Indeed, a total of 22 languages are used in the channels broadcast by EUTELSAT.

It has eventually become clear that the principal role of the EUTELSAT satellites is in television distribution, in particular DTH-TV, and television now forms the majority, perhaps more than 70–80 per cent, of the traffic carried by the organisation's satellites. A quick estimate of EUTELSAT's transponder loading in the summer of 1998 shows that it is supporting around 65 analogue, 49 MCPC and 31 SCPC MPEG-2 encoded digital entertainment television carriers, a number of analogue and digital television feeds, large numbers of sub-carriers supporting radio channels and a mix of analogue and digital television carriers for the EBU on EUTELSAT II-F4M. In terms of television channels, it is estimated that EUTELSAT currently supports 75 analogue and nearly 300 digital channels. Many of the digital channels are incorporated in multiplexed packages or bouquets and they are increasing at a very rapid pace. The number of carriers supporting digital channels is already substantially greater than the number offering analogue services.

Revisiting the focus of this Chapter, various multimedia trials and related commercial developments are now being supported in Europe using both EUTELSAT and ITALSAT space segments.

Hughes Olivetti Telecom has already launched the DirecPC service on HOT BIRD 3 supporting a consumer-oriented interactive Internet service using a 400 kbit/s link in the forward direction and a telephone link for the return path. EUTELSAT is also a partner in a 13-member European-Commission-sponsored consortium developing the interactive satellite information system (ISIS) multi-media experiment and system demonstrator. Ku-band capacity from one of the EUTELSAT satellites at 13.0 °E is to be used for the DVB-format 34 Mbit/s downlink signal, and Ka-band capacity from the adjacent ITALSAT satellite located at 13.2 °E will be used for the uplink. The project may be seen as a fore-runner to the commercial EuroSkyWay system being developed by Alenia Aerospazio for deployment in 2001. The ISIS project is part of the EC Interactive Digital Multimedia Services ACTS programme.

EUTELSAT has also developed and deployed a digital multiplex system for HOT BIRDs 4 and 5. With this system, known as SKYPLEX, the satellite can perform all the necessary onboard processing of FDMA/SCPC data signals received from users or clients on the ground, retransmitting data in the DVB MCPC multiplex format. Data rates in the forward link will be up to 36 Mbit/s, conforming to the DVB specification (see Chapter 18). On the uplink the SKYPLEX multiplexer can assemble up to six low-bit-rate carriers at 6 Mbit/s each. Lower input bit rates will be available with HOT BIRD 5. The advantage of the technique is that low-cost and easily located resources, such as SNG equipment, can be used to uplink the low-bit-rate carrier, and the multiplexed wide-bandwidth signal can be received by low-cost DTH-TV receivers with rela-tively low phase-noise performance local oscillators.

Finally, EUTELSAT has designed a new range of satellites, the W series, to replace the EUTELSAT-2 generation. The W-series satellites have 24 (32 in the case of W4) transponders and will be launched in late 1998/early 1999, providing telephony and television services across Europe and Central Asia with two steerable and one fixed antenna beam. The steerable beams will also permit connectivity with Africa and Latin America. (The first satellite in the series, W2, was launched in October 1998.)

By the year 2000, EUTELSAT is planning to have around 250 transponders available from the existing EUTELSAT II, HOT BIRD and the new satellites under construction to meet the expected demand for digital television and multi-media services.

3.2.3 INMARSAT

INMARSAT was originally established in 1979 as an international treaty orga-nisation out of the need to provide reliable, high-quality, maritime mobile communications. High-frequency radiotelex and telegraphy was, and remained until quite recently, the most commonly installed and lowest-cost system of choice for many merchant ship operators.

INMARSAT's structure is broadly similar to that of INTELSAT, with an Assembly of Parties and a Council with functions similar to a Board of Directors. The INMARSAT Directorate is similar to the INTELSAT and EUTELSAT executive organs, performing the day-to-day management and operations for the organisation.

A very important feature addressed by many of INMARSAT's operations has been the safety communications needs of the aeronautical, maritime and land markets it serves. An emergency reporting capability has been a feature of nearly all the services developed by the organisation, and INMARSAT has close links with the International Maritime Organisation, IMO, making

INMARSAT systems part of the new Global Maritime Distress and Safety System (GMDSS). The International Civil Aviation Organisation, ICAO, has also incorporated many elements of the INMARSAT aeronautical system into its Standards and Recommended Practices (SARPS). INMARSAT systems contain emergency-reporting features which are approved for safety at sea and in the air by these organisations and are unique satellite communications services not as yet fully supported by any other operational commercial mobile satellite networks.

INMARSAT's revenue from its operations in 1997 was approximately US$384 million, with the majority coming from the INMARSAT-A telephony service. In 1997 the US Signatory, COMSAT, had the largest Signatory investment share, at approximately 23%, followed by the UK Signatory, BT, Japan (Signatory KDD Co. Ltd.), Norway (Telenor Satellite Services AS), Greece (Hellenic Telecommunications Organisation (OTE)), France (France Telecom) and Germany (Deutche Telekom AG) at approximately 8.4%, 7.59%, 6.81%, 5.4%, 5.11% and 4.29%, respectively.

The INMARSAT Assembly voted in April 1998 to amend the INMARSAT convention and operating agreement sanctioning the structural change proposed by the INMARSAT Council to meet the challenge of the competition which it is soon to face in its global mobile satellite communications market from the new LEO/MEO networks. INMARSAT will now become a public limited company on 1 April 1999. The new commercial company will undertake all of INMARSAT's existing operations with governmental oversight of the organisation's obligations to provide safety-of-life and other public services. The privatised INMARSAT is to seek public limited company status within two years following a public share offering.

This action was preceded by the successful spin-off of an affiliate company, ICO Global Communications, in early 1995. ICO is a completely separate private company, of which INMARSAT owns around ten per cent equity and has exclusive rights to its wholesale services provided for aeronautical and maritime communications. ICO was formed to develop new mobile satellite systems based on satellites in nongeostationary orbits and to compete directly with the increasing number of such systems being developed worldwide.

Returning again to our theme, INMARSAT is also addressing the multimedia market and is planning to develop new satellite networks capable of offering broadband (in relatively low-speed terms compared to the speeds being offered by the fixed satellite service operators, i.e. up to 144 kbit/s) mobile services such as video, e-mail etc. via the Horizons project. Horizons is a mobile multimedia system supported by a fourth-generation INMARSAT constellation of four geostationary satellites with a number of high-power spot beams, probably several times greater capacity than is currently provided by INMARSAT's recently launched third-generation satellites. The start of the services is planned for 2002.

For further information see the INMARSAT worldwide web site at www.inmarsat.org.

3.3 Private satellite networks

As with so many technologies, the commercial side of the private satellite communications business started first in the US with the development of domestic satellite networks for a wide range of services by a number of large operators, such as AT&T, GE Americom, GTE and Hughes Communications Inc. Television distribution by satellite to the C-band television receive-only (TVRO) market started in the US and provides broadcast-quality analogue television signals to cable-TV (CATV) head ends for distribution over their networks throughout the USA via 4.5 m diameter antennas. However, it was not long before technology improvements in the noise temperature of the C-band LNAs/LNBs enabled individuals to install more modest antennas, perhaps < 3 m diameter, and receive the television signals in their backyards, and DTH-TV was born.

The US also saw the start of the burgeoning so-called small-dish business systems, beginning with Satellite Business Systems (SBS) formed in 1975 which offered private circuits for corporate data and television networks based on customer-premises terminals. SBS, the world's first Ku-band satellite network, launched six satellites in all but was then acquired from its principals IBM, COMSAT General and Aetna Life & Casualty by MCI in 1986. Two of the surviving SBS satellites are now owned and operated by PanAmSat.

In recent years most parts of the satellite business have become increasingly competitive and similar to any other commercial enterprises, funding investment and operations from shareholder capital and operating revenue, respectively, while providing a profit to investors.

In Europe, the private company SES ASTRA, founded in 1985 with a 20 per cent holding by the Luxembourg government, launched its first satellite in 1988 with sixteen 45-watt output transponders, more than twice the power of the first ECS satellites. The satellites were only ever intended for DTH-TV operations. The early demise of the UK DBS BSB company and its take-over by Sky, and the subsequent failures of some of the other European DBS projects, inevitably led to the present situation where SES ASTRA is the largest commercial DTH-TV transponder-leasing business in Europe. As of Autumn 1998 the company has eight satellites in orbit, ASTRA 1A-G and ASTRA-2A, and its signals are currently received in over 70 million homes throughout Europe by a combination of SMATV, CATV and DTH-TV delivery. Over 25 million households receive ASTRA signals via SMATV or DTH-TV systems. Seven satellites are in position at 19.2 °E, and ASTRA 2A is in position at 28.2 °E broadcasting DVB DTH-TV signals generated by its largest customer, BSkyB, to Sky digital subscribers in the UK and Ireland. ASTRA 2B is also under construction, to be launched in the first half of 1999. The two satellites at 28.2°E will provide the digital carrier capacity required for the full service. Digital services are also supported on several satellites at the 19.2 °E slot. A count of the ASTRA satellite network transponder loading in Autumn 1998 reveals that there are now 118 transponders at 19.2 °E and 28 at 28.2 °E with a total of more than 90 analogue and 320 digital television channels transmitted in 14 European languages. On this basis the ASTRA network is supporting slightly more channels than the EUTELSAT network and both networks now cover the whole Ku-band spectrum (10.7–12.75 GHz) from a single orbital location.

So far as multimedia is concerned, SES ASTRA has developed the ASTRA-Net multimedia service from European Satellite Multimedia Services using the DVB format. ASTRA 1H, to be launched in early 1999, and 1K, to be launched in 2000, support return-path transmission via a Ka-band payload using the ASTRA Return Channel System (ARCS), permitting the development of a bidirectional asymmetric multimedia service. Further information about SES and its ASTRA satellites can be found at its worldwide web site www. astra-net.com.

BSkyB leases 18 transponders from SES ASTRA and the company already has over six million viewers in the UK and a ubiquitous UK high-street presence in retailing DTH satellite services. In a few years' time it will be joined by other retail operations, for mobile satellite services for example, other DTH service operators, multimedia satellite service operators and perhaps even global navigation satellite service companies.

BSkyB's major shareholder, the holding company News Corp. Ltd, with a 40 per cent share in BSkyB, is the world's largest commercial user of satellites with an annual operating revenue of over A\$14 billion for 1997. The chairman and chief executive is the Australian Rupert Murdoch. News Corp. Ltd now has satellite television broadcasting interests in six markets worldwide, including the UK and Ireland (BSkyB), the USA (Primestar), Asia (STAR TV), India (ISkyB), Japan (JSkyB) and Latin America (Sky Latin America). BSkyB has joined forces with BT, the Midland Bank and Matsushita Electric in the UK to launch British Interactive Broadcasting, BIB, to offer interactive multimedia services to the home. More details of News Corp. Ltd can be found at www.news-corp.com.

An interesting feature of all the above DTH-TV systems is that very few satellite operators conform to the precise parameters of the European frequency allocation plan agreed at the 1997 WARC for the broadcasting satellite service, including the 11.7–12.5 GHz channel frequencies, downlink beams, orbital slots and downlink EIRPs. For example, ASTRA-1F and 1G at 19.2 °E, ASTRA-1D at 28.2 °E and HOT BIRD's 2 and 3 at 13 °E now largely use digital carriers in the WARC 77 BSS band because of the attractiveness of multiplexing a number of channels into a single 27 MHz bandwidth. Some of the early BSS satellites, such as SIRIUS, TV-Sat, TDF-2 and the renamed BSB Marco-Polo satellite, Thor, are still operating although most, apart from SIRIUS, have been moved from their original WARC 77 orbital slots. The original BSS plan is now largely out of date primarily because of improvements in technology (e.g. DVB format signals), the cost of the high-power satellites required and commercial changes in the way in which satellite television broadcasting has developed in Europe. WRC 97 advocated a new plan for BSS in ITU region 1, i.e. Europe, Africa and the Middle East, that addresses, among other areas, the technical parameters of the downlinks, such as bandwidth, EIRP etc., and the addition of assignments for further countries. ITU-R is actively studying ways of revising the BSS plan in response to the resolutions of WRC 97. However, there are still further issues to be resolved, such as the number of channels to be preassigned to each country; ten channels were considered at WRC 97 but this is to be studied and a further conference will be convened in 2000 to reconsider the parameters for the BSS band.

3.4 Ka-band satellite systems

A not inconsiderable number of private companies from Europe, North America and the Pacific Rim countries are currently preparing plans for multi-media-capable satellite networks using Ka-band platforms addressing global, regional and domestic markets. At least 13 new Ka-band systems have been approved by the US FCC and the ITU has received dozens of Ka-band filings from administrations around the world.

The attractions of Ka-band are that it offers:

* allocations agreed at WRC 95;
* greater unused bandwidth than the congested, lower frequency, C- and Ku-bands;
* multiple narrow spot beams providing high PFDs and hence small earth stations;
* potential for onboard processing and switching between spot beams as a result of narrow spot beams;
* extensive frequency reuse and hence even greater effective bandwidth;
* a technology that can meet the perceived capacity demand for multimedia services.

In the rush to exploit the new territory of Ka-band, i.e. the frequencies between 17 and 30 GHz, it may be recalled that the bands above 14 GHz have been used before in civilian satellite communications systems both experimentally and for real traffic-carrying applications.

Japan has been in the forefront of exploiting the higher frequency bands since 1977 with its CS- and N-STAR series of satellites used by NTT for national telecommunications. These bands were chosen because the country has dense terrestrial microwave networks and wished to avoid intersystem interference.

In North America and Europe there have been several experimental satellites, such as NASA's ACTS in the US and ESA's Olympus satellite in Europe.

ITALSAT 1 and 2 are examples of European operational Ka-band domestic multibeam satellites built entirely under Italian management at a total cost of around $660 million, mostly funded by the Italian Space Agency. The ITALSAT network provides a national voice telecommunications service to Italy, Sardinia and Sicily operated by Telecom Italia.

In Europe a new European experimental satellite programme, Artemis, will use Ka-band and the SES Astra-1H and -1K satellites will carry two Ka-band transponders in the 29.5–30 GHz band for return links.

The planned and proposed commercial systems comprise GEO, LEO and MEO satellite mixes and feature most of the major US and European satellite manufacturers in consortia. More details of these systems are explored in Chapter 25 but suffice it to write here that plans are well advanced for such multimedia satellites to be operational early in the twenty-first century.

As a footnote, it is worth noting that Ka-band is now not the top limit of the frequency spectrum being addressed by commercial satellite companies. In autumn 1997 several US operators submitted plans for GEO, LEO and MEO satellite mixes in V-band, i.e. with uplinks and downlinks in the 50 and 40 GHz bands, respectively. These applications amounted to several hundred satellites and may involve investments of over US$30 billion, should they ever get to the stage of being realised.

3.5 Starting up a satellite business

So let us imagine a new, multimedia, Ka-band geostationary satellite network is to be constructed to provide a broadband forward link, say up to 6 Mbit/s per channel, and a return link of 384 kbit/s, similar to the Hughes SPACEWAY design. SPACEWAY's initial global network of Ka-band satellites is intended to provide two-way interactive services. (To find out more about the Hughes SPACEWAY system, see Chapter 25 and the worldwide web sites at www.spaceway.com and www.hcisat.com.)

A list of issues, some or all of which will need addressing in starting up a new multimedia satellite service, is given below:

- the business plan,
  market research,
  risk analysis,
  finance;
- technical design;
- scheduling;
- regulatory and licensing;
- frequency coordination;
- security and conditional-access schemes;
- sales, advertising and marketing;
- project management;
- administration;
- billing;
- service launch;
- training;
- environmental issues;
- testing and commissioning;

- network management;
- service provision and service centres;
- operations and maintenance.

All of these issues are critically important to the success of a new venture. Some, such as the business plan, will probably be completed by the core business team heading up the venture; other, specialised tasks, such as the market research, technical system design, frequency coordination activities, conditional access system design, training, operations and maintenance, advertising and marketing may be carried out by specialist consultants or contractors.

Some of these issues may pose greater challenges than others, depending upon the nature of the satellite business, but all will be involved if the business is one which requires the design, development, procurement, launch and operation of new space, ground and user segments.

3.6 Trade issues

In our plan to set up a global multimedia Ka-band satellite network we will need access to as many markets worldwide as possible. Until recently this has been a difficult task to achieve as many states have had restrictions on allowing foreign companies to provide satellite services in their countries. There have also been restrictions placed on the amount of foreign investment permitted in local satellite operating companies.

Although there is less difficulty in obtaining agreement to operate private networks not connected to the local public networks in other countries, there have been many difficulties in obtaining approval and licences to interconnect to the fixed public networks. A particular difficulty is obtaining approval for the satellite distribution of one-way broadcast television in many countries. The exclusion of broadcasting services is typically associated with copyright, language and cultural issues of the broadcast content. Broadcast material is usually purchased with the right to broadcast it to specific countries and the price is determined by the number of persons likely to receive it. Furthermore, the broadcasting across national boundaries of material which is legal in one country but not in another is often a delicate matter. It is as a result of these considerations that sophisticated conditional-access schemes have been developed by the satellite broadcasters to limit reception of channels carrying copyright material to specific domestic markets. The satellite service provider chosen to broadcast the signals is also an important issue.

In all European Community countries public telecommunications networks and services have been subject to liberalisation and competition since 1 January 1998. Liberalisation and free competition in the provision of European telecommunications services over satellite links has been a fact since November 1994.

Elsewhere in the world the picture is less clear, but a recent World Trade Organisation (WTO) agreement on telecommunications, the 'Fourth protocol to the general agreement on trade in services', details commitments from a large number of countries worldwide constituting the greater part of the world market in telecommunications. Some of these countries have opened their markets in whole or in part to foreign satellite communications operators and service providers. Some have delayed implementing free access for a number of years, some have specifically restricted satellite DTH-TV and DBS provision by foreign entities and some have set limits on the extent of foreign investment in domestic telecommunications operations. However, the agreement is apparently an improvement on the previous situation. With the possibility of access to foreign markets improving over the coming years, particularly as more and more nations develop a satellite communications industry, there is hope that a free market in satellite communications of all types will emerge in the early years of the 21st century.

3.7 The business plan

The business plan will essentially define the business objectives from a service point of view, giving the basic reasoning for starting up a new satellite venture and the strategy to be followed. To begin with we may have just an idea that a global satellite-based multimedia service could be profitable. These ideas have to refined into an appropriate plan based on hard facts. Before committing funds we have to work out the risk factors involved and the basic economics of the system.

The plan is also needed to enable the company to evaluate and monitor the financial outcome of the project against the chosen strategy, to assess the results of any changes in strategy and to ensure that all necessary resources will be available to implement the plan.

A typical business plan might include the following:

- mission statement;
- objectives following on from this (SMART);
- marketing audit,
  internal audit (product strengths/weaknesses, structure of the organisation, its strengths and weaknesses),
  external audit;
- SWOT analysis (key points from marketing audit);
- strategic marketing plan addressing
  performance of the product,
  planning,
  product pricing and profitability,
  product positioning,
  product launch activities,
  launch timing,
  advertising and promotion.

3.7.1 Mission statement

A mission statement is a common feature of modern business practice and will indicate in a succinct form the focus of the activity, i.e. in our case to be the world's leading global satellite multimedia network provider offering a full range of high-quality services to our customers worldwide with the financial objective of increasing shareholder value while providing value for money and remunerating our people fairly in line with best practice.

3.7.2 Objectives

In today's terminology the business objectives need to be SMART, i.e. specific, measurable, achievable, realistic and timed.

3.7.3 Marketing audit

Internal factors — the company environment

Internal, as well as external, factors related to the anticipated markets need to be considered. An analysis of the internal environment of the company needs to be performed to assess how the company will be able to meet the challenge of the project, in terms of company size, internal auditing and quality-control procedures, project management, competences and technical strength. Clearly, previous experience in the field will be a considerable help in ensuring a healthy start to the project.

External factors — market research

Arguably one of the hardest aspects of planning a new multimedia satellite service will be carrying out the initial market research. In the author's experience this factor is perhaps the most important item to get right before any large sums are committed to the project and if it is wrong then the road to profits may be long.

Our hypothetical multimedia service is likely to have global reach. It will therefore need to be targeted and marketed to many different countries worldwide, possibly in up to 200 markets. The service is to be consumer-oriented, i.e. will be marketed to individuals, but corporate users will also be likely customers.

It is almost inevitable that most of the market research will need to be carried out by specialist consultancies. In selecting these it may be important to choose a consulting company that has its base in the home country, e.g. the UK, and having in-country branches or associate consultants familiar with the company's formal methodologies in the various major target countries. This will make the process of collecting primary research data more efficient and integrable in the final grossing-up exercises when the total global market is calculated.

The market research may be performed in several phases, starting with desk research into the target markets using information from, in the case of the UK, the Department of Trade and Industry, existing research papers on similar products, databases etc. An analysis of competing systems and networks will need to be carried out for any service providers which are offering similar services by whatever means, e.g. interactive multimedia services by the various satellite-based techniques discussed in Section 3.4 above and terrestrial systems.

During the early stages of the research it will be necessary to examine the so-called SLEPT factors of our intended markets, i.e. the social/cultural, legal, economic, political and technical factors. The following lists some of the factors, some more relevant than others to the multimedia project, which may need consideration and study:

(i)   Social and cultural

- how society is structured;
- education levels;
- how family life is organised and how the service will fit into the domestic and/or SOHO environment;
- demographics, i.e. the geographical distribution of different segments of the market;
- language;
- values and beliefs, religion, censorship etc.;
- existing media in the target country, e.g. newspapers, TV.

(ii)   Legal and institutional

- operating restrictions;
- licensing regulations;
- spectrum-allocation policy.

(iii) Economic

- GDP;
- state of economic development;
- percapita income;
- disposable income;
- stability of currency and exchange rate;
- how consumers pay for goods and services.

(iv)   Political

- political risk factors, e.g. stability of government etc. (very important!);
- local government attitude towards the foreign satellite company transmitting multimedia signals into its territory;

- trade barriers;
- tariff barriers, e.g. import taxes, and nontariff barriers, e.g. import quotas.

(v) Technical

- how the service could be used in-country;
- type of infrastructure available for support system.

Primary research may then be carried out in one of the later phases of the work, it being more costly than desk studies, and this is where the consultants will actually ask people about their requirements and interests by various methods — telephone interviews, face-to-face meetings, focus groups etc. In this later research phase it will be possible to ask interviewees for their opinions about the existing multimedia services which they have encountered, e.g. fixed-line Internet service providers etc., and explore the acceptability of various pricing schemes, price thresholds and trade-offs etc.

3.7.4 Strategic marketing plan

The final phases of the market research will be to review all the research data and conclusions to create the strategic marketing plan. This will provide a clear picture of market segmentation and allow us to determine whether there are different service packages and pricing schedules required by the different market segments. Our multimedia service has already been targeted at two such segments, individual and corporate users, but there may be many others — educational at various levels, institutional, e.g. libraries, government departments, police etc. During this part of the work it will be necessary to ensure that each segment is of a size which is measurable, economically viable, possesses unique selling points, accessible and sustainable before committing marketing effort.

Market research will thus show whether we need to take a standardised or differentiated policy to products, pricing distribution, promotion etc. and whether there will need to be different packaged services for the different countries, with different pricing and payment schemes and promotion techniques.

During the final stages of analysis, the sensitivity of the grossed-up customer base should be probed for variability in numbers, pricing and as many other variable factors as can be determined from the data. Testing the robustness of the total market to change in this way may yield some last-minute surprises!

There may be several revisions of the plan as market research data becomes available. The principal task of the plan, once the business objectives have been stated, will be to estimate the costs and revenue for a certain operating period, perhaps 10–15 years, depending on the lifetime of the satellites.

3.7.5 SWOT analysis

All business plans will include a section analysing the strengths, weaknesses, opportunities and threats of the proposed venture, a SWOT analysis, and will endeavour to quantify the financial risk posed to the project by hypothetical events which may threaten its viability.

The most obvious threats to any venture will be either from unforeseen competition, unpredicted growth in the competition from existing or new operators or a downturn in the target country's economy. A further risk to the viability and hence survival of the venture may be that it could be too successful and demand an expansion in capacity which simply cannot be met by increased resources, whether financial or system resources such as additional space segment. Such a shortfall in resources to meet unforeseen growth in the market might ultimately lead to failure if new competition capable of meeting the real market demand should emerge and attract all the custom.

A global multimedia satellite project illustrates at least two new risks that are posed to companies proposing global ventures. First, in order to work, globalisation requires access to as many markets worldwide as possible. To operate in each and every country in which we wish to operate and to offer our multimedia services requires the cooperation and approval of governments. In most if not all cases operating licences will be required from the administrations concerned and local regulations will need to be observed.

Secondly, with so many new, very capital-intensive, satellite ventures being proposed there is a real risk of a shortage of capital; the smallest single-satellite network will require US\$200–300 million or more, and multisatellite global systems several billions. Although there is apparently little shortage of investors happy to pour money into these projects at present, not all may eventually survive and a global recession may reduce the number of investors dramatically.

This highlights again the importance of understanding the market and getting one's product in front of customers in a timely and appropriate fashion by smart marketing—the customers need to be convinced that they need to buy your product.

One possible approach to minimising risk in a project of this nature is to organise a two-phase project, or possibly even more phases, with the deployment of additional satellites triggered by market response. Contracts and financing could be organised to be conditional upon triggers such as the number of subscribers or demand for transponders accompanied by clear evidence of growth in demand. This approach could almost take the form of a trial phase, to be followed by a full service when utilisation has reached a certain critical so-called take-off point. I cannot wholeheartedly recommend this approach because it is accompanied by some significant problems where a market is stimulated in anticipation of a service, with manufacturers funding development and gearing up production etc. only to discover that the network operator decides to terminate the business. This could certainly leave a bad taste in the mouths of the consumers and manufacturers who lost money, and may even lead to litigation. Something similar occurred when BSB started broadcasting DBSTV from the Marco Polo satellites, although in this case the major problems were delays and difficulties with the supply of receivers, allowing the main competitor, Sky Television, to take the market from beneath BSB's nose!

3.8 Finance

Anyone contemplating the start-up costs of the new multimedia satellite venture is going to need deep pockets; the financing of new satellite systems is fundamentally different from terrestrial systems in that the majority of the capital costs, however they are funded, are likely to be incurred in the early days of the project, with the procurement and placement of the space and ground-control segment providing immediate (after testing and commissioning) coverage for the user segment. Terrestrial systems, on the other hand, tend to evolve more slowly, with the installation schedule arranged over time to more accurately track the take-up of the service. (Examples of this might include terrestrial cellular radio networks).

However, the revenue from operating and selling the service will take time to grow. If the marketing calculations are inaccurate it may take several years longer to reach the break-even point than expected, leading to nonexistent dividends to shareholders and extensions to the periods of any loans taken out to finance the project, with the consequent unpredicted interest charges. So, again, it can be seen how important accurate market projections will be to the financial viability of the project.

3.8.1 Project costs

The costs associated with starting a multimedia service from scratch may include the following items:

(i)   Planning and tendering phase

- system design and specification drafting consultancy fees;
- tender evaluation consultancy fees;
- market research consultancy fees;
- spectrum management consultancy fees.

(ii)   Development phase

- satellite(s);
- uplink and TT&C earth station(s).

(iii) Procurement phase

- satellite(s);
- ground-control and networking facilities;
- progress monitoring consultancy fees;
- marketing, advertising and promotions;
- content providers;
- recruitment;
- training.

(iv)   Launch phase

- launcher;
- insurance.

(v)   Commissioning phase

- acceptance testing.

(vi)   Operational phase

- O& M.

(vii) Taxation.

(viii) Decommissioning phase.

There will be a possible trade off between procuring, launching and operating dedicated satellites for our multimedia venture compared with the costs of leasing or purchasing outright transponders from another space segment provider. This trade-off may be calculated for a typical satellite lifetime of 12 years to find which option will provide the greater return.

3.8.2 *User equipment costs*

Notice that the above list excludes the costs involved in the development and procurement of user equipment. In the case of our multimedia project, this equipment may consist of a small, <1 m diameter, low-power Ka-band consumer VSAT capable of being interfaced to a PC with multimedia capabilities. The entrepreneur has several options with regard to user equipment. In any event, the system design will have to include a full specification for the user segment to ensure compatibility with the space segment.

Several possible routes to the production of end-user equipment can be envisaged:

(a)   The system owner and operator assumes full responsibility for the specification, design and production of end-user equipment and sets up a distribution and dealer network to sell the units to customers.

(b)   Full hardware and software designs are licensed by the system owner to multiple third-party manufacturers which then distribute the equipment to dealers that supply consumers.

(c)   The system owner makes full hardware and software designs available to third-party manufacturers which then distribute the equipment to dealers that supply consumers.

The choice between these options is a difficult one for the system owner/network operator and is another factor which may emerge from the later stages of the market research activity, as discussed above. Clearly, there is some potential for the system operator to make some additional revenues from either manufacturing terminals or licensing designs to third parties. There is a risk of losing the market if the supply of terminals becomes too proprietary and limited. It is essential, as demonstrated by the BSB history, to get reliable, low-cost, user equipment developed in a timely fashion and available in sufficient quantity to meet demand. Alternatively, if the terminal specification is placed in the public domain, there is the risk that competitors will get a free ride, but at the same time there could be the PC-compatible effect where everyone starts to build products, add-ons etc. and the market could grow dramatically as it has done with PC equipment.

3.8.3 *Insurance costs*

A very important cost item, both during the launch phase and subsequently, is the cost of insurance for satellites and launchers. New, unproven launchers are obviously a target for very heavy insurance premiums because of the lack of any track record. The cost of insuring a satellite and launch, plus the first year of operation, is typically between 15 and 20 per cent of the capital cost of the satellite and launcher, i.e. US$45–US$60 million for each of the multimedia service's Ka-band satellites. Although Ka-band has been used relatively extensively now, it is still possible that the technological risk in using the higher frequencies between 17 and 30 GHz could lead to higher insurance premiums, another factor to consider in our SWOT analysis! The insurance premium to cover subsequent years of operation will be significantly less, perhaps two per cent of the cost of the satellite per annum, say $5 million for each Ka-band craft.

3.8.4 *Financing the project*

Financing the project will depend again on whether we use a new satellite system or whether we buy or lease transponder capacity on other company's planned or existing satellites.

Several methods of financing could be considered:

- debt financing;
- equity financing;
- joint venture partners;
- leasing.

Debt financing will be feasible if we can demonstrate that the traffic to be generated by our multimedia network is of sufficient volume to service the debt, including the interest charges and repayment of capital. If we will need to obtain funds on the commercial capital market we will have to provide very sound and well researched marketing figures to convince lenders of the viability of the project.

Equity financing is perhaps one of the more attractive options to secure funding for the project. A public offering could, in a climate where high-technology stocks are at an all-time high and there is a clamour for more and more new offerings, be a very attractive route to acquiring a large proportion of the necessary capital. There are some doubters who believe that the market for satellite communications initial public offerings is beginning to weaken, which is perhaps not surprising considering the number of systems that have been up for offer in recent years.

Joint ventures with several partners can also be an attractive way of funding and managing a complex project like a new satellite communications system. The business could benefit greatly by bringing in partners with specific competences, such as user-equipment manufacturers, terrestrial-network operators, and service and content providers, as our multimedia system will need to be able to offer content.

Leasing is a very important alternative financing approach where it is possible to defer the substantial capital expenditure over a greater period than would be the case if outright purchase of the major items of capital expenditure were to be considered.

During our financial analysis of the viability of the multimedia satellite project we will need to use some financial decision-making techniques to confirm the financial viability of our project, and possibly compare the financial performance of a number of options. As remarked before, the options might be whether to build a new system or to lease transponders on satellites owned by third parties, i.e. to use 'satellites of opportunity'.

The procedures involved all take into consideration the value of money over time, the fact that any financial decision involving spending large sums of money at the start of a project and receiving the financial benefit in the future must take into account the falling value of money with the passage of time. Therefore, we have to choose a method of depreciating the value of money that makes sense in the business environment in which we are operating. For instance, we could invest the money we might have available to build our system in a building society and receive interest without doing anything at all! As we probably all know, this is a relatively safe investment but regrettably produces, generally, the smallest rewards. So we use a concept termed the discount rate which reflects the return which we might expect to receive on our funds and we will normally expect this figure to be well in excess of prevailing interest rates. A figure of 15 per cent might be typical, which was in 1997 12.5% greater than the UK government's target for inflation.

We can apply this discount rate to calculate the present value of all expenditure made and all revenues received every year in the future for a period which might match the lifetime of the satellites, for instance 15 years for the latest designs.

Thus, present value, P, of a sum, S, received at the end of each year $1\ldots n$ using a discount rate i in decimal form, i.e. for a discount rate of 15% $i = 0.15$, is given by the following expression:

$$P = \frac{S}{(1 + i)^n}$$

This enables each item of expenditure and income to be directly comparable and the financial status of the project can then be calculated. We can use the discounted cash flows in two ways to show the returns for the project: net present value (NPV) and internal rate of return (IRR).

When using the NPV method we compare the present value of the cash inflows with the present value of the cash outflows over a period of time. If the net present value, i.e. the difference given by (present value of cash inflows minus present values of cash outflows) over a period is positive at a particular point in time then we can state that the project shows a positive cash-flow return at that point.

As an example, we can calculate the NPV each year for the 17-year period of the example Ka-band multimedia project, assuming some costs and revenues as given below, and limiting the project to, say, two colocated Ka-band satellites constructed, launched, insured and in orbit at a total cost of US$350 million. The satellites will be launched in two phases to reduce initial setup costs and to cater for increased demand in the later years of the service. After calculating the NPV each year, we can then sum the NPV for each period from the base year to any chosen year to obtain the NPV of the project to date.

It is assumed for simplicity that the satellites are used to provide demand-assigned primary-rate (2 Mbit/s) carriers for Internet service providers, Intranets etc. This is obviously a very unrealistic case, as other bearer rates and access techniques would probably be used to address this type of market. In a realistic case there will also be various types of carrier to accommodate the flexibility demanded by multimedia customers. We may assume that the multibeam satellite can support a frequency reuse factor of about three, e.g. supporting about nine to 12 beams covering the areas within the coverage region with the greatest traffic potential. Each beam will have a sufficient number of transponders to cover half the available downlink bandwidth at Ka-band, i.e. 400 MHz. Assuming demand-assigned access techniques, the two satellite systems may support about 1500 2 Mbit/s half circuits. For simplicity, these could generate a maximum effective revenue of US$900 million/annum, assuming a US$50 000 per-month charge per equivalent 2 Mbit/s leased half circuit. Of course, we need to develop a number of very much more sophisticated tariffs more suitable for timeshared use by many more customers than would be able to use the above equivalent leased half circuits, probably by a factor of at least 500–1000. We would then be able to charge an initial, one-off connection fee of, say, US$1000 and a monthly subscription fee of US$50–100 with download limits of perhaps 100–1000 megabytes per month per subscriber.

Our NPV calculations, shown in Table 3.1 using the formula given above, demonstrate that the project comes into positive cash flow by around year 6.5, indicated by the point at which the NPV becomes positive.

By repeating this exercise on a *what if?* basis it will be possible to vary the discount rate, or even develop a new project, perhaps based on leasing transponders, to calculate which project will produce the earliest positive cash flow.

Another method more favoured in the financial community is that of the internal rate of return, or IRR. The IRR technique enables the calculation and comparison of the returns achievable for alternative projects. Here one does not select the discount rate in advance but instead one computes the discount rate,

Table 3.1   NPV and IRR calculations for hypothetical multimedia satellite communications (all figures in US$000s, discount rate 15%)

							Year										
	-3	-2	-1	0	1	2	3	4	5	6	7	8	9	10	11	12	13
Two satellites constructed, launched and insured (350000)	25000	75000	75000			75000	75000										0
Uplink earth stations (30000)			2000	5000		5000	2000										
Network control centre (20000)			1500	3000		3000	1500										
Prelaunch operations (20000)		2000	5000														
Advance marketing (10000)		1000	5000														
Interest charges (5000)		3000	9360	10620	960		9960	9420									
Working capital (15000)		4000	5000	5000	1000												
Capital required (450000)		85000	102860	23620	1960	83000	88460	9420	0	0	0	0	0	0	0	0	0
Sources of cash																	
Internal funds (400)	400																
Debt (225000)	225000																
Equity (524600)				300000	224600												
Total sources of cash (750000)	225400			300000	224600	0	0	0	0	0	0	0	0	0	0	0	0
Income																	
Rental revenue				90000	375000	562500	675000	825000	900000	900000	862500	712500	487500	337500	240000	150000	0
Cost of revenue collection				9000	37500	56250	67500	82500	90000	90000	86250	71250	48750	33750	24000	15000	15000
Gross margin				81000	337500	506250	607500	742500	810000	810000	776250	641250	438750	303750	216000	135000	-15000
Operating expenses																	
NCC O&M				3500	7500	7875	8269	8682	9116	9572	10051	10553	11081	11635	12217	12828	0
uplink ES O&M				12500	23000	24150	25358	26625	27957	29354	30822	32363	33981	35681	37465	39338	0
marketing and sales				12600	52500	78750	94500	115500	126000	126000	120750	99750	68250	47250	33600	21000	0
technical support				8000	33333	50000	60000	73333	80000	80000	76667	63333	43333	30000	21333	13333	0
administration				3333	13888	20831	24998	30553	33330	33330	31941	26386	18054	12499	8888	5555	2778
bad debts				3333	13888	20831	24998	30553	33330	33330	31941	26386	18054	12499	8888	5555	5555
depreciation				41667	83333	83333	83333	83333	83333	83333	83333	83333	83333	83333	83333	83333	83333
decommissioning																	350000
Total expenses				84933	227441	285771	321454	368579	393066	394920	385505	342105	276086	232896	205724	180942	441666
Operating income				-3933	110059	220480	286046	373921	416934	415080	390745	299145	162664	70854	10276	-45942	-456666
Taxes at 25%				-983	27515	55120	71511	93480	104234	103770	97686	74786	40666	17713	2569	-11485	-114166
Net income		0	0	-2950	82544	165360	214534	280441	312701	311310	293059	224358	121998	53140	7707	-34456	-342499
Yearly NPV rel. base year 0	-40708	-117647	-121012	-26570	70073	62276	82896	154957	155467	134588	110172	73343	34679	13136	1657	-6440	-55666
NPV rel. base year 0	-40708	-158355	-279367	-305937	-235864	-173588	-90692	64265	219732	354320	464492	537835	572514	585650	587307	580866	525201
IRR (%)								4.41	11.44	15.19	17.29	18.32	18.71	18.82	18.84	18.80	18.49

on an iterative basis, which reduces the NPV to zero. In the example shown in Figure 3.1 a built-in spreadsheet function IRR(...) has been used, which is a lot easier than trying various rates until one gets the figures to work!

A barchart illustrating how the revenue and costs vary over the whole project life cycle is shown in Figure 3.1. Figure 3.2 then shows in graphical form the annual NPV (i.e. the cash flow in any particular year, discounted to year 0),

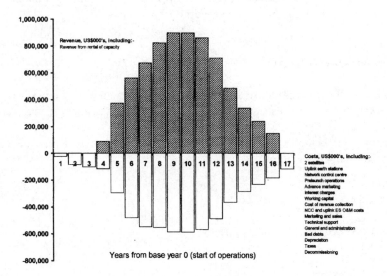

Figure 3.1   Revenue and costs for two-satellite Ka-band multimedia project

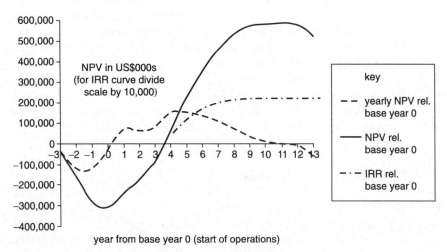

Figure 3.2   Annual NPV, cumulative NPV and IRR for two-satellite Ka-band multi-media project

NPV for the project to date and the IRR for the project to date for the whole 17-year lifecycle. These curves enable the variation in NPV and IRR to be shown throughout the life of the project, although conventionally NPV and IRR are quoted as single figures for the whole project from start to finish. For the multimedia project these figures are an NPV of US$376 million and an IRR of 42 per cent.

Alternative projects can then be compared and the one achieving the highest IRR is likely to be the better option. Currently banks lending finance on projects would be looking for IRRs exceeding 30 per cent.

3.9 Technical design

The technical-design factors are, of course, a major element of the specification of a new satellite service, and will principally address the technical specification of all components of the complete system.

If the business being proposed is a complete satellite system then the factors to be considered will include:

- orbital parameters;
- coverage;
- access technique;
- signal in space;
- payload;
- need for ISLs;
- ground networks.

The two possible options when starting a new multimedia satellite system could be to either design and procure a complete satellite system and launch it or to purchase outright or lease satellite transponders from a third party. The first of these is likely to offer much greater flexibility in the choice of system-design parameters than the latter course, such as coverage, number of beams and satellite EIRP, although it will involve more of the decision-making steps outlined in Section 3.5 and thus, perhaps, expose the entrepreneur to greater risk of failure. On the other hand, there may be the opportunity for the business to be a provider of leased capacity to others by deliberately oversizing the system, leading to additional revenue streams and some diversification.

3.10 Frequency coordination

Overcrowding of the geostationary orbit, which is still the orbit of choice for many applications, is a growing problem. It has been getting worse in some parts of the orbit since the late 1980s as the growth in applications for orbital slots has increased. This is the flip side of the rapid surge in interest in satellite telecommunications now that globalisation of telecommunications has become very big business. The situation has been worsened in recent years with many so-called paper-satellite applications made under the ITU procedures. Any ITU member state can make an application, which has to be accepted as genuine with the applicant having every intention of proceeding to realise the system. WRC 97 produced a procedure to address the problem of these systems failing to materialise. This will require administrations to provide regular evidence of progress in the establishment of the satellite system, including implementation dates, contractors, delivery dates etc. Further measures were also considered at WRC 97, including reducing the total timescale between initial applications, the advance publication and the date of bringing a network into use.

The desired orbital position of our multimedia satellites must be decided at a very early stage in the project, as the coordination process can be very lengthy and may take several years to complete. Coordination is essential from the system operation point of view, but also has vital commercial implications because only when the process has been completed will we know whether we can implement the network as originally conceived. At various points during the coordination process we may have to reduce the proposed number of channels, modify their characteristics, change the orbital positions of our satellites, move the beams to different locations etc. This is very necessary work and fully in the interests of fitting our networks in with our neighbours to reduce intersystem interference to mutually acceptable levels.

The coordination process (see Chapter 4 for details) must also be close to completion before we can finally sign any contracts to procure the satellites, since any of the above changes, if introduced during the design stage, or worse during production, could dramatically increase the cost of the contract.

Satellite coordination is an international notification and negotiation process carried out under the International Telecommunications Union (ITU) Radio Regulations with the aim of ensuring that all existing and proposed satellite networks will operate together without causing unacceptable interference.

The notification or filing process requires national administrations, which are governmental bodies and members of the ITU, such as the US FCC, to submit orbital slot details and a simplified frequency plan for each proposed satellite network (or amendments/modifications/additions/deletions to previous filings). A network in this context refers to each satellite antenna beam. Each network is further characterised by its intended service description, frequency band of operation, beamwidth, polarisation, steerability, beam pointing, associated transmitter carrier power, transponder strapping details and earth-station characteristics. These details are sent to the ITU on behalf of the applicants by their national member administrations in a first step referred to as 'Advance publication of Appendix 4 information'.

The Appendix 4 details are published and circulated to all ITU member administrations on a weekly basis. This enables experts in the administrations to check the advance details in case any of the proposed networks pose a possible source of interference to their own current or planned networks, particularly if these networks have priority over the circulated filing. Member administrations have four months in which to respond, both to the ITU and to the administration submitting details, with comments about the proposed networks. For example, they may find that either their existing or prior satellite network applications will suffer from unacceptable interference. The ITU also performs preliminary thermal-noise-based interference calculations on the preliminary information and informs any member administration which it finds may experience interference, as well as notifying the administration responsible for the filing.

Not less than six months after the first publication of the Appendix 4 data the administration proposing the new network provides greater, more specific, detail about its proposal in an Appendix 3 submission. The Appendix 3 data includes details of actual channel frequencies, modulation formats to be used etc. This is then sufficient information to enable a variety of detailed wanted carrier-to-unwanted interference ratio (C/I) calculations to be made by other administrations. These may be compared with figures for the minimum protection ratios to apply to the potentially interfering signals, based on the types of modulation, carrier power, location of receiving stations with respect to antenna beam patterns etc.

The key to the submission process is the fact that the filing of the Appendix 3 data effectively sets the priority date of the applicant's submission relative to any subsequent submissions from other administrations. Applicants having a later priority date have the responsibility of coordinating with all earlier applicants where there is the possibility of unacceptable interference, whether these networks are in operation or are just paper proposals.

After the filing of the Appendix 3 data the administration then proceeds to negotiate with all those other administrations, in a series of bilateral, and sometimes multilateral, coordination meetings to resolve any instances of unacceptable interference, with a view to finally notifying the ITU that all these instances have been satisfactorily resolved. Until this is done the administration may not operate the network, although it is possible to bring into service those beams and channels which have been coordinated in advance of the complete network if this is deemed necessary.

3.11 Billing

Accurate and timely billing is absolutely crucial to the success of any business venture and nothing could be more relevant than the case of satellite communications systems.

The method of payment, and billing systems required to operate the payment schemes, should both receive the highest possible attention during all phases of the development of the business.

Secure and accurate computerised billing systems will need specifying and procuring to meet the needs of the business. Most importantly, they will need to be sized to meet the anticipated demand for the service from start-up to some future, hopefully, higher capacity. The sizing of the billing system is, of course, one of the many variables which will need to be estimated at the outset in the business plan. Decisions will be needed as to whether to invest in a very much larger system than required at the outset to cope with the highest estimate for demand, or whether to specify a system which may be grown, perhaps in a modular fashion, as demand grows.

There are also decisions to be made about the methods of payment to be adopted for each sector of the market being addressed. In our example of a multimedia service there may be several payment options, perhaps depending on whether the customer is in the corporate or private sector.

3.12 Operations and maintenance

If our new multimedia system does include the deployment of new space segment and is not hosted in existing transponders there will be at least two operation and maintenance teams and associated ground control and uplinking equipment required to support the space segment and the service. One team will be required to operate and maintain the space segment and a second to perform similar activities for the satellite and terrestrial-network transmission aspects of the service.

The quality and professionalism of the operations and maintenance (O&M) teams will be vital to the success of the whole operation.

The space-network O&M team will be responsible for the monitoring and control of the satellites and communicating with the spacecraft platform telemetry, tracking and control (TT&C) subsystem via dedicated up- and downlink channels. As well as maintaining a 24-hour watch on the onboard housekeeping systems, monitoring the thermal state of the satellite, battery and power status and a host of other variables, this team will be responsible for executing the orbital manoeuvres required to maintain the satellites in their orbital positions. This is usually defined by tight limits on the maximum allowable x, y and z-excursions. These excursions are caused by a variety of orbital perturbations such as those caused by the earth's inhomogeneous gravity field, gravitational effects from the sun, moon and planets, solar radiation pressure etc. These all cause precession of the orbital plane around the earth's north-south axis and precession of the orbit perigee in the plane of the orbit.

The space-network O&M team will also need to monitor the satellite for signs of any interference from any terrestrial sources, such as errant transmitters or even intentional interference, as well as monitoring and controlling the satellite transponders, for example monitoring the loading of the transponders and perhaps changing the transponder gain settings.

The transmission team will be responsible for the operating the networking and uplinking equipment, probably located at the earth station, necessary to maintain the service on a daily basis. They will require the support of as much sophisticated test equipment as possible, including a complete simulation facility to enable operational problems to be simulated in an offline environment. The networking and uplink equipment will need to be based on some form of redundant configuration, duplicating highly vulnerable items such as LNAs, HPAs, up- and downconverters etc. on a one-to-one basis and other, more numerous, items such as channel units and modems, perhaps replicated on a one-for-N scheme.

It will be essential to provide an interface between the O&M people and end users. Such an interface may take the form of a service desk or customer-support centre, which would log all queries and complaints and filter them before contacting the O&M team if necessary. The customer-support centre will maintain records of the action taken to rectify any problems or customer feedback on the service and produce statistics for management to enable the quality of service to be constantly monitored.

Of course, O&M may well be contracted to a network operator which already has a suitable earth station housing the necessary equipment, or space to install new racks to support the service. This would certainly be the case at present if the service is to be provided using space segment provided by one of the consortia but not necessarily so if the space segment is either procured as part of the venture or is obtained from a private operator.

3.13 Acknowledgments

All trademarks quoted are the property of their respective owners and used with acknowledgment. The Internet worldwide web resources quoted are current at the time of writing, October 1998.

Chapter 4

Radio regulatory considerations relating to satellite communications systems

P.T. Thompson and A.G. Reed

4.1 Introduction

The operation of any radio system may cause harmful interference to another. Consequently, there is a need to have some form of management of the spectrum and orbit resources in order to ensure appropriate quality of service for existing operations, at the same time permitting access to these resources for newcomers to the field. The International Telecommunications Union (ITU) was established to facilitate this by means of a set of procedures detailed in the radio regulations (RR). We examine here the obligations of those wishing to establish satellite communications systems under such regulations. This is an extensive and continually evolving subject and thus it is only possible to outline the key principles involved. Further specific details can be obtained from the ITU and its radio regulations.[1]

Provisions for space-based telecommunications were first introduced into the International Telecommunications Union (ITU) radio regulations in 1963 at a specially convened extraordinary administrative radio conference; these provisions have since been extended and refined at subsequent conferences. In 1971 a world administrative radio conference (WARC) enhanced the space-oriented regulations, and broadcasting satellite services were specially handled at another WARC in 1977 (for ITU regions 1 and 3) and at a regional ARC in 1983 (for region 2).

A more general WARC in 1979 undertook revision of the radio regulations in total and hence the space services were considered again. During 1985 the first session of a two-session WARC relating to space services was held and the second was held in late 1988. Subsequent WARCs and world radiocommunications conferences (WRCs) have addressed special aspects of satellite communications including mobile services, high-definition TV, radiolocation services and non-geostationary orbit services.

4.2 The nature of satellite services

Within the ITU radio regulations fixed satellite services (FSS) are defined as:

> 'A radiocommunication service between earth stations at specified fixed points when one or more satellites are used; in some cases this service includes satellite-to-satellite links, which may be effected in the intersatellite service; the fixed satellite service may also include feeder links for other space radiocommunication.'

Thus, although this definition is quite broad, it does not include mobile satellite services (MSS) or broadcasting satellite services (BSS) and these have their own definitions.

The general scope of the definition of the FSS is such that the nature of the satellites and earth-station equipment, modulation methods and interference aspects of the service can be very diverse. For example, today we have mammoth 33 m diameter earth stations with large capacity and very small aperture terminals (VSATs) with antenna diameters of under 1 m and capacity adequate for low-volume traffic, almost a personal communicator! To accommodate these wide-ranging systems there is a need for a flexible frequency-management regime that is simple to apply.

Mobile satellite services have a smaller degree of variation in their parameters but, owing to the fact that the mobile terminal may be located on a variable basis over a wide geographical area, such services are normally not in frequency bands shared by fixed links.

The broadcasting satellite service also operates to a wide geographical distribution of receiving terminals. In this case interference from fixed links into the TVRO terminals may be a problem, but such terminals cannot cause interference into fixed links as they do not transmit. The frequency-management regime for such services may be less flexible than that required for the FSS.

4.3 Objectives of frequency/orbit management

It is clear that the spectrum and orbit resources are limited. However, unlike many other natural resources, they are not expendable; if all transmissions were to cease tomorrow there would still be exactly the same resources left. This feature lies at the background of all thinking related to spectrum management. The management regime can be structured to recognise the nondestructive nature of the resource and can be evolutionary to deal with the constantly changing patterns of communications. This strength of the flexible regulatory regime used for managing the spectrum resource is most apparent for the FSS where orbit resource management is also a factor of importance. Another feature of the resource-management approach currently in use is its ability to handle evolving technologies, especially those which effectively increase the utilisation of the limited resources. What then are the objectives of such regulatory machinery?

One objective is to provide a mechanism whereby an internationally agreed set of rules can be applied in order to promote harmonious use of the orbit/spectrum resource by all nations, but still leaving final and ultimate authority on a country's communications to each individual national administration.

Another objective is to ensure that services which have been through the appropriate procedures are recognised as operating, and that a level of protection is afforded to them such that they do not suffer unacceptable interference. Furthermore, these operating systems are obliged not to cause unacceptable interference to other occupants of the spectrum or orbit and to consider new networks as their requirements arise.

Yet another feature of the regulations is that of ensuring that the most advantageous frequencies are allocated to the satellite services while at the same time maximising spectrum sharing with other services (thereby enhancing the resource utilisation).

The rapid growth of satellite systems has led to such concentrated use of some parts of the resources that concern has rightfully been raised by those countries that have not yet been in a position to introduce their own satellite systems. Their concern is that when they are in a position to operate such systems the occupancy of the orbit/spectrum will be such that they will not be able to gain entry. As a consequence of such concerns another objective of the regulatory machinery is to ensure equitable access to the radio spectrum and to the orbits for satellite services.

The radio regulations detail a number of general rules for the assignment and use of frequencies. Some of these are of particular importance to satellite services:

1 Members shall endeavour to limit the number of frequencies and the spectrum space used to the minimum essential to provide in a satisfactory manner the necessary services. To that end they shall endeavour to apply the latest technical advances as soon as possible.

2 Members undertake that in assigning frequencies to stations which are capable of causing harmful interference to the services rendered by stations of another country, such assignments are to be made in accordance with the table of frequency allocations and other provisions of the regulations.

3 Any new assignment or any change of frequency or other basic characteristic of an existing assignment shall be made in such a way as to avoid causing harmful interference to services rendered by stations using frequencies assigned in accordance with the table of frequency allocations and with the other provisions of the

regulations, the characteristics of which assignments are recorded in the master international frequency register.

Another rule states:

> 'Where, in adjacent regions or subregions, a band of frequencies is allocated to different services of the same category, the basic principle is the equality of right to operate. Accordingly, the stations of each service in one region or subregion must operate so as not to cause harmful interference to services in the other regions or subregions.'

For satellite systems, these rules are fundamental to ensuring that the required objectives are met.

4.4 Frequency allocations

A band of frequencies prescribed for use by a satellite service is said to be allocated to that service. There are several degrees of allocation:

a Primary (and permitted) allocations, which are entitled to protection against unacceptable interference from all other services having allocations in the band. There can be more than one primary allocation in a frequency band and this gives rise to the need to have sharing procedures and criteria. Exclusive allocations in which the satellite service is the only service in a particular band mean that interservice interference will not occur. Such allocations are rare.
b Secondary allocations which do not enjoy the protected status but can operate under restricted conditions.
c Footnote allocations which, for a variety of reasons, are treated as additions or alterations to the normal, or table, allocations and may be primary, permitted or secondary or specifically constrained with regard to their use.

Article S5 of the radio regulations contains all of the frequency allocations. For most satellite services the allocations are predominantly in the microwave bands. The reader is referred to the radio regulations for more specific data[1].

4.5 Frequency management regimes

In each country it is the responsibility of the government-based administrative member of the ITU to undertake the appropriate management activities on behalf of potential satellite operators or users. This is implemented in various ways in different countries but the key aspect is a government oversight and involvement in the process.

Several management regimes exist in the radio regulations and are generally differentiated by the nature of the service being managed. All have some defects and meet the range of frequency-management objectives in varying degrees.

4.5.1 A priori planning

This approach establishes a rigid plan in advance of service operation for a specific service type, and has been applied to the broadcasting satellite service. A priori plans were established in 1977 and 1983 for the BSS, recognising the technologies that were considered feasible at that time. These plans are detailed in appendices S30 and S30A of the radio regulations.

Such regimes have not proved to be efficient in terms of utilisation of the orbit and spectrum, but they do give assurance regarding access to such resources. It is interesting to note that in late 1977 allocations of five channels each for BSS were made to 196 countries, but in late 1997 (20 years later) only 32 countries (16 per cent) had applied to take up their allocations and fewer than half of these had actually been used. Possible reasons for this apparent inefficiency are:

- plans are now incompatible with subsequent technological developments;
- the inflexibility militates against multipurpose satellites and their cost effectiveness;
- national beam-only satellites are uneconomic for many countries.

This plan is now under active consideration by the ITU in terms of improving its effectiveness.

4.5.2 Allotment plans

In this arrangement, which has been restricted to two particular segments of the spectrum which were less occupied than elsewhere, each country's guarantee of access will be obtained by means of a predetermined allotment. This plan needs to have some degree of flexibility as well as a cast-iron guarantee. Thus, it is more flexible than the *a priori* plan used for the BSS, a feature considered necessary if the FSS is not to be constrained in terms of future applications.

This approach has been adopted for national systems and potentially subregional systems, thereby exploiting the technical advantages of noncocoverage satellite antenna beams. WARC 85/88 defined the concept of an FSS allotment plan, and its main flexibility is provided by means of the allocation being made for a predetermined orbital arc within the geostationary orbit. The specific orbit location of a satellite within the predetermined orbital arc is determined when the system goes into service.

The FSS allotment plan was developed on the basis of at least one allotment per country from one orbital position within a predetermined orbital arc. The frequency bands allocated for FSS allotment planning are:

- 4.5−4.8 GHz and 6.725−7.075 GHz;
- 10.7−10.95 GHz, 11.2−11.45 GHz and 12.75−13.25 GHz.

Details of the plan are given in appendix S30B of the radio regulations.

In late 1988 allotments were made to 203 countries but in late 1997 (almost a decade later) only 15 (seven per cent) had applied to take up their allotments. Some members of the ITU are pressing for this more flexible regime to be reviewed, but this is unlikely to happen quickly as many are content with the guarantee of access that this scheme provides.

4.5.3 Coordination

Most satellite services are managed under a flexible-management regime within the radio regulations, known generally as coordination. The procedures comprise two major phases and have to be completed before an administration can be assured of the protection of its network. Phase 1 is the advance publication and forms a mechanism for informing others of the intent to establish a new network (or change an old one). This information phase may not be initiated earlier than five years and no later than two years before the intended commencement of use of the new network. The ITU circulates advance-publication material to all administrations. Thus, all administrations that may have an interest in the matter have the opportunity to see what is going on.

A simple interference assessment is made between the new published network and any existing or planned networks in the same frequency band, and the potential for unacceptable interference assessed (using appendix S8* of the regulations). Figure 4.1 depicts the interference paths considered in the analysis. If such an analysis indicates a need to consider methods of alleviating interference then discussions are initiated.

The ITU may help in resolving problems, but other than that it need only be informed of the status of the coordination at six-monthly intervals. All other actions are undertaken by the administrations concerned.

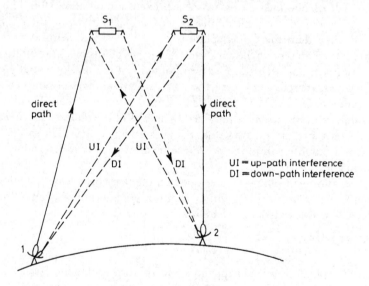

Figure 4.1   Interference from (to) systems using adjacent satellites

---

*Prior to WRC 95 this was known as Appendix 29 of the radio regulations.

This second and more formal phase is similar to the first phase but has three notable differences:

(i)   The coordination of space and earth stations with respect to interference compatibility with other networks takes specific earth-station locations into consideration.

(ii)   In frequency bands shared with terrestrial services the specific earth stations must also be coordinated with stations of these terrestrial services.

(iii)   Whereas phase 1 examination may have used only spectral emission-power densities in the assessment, actual transmission characteristics must now be taken into consideration.

Thus, the detailed second phase (coordination) needs to go beyond the simplistic interference analysis used in phase 1. Various ITU recommendations and reports give some guidance as to how to do this, but the matter is not specifically detailed in the radio regulatory text enabling the latest interference-analysis approach to be exploited. Furthermore, the precise operational role of the networks may be exploited in order to minimise interference by giving carriers a suitable small frequency offset or avoiding putting incompatible carriers near to each other in frequency. It is important that this flexibility be present in order to permit the establishment of as many networks as possible.

If potential, or actual, conflicts with other prior assignments have been resolved then the system under consideration can be notified, that is, submitted for recording in the master international frequency register (MIFR).

Historically, the ITU has been the guardian of the MIFR which contains the records of frequency assignments for the radio services. These assignments are the output of the procedures, and are sets of data characterising individual transmitting or receiving facilities. The assignments are made by individual administrations, but only after a formal procedural and technical examination by the ITU are they included in the MIFR (when all potential, or actual, conflicts with other prior assignments have been resolved).

Thus, the coordination procedure is a carefully staged process with detailed actions and schedules which are particularly important for networks using the geostationary satellite orbit. Further details on the interference assessment undertaken in the two phases are given in Chapter 7.

4.5.4 Coordination of earth stations

When a network is coordinated as a whole, the task is undertaken using typical earth-station parameters. In the event that the parameters of a particular earth station are different, there is no problem provided that the interference it may cause is no worse than that which was calculated using the original figures. Furthermore, no greater protection from interference can be expected than that originally afforded the station.

4.6 Responsibilities of large satellite organisations

Members of organisations such as INTELSAT, EUTELSAT or INMARSAT which operate in the FSS bands all have specific conditions imposed upon them, by rules adopted by those organisations in order that a well organised and disciplined network is maintained. Most such organisations mandate that their members safeguard the organisation from technical and economic harm by means of articles and agreements which member states sign. Detailed technical coordination of the organisations' networks is carried out in a manner very similar to that of the ITU Radio Regulations. In addition, technical coordination of the organisations' networks is undertaken in the ITU via a national administration, usually the administration of the country in which the headquarters of the organisation is located. By means of internal procedures each organisation ensures that its members adhere to the parameters of the coordinated network. It is the responsibility of the member to coordinate its earth station within the ITU procedures and to operate that station within the coordinated parameters.

4.7 Coordination with terrestrial services

When the frequency bands are shared with terrestrial systems such as radio-relay links or radar systems a number of interference conditions need to be investigated, namely:

- signals from the satellites must not cause unacceptable interference to the receivers of the terrestrial service;
- signals from the satellite earth station must not cause unacceptable interference to the receivers of the terrestrial service;
- signals from the terrestrial stations must not cause unacceptable interference to the receivers of the satellite earth stations;
- signals from the terrestrial stations must not cause unacceptable interference to the satellite receivers.

As with the interference between two satellite networks, there is a two-stage process for addressing interference between terrestrial and satellite systems.

The first stage of this procedure is aimed at determining an area around the earth station, known as the coordination area, within which the presence of a terrestrial station could lead to unacceptable interference. A simplistic model is adopted similar to that used for the satellite network procedures. However, to determine the contour of the coordination area several additional factors must be taken into account:

(i)   The proximity of the earth can introduce effects which influence the transmission of the radio-relay-link signals beyond the horizon. These signals are not cut off immediately by the intersection of the earth, and thus detailed knowledge of the diffraction over the horizon is required. The presence of hills and mountains can make this relatively complicated.

(ii)   The changing value of the radio refractive index of the air with height and temperature gives rise to a beam-bending effect and this can cause propagation of signals far beyond the horizon. Furthermore, this bending effect is variable depending upon the weather conditions and situations can occur where the refractive index of the atmosphere is such that the signal can be trapped in a duct and propagated for long distances with little loss. This is known as duct propagation.

(iii)   Scattering from raindrops in a common volume where the signals can be directed from one interfering source into the radio beam of another system can give rise to rain-scatter interference. The level of such interference is dependent upon the rain intensity, the particular geometry involved and the various antenna characteristics as well as the signal type.

These factors are used in determining the coordination area of the earth station and, in general, rain-scatter considerations determine most of the area except in the region of the earth-station main beam. The calculation of the coordination area is performed using the detailed procedure described in appendix S7* of the radio regulations. It should be noted that the coordination area can extend across national boundaries and therefore requires international cooperation in order to address the situation.

If there are radio-relay links within the coordination area then a second stage of assessment and resolution is required. This stage reviews the situation in much more detail and is similar to that used for the satellite network coordination, except that various propagation factors have now to be taken into account. In many cases it is found that, owing to the specific pointing of the signal paths, features such as rain scatter and duct propagation are not critical and that the services can coexist. Again, specific interference qualities of the signal types are now investigated and the analysis can be complex to conduct. The path profile of the terrain can also be taken into account. Techniques such as positioning the earth station in a natural bowl or behind a mountain can ease the interference situation as can locating the earth station away from areas of high population where fewer radio-relay systems are likely to be operating.

With the use of small earth stations it becomes practical to shield the site from interference and reductions of interference of the order of 20 dB or so are possible.

4.8 NonGSO systems

Since 1992 there has been an increased interest in operating constellations of satellites in nongeostationary orbits. These low-earth orbit (LEO) and medium-

---

* Prior to WRC 95 this was known as Appendix 28 of the radio regulations.

earth orbit (MEO) systems may be interfered with or cause interference to GSO systems. Much effort has been expended in understanding how to regulate this situation. Historically, such satellites were required to cease transmission as they crossed the GSO in order to eliminate harmful interference. This is quite an onerous situation and the regulations are being changed to allow alternative approaches where feasible. Thus there is a need to consider short-term interference limits and it is true to say that some objection has been created to accepting such limits, from both sides. The situation is made worse by the complexity of the methods used to calculate short-term interference limits on a carrier by carrier case. For example:

- Different approaches are required when the satellites employ onboard processing;
- There is a need to consider the influence of fade countermeasures;
- How do you define 'short-term interference'?

In general, it is possible to conduct the necessary analysis only by means of sophisticated computer simulation models, an increasing trend in frequency management.

WRC 95 and WRC 97 have set the foundations for handling nonGSO systems, albeit incomplete and complex in nature. This is an area that will

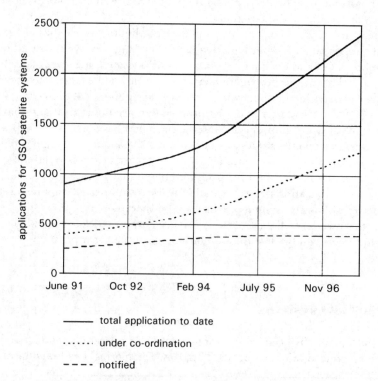

Figure 4.2   Escalation in demand for spectrum for GSO satellite services

receive much more attention in the future as the demands are significant and the investments in systems very high.

4.9 The workload of the ITU satellite systems frequency management regime

It is interesting to examine the workload of the ITU satellite frequency management regime especially from the viewpoint of coordination. Figure 4.2 shows the situation regarding GSO systems for the period June 1991 to September 1997. It is of concern both that many advanced publications are still being submitted and that the number of systems reaching the notified stage is flattening out.

Figures 4.3 and 4.4 indicate the applications for GSO and nonGSO systems, respectively. For the latter the dominant requirement is in the 2 GHz bands (although interest in higher bands is now growing rapidly) and the former shows demand for almost all areas of the GSO.

Figure 4.5 indicates the countries which have applied for GSO systems and Figure 4.6 gives similar data for nonGSO systems. It is apparent that a small number of countries apply for the majority of the systems.

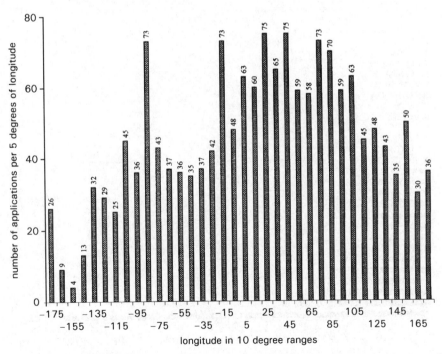

Figure 4.3   Use and potential use of the GSO—applications for spectrum as of September 1997

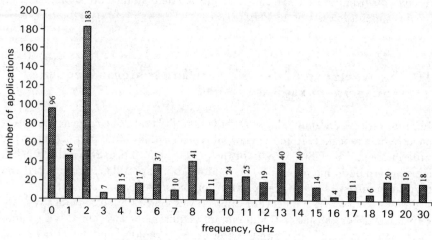

Figure 4.4   NonGSO applications for spectrum as of September 1997

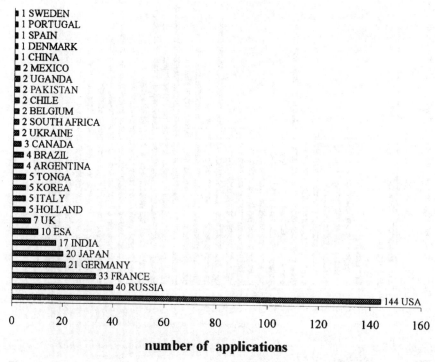

Figure 4.5   Applications for nonGSO systems as at September 1997

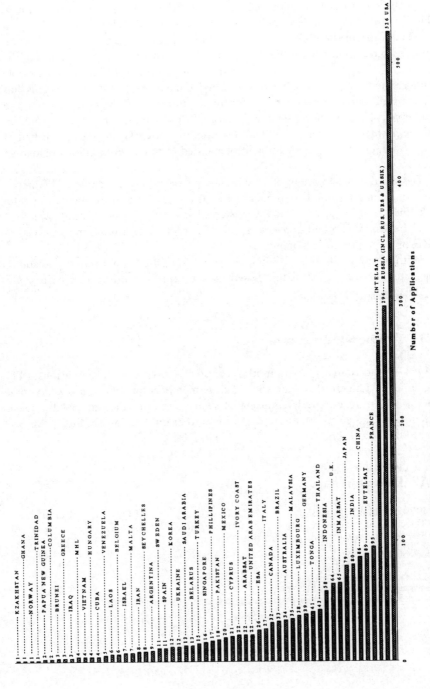

Figure 4.6   Applications by country for GSO systems

4.10 Conclusions

An overview of the objectives of resource management for both the spectrum and orbit resources has been given and some of the technical features examined. Three regulatory regimes have been addressed and their values indicated.

The nature of the FSS is such that a wide range of parameters exists and no simple rigid regulatory mechanism would suit. Instead, as a consequence of the evolution of the regulatory machinery being matched to the objectives, a flexible management regime has been established, known as coordination. This has served well to date but scope exists for further improvements given the emerging technology and changing service patterns. In particular, the practical solution of the problem of 'paper satellites' (i.e. satellite networks advance published as 'place holders' but unlikely to be implemented in practice) is still a worthwhile goal.

Despite the development of very high-capacity fibre-optic cables, ITU activities in managing satellite frequency bands are growing at an increasing rate. The increased interest in nonGSO systems has required a substantial volume of new thinking and regulation to be applied in order to continue to meet the objectives of frequency management.

With national administrations at its heart, the ITU is in a sound position to meet the continuing challenge posed by the rapid evolution of satellite communications.

4.11 Acknowledgments

Much of the material used in this chapter is extracted from various ITU documents. The efforts of the ITU and its members have been a fundamental factor in setting our understanding of such matters to the current-day level. These remarkable efforts and the quality of their publication are fully acknowledged by the authors.

4.12 References

1 ITU Radio Regulations, available from the ITU at: International Telecommunications Union (ITU), Place des Nations, CH-1211 Geneva 20, Switzerland. www.itu.int/publications/

Chapter 5

Introduction to antennas

B. Claydon

5.1 Introduction

Antennas employed in space communications are key components providing the vital links between the ground and the spacecraft.

The fundamental principles of antenna theory are based on classical electromagnetic field theory and are discussed in detail in many major text books.[1,2] The basic starting point for the theory is Maxwell's equations for a homogeneous isotropic medium which relate electric and magnetic fields to source currents. From Maxwell's equations, the wave equation can be derived which describes the propagation of an electromagnetic wave in a nondispersive homogeneous isotropic medium. The performance of an antenna configuration is found from solution of the wave equation, generally using approximate techniques.

Antennas can be broadly classified by the frequency spectrum in which they are commonly applied, or by their basic mode of radiation. The most important type for space communications is the aperture antenna which includes horns, reflectors and lenses (although the latter are not usually employed in this particular application). The aperture antenna will form the basis of the present discussion. A further antenna type, namely the array antenna, is used in some space applications, for example mobile communications.

5.2 Basic aperture antenna definitions and relationships

5.2.1 Principle of reciprocity

The principle of reciprocity is of fundamental importance in antenna theory and practice since the properties of a particular antenna can be determined either by analysis or measurement, with the antenna as a transmitter or as a receiver. In practice, for computational analysis the antenna is generally assumed to be in the transmit mode, whereas for measurements the antenna is assumed to be receiving.

5.2.2 Antenna radiation pattern

A transmitting antenna does not radiate uniformly in all angular directions, nor does a receiving antenna detect energy uniformly from all directions. This directional selectivity of an antenna is characterised in terms of its radiation pattern. This pattern is a plot of relative strength of radiated field, amplitude and phase, as a function of the angular parameters, θ and ϕ, of a spherical coordinate system for a constant radius, r. The amplitude of this pattern is most important and can be expressed as a relative power or field pattern (normalised to unity maximum) or a logarithmic decibel pattern (with a maximum of 0 dB).

The antenna radiation pattern, which typically comprises a main beam and a sidelobe structure, is commonly depicted as a two-dimensional plot as illustrated in Figure 5.1. For antennas with dimensions small compared to the wavelength, a polar plot, Figure 5.1*a*, is often employed. As the antenna dimension increases, a more detailed picture of the radiation pattern is achieved using a cartesian plot, Figure 5.1*b*, since the width of the main beam decreases and the periodicity of the sidelobe region increases. Such a representation is generally employed for earth-station antennas. Satellite antenna radiation patterns are often expressed as isogain contour plots which are superimposed on the coverage area, as illustrated in Figure 5.1*c*. In this case only that portion of the radiation pattern which is actually incident on the earth's surface is of interest.

The peak of the main beam represents the highest level of field strength and approximately 70 per cent of the radiated energy is enclosed in the main-beam region. The sidelobe region represents a potential source of interference into the communications link and for this reason is generally required to be of low level.

5.2.3 Antenna half-power beamwidth

The angular width of the main beam of the antenna radiation pattern is characterised by the half-power beamwidth (HPBW). This is defined as the full angular width between the two points which are 3 dB below the main-beam peak.

The HPBW is dependent on the illumination distribution in the antenna aperture and the aperture dimension in the plane in which the pattern is measured. For a circular aperture with diameter D/λ in wavelengths:

$$\text{HPBW} \simeq \frac{N\lambda}{D} \text{ degrees} \tag{5.1}$$

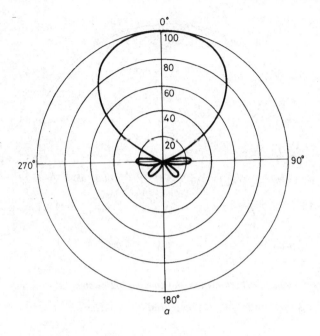

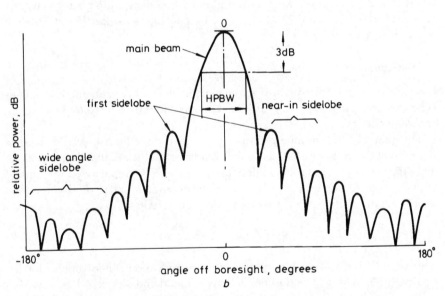

Figure 5.1   *Various methods of representing an antenna radiation pattern*

      a   Polar plot of antenna radiation pattern

      b   Cartesian plot of antenna radiation pattern

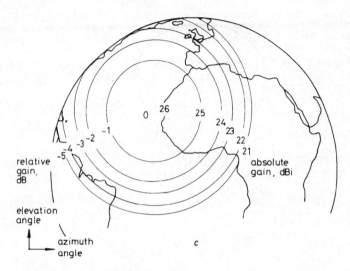

Figure 5.1   Various methods of representing an antenna radiation pattern

c Isogain contour plot of antenna radiation pattern

where N is dependent on the aperture illumination distribution being typically 58 when it is uniform and 73 when tapered according to the law $(1 - r^2)$, r being the normalised radial parameter.

5.2.4 Gain, directivity and efficiency

Gain and directivity are quantities which define the ability of an antenna to concentrate energy in a particular direction and are directly related to the antenna radiation pattern.

The gain, $G(\theta, \phi)$, of an antenna in a specified direction (θ, ϕ) is defined as the ratio of the power radiated per unit solid angle, $P(\theta, \phi)$, in the direction (θ, ϕ) to the power per unit solid angle radiated from an isotropic antenna supplied with the same total power, P_T. (The isotropic antenna is a hypothetical antenna which radiates uniformly in every direction.) Therefore:

$$G(\theta, \phi) = \frac{P(\theta, \phi)}{P_T/4\pi} \tag{5.2}$$

This quantity is an inherent property of the antenna and includes ohmic and dissipative losses arising from the conductivity of metal and dielectric loss.

The directivity, $D(\theta, \phi)$, of an antenna does not include these dissipative losses and is defined as being the ratio of $P(\theta, \phi)$ to the power-per-unit solid angle from an isotropic antenna radiating the same power, P_R. Therefore:

$$D(\theta, \phi) = \frac{P(\theta, \phi)}{P_R/4\pi} \tag{5.3}$$

The ratio of $G(\theta, \phi)$ to $D(\theta, \phi)$ is termed the radiation efficiency of the antenna.

The value of $G(\theta, \phi)$ where maximum radiation occurs is simply called the gain of the antenna and in most cases is expressed in dBi (decibels relative to isotropic). This value corresponds to the peak of the main beam of the radiation pattern which is generally in the direction $(0, 0)$ referred to as the antenna boresight direction. This value of gain, G, is related to physical aperture area, A, of the antenna by the expression:

$$G = 10 \log_{10} \left(\frac{4\pi}{\lambda^2} \eta A \right) \text{dBi} \tag{5.4}$$

where λ is the operational wavelength and η the antenna efficiency. In the case of a circular antenna aperture of diameter D, the gain is expressed by:

$$G = 10 \log_{10} \left[\eta \left(\frac{\pi D}{\lambda} \right)^2 \right] \text{dBi} \tag{5.5}$$

The antenna-efficiency factor, η, used in the above expression is always less than unity although it is more often expressed as a percentage. For a practical antenna configuration, the efficiency factor comprises several components which may, for example, include:

$\eta_I =$ illumination efficiency, which accounts for the nonuniformity of the illumination and phase distributions in the antenna aperture;

$\eta_S =$ spillover efficiency, which represents the ratio of the total power in the antenna aperture to the total power radiated by the primary feed horn;

$\eta_B =$ blockage factor, which allows for the incomplete utilisation of the antenna aperture owing to the blocking effect of supports, subreflector, etc.;

$\eta_E =$ losses due to manufacturing and alignment errors, which include variations from a uniform phase distribution in the antenna aperture owing to irregularities in the reflector profiles;

$\eta_O =$ ohmic losses in the primary feed chain.

5.2.5 Antenna noise temperature

In satellite communications, the noise caused by the thermal loss of the ground and the atmosphere is received via the sidelobes of the antenna and degrades the overall receive-band performance of the antenna. The antenna noise temperature in degrees Kelvin at an antenna elevation angle of θ_0, can be expressed as:

$$T_A(\theta_0) = \frac{1}{4\pi} \int_0^{2\pi} \int_0^{\pi} T^1(\theta_0, \theta, \phi) G(\theta, \phi) \sin \theta \, d\theta \, d\phi \tag{5.6}$$

where $T^1(\theta_0, \theta, \phi)$ is the brightness-temperature function $T(\theta)$, transformed into the coordinate system of the antenna gain function, $G(\theta, \phi)$. The brightness-temperature function can be found from consideration of atmospheric attenuation under clear-sky conditions at the frequency of interest, and is usually determined from tables.

An estimate of the antenna noise temperature can be found by considering the overall antenna efficiency factor, η, and its various contributions, η_I, η_S, η_B, η_E and η_O defined in the previous Section. The noise temperature at an elevation angle of θ_0 is given by[3]:

$$
\begin{aligned}
T_A(\theta_0) \approx\ & \eta T(\theta_0) + (1 - \eta_I) T(\theta_0) \\
& + (1 - \eta_S) \cdot \tfrac{1}{2} \cdot [T(\theta_0) + T(0)] \\
& + (1 - \eta_B) \cdot \tfrac{1}{2} \cdot [T(\theta_0) + T(0)] \\
& + (1 - \eta_E) T(\theta_0) \\
& + (1 - \eta_O) \cdot 290
\end{aligned}
\tag{5.7}
$$

5.2.6 Reflection coefficient, voltage standing-wave ratio and return loss

For satellite communications applications, a conjugate match is desired between the antenna and the transmission line or device to which it is connected. When such a match does not exist, some of the available power is lost by reflection. The ratio of the reflected power to the incident power is related to the voltage reflection coefficient, Γ, by the expression:

$$
|\Gamma|^2 = \frac{\text{reflected power}}{\text{incident power}}
\tag{5.8}
$$

The voltage standing wave ratio (VSWR) is related to the voltage reflection coefficient by:

$$
\text{VSWR} = \frac{1 + |\Gamma|}{1 - |\Gamma|}
\tag{5.9}
$$

The return loss is the logarithmic form of the voltage reflection coefficient and is given by:

$$
\text{return loss (dB)} = 20 \log |\Gamma|
\tag{5.10}
$$

Typical values for an earth-station antenna might be: a voltage standing wave ratio of 1.2:1, a voltage reflection coefficient of 0.091 and a return loss of -20.8 dB.

5.2.7 Polarisation

Both the antenna and the electromagnetic field received or transmitted have polarisation properties. The polarisation of an electromagnetic wave describes the shape and orientation of the locus of the extremities of the field vectors as a function of time. A wave may be described as linearly polarised, circularly polarised or elliptically polarised.

Linear polarisation is such that the E-field, electric field, is oriented at a constant angle as it is propagated. The angle may be arbitrary but often for convenience is defined as being either vertical or horizontal.

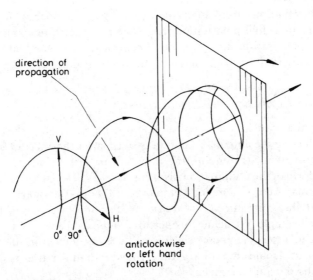

Figure 5.2   Schematic of left-hand circular polarisation

Circular polarisation is the superposition of two orthogonal linear polarisations, for example vertical and horizontal, with a 90° phase difference. The tip of the resultant *E*-field vector may be imagined to rotate as it propagates in a helical path as illustrated in Figure 5.2. If the wave intersects a stationary plane normal to the direction of propagation, then the circular path which the rotating *E*-vector traces out on the plane defines the hand of polarisation. In Figure 5.2, a left-hand circularly-polarised (LHCP) wave is shown since the circular path, viewed as the wave propagates away from the observer, describes an anticlockwise rotation. A clockwise rotation would indicate a right-hand circularly-polarised (RHCP) wave.

An elliptically-polarised wave may be regarded as either the resultant of two linearly-polarised waves or as the resultant of two circularly-polarised waves with opposite directions.

5.2.8 *Crosspolarisation and polarisation discrimination*

The crosspolarisation of a source is becoming of increasing interest to satellite communication antenna designers.

In the case of an antenna transmitting, or receiving, a linearly-polarised field, the crosspolar component is the field at right angles to this copolar component. For example, if the copolar component is vertical, then the crosspolar component is horizontal.

Circular crosspolarisation is that of the opposite hand to the desired principal or reference polarisation. Impure circular polarisation is, in fact, elliptical. The level of impurity is measured by the ellipticity and known as the axial ratio.

A detailed definition of crosspolarisation has been considered by Ludwig.[4] Of importance in a dual polarisation frequency reuse satellite communication system is the polarisation discrimination between the copolar and crosspolar signals, especially in the antenna main beam region as illustrated in Figure 5.3.

5.2.9 Aperture distribution and illumination taper

For an antenna which generates a single focused beam, the principal parameter affecting the antenna radiation pattern, after the aperture size, is the aperture illumination distribution; the radiation pattern being essentially the Fourier transform of the illumination distribution.

Since the majority of antennas employed in both the space and earth segment of satellite communications utilise circular apertures, we will consider this case as an example. We define an aperture illumination distribution $E_a(r, \psi)$ which is dependent on the polar coordinates (r, ψ) as defined in the geometry shown in Figure 5.4. The amplitude of the far-field radiation pattern at the point (θ, ϕ), normalised to unity, is given by:

$$E(\theta, \phi) = \frac{1}{\pi a^2} \int_0^{2\pi} \int_0^a E_a(r, \psi) \exp[-jkr \sin \theta \cos(\phi - \psi)] r \, dr \, d\psi \qquad (5.11)$$

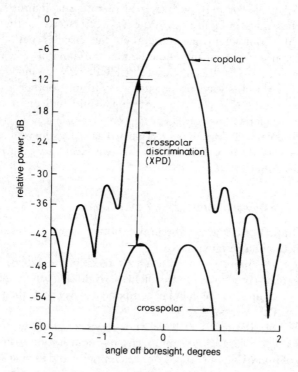

Figure 5.3   *Polarisation discrimination*

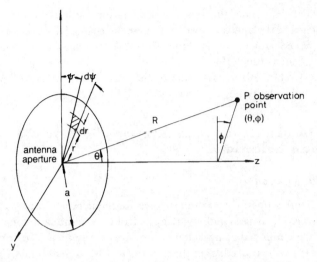

Figure 5.4 Geometrical parameters relating to far-field radiation pattern of an antenna

For circularly symmetric aperture illumination distributions, this expression reduces to:

$$E(\theta) = \frac{2}{a^2} \int_0^a E_a(r) \cdot \mathcal{J}_0(kr \sin \theta) \cdot r \, dr \qquad (5.12)$$

where $\mathcal{J}_0$ is the Bessel function of the first kind and order zero.

In general, the integral above has to be evaluated numerically, although, in certain cases it is possible to find an analytic solution.

A most useful aperture distribution for circular aperture antennas is:

$$E_a(r) = (1 - r^2/a^2)^n + b \qquad (5.13)$$

where n is an integer and b the level of the field at the edge of the aperture. The antenna radiation pattern resulting from this distribution can be written as:

$$E(u) = 2b \frac{\mathcal{J}_1(\pi u)}{\pi u} + n! \frac{\mathcal{J}_{n+1}(\pi u)}{(\pi u/2)^{n+1}} \qquad (5.14)$$

where

$$u = \frac{2a}{\lambda} \sin \theta$$

$\mathcal{J}_n(\pi u)$ being the Bessel function of the first kind and order n.

For the special case of uniform illumination in the aperture, $b=0$ and $n=0$ and the antenna radiation pattern becomes:

$$E(u) = \frac{2\mathcal{J}_1(\pi u)}{\pi u} \qquad (5.15)$$

For a reflector antenna, the aperture distribution and the illumination taper (the relative level of the aperture distribution at the edge of the aperture to that at the centre) are controlled by selection of the feed horn radiation pattern and the reflector geometry such that the reflector edge angle intersects the feed pattern at the desired relative level. In general, the ratio between centre and edge power densities will lie between −3 dB and −30 dB, depending on the application.

5.3 Typical antenna configurations for satellite communications

5.3.1 Horn antennas

Horn antennas operating at microwave frequencies are used to produce wide beam coverage, such as global coverage from a satellite antenna, but more frequently are employed as feeds for reflectors when higher gain and/or narrower beamwidths are required, as in the case of earth-station antennas.

5.3.1.1 Simple pyramidal and conical horn antennas

The simplest forms of horn antenna are open-ended waveguides of either circular or rectangular cross-section. However, to provide a narrower beamwidth, a flared section of waveguide is used resulting in a radiating aperture several wavelengths wide and giving rise to a good match between waveguide impedance and free space.

By choosing the aperture dimensions of a rectangular apertured horn with fundamental TE_{01} mode excitation such that the beamwidths are equal in the two principal planes (E- and H-plane), a radiation pattern with good circular symmetry can be achieved. Typically, the full −10 dB beamwidths in the two planes are given by:[2]

$$E\text{-plane: } \theta^E_{-10} \approx \frac{88\lambda}{b} \text{ degrees, } \frac{b}{\lambda} < 2.5 \qquad (5.16a)$$

$$H\text{-plane: } \theta^H_{-10} \approx 31 + 79\frac{\lambda}{a} \text{ degrees, } 0.5 < \frac{a}{\lambda} < 3 \qquad (5.16b)$$

where a and b are the aperture dimensions in the two planes. Such a horn, unfortunately, suffers a high crosspolarised component in the diagonal 45° planes.

More frequently, horns with circular apertures are employed in satellite communications applications. The radiation pattern of such a conical horn with fundamental TE_{11} mode excitation and small flare angle is given by simple formulae involving Bessel functions, *viz*:

$$E\text{-plane: } F_E(u) = E_0 J_1(u)/u \qquad (5.17a)$$

$$H\text{-plane: } F_H(u) = E_0 \frac{J'_1(u)}{1 - \left(\dfrac{u}{1.841}\right)^2} \qquad (5.17b)$$

where a is the aperture radius, E_0 is constant and $u = (2\pi a/\lambda)\sin\theta$.

For large aperture diameters, the *E*- and *H*-plane pattern beamwidths differ, implying that a high crosspolar component exists in the diagonal plane. The *E*-plane pattern also suffers from high sidelobe levels.

For smaller aperture diameters, the radiation-pattern performance of conical horns is very dependent on the flange surrounding the aperture.

5.3.1.2 Dual-mode conical horns

The dual-mode concept, originally proposed by Potter,[6] relies upon the use of the higher-order TM_{11} mode to symmetrise the aperture distribution of large-aperture conical horns excited in purely TE_{11} mode. The presence of the TM_{11} mode does not affect the *H*-plane radiation pattern but, when properly phased and combined with the fundamental TE_{11} mode, improves the *E*-plane radiation characteristics. The resultant radiation fields exhibit near axially symmetric copolar patterns with low sidelobes, low crosspolarisation and a true phase centre.

The overall bandwidth of the dual-mode horn is limited by the need to maintain the correct phase relationship between the two waveguide modes at the horn aperture. In the original Potter design, the TM_{11} mode is generated by a step caused by an abrupt change in the horn radius and the correct phase relationship achieved by a straight section immediately following the step. The design is, therefore, frequency dependent with crosspolar suppression of better than -30 dB being realisable over typically five per cent bandwidth only.

5.3.1.3 Hybrid-mode corrugated conical horn

Corrugated conical horns are capable of providing radiation fields with circularly symmetric copolar patterns, low levels of crosspolarisation and low sidelobes over relatively broad frequency bandwidths. The fundamental mode associated with the corrugated structure is the HE_{11} mode which can be considered as a hybrid of the TE_{11} and the TM_{11} modes of smooth-wall waveguides. In contrast to the dual-mode smooth-wall configurations, however, the TE and the TM components in the HE_{11} mode have the same cut-off frequency and the same phase velocity. They, therefore, remain in the correct phase relationship along the guide, independent of frequency, so that the bandwidth limitation of dual-mode horns is removed.

Two distinct versions of hybrid-mode horns are common: one in which the flare angle is large and the horn is relatively short, the other in which the flare angle is very gradual so that the horn may be long, Figure 5.5. The first of these is often termed the scalar horn. Its distinguishing feature is that the large flare angle causes the spherical wave in the horn aperture to depart from a plane wave by half a wavelength or more. Because of this, the shape of its radiation pattern depends very little on frequency, being determined almost entirely by the flare angle. Over at least a 1.5 : 1 frequency range the width of the pattern at the -15 dB level is about equal to the total flare angle in the horn. The *E*- and *H*-plane phase centres are coincident being located at the apex of the flare, and

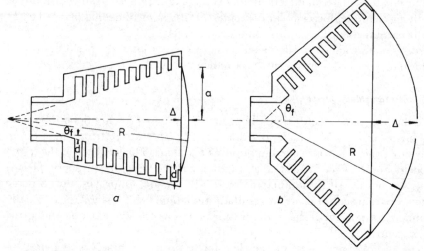

Figure 5.5   Geometry of the corrugated horn

    a Narrow flare
    b Wide flare

there is virtually no sidelobe structure to the pattern. These desirable properties hold for semiflare angles of up to 70° and for horn apertures exceeding about 2.5 λ in diameter. Thus such a scalar horn is well suited for prime-focus reflector applications.

The other version of the corrugated configuration is exemplified by a conical horn with a flare angle sufficiently small that the spherical wave in the aperture is within $\lambda/5$ or less of a plane wave. The bandwidth is thus limited by this condition and by the additional circumstance that the phase centre is now located between the apex and the aperture, its exact position depending on the frequency.

In all cases, the crosspolar bandwidth of corrugated horns is limited, first by the departure of the HE_{11} mode from the so-called purely balanced hybrid condition and, secondly, by the excitation of higher-order modes along the length of the horn. The dependency of these factors upon the horn design parameters such as the corrugation depth, period, horn flare angle and the aperture dimensions are generally well understood, however, and relatively broadband performances can be realised with good design practice.

Considering the case of conventional single-depth corrugated horns with moderate (<15°) flare angles and aperture dimensions in excess of about 3λ, the present state-of-the-art suggests bandwidths of typically 15 per cent over which the crosspolar radiation fields can be suppressed to below −35 dB relative to the copolar peak. Wider bandwidths of typically 35–40% can be realised with wide flare angle horns.

Corrugated elliptical and rectangular aperture horns offer a means of realising elliptical beam patterns with low crosspolar characteristics. Their use in space applications to date has been very limited.

5.3.2 Reflector antennas

When higher gain and narrower beamwidths are required from an antenna it is necessary to increase the aperture dimension and simple horn antennas become impractical. For these applications a passive reflecting surface is utilised which is illuminated by a smaller primary feed horn.

5.3.2.1 Prime-focus paraboloidal reflector antenna

The simplest form of reflector antenna is the axisymmetrical paraboloid with a feed horn located at its focal point, as illustrated in Figure 5.6. A spherical wave emanating from the feed horn at the focus is converted into a plane wave radiated from the aperture of the antenna. The antenna is characterised by its diameter, D, and focal length, F, with the reflector curvature being a function of the ratio $F:D$.

Such a configuration can be employed for antenna diameters typically of up to 3 m. For larger diameters the necessary long waveguide run from the feed at the focus to a conveniently located electronics box, generally at the rear of the reflector, is undesirable.

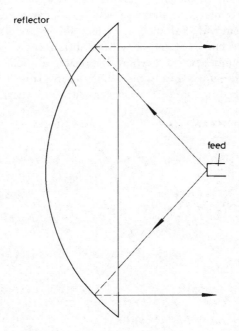

Figure 5.6 Axisymmetric single reflector system

5.3.2.2 Dual-reflector axisymmetric reflector antennas

For larger reflector antenna diameters, a more compact axisymmetric configuration can be realised by the introduction of a smaller subreflector as illustrated
in Figure 5.7. The primary feed horn is now located near the rear of the main
reflector, eliminating the need for long and lossy waveguide runs to the electronics box.

Two types of subreflector configuration are possible:

(i)   The Cassegrain arrangement where the subreflector is a section of hyperbo-
       loid situated within the focus of the paraboloidal main reflector.

(ii)   The Gregorian arrangement where the subreflector is a section of an ellipsoid
       located outside the focus of the paraboloidal main reflector.

Both systems result in similar RF performance characteristics, although the
Cassegrain is more commonly employed in earth-station applications.

5.3.2.3 Asymmetric (or offset) reflector antennas

Axisymmetrical reflector antennas, such as those described in the previous
Sections, have been favoured because of their straightforward geometry and
mechanical simplicity. However, the electrical performance of an axisymmetrical antenna is limited by the effects of aperture blockage caused by the
primary feed horn and, in dual-reflector systems, the subreflector. Although the
geometry of an asymmetric, or offset, reflector is less straightforward than that
of its axisymmetric counterpart, and various effects due to the asymmetry must
be considered, the removal of all blockage effects brings about major improvements in both antenna efficiency and sidelobe performance.

Similar to its axisymmetric counterpart, the offset parabolic reflector can be
realised as a single reflector fed from the vicinity of its prime focus, as shown in
Figure 5.8, or arranged in a dual-reflector configuration where the main offset

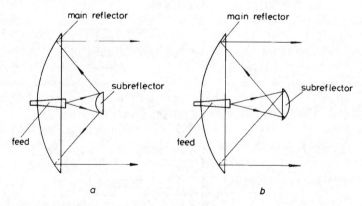

Figure 5.7   Axisymmetrical dual-reflector systems

       a Cassegrain
       b Gregorian

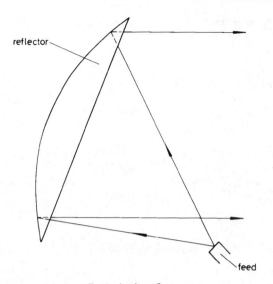

Figure 5.8   Asymmetric (or offset) single-reflector system

reflector is illuminated by the combination of a primary feed horn and subre-flector as shown in Figure 5.9. The latter can be arranged such that the axes of the main reflector and subreflector are coincident or tilted with respect to each other so that no blocking of the optical path occurs either by the primary feed or subreflector. The dual-offset configuration can be realised in either a Cassegrain type in which the subreflector is a section of a hyperboloid, or a Gregorian type in which the subreflector is a section of an ellipsoid with both focii situated between the two reflecting surfaces.

5.3.3 Array antennas

Array antennas comprise a series of discrete radiating elements fed from a single input source. The type of radiating elements depends on such parameters as the application, frequency bands and polarisation and, typically, could be dipoles, waveguide slots, horns or microstrip patches. A beamforming network is used to distribute the input power either uniformly or, with a prescribed weighting factor, to each of the elements. The beamforming network also ensures that each element is phased correctly. For a beam to radiate normal to the plane of the array, each element should be in phase with the others. By creating a linear phase tilt in a particular plane across the array, the radiated beam can be made to scan in that plane. An antenna in which the direction of the radiated main beam is controlled by changing the phase distribution applied to the individual elements is known as a phased array.

The secondary radiation pattern associated with the array comprises the product of the element pattern and the array factor. The latter is solely a function of the array geometry, i.e. the element spacing and the array lattice structure, typically triangular or square. If the element spacing is too large,

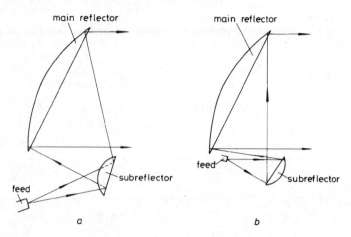

Figure 5.9   Asymmetric (or offset) dual-reflector systems

    a Cassegrain
    b Gregorian

there will be more than one direction in which the radiation from all the elements adds in phase, giving rise to extraneous main beams known as grating lobes. To avoid grating lobes, the element spacing, d, must be related to the maximum scan angle, $\pm\theta_s$, of the main beam by the inequality:

$$\frac{d}{\lambda} < \frac{1}{1 + \sin\theta_s} \tag{5.18}$$

For many applications, this criterion for the element spacing may not be compatible with the element type, and the resultant grating lobe will give rise to a reduction in gain and be a potential source of interference. Several techniques are available for suppressing grating lobes.[7]

5.4 References

1 RUDGE, A.W., MILNE, K., OLVER, A.D., and KNIGHT, P.: 'The handbook of antenna design—Volume 1'. IEE Electromagnetic Waves Series 15 (Peter Peregrinus Ltd, 1982)
2 SILVER, S.: 'Microwave antenna theory and design' (McGraw-Hill Book Company, 1949)
3 PRATT, T. and BOSTIAN, C.W.: 'Satellite communications' (John Wiley, 1986)
4 LUDWIG, A.C.: 'The definition of cross-polarisation', *IEEE Trans. Antennas Propag.*, 1973, **21**, pp. 116–119
5 CCIR: 'Handbook on satellite communications' (International Telecommunication Union, 1985)
6 POTTER, P.D.: 'A new horn antenna with suppressed sidelobes and equal beamwidth', *Microwave J.*, 1963, **6**, pp. 71–78
7 MAILLOUX, R.J.: 'The handbook of antenna design—Volume 2'. IEE Electromagnetic Waves Series 28 (Peter Peregrinus Ltd, 1986)

Chapter 6

Propagation considerations relating to satellite communications systems

P.T. Thompson

6.1 Introduction

A number of factors resulting from changes in the atmosphere have to be taken into account when designing a satellite communications system in order to avoid impairment of the wanted signal. Generally, a margin in the required carrier-to-noise ratio is incorporated to accommodate such effects.

6.2 Radio noise

Radio noise emitted by matter is used as a source of information in radioastronomy and in remote sensing. Noise of a thermal origin has a continuous spectrum, but several other radiation mechanisms cause the emission to have a spectral-line structure. Atoms and molecules are distinguished by their different spectral lines.

For other services such as satellite communications noise is a limiting factor for the receiving system; generally, it is inappropriate to use receiving systems with noise temperatures which are much less than those specified by the minimum external noise.

From about 30 MHz to about 1 GHz cosmic noise predominates over atmospheric noise except during local thunderstorms, but will generally be exceeded by man-made noise in populated areas. Above 1 GHz, the absorptive constituents of the atmosphere (water vapour, oxygen, hydrometeors) also act as noise sources.

In the bands of strong gaseous absorption, the noise temperature reaches maximum values of some 290 K. At times, precipitation will also increase the noise temperature at frequencies above 5 GHz.

Figure 6.1 gives an indication of sky noise at various elevation angles and frequencies.

The sun is a strong variable noise source with a noise temperature of about 10^6 K at 30 MHz and of at least 10^4 K at 10 GHz under quiet sun conditions (see Figure 6.2); large increases occur when the sun is disturbed. In general, the sun interferes with the wanted signal for short periods (<10 min) about 8 days a year. The noise temperature of the earth as seen from space is about 290 K.

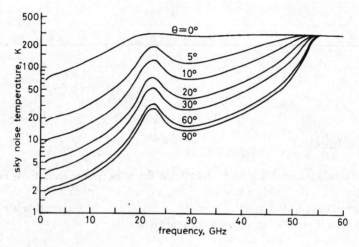

Figure 6.1   Sky-noise temperature for clear air (θ is the elevation angle)

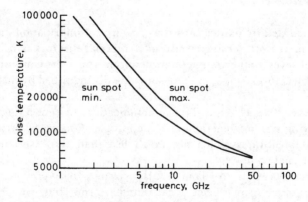

Figure 6.2   Noise temperature of the sun

The radio noise temperature observed with an antenna having an infinitely narrow beam, pointed towards the zenith but away from any localised noise source, is about 3 K for frequencies from about 2 GHz to about 10 GHz. For decreasing elevation angles the noise temperature increases, reaching about 80 K when the antenna is pointed at low angles in clear-air conditions. The above is normally adequate for system design but occasionally a more detailed analysis may be required where noise contributions through both the main beam and the sidelobes (including the contributions from the ground) will need to be considered.

6.3 Ionospheric effects

Although radio waves at frequencies greater than 100 MHz can normally pass through the ionosphere between earth and space, they may be modified by the presence of free electrons and the earth's magnetic field. Effects due to bulk ionisation include absorption, propagation time delay, refraction and polarisation rotation, all of which vary with position on the earth and with time according to the various long and short-period changes of the ionosphere. The existence of small-scale irregularities in the electron concentration also gives rise to relatively rapid fluctuations of a number of signal parameters, these fluctuations being collectively known as scintillation.

All of the ionospheric effects mentioned above show a progressive decrease with increasing radio frequency.

Table 6.1 gives estimates of maximum values of various ionospheric effects at a frequency of 1 GHz, together with their respective frequency dependences.

Table 6.1   Estimated maximum values of ionospheric effects for elevation angles of about 30° one-way traversal ($TEC = 10^{18}$ electron$/m^2$)

Effect	Magnitude at 1 GHz	Frequency dependence
Faraday rotation	108°	$1/f^2$
Propagation delay	0.25 μs	$1/f^2$
Refraction	< 0.17 mrad	$1/f^2$
Variation in the direction of arrival	0.2 min of arc	$1/f^2$
Absorption (polar-cap absorption)	0.04 dB	$\simeq 1/f^2$
Absorption (auroral and polar-cap absorption)	0.05 dB	$\simeq 1/f^2$
Absorption (mid-latitude)	< 0.01 dB	$1/f^2$
Dispersion	0.5 ns/MHz	$1/f^3$
Scintillation	see Section 6.1.2.3	no simple dependence

The figures relate to one-way traverse of the ionosphere at an elevation angle of 30°, the vertical total electron content (TEC) being 10^{18} electron/m^2.

6.3.1 Ionospheric absorption

At VHF and above the absorption at a frequency f on an oblique path with angle of incidence i at the ionosphere is in general proportional to $(\sec i)/f^2$. At middle latitudes, the absorption for a one-way traverse of the ionosphere at normal incidence is generally less than 0.1 dB at frequencies greater than 100 MHz. Enhanced absorption can occur as a result of solar flares, and during auroral and polar-cap events, but an upper limit of about 5 dB should be expected.

6.3.2 Ionospheric effects depending on total electron content

The total electron content (TEC), which is the integral of the electron concentration along the ray path, expressed in units of electrons/m^2, is significant in the determination of phase path, group delay, dispersion, refraction and Faraday polarisation rotation of transionospheric signals. These parameters are all directly proportional to either TEC or, in the case of Faraday rotation, to the integrated product of the electron concentration and the component of the magnetic field along the propagation path.

Anisotropy of the ionosphere produces Faraday rotation of the plane of polarisation of a linearly polarised wave. The rotation at a given radio frequency exhibits fairly predictable diurnal, seasonal and solar cyclical variations which may be compensated for by adjustment of the polarisation tilt angle of the earth-station antenna. However, large deviations from the regular behaviour can arise for small percentages of the time, and cannot be predicted in advance. Statistical data on TEC variations experienced for small time percentages are lacking, existing data indicates peak Faraday rotations of 9° at 4 GHz, 4° at 6 GHz and 1° at 12 GHz.

6.3.3 Ionospheric scintillation

Amplitude, phase, polarisation and direction-of-arrival scintillations may represent practical limitations to space communication systems. They are produced as radio waves propagate through electron-density irregularities in the ionosphere.

Scintillations are commonly observed at VHF and may occur at frequencies at least up to 7 GHz. Scintillation activity is mostly a night-time phenomenon. It is most severe in the vicinity of the geomagnetic equator and at high altitudes, least severe at mid-altitudes and increases with solar and geomagnetic activity. For example, scintillations with peak-to-peak amplitude fluctuations exceeding 10 dB at 154 MHz have been observed for 7 per cent of the time at night at Huancayo, an equatorial station; at Narssarssuaq, an auroral station, the corresponding time percentages were 0.9% and 8.4% under quiet and disturbed magnetic conditions, respectively.

6.4 Tropospheric effects

In addition to the thermal-noise radiation from any absorbing constituent in troposphere, the following factors need to be considered:

- attenuation by gases, clouds and precipitation, sand and dust storms;
- various aspects of refraction (including scintillation, loss of antenna gain and possible limitations in bandwidth);
- depolarisation by precipitation.

In general, the effects become increasingly important above about 3 GHz and at low elevation angles. Above 40 GHz, especially in the region of the absorption lines of water vapour and oxygen, significant attenuation can occur even in clear conditions.

6.4.1 Attenuation due to precipitation and clouds

The principal factor when considering attenuation caused by precipitation and clouds is rain, and the attenuation it causes increases with rainfall rate, with frequency (up to about 100 GHz) and with decreasing elevation angle. Rain attenuation can normally be neglected at frequencies below about 5 GHz.

Snow, especially dry snow, is much less serious than heavy rain, but melting snow can cause significant attenuation. Also, snow on the antenna and feed can be more serious than heavy rain.

The effect of clouds is small compared with that of rain and is only a serious factor at frequencies well above 30 GHz. The attenuation is proportional to the liquid-water content, which will vary with cloud type. Thunderstorm clouds (cumulo-nimbus) cause the maximum attenuation.

To predict attenuation due to precipitation along a slant path, it is necessary to obtain information on its distribution in space and time. A general method developed by the ITU for calculating attenuation is presented below.[1]

If reliable long-term statistical attenuation data are available, which were measured at an elevation angle and frequency (or at frequencies) different from those for which a prediction is needed, it may often be preferable to scale this information to the elevation angle and frequency in question rather than using the general prediction formula. Frequency scaling is discussed in more detail in Reference 1.

To calculate the long-term statistics of the slant-path rain attenuation at a given location, the following parameters are required:

$R_{0.01}$ = point rainfall rate for the location for 0.01 % of an average year (mm/h)
h_S = height above mean sea level of the earth station (km)
θ = elevation angle (degs)

ϕ = latitude of the earth station (degs)
f = frequency (GHz)

The latest method detailed by the ITU[1] consists of the following steps 1 to 7 to predict the attenuation exceeded for 0.01% of the time, and step 8 for other time percentages.

Step 1: The effective rain height, h_R, is calculated from the latitude of the station ϕ:

$$h_R \text{ (km)} = 5.0 - 0.075(\phi - 23) \quad \text{for} \quad \phi > 23° \text{ northern hemisphere}$$
$$= 5.0 \qquad\qquad\qquad \text{for} \quad 0° \leqslant \phi \leqslant 23° \text{ northern hemisphere}$$
$$= 5.0 \qquad\qquad\qquad \text{for} \quad 0° > \phi \geqslant -21° \text{ southern hemisphere}$$
$$= 5.0 + 0.1(\phi + 21) \quad \text{for} \quad -71° \leqslant \phi < -21° \text{ southern hemisphere}$$
$$= 0.0 \qquad\qquad\qquad \text{for} \quad \phi < -71° \text{ southern hemisphere}$$

$$(6.1)$$

Step 2: For $\theta > 5°$ the slant-path length, L_s, below the rain height is obtained from:

$$L_s = \frac{(h_R - h_s)}{\sin \theta} \text{ km} \tag{6.2}$$

For $\theta < 5°$ a more accurate formula should be used, that is:

$$L_s = \frac{2(h_r - h_s)}{(\sin^2 \theta + 2(h_R - h_s)/R_e)^{1/2} + \sin \theta} \text{ km} \tag{6.3}$$

where R_e = radius of the earth = 6378 km

Step 3: The horizontal projection, L_G, of the slant-path length is found (see Figure 6.3):

$$L_G = L_s \cos \theta \text{ km} \tag{6.4}$$

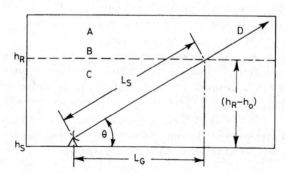

Figure 6.3   *Schematic presentation of an earth-space path giving the parameters to be input into the attenuation prediction process*

A: frozen precipitation
B: rain height
C: liquid precipitation
D: earth-space path

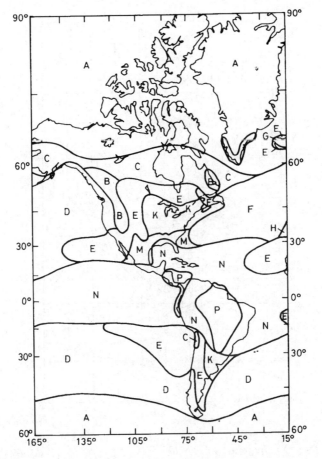

Figure 6.4   *Rain climatic zones (15° W – 165° W)*

Step 4: Obtain the rain intensity, $R_{0.01}$, exceeded for 0.01 % of an average year (with an integration time of 1 min). If this information cannot be obtained from local data sources, an estimate can be obtained from the maps of rain climates given in Figures 6.4, 6.5 and 6.6 together with Table 6.2.[2]

Step 5: The reduction factor, $r_{0.01}$, for 0.01 % of the time can be calculated from:

$$r_{0.01} = \frac{1}{(1 + L_G/L_0)} \tag{6.5}$$

where $L_0 = 35 \exp(-0.015\ R_{0.01})$ for $R_{0.01} \leqslant 100$ mm/hr; if $R_{0.01} > 100$ mm/hr then calculate L_0 with 100 mm/hr instead of $R_{0.01}$.

Step 6: Obtain the specific attenuation, Υ_R, using frequency-dependent coefficients derived from Table 6.3 (also see Reference 3) and the rainfall rate, $R_{0.01}$, determined from step 4:

$$\Upsilon_R = k(R_{0.01})^\alpha \text{ dB/km} \tag{6.6}$$

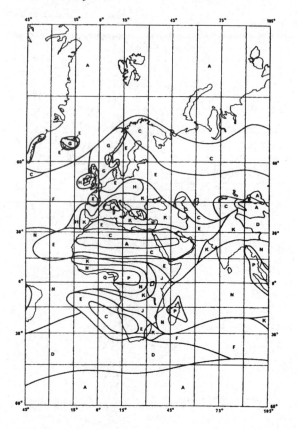

Figure 6.5   Rain climatic zones (45° W – 105° E)

To use Table 6.3 coefficients additional steps are needed to arrive at k and α as follows: Based on the assumption of spherical drops, values of k and α have been calculated at a number of frequencies between 1 and 1000 GHz for several drop temperatures and drop-size distributions. Values of k and α for a Laws and Parsons drop-size distribution and drop temperature of 20°C have also been calculated by assuming oblate spheroidal raindrops aligned with a vertical rotation axis and with dimensions related to the equivolumic spherical drops. These values, which are appropriate for horizontal and vertical polarisations, are presented in Table 6.3 (denoted k_H, α_H and k_V, α_V). Values of k and α at frequencies other than those in Table 6.3 can be obtained by interpolation using a logarithmic scale for frequency and k, and a linear scale for α.

For linear and circular polarisation, the coefficients in eqn. 6.6 can be calculated from the values of Table 6.3 using the approximate equations given below:

$$k = [k_H + k_V + (k_H - k_V)\cos^2\theta\cos 2\tau]/2 \tag{6.7}$$

$$\alpha = [k_H\alpha_H + k_V\alpha_V + (k_H\alpha_H - k_V\alpha_V)\cos^2\theta\cos 2\tau]/2k \tag{6.8}$$

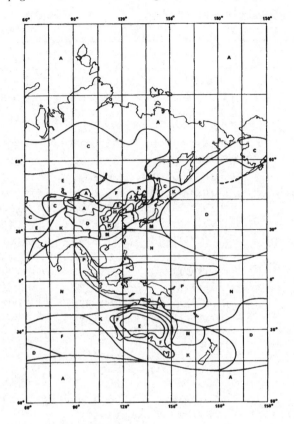

Figure 6.6 Rain climatic zones (60° E − 150° W)

Table 6.2   Rainfall intensity exceeded (mm/h) for specified time percentages

% of time	Rain climatic zone														
	A	B	C	D	E	F	G	H	J	K	L	M	N	P	Q
1.000	–	0.5	0.7	2.1	0.6	1.7	3	2	8	1.5	2	4	5	12	24
0.300	0.8	2	2.8	4.5	2.4	4.5	7	4	13	4.2	7	11	15	34	49
0.100	2	3	5	8	6	8	12	10	20	12	15	22	35	65	72
0.030	5	6	9	13	12	15	20	18	28	23	33	40	65	105	96
0.010	8	12	15	19	22	28	30	32	35	42	60	63	95	145	115
0.003	14	21	26	29	41	54	45	55	45	70	105	95	140	200	142
0.001	22	32	42	42	70	78	65	83	55	100	150	120	180	250	170

Table 6.3   Regression coefficients for estimating specific attenuation in eqns. 6.6–6.8

Frequency (GHz)	k_H	k_V	α_H	α_V
1	0.0000387	0.0000352	0.912	0.880
2	0.0001540	0.0001380	0.963	0.923
4	0.0006500	0.0005910	1.121	1.075
6	0.0017500	0.0015500	1.308	1.265
7	0.0030100	0.0026500	1.332	1.312
8	0.0045400	0.0039500	1.327	1.310
10	0.0101000	0.0088700	1.276	1.264
12	0.0188000	0.0168000	1.217	1.200
15	0.0367000	0.0335000	1.154	1.128
20	0.0751000	0.0691000	1.099	1.065
25	0.1240000	0.1130000	1.061	1.030
30	0.1870000	0.1670000	1.021	1.000
35	0.2630000	0.2330000	0.979	0.963
40	0.3500000	0.3100000	0.939	0.929
45	0.4420000	0.3930000	0.903	0.897
50	0.5360000	0.4790000	0.873	0.868
60	0.7070000	0.6420000	0.826	0.824
70	0.8510000	0.7840000	0.793	0.793
80	0.9750000	0.9060000	0.769	0.769
90	1.0600000	0.9990000	0.753	0.754
100	1.1200000	1.0600000	0.743	0.744
120	1.1800000	1.1300000	0.731	0.732
150	1.3100000	1.2700000	0.710	0.711
200	1.4500000	1.4200000	0.689	0.690
300	1.3600000	1.3500000	0.688	0.689
400	1.3200000	1.3100000	0.683	0.684

where θ is the path elevation angle and τ is the polarisation tilt angle relative to the horizontal ($\tau = 45°$ for circular polarisation).

Step 7: The attenuation exceeded for 0.01 % of an average year may then be obtained from:

$$A_{0.01} = \Upsilon_R L_S R_{0.01} \ \text{dB} \tag{6.9}$$

If alternative attenuation statistics are required then the following steps may be applied:

Step 8: The attenuation to be exceeded for other percentages (P) of an average year, in the range 0.001 % to 1.0 % (1 % often being appropriate for broadcasting services), may be estimated from the attenuation to be exceeded for 0.01 % of an average year by using:

$$A_p = A_{0.01} 0.12 p^{-(0.546+0.043 \log P)} \ \text{dB} \tag{6.10}$$

System planning often requires the attenuation value, exceeded for the time percentage p_w of the worst month. For a definition of the worst month see Reference 4.

Using the following approximate relation (which relates to the case for global planning purposes), an annual time percentage, p, corresponding to p_w may be obtained:

$$p = 0.3p_w^{1.15} \ (\%) \tag{6.11}$$

Consequently the attenuation exceeded for this annual time percentage, p, calculated using the eight-step method above may be taken as the attenuation value exceeded for $p_w\%$ of the worst month.

6.4.2 Atmospheric absorption

As well as attenuation caused by rain and other particles the gases present in the atmosphere cause attenuation especially at low elevation angles. Figure 6.7 depicts typical values for this atmospheric attenuation for which an additional link margin must be provided. Reference 1 gives details on the calculation of gaseous absorption should it be required.

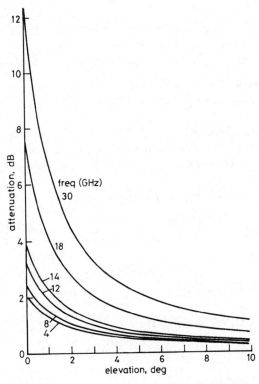

Figure 6.7   Attenuation due to atmospheric absorption for water vapour density
$p = 5g/m^3$

6.4.3 Site diversity

The small geographical extent of high-rain-intensity rain cells can be advantageous in combating signal fading by appropriately combining signals from two or more earth stations which are located some reasonable distance apart. Two methods of indicating the effectiveness of such space diversity are presented by the ITU. The first is called the 'diversity-improvement factor' and is given as the ratio of the single-site time percentage and the diversity-based time percentage. The second indicator is termed 'diversity gain' and is the difference (in dB) between the single-site and diversity-based attenuation values for the same percentage time.

It can be shown that the diversity-improvement factor, I, is approximately given by:

$$I = p_1/p_2 = 1 + 100\beta^2/p_1 \tag{6.12}$$

where p_1 and p_2 are the respective single-site and diversity-based time percentages and β is a parameter based on the link characteristics. Based on the results of extensive measurement programmes it is found that between 10 and 20 GHz:

$$\beta^2 = 10^{-4}d^{1.33} \tag{6.13}$$

where d is the distance between the diversity stations in km. Figure 6.8 gives p_2 against p_1 for a range of percentage times and distances.

Calculation of diversity gain can be undertaken as follows: Given that

d = separation between the two sites (km)
A = path rain attenuation for a single site (dB)
f = frequency (GHz)
θ = path-elevation angle (degs)
ψ = angle (degrees) made by the azimuth of the propagation path with respect to the baseline between the two sites chosen such that $\psi < 90$ degs

Step 1: The gain contributed by the spatial separation is:

$$G_d = a(1 - e^{-bd}) \tag{6.14}$$

where

$$a = 0.78A - 1.94(1 - e^{-0.11A})$$
$$b = 0.59(1 - e^{-0.1A})$$

Step 2: The frequency-dependent gain is:

$$G_f = e^{-0.025f} \tag{6.15}$$

Step 3: The elevation-angle term is:

$$G_\theta = 1 + 0.006\theta \tag{6.16}$$

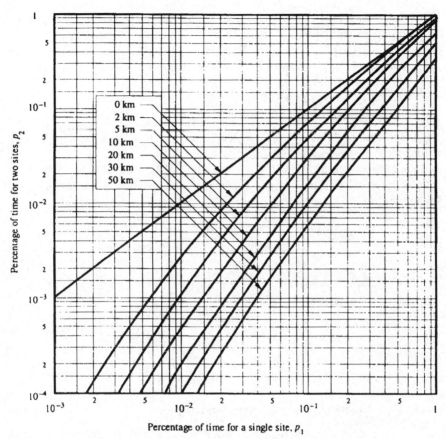

Figure 6.8   Relationship between % of time with and without diversity for the same attenuation

Step 4: The baseline-dependent term is:

$$G\psi = 1 + 0.002\psi \qquad (6.17)$$

Step 5: The net diversity gain is:

$$G = G_d \cdot G_f \cdot G_\theta \cdot G_\psi \text{ dB} \qquad (6.18)$$

6.4.4 Depolarisation

Depolarisation effects on earth-space paths can be due to both rain and ice (as well as antenna imperfections). In the following ITU calculation method a simplified approach for calculating the depolarisation based on rain effects and then correcting for ice is adopted.

To calculate long-term statistics of depolarisation from rain attenuation statistics the following parameters are needed:

A_p: the rain attenuation (dB) exceeded for the required percentage of time, p, for the path in question, commonly called copolar attenuation, CPA

τ: the tilt angle of the linearly polarised electric-field vector with respect to the horizontal (for circular polarisation use $\tau = 45°$)

f: the frequency (GHz)

θ: the path-elevation angle

The method described below to calculate XPD (crosspolar discrimination) statistics is based on the fact that there is a high correlation with rain-attenuation statistics.

The method is valid for $8 \leqslant f \leqslant 35$ GHz and $\theta \leqslant 60°$ and consists of the following steps 1 to 8 to compute the respective contributions due to separate factors (for $f < 8$ GHz see later):

Step 1:   Calculate the frequency-dependent term:

$$C_f = 30 \log f \quad \text{for } 8 \leqslant f \leqslant 35 \text{ GHz} \tag{6.19}$$

Step 2:   Calculate the rain-attenuation-dependent term:

$$C_A = V(f) \log(A_p) \tag{6.20}$$

where:

$$V(f) = 12.8 f^{0.19} \quad \text{for } 8 \leqslant f \leqslant 20 \text{ GHz}$$
$$V(f) = 22.6 \quad\quad\quad \text{for } 20 < f \leqslant 35 \text{ GHz}$$

Step 3:   Calculate the polarisation-improvement factor:

$$C_\tau = 10 \log[1 - 0.484(1 + \cos(4\tau))] \tag{6.21}$$

Step 4:   Calculate the elevation-angle-dependent term:

$$C_\theta = -40 \log(\cos \theta) \quad \text{for } \theta < 60° \tag{6.22}$$

Step 5:   Calculate the canting-angle-dependent term:

$$C_\sigma = 0.0052\sigma^2 \tag{6.23}$$

σ is the effective standard deviation of the raindrop-canting-angle distribution, expressed in degrees. σ takes the value $0°$, $5°$, $10°$ and $15°$ for 1%, 0.1%, 0.01% and 0.001% of the time, respectively.

Step 6:   Calculate rain XPD not exceeded for $p\%$ of the time:

$$XPD_{rain} = C_f - C_A + C_\tau + C_\theta + C_\sigma \text{ dB} \tag{6.24}$$

Step 7:   Calculate the ice-crystal-dependent term:

$$C_{ice} = XPD_{rain} \times (0.3 + 0.1 \log p)/2 \text{ dB} \tag{6.25}$$

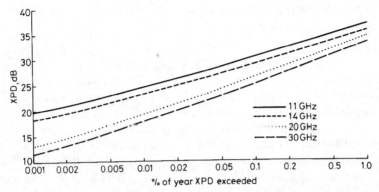

Figure 6.8   *Typical XPD values for various % times (latitude 50°N, elevation angle 30°, rain climate E, circular polarisation)*

Step 8: Calculate the XPD not exceeded for $p\%$ of the time and including the effects of ice:

$$XPD_p = XPD_{rain} - C_{ice} \text{ dB} \qquad (6.26)$$

Figure 6.8 indicates typical results using the above method.

A design approach based on joint probability cumulative distributions of XPD and A_p is preferable, particularly for earth-space paths where the variability in XPD for a given A_p is large for the low attenuation margins normally employed. It should be noted, however, that the use of an equiprobability relation between XPD and A_p for outage calculations may give the same results as the use of joint probabilities if it is applied to the calculation of fading margins in systems using dual polarisation.

For the frequency band 4–6 GHz, where attenuation is low, A_p statistics are not useful in predicting XPD statistics. For frequencies below about 8 GHz, it may be more useful to employ relationships between XPD, point rain rate and effective path length.

6.4.4.1 Frequency and polarisation scaling of XPD

Long-term XPD statistics obtained at one frequency and polarisation tilt angle can be scaled to another frequency and polarisation tilt angle using the semi-empirical formula:

$$XPD_2 = XPD_1$$
$$- 20\log\{f_2/f_1(1 - 0.484(1 + \cos 4\tau_2)^{1/2}/(1 - 0.484(1 + \cos 4\tau_1)^{1/2}\}\text{dB}$$
$$\text{for } 4 \leqslant f_1, f_2 \leqslant 30 \text{ GHz} \quad (6.27)$$

where XPD_1 and XPD_2 are the XPD values not exceeded for the same percentage of time at frequencies f_1 and f_2 and polarisation tilt angles τ_1 and τ_2, respectively.

6.4.5 *Tropospheric refraction and scintillation effects*

The refractive index of the atmosphere is variable in both space and time, and the effective path length through the lower layers of the troposphere depends on the angle of elevation of the antenna beam at the earth station. The irregular refractive-index structure of the troposphere can cause signal fluctuations (scintillation) and a loss in antenna gain owing to phase incoherence across the aperture. For the former effect, measured data at elevation angles of 20°–30° and at 23 GHz indicate that, in temperate climates, the fluctuations are of the order of 1 dB (peak-to-peak) in clear sky in summer and 0.2–0.3 dB in winter. Clouds can produce fluctuations of 2–4 dB (occasionally 6 dB). Fading rates of about 1 Hz are typical but components of at least 10 Hz have been detected. A slower fading rate, at about 0.01 Hz, has also been observed. This is probably due to angle-of-arrival variations caused by relatively large-scale changes in refractive index. A detailed method of calculating tropospheric scintillation from meteorological data is given in Reference 1. This requires a knowledge of average surface ambient temperature at the site for a period of one month or longer along with a knowledge of the average surface relative humidity for the same period.

6.4.6 *Shadowing and multipath effects*

Link margins must be included in certain satellite communications applications to account for shadowing and/or multipath effects (more details will be found in Chapter 19 and an overview only is given here). As an alternative to simply using power margins, certain modulation techniques designed to combat multipath effects may be used. Applications include maritime mobile, land mobile and aeronautical mobile satellite systems, generally using low G/T mobile earth stations and operating, typically, below about 3 GHz. Three specific ITU recommendations cover the calculation of signal impairments for these three cases (given in References 5, 6 and 7). These References provide formal methodology for calculating signal impairments for the different services and in many cases refer back to the techniques detailed above but supplemented with service-specific situations. The following gives a brief review of these more specific aspects. Should more precise values be required then the reader will need to perform the calculations detailed in the above References.

For maritime applications, multipath effects predominate at low elevation angles and increase in severity as the mobile earth-station antenna gain decreases. For a maritime mobile earth-station antenna gain of 20 dB, an elevation angle of 5° and a rough sea, multipath fading of 2.7 dB and 4.8 dB are not exceeded for 95% and 99% of the time, respectively; decreasing to 0.5 dB and 0.6 dB, respectively, at an elevation angle of 20°. For a smooth sea and an elevation angle of 10°, the corresponding values are 3.2 dB for 95% of the time and 5.4 dB for 99% of the time (these figures relating to operation in the 1600 MHz band).

For land mobile satellite application, both shadowing and multipath effects must be taken into account. For a suburban/rural environment, link margins of the order of 10 dB to 15 dB are required for 90% of locations for 90% of the time at an operating frequency of 860 MHz. Significantly, greater link margins, of the order of 30 dB, are required to provide the same performance in an urban environment.

Satellite sound-broadcasting systems operating in the 500 MHz to 2 GHz region of the spectrum will, like the land mobile satellite systems, experience shadowing and multipath effects. Results of measurements conducted in Europe are of the order of 15 dB. The effects of shadowing and multipath may necessitate that between 5 dB and 30 dB link margins be required for rural and urban areas, respectively. Additional studies and experimental data are required, particularly since there are differences in terrain and building features in different countries.

6.5 Acknowledgments

Much of the material presented in this chapter is extracted from various ITU documents. The efforts of the ITU and its members have been a fundamental factor in setting our understanding of such matters to the current-day level. These remarkable efforts and the quality of their publications are fully acknowledged by the author.

6.6 References

1 Rec. ITU-R P.618-5
2 Rec. ITU-R P.837-1
3 Rec. ITU-R P.838
4 Rec. ITU-R P.841
5 Rec. ITU-R P.680-2
6 Rec. ITU-R P.681-3
7 Rec. ITU-R P.682-1

All references are ITU-R texts and are available from: International Telecommunications Union (ITU), Place des Nations, CH-1211 Geneva 20, Switzerland. Latest details on these documents are also available on the worldwide web at www.itu.int/publications/itu-r/iturrec.htm

Chapter 7

Interference considerations relating to satellite communications systems

P.T. Thompson

7.1 General

Congestion of the radio-frequency spectrum has necessitated the sharing of frequencies between a number of services, involving both terrestrial and space systems. The possibilities of interference arise therefore between terrestrial and satellite systems and between systems of the same type. Thus, the designer needs to take account of potential interference when designing systems. Clearly, due account needs to be taken of existing, planned and future systems when allowing for interference in the system design.

In order to estimate mutual interference between different radio systems it is necessary to know the statistical distribution of the difference in decibels of the level of the interfering signal and the level of the wanted signal. For most radio systems, except transhorizon radio-relay links, the wanted signal level may be considered to be approximately constant, although multipath fading of the wanted signal can cause complications. Therefore, it is essential for the engineer to have statistics describing the occurrence of the interference, the variability of which is due to the existence of several different propagation mechanisms, each of which is itself subject to variations arising from the nature of the tropospheric propagation medium. Changes in the medium are a function of climate, and it is necessary to take into consideration both the effects due to the propagation characteristics in clear air and phenomena which may arise owing to the presence of hydrometeors in the propagation path.

In system planning it is generally required to estimate the level of interference not to be exceeded for both the small-time percentage (<1% of the time), and for about 50 per cent, the propagation mechanisms mainly to be considered are:

- line-of-sight;
- diffraction over isolated obstacles;
- diffraction over irregular terrain;
- tropospheric forward scatter.

At time percentages below about ten per cent, additional mechanisms require consideration for the evaluation of the stronger interference levels. These are:

- superrefraction and ducting;
- scatter from hydrometeors (mainly rain);
- reflections from aircraft.

In considering satellite-communications systems design there are cases which need specific consideration:

- interference from other satellite networks into the one being designed;
- interference from the network being designed and into other networks;
- interference from terrestrial networks (including radar) into the network being designed;
- interference from the new satellite network into terrestrial services.

Wherever possible the ITU has arranged the sharing of services such that they are reasonably compatible and can be coordinated. This last term relates to the process of achieving agreement between the operators of the different systems on the acceptability, or otherwise, of interference.

There are well laid down rules on how to coordinate systems and these are contained in the radio regulations of the ITU [1]. Consequently, it is not proposed to give specific numerical rules for calculating such interference cases here but to concentrate on the factors to be taken into account, thus giving the satellite system designer a feel for the magnitude and scope of the problem. It should also be noted that from time to time the ITU updates the coordination procedures and the latest set should be used in detailed analysis [1].

7.2 **Interference between satellite networks**

Interference analysis of separate systems is a complex procedure which depends both on the parameters of the various systems and the impact which interference has on the wanted signal. The deterioration of a wanted signal by an interferer can be determined both analytically and subjectively. The latter is often needed if the interferer causes intelligible signals to appear in the wanted signal, such as one television signal interfering with another television signal.

If the interference is not correlated with the wanted signal it is generally assumed to be noise like and account is taken of it in the satellite system design by having an allowance of extra noise in the link budget. Thus, the critical factor in the link budget is carrier-to-noise level (C/N) which gives the required quality of performance at system threshold. The actual C/N must be increased to account for interference as well as propagation effects.

In considering interference in satellite systems account should also be taken of interference from other transponders on the same satellite, especially if dual polarisation or multiple beams are used where several additional interference entries would exist.

Figure 7.1 depicts the various signal and interfering paths associated with two adjacent satellites. Clearly, both up-path and down-path interference need to be evaluated.

7.2.1 Basic first-stage analysis

Owing to the many complex parameters associated with interference assessment the ITU has adopted a rather simplistic model from which to determine the worst-case potential for interference. This is described in appendix S8 of the Radio Regulations.* If when using this model the threshold of acceptable interference is exceeded then the network operators must get together and find ways of performing a more detailed analysis and explore methods of mutually agreeing to acceptable conditions for service to be effected.

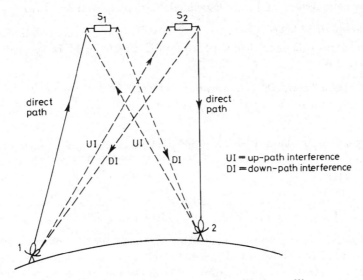

Figure 7.1   Interference from (or into) systems using adjacent satellites

---

*Prior to WRC 95 this was known as Appendix 29.

The simplistic model referred to above assumes that the interferer is operating at the same frequency as the wanted signal, that the interference is noise like and that the analysis can be performed using power spectral density figures, i.e. the power per Hertz of wanted signal against the power per Hertz of the interferer.

The criterion used to indicate that the interference should be analysed in a more detailed manner is that of an increase in apparent noise temperature of the signal being interfered with by more than six per cent of the noise already existing in that system. This so called ΔT method is a reasonably good indicator of potential interference which can be applied independent of precise signal characteristics. It is valid for FM and PSK systems alike.

The reader is referred to appendix S8 of the ITU Radio Regulations for further details of this first-stage procedure.

Should the increase in noise level be less than six per cent then all is well and the system should work without problems.

7.2.2   Detailed analysis

At this stage a detailed examination of the interference situation is conducted taking account of the precise frequency plans of the networks, the types of signal that are interfering, the relative powers of the transmitted and received signals, the known levels of coupling between signal types, the polarisation properties of the signals, the acceptable levels of carrier interference of the systems and any possible methods of reducing interference.

The following information gives a feel for the types of signal currently being used in satellite systems and their associated interference characteristics.

Wide-deviation FDM/FM: covers most of the more long standing satellite transmissions; bandwidths range from about 1.25 MHz up to 36 MHz; spectrum has gaussian shape.

Narrow-deviation FDM/FM: bandwidths typically 20 MHz or 30 MHz; non-gaussian spectrum often with carrier spike, especially during light loading conditions.

FM TV carriers: complex spectrum dependent on modulation; spectrum rectangle when energy dispersal only modulating signal; bandwidths typically 17.5 or 30 MHz.

Digital PSK carriers: spectrum is truncated or modified sin x/x; line spectrum if no energy dispersal; otherwise relatively uniform and well spread.

A uniform and well spread spectrum can be advantageous and most signals employ some form of energy dispersal which spreads the energy over the allocated bandwidth thereby avoiding peaks in the spectral density. Naturally, these energy-dispersal signals are normally removed from the signal upon reception, thus restoring the signal to its original form. The following indicates the various energy-dispersal systems employed in current systems:

FDM/FM: a low-frequency triangular-wave energy-dispersal signal is applied to the baseband at a frequency well below the lowest baseband frequency of 4 kHz. Various frequencies are used to avoid interactions. It is unnecessary to remove the dispersal signal at the receiver. The deviation due to the dispersal signal is chosen so that it maintains the power spectral density under light loading conditions to within 2 dB of that applying full-load conditions.

FM/TV: a symmetrical triangular waveform is added to the TV baseband signal at a frequency equal to the frame rate or half the frame rate and synchronised with the TV signal. The signal is removed at the receiver usually by means of a black level clamp. The maximum peak-to-peak dispersal deviation is 4 MHz, but 2 MHz is more typical. For direct broadcasting satellites the figure is 600 kHz. Line-rate energy dispersal has also been considered in order to give added protection for interference to SCPC.

SCPC/FM: usually no energy dispersal is applied.

PSK: to avoid strong line components in the spectrum of a digital signal under certain traffic conditions digital-energy dispersal is performed by applying a pseudorandom (PR) scrambler to the transmitted digital baseband signal. The scrambling is removed at the receiver. The worst spacing between the lines in the resultant spectrum is then the reciprocal of the reception rate of the PR pattern. For example a $2^{15} - 1$, PR — scrambler would repeat approximately every 32 000 bits. In a 120 Mbit/s system this would produce lines at 4 kHz spacing. If it is a four-phase PSK system and the scrambling is applied separately to the two orthogonal streams, the line spacing would be 2 kHz.

As well as energy-dispersal techniques there are currently under study various forms of interference canceller which could be used to assume a higher tolerance to interference.

In the more detailed considerations addressed in this case account can be taken of the impact of the specific modulation characteristics of both the interferer and the wanted signal. Such an analysis is rather complex and beyond the scope of consideration here, however the interested reader is referred to the ITU documentation on the subject, which is kept current and details most of the signal formats in use today. It should be noted that for digital signals the measure of interference is that of an increase in bit error rate (BER) and that some digital techniques have a high tolerance to interference.

7.3 Interference with terrestrial networks

When the frequency bands are shared with terrestrial systems such as radio-relay links or radar systems a number of interfering conditions need to be investigated namely:

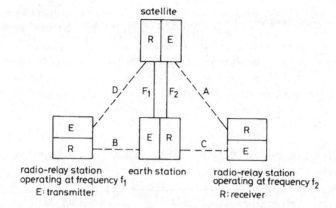

Figure 7.2   Interference paths between systems in the fixed satellite service and terrestrial radio services (The frequencies shown are in the bands shared between terrestrial radio services and fixed satellite service, allocated to earth-to-space transmission (F_1) and space-to-earth transmission (F_2)).

—————   wanted signal
- - - - - -   interfering signal

⦁   the signal from the satellites must not cause unacceptable interference to the receivers of the terrestrial service, as in A in Figure 7.2
●   the signals from satellite earth stations must not cause unacceptable interference to the receivers of the terrestrial service, as in B in Figure 7.2
●   the signals from terrestrial stations must not cause unacceptable interference to the receivers of satellite-system earth stations, as in C in Figure 7.2
●   the signals from terrestrial stations must not cause unacceptable interference to the satellite receivers, as in D in Figure 7.2.

As with interference considerations for interference between two satellite networks as detailed in Section 7.2 there exists in the case of interference between terrestrial and satellite systems a two-stage coordination approach.

7.3.1 Basic first-stage analysis

This procedure is aimed at determining an area around the earth station, known as the coordination area, within which the presence of a radio-relay link station could lead to unacceptable interference conditions. A simplistic model of overlapping frequency bands etc. is adopted in a similar manner to that used in Section 7.2.1 above. However, in order to determine the outer contour of the coordination area several additional factors now need to be taken into account.

The proximity of the earth can introduce effects which influence the transmission of the radio-relay link signals beyond the horizon. These signals are not cut off immediately by the intersection of the earth, and thus detailed knowledge of the diffraction of the earth's surface is required. The presence of hills and mountains makes this a reasonably complex feature.

The changing value of the radio refractive index of the air gives rise to a beam-bending effect and this can cause the propagation of signals far beyond the horizon. Furthermore, this bending effect is variable upon the weather conditions and situations can occur where the refractive-index gradient is so configured that the signal can be trapped in a duct and propagated long distances with little loss. This is known as duct propagation.

Irregularities of a fine nature in the refractive index of the atmosphere can give rise to scattering in the troposphere. In general, the scattered signal is very weak and confined to great circle paths and is the mechanism used to effect transmission in high-power tropospheric-scatter communications systems.

Finally, scattering from rain drops in a common volume where the signal can be directed from interfering source into the radio beam of the other system can give rise to rain-scatter interference. The level of such interference is dependent upon the exact geometries involved and the various antenna characteristics as well as the signal type.

These factors are used in determining the coordination area of the earth station and are discussed in greater detail in Reference 2. In general, rain-scatter considerations dictate most of the area except in the region of the earth-station main beam.

Figure 7.3 indicates the coordination area of a representative BT earth station as calculated using the method described in appendix S7 of the radio regulations.*

It should be noted that the coordination area can extend to cover other nations and at that stage the matter can require significant international co-operation.

If radio-relay links exist within the coordination area then detailed interference assessments need to be made and agreements reached between the operators.

7.3.2 Detailed analysis

This stage involves a much more detailed review of the situation in a manner similar to that detailed in Section 7.2.2 but including the various propagation factors outlined above. In many cases it is found that, owing to the specific pointing of the signal paths, features such as rain scatter are not critical and that the services can coexist.

Again, specific interference qualities of the signals and the precise acceptable level of carrier-to-interference ratio are used in the analysis which can take a

---

*Known prior to WRC 95 as Appendix 28.

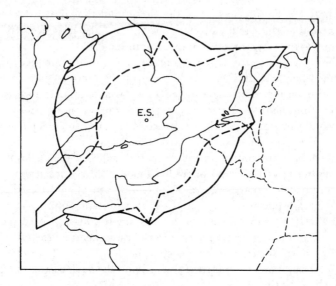

Figure 7.3   Typical earth-station coordination area (ITU RR appendix S7)

- - - - - - mode 1 contour	earth station: London
———— coordination contour	satellite: INTELSAT VF4
	:325.5°E
(Note: mode 1 does not include rain scatter)	TX freq. :14.25 GHz

considerable time to conduct. The path profile of the terrain between the inter-ferers can also be taken into account.

Techniques such as locating the earth station in a natural bowl or behind a mountain can ease the interference situation as can locating the earth station away from high population density areas where radio-relay systems may be more intensely used.

With the use of smaller earth stations it becomes practical to shield the site and reductions in interference of the order of 20 dB can be achieved.

7.4 Acknowledgments

Much of the material presented in this chapter has been extracted from various ITU documents. The efforts of the ITU and its members have been a funda-mental factor in setting our understanding of such matters to the current day level. These remarkable efforts and the quality of their publication are fully acknowledged by the author.

7.5 References

1 ITU Radio Regulations
2 Rec. ITU-R P.620-3

Available from the ITU at: International Telecommunications Union (ITU), Place des Nations, CH-1211 Geneva 20, Switzerland. Latest details on such documents are also available on the worldwide web at www.itu.int/publications/itu-r/iturrec.htm.

Chapter 8
Satellite access techniques

T. Tozer

8.1 Introduction

A satellite communication system will have a number of users operating via a common satellite transponder, and this calls for sharing of the resources of power, bandwidth and time. Here we describe these techniques and examine their implications, with emphasis on principles rather than detailed structure or parameters of particular networks, which tend to be very system specific.

The term used for such sharing and management of a number of different channels is multiple access. Each resource is limited, and ultimately relates to cost or revenue; their efficient use is important, as is meeting the needs of the users' traffic demands.

There are four fundamental techniques of multiple access:

- frequency-division multiple access (FDMA);
- time-division multiple access (TDMA);
- code-division multiple access (CDMA);
- packet (or random) access.

Each technique has its advantages and disadvantages, and in practice hybrid schemes are likely to be employed, having features of each. It should also be appreciated that a satellite transponder may be serving more than one network simultaneously, together with a variety of modulation schemes, signal powers and characteristics.

An outline description of spread spectrum is also included here as a modulation technique allowing multiple access capability in the form of CDMA.

8.2 Network architectures

A group of satellite terminals may be regarded as constituting a network, and the arrangement of their communication links as the network architecture or configuration. The architecture of a satellite communication network is different from most terrestrial networks, as it has all its links passing through a common satellite transponder, which will see all the earth terminals within its coverage area. The network may therefore in one sense be regarded as a star topology with the satellite transponder as the central node. However, the transponder is generally transparent, and it is the logical connectivity and topology of the users which is of interest, and this will be either a hub configuration, having a large hub or anchor station, or a mesh configuration where there is effective equity between all terminals. As well as the user links, there may be a distinct network management function handled from a hub station (NMCC—network management and control centre), also representing a star configuration.

The simplest satellite communications configuration to be considered is point-to-point between two dedicated earth terminals (Figure 8.1a). We may term the combination of up- and downlink a hop. As a one way, or simplex, hop this has sole call upon the entire transponder power and bandwidth. Different up- and downlink frequencies will be used, and the link budget calculation is straightforward. The two earth terminals need not necessarily have similar antenna apertures, for example one might be a large ground station (e.g. hub) and the other a smaller VSAT (very small aperture terminal). A practical system may be two way or duplex, and, if the same transponder is required to handle each direction of traffic simultaneously, then nonoverlapping frequencies may be allocated and the downlink transmit power shared between each direction according to relative strengths at the satellite. This represents a basic form of frequency-division multiple access (FDMA).

Where there is one transmit station and several receive terminals, we have a point-to-multipoint configuration (Figure 8.1b). Outgoing traffic from the large hub station may be selected for a particular receive terminal by means of choice of frequency or time slot, or by digital addressing. If, however, the traffic is

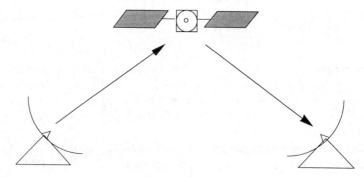

Figure 8.1a   Point-to-point network architecture

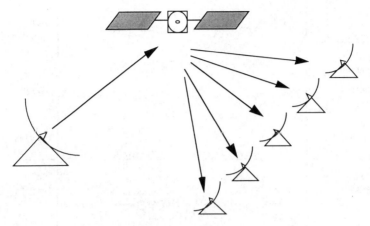

Figure 8.1b   Point-to-multipoint network architecture

common to all destinations, this is referred to as a broadcast network. The converse is multipoint-to-point, where several terminals are transmitting to one receive terminal.

The general case with a number of users is multipoint-to-multipoint (Figure 8.1c). If there is not a single dominant hub, we have a mesh network.

A communications network will also require a set of protocols for access into the network itself by the user. Protocols are sets of rules for the reliable establishment, maintenance and termination of communications. They may be system specific, although there are a number of international standards such as TCP/IP or other protocols based upon a general hierarchical structure such as the ISO-OSI seven-layer model (Figure 8.2).

In principle, these network-access protocols are distinct from the satellite multiple-access schemes described here, although the one will impact upon the other. A link via the satellite will provide the means of data transmission, and thus represent level 1, and perhaps levels 2 or even 3, of the OSI model,

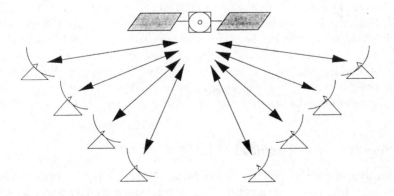

Figure 8.1c   Multipoint-to-multipoint network architecture

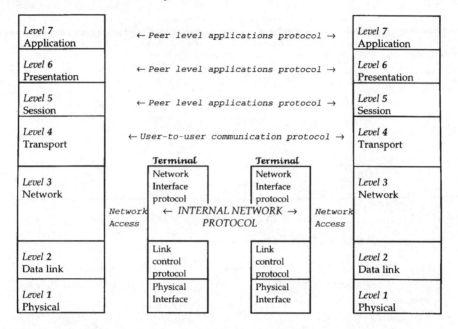

Figure 8.2   ISO-OSI model

depending upon the access scheme. It is the efficient interworking between the internal multiple-access protocols of the satellite network and the data protocols employed externally by the users which may represent the engineering challenge.

8.3 Traffic multiplexing

A number of user channels may be combined terrestrially at or before an earth station or terminal, for transmission with a common RF antenna as a single composite uplink. This is known as multiplexing, and it is helpful to deal first with this concept before examining the problem of combining signals at the satellite. If the carrier transmitted from a terminal represents only a single user channel, it is designated SCPC (single channel per carrier), and such is generally the case with a very small terminal (VSAT) uplink. If the carrier represents a number of multiplexed users it is designated MCPC (multiple channel per carrier), and this may apply to a large terminal or hub-station uplink.

The principal forms of multiplexing carried out at a terminal are described below.

8.3.1 Frequency-division multiplex (FDM)

FDM involves each individual channel being shifted in frequency and the frequency spectra placed contiguously. The original channels may be recovered

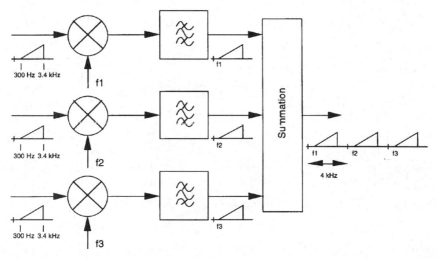

Figure 8.3   FDM telephony illustration (transmit end only)

by demultiplexing, with filtering and frequency translation back to baseband. Figure 8.3 illustrates a typical traditional example of a multiplexer for analogue telephony; FDM has long been established for this purpose. Essentially, each user is single-sideband (SSB) modulated onto a different carrier frequency, and these are summed together. Small guard bands are inserted between each channel to prevent frequency overlap or aliasing, and to facilitate demultiplexing. For N users the overall bandwidth will be approximately $N \times$ that of a single channel (usually 4 kHz for analogue speech, including a guard band).

(As an aside, one rather different modulation and multiplexing format used in some VSAT systems is FM^2 (FM squared). This involves a number of baseband user signals being FM modulated onto separate carriers prior to FDM multiplexing; the composite signal then FM modulates the final carrier. Typical bandwidths prior to multiplexing might be of the order of $20-30$ kHz and the final carrier bandwidths up to a few tens of MHz. The advantage of such schemes is that simple low-cost demodulators can be used, akin to domestic FM radio receivers, and the received multiplexed signal can be readily distributed terrestrially for individual channels to be taken off in stages. Enhancements to this concept include FM^3, where a further stage of multiplexing and modulation is employed.)

8.3.2 Time-division multiplex (TDM)

TDM involves each individual channel transmitted as a digital word, or slot, within a high-speed digital frame. Selection of the appropriate time slots in the frame permits reconstruction of the original channel, and the resultant transmission bit rate is approximately $N \times$ that of a single channel. Figure 8.4 illustrates the concept of TDM, showing both multiplexer and demultiplexer. One or more

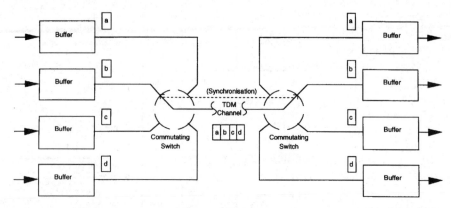

Figure 8.4   TDM concept (transmitter and receiver)

words within the TDM data frame may be allocated for synchronisation purposes (including clock or carrier recovery), or for carrying signalling information.

TDM is well established for digital-telephony transmission, as a series of multiplexing hierarchies with data rates up to around 2.5 Gbit/s, and is widely used for fibre-optic, microwave and satellite links. Owing to limitations in transponder bandwidths (and frequency allocations), satellite systems tend to employ lower data rates, with 140 Mbit/s being a typical upper limit. Such data rates, however, call for large terminal antennas, and small terminal systems may operate at no more than a few Mbit/s.

Some form of carrier modulation is required to transmit a multiplexed signal. Although an FDM signal could be transmitted directly (shifted to an appropriate frequency band), it is usual to frequency modulate (FM) the entire FDM block; this provides a constant envelope signal and may make optimum use of the wide transponder bandwidth. TDM uses digital-modulation schemes such as binary or quaternary phase-shift keying (BPSK or QPSK), or higher order schemes, although other modulation forms such as noncoherent FSK (frequency-shift keying) may be used.

8.4 Multiple access, and assignment strategies

Several uplink carriers from different earth stations may need to share a satellite transponder's resources of power and bandwidth, and also share common transmission time. This represents multiple access, and needs to be performed in a manner which ensures an equitable and efficient usage of these parameters coupled with minimisation of mutual interference. The management and allocation of accesses may be performed on a fixed-assigned, demand-assigned or random basis.

Fixed, or preassigned, multiple access is a simple technique where FDMA frequency slots or TDMA time slots are assigned by the network-control station for a lengthy or indefinite period to a particular terminal. The allocation may be varied manually according to average traffic demands or expectations. It is an efficient method for MCPC where large numbers of multiplexed speech channels are involved, giving relatively small statistical fluctuations, and is used for many commercial (large terminal) fixed-route telephony systems.

Demand-assigned multiple access (DAMA) involves network capacity being automatically allocated in the form of FDMA or TDMA slots as demanded. In this way, power and bandwidth may be reassigned where they are needed, and overall network capacity effectively optimised. It is efficient where traffic fluctuations are proportionately large, such as thin-route telephony with only a few speech channels, or for bursty data communications.

A mechanism must be provided for requesting service and for handling the assignment, and this can represent an overhead of capacity and complexity. DAMA schemes are also known as dynamic, reservation or controlled access. Such schemes call for considerable intelligence (i.e. software) to manage, and may be achieved under the control of a central hub station (dedicated or shared), or may be distributed between all user terminals. Distributed DAMA can reduce call setup times, and may also provide a measure of redundancy, e.g. against hub failure. Future satellite onboard processing techniques may provide some of this intelligence in the satellite itself, with resultant system benefits. Some DAMA systems are discussed further below.

Random access represents an alternative approach to providing capacity as it is required. Schemes are based upon packets of data in the form of bursty transmissions having a low duty cycle, and hence low average traffic rate. Some collisions between packets may be expected to occur, calling for retransmission (see below). There are various techniques, many of which may form part of a hybrid access scheme, e.g. to reserve TDMA capacity.

8.5 Frequency-division multiple access (FDMA)

8.5.1 Description

FDMA is a traditional and popular technique whereby several earth stations transmit simultaneously, but on different preassigned frequencies, into a common transponder. The transmitted carriers may range from an SCPC access, such as a single voice or data channel using FM, BPSK or QPSK, to an MCPC signal such as FDM/FM, FM/TV or a digitally modulated carrier.

It is important to recognise first that a single transponder may well carry a number of quite disparate groups of traffic, possibly even from different networks, and the transponder bandwidth (BW) will be coarsely preassigned by means of frequency division. Figure 8.5 shows a crude example of FDMA over a 72 MHz wide transponder; here a number of different types of signal are sharing the transponder bandwidth. The analogy with FDM is apparent, and

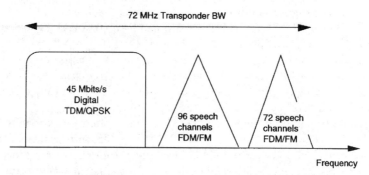

Figure 8.5   Example of FDMA allocations over a 72 MHz wide transponder

guard bands need to be allocated between accesses to reduce adjacent channel interference.

Preassigned FDMA is attractive because of its simplicity and cheapness. Analogue FDM–FM–FDMA has a well established history for high-capacity telephony links between large ground stations, and SCPC–FDMA is commonly used for very thin route telephony, VSAT systems and mobile services. Virtually all services are now digital, e.g. TDM/QPSK/FDMA.

A common scheme is the INTELSAT IDR (intermediate data rate) service, with carriers providing trunk communications at bit rates between 64 kbit/s and 45 Mbit/s. 2 Mbit/s IDR is widespread, comprising 32×64 kbit/s bearers, with QPSK modulation. INTELSAT also offers IBS (INTELSAT business services) with data rates from 64 kbit/s to 8 Mbit/s—again as TDM/QPSK/FDMA—and this is similar to EUTELSAT's SMS (satellite multiservices) system. Future IDR schemes are expected to employ 8PSK with TCM (trellis-coded modulation), which may provide around a 25 per cent saving in bandwidth for no increase in power.

Commonly-used digital FDMA systems (e.g. IDR, IBS, SMS) make use of FEC (forward error correction) encoding, typically rate 3/4 and rate 1/2, together with Viterbi soft-decision decoding. Options are available for additional (outer) encoding using RS (Reed–Solomon) block coding, and this is standard with TCM/IDR. For transmission rates up to around 10 Mbit/s digital modems tend to be referred to as channel units, being characterised by the ability to vary the transmit and receive carrier frequencies anywhere within the IF (intermediate frequency) band (e.g. 70 MHz $\pm$ 8 MHz, or 140 MHz $\pm$ 36 MHz). A typical channel unit comprises: terrestrial interface; FEC codec; scrambler/descrambler (for energy dispersal); frequency synthesiser; modulator; demodulator.

One illustrative, although dated, example of FDMA is the INTELSAT preassigned SCPC telephony service, having 36 MHz wide transponders apportioned into 800 frequency slots with 45 kHz centre-to-centre spacing. Each slot can be occupied by a single carrier such as a 64 kbit/s QPSK voice channel, and slots are coupled in pairs to provide each direction of speech.

8.5.2 Transponder effects

The majority of satellite transponders are transparent; that is, they amplify the uplink signal, shift it in frequency and transmit it as the downlink. Typical transponder bandwidths are a few tens of MHz, with RF output powers of a few tens of watts. Travelling-wave-tube amplifiers (TWTAs) are generally used for the power output stages, although solid-state amplifiers are finding increasing usage, especially where low-power spot beams are served.

All amplifiers possess limited output power, giving rise to saturation performance, which is well characterised in the case of TWTAs. The effect in terms of the carrier-to-noise density for up- and downlink signals, including intermodulation products (IPs or IMPs), is shown in Figure 8.6. These IPs arise at the output owing to a number of different frequencies being present combined with the nonlinearities of the transponder. Access by multiple carriers, as in FDMA, can give significant problems with IPs, and hence a few dB of backoff from saturation is required. The resultant reduction in downlink EIRP may represent a significant penalty, especially with downlinks to small terminals.

It is the third-order IPs which give most problems. These are frequency components arising from two or more carriers within the transponder passband, as a result of a cubic term in the polynomial representation of the nonlinear TWTA transfer characteristic. Such signals may themselves fall back into the same passband. For example, consider two carriers at f_1 and f_2: one third-order product is $(2f_2 - f_1)$, which represents a frequency that is effectively a reflection

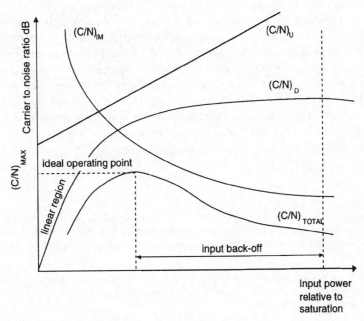

Figure 8.6   Characteristics of satellite transponder

of f_1 the other side of f_2, and which may fall directly upon another channel, f_3. This adds intermodulation noise to that signal which may be a problem, and calls for inclusion of a C/I_0 term in the link budget (see Chapter 11).

Provided that the transponder is operating in the linear region, the downlink output power for each carrier is simply proportional to its uplink power, and link budget calculations for each downlink carrier are based upon the *pro rata* apportionment of the total (backed-off) downlink power. However, as carrier accesses come and go, the TWTA operating point will vary. Some monitoring may be required to maintain the degree of backoff, and careful control of uplink transmit powers is necessary in most FDMA systems, both to keep IPs within limits and to ensure equitable distribution of the finite total downlink power.

Virtually all satellite systems operate in a backed-off mode, except perhaps where a single-carrier TDMA link is concerned. If a transponder is operating in the saturated region as opposed to the linear backed-off region, the distribution of powers may be further slightly affected by small signal suppression* (SSS).[1]

For some very low-power uplinks, the signal strength at the satellite may be small compared to the total thermal noise at the satellite front end. In extreme situations, e.g. low-power handheld terminals into a global or very wide-coverage antenna, transponded thermal noise may represent the predominant component of downlink power, and this will need to be taken into account when determining the available downlink EIRP for the wanted signal.

Because satellite downlink power is shared *pro rata*, advantage can be gained by removing accesses during periods when no information is being conveyed. This will not only allow more EIRP for the residual signals, but may also reduce IPs. Thus, systems handling intermittent digital traffic should endeavour to turn off the transmitted carrier when not required. An implementation of this strategy is voice activation, a technique whereby activity on a speech channel is detected and used to turn the carrier on and off, and can be used with some SCPC transmissions. A speech channel involves information in each direction for only about 40 per cent of the time. Where a large number of voice-activated channels are sharing transponder power in FDMA, the overall fluctuation in total power is within statistical limits and there is a net gain in useful downlink EIRP of up to a few dB. Such schemes may be particularly applicable to small-terminal or mobile satellite communications.[2]

---

*Small-signal suppression arises when more than one signal shares a common limiting amplifier, and the analysis of the nonlinear situation can be very complex, especially where several signals are concerned. Two particular extreme cases are worth noting however:

(i) A single-carrier signal with power which is far below the accompanying wideband gaussian noise: the carrier suffers additional SSS of about 1 dB over and above the simple power apportionment between the signal and the noise.

(ii) A single-carrier signal in the presence of another much higher power carrier: the carrier suffers additional SSS of up to about 6 dB.

8.5.3 Demand-assigned FDMA

Many SCPC users are unlikely to require continuous service, and large fluctuations in traffic load can arise. In order to use bandwidth efficiently demand assignment is commonly used, whereby a channel between two terminals is assigned only when communication is requested. This involves an exchange of information between terminals to assign channels; the detailed protocols, however, would be specific to a particular system.

A classic and illustrative example of FDMA DAMA is the SPADE system which was evolved to provide economical links over thin telephony routes within developing countries using medium to large terminal systems but where the average traffic per channel may be small.

SPADE (single-channel-per-carrier PCM multiple-access demand assignment equipment) was based upon the INTELSAT SCPC FDMA format described above. It is a distributed form of SCPC DAMA, without the need for a central control station. In addition to approximately 800 frequency slots, it includes a common signalling channel (CSC) which facilitates the demand assignment. The frequency plan is illustrated in Figure 8.7.

When a call request arises, an earth station chooses a pair of frequency slots for the duration of the call. Information about the slots chosen is exchanged on the CSC, which has a bandwidth of 160 kHz. A separate multiple-access scheme is required for the CSC itself, and this is TDMA with one time slot preallocated to each earth station in the system. Each station maintains a list of the available frequency slots, and chooses a pair at random. Detailed protocols over the CSC provide for signalling information exchange, and at the termination of a call the slots are returned to a common pool.

Part of the early problem with SPADE was that it required a minicomputer (in the days before microprocessors) and was ahead of its time; operating difficulties led to its being discontinued by INTELSAT in around 1992. However, its principles are still of interest and newly emerging DAMA schemes are remarkably similar in concept.

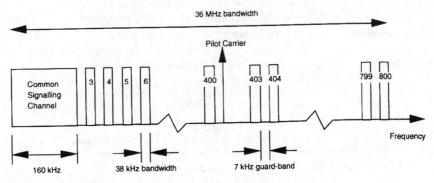

Figure 8.7   Channel allocations in the SPADE system

Various alternative schemes may be envisaged based upon centralised control, and these would be particularly applicable to mobile satellite services.

Commercial VSAT and other satellite networks may employ their own proprietary versions of DAMA, all heavily dependent upon software control. INTELSAT is currently establishing a new SCPC DAMA scheme, initially for the public-switched network, which is similar to SPADE, but with simpler terminal equipment. It will be SCPC but with a channel bit rate which can be varied; ultimately the aim is to provide bandwidth on demand, which is particularly useful for corporate networks requiring mixed service (e.g. speech, data, videoconferencing). The INTELSAT DAMA system consists of a number of traffic earth stations under the control of at least one network management and control centre (NMCC). Initially, all terminals need visibility of the frequency band allocated to DAMA, i.e. the service must be via a global beam or a looped-back hemi/zone beam. The wide range of choice of bit rate (16 kbit/s to 8 Mbit/s) means that a fully utilised DAMA network could well comprise the full range of antenna sizes.

8.6 Time-division multiple access (TDMA)

8.6.1 Description

In TDMA each user is allocated a periodic time slot, within which a burst of information is transmitted, sharing a common carrier frequency with other users. The successive bursts form a multiplexed TDMA frame at the satellite (akin to TDM). The basic concept is illustrated in Figure 8.8. Clearly, bursts need to be transmitted at an appropriate time such that no overlap occurs at the satellite, and hence on the downlink, and this calls for careful synchronisation. A small guard time may separate each burst. The simplest systems have a fixed assignment of slots (and are sometimes referred to as F-TDMA), although demand-assigned schemes (see below) can be very flexible, with allocation of bursts and burst lengths as required.

Only one TDMA carrier is active at a given time, and thus the full satellite downlink power is available for that access. Provided that there are no other groups of users sharing the transponder by way of frequency division, intermodulation products can be virtually absent as only one signal is present, and hence the transponder may not require to be backed off. A transponder can thus

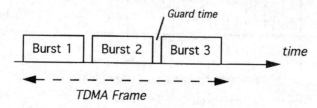

Figure 8.8   Outline concept of TDMA frame format

provide higher overall EIRP with TDMA than with FDMA. Care must be taken, however, with intersymbol interference (ISI), which can arise owing to the combination of filtering and TWTA nonlinearity; for this reason a small backoff is normally used (typically 2 dB input BO, 0.5 dB output BO).

The length of a TDMA frame depends on the type of traffic being carried. As traffic in a TDMA system is transmitted in bursts, data needs to be buffered at the ground station. TDMA has been well established for PCM (pulse code modulation) speech, where the baseband is sampled at 8 kHz with typically eight bits per sample, giving 64 kbit/s per (one-way) channel. Overall frame durations may be several ms; longer frame durations may be limited by the need to keep the end-to-end latency (delay) within comfortable limits.

One disadvantage of TDMA is that each uplink transmit station requires sufficient peak power to support the high instantaneous channel data rate, albeit for a short duty cycle. This may represent a technology challenge, especially for small handheld TDMA terminals, and is in contrast to FDMA which has continuous low-data-rate transmission. A further feature of TDMA is that the satellite downlink power is simply wasted during any empty slots, when the network is lightly loaded, whereas with FDMA the power may be shared between those carriers which are present at any time.

A TDMA network is complex to implement, and requires dissemination of a time reference. For a large, well ordered network this is practical and straightforward, and the all-digital nature interfaces well with sophisticated network management systems. A fixed assigned TDMA scheme will have a well defined capacity, but may need to be reconfigured to accommodate any changes in the number of links.

TDMA frames must be structured so that overlap at the satellite does not occur between bursts from different earth stations. Such synchronisation may be achieved by one earth station acting as a reference station, which then transmits reference bursts used by all the other stations to achieve timing synchronisation. After this initial reference burst, each earth station transmits its own traffic bursts, which together comprise the frame. Overlap is further prevented by guard times between bursts; this takes into account the different distances between the satellite and the earth stations as well as variations in the timing accuracies of the different stations.

TDMA systems developed for telephony will have an initial reference burst incorporating a carrier and clock-recovery period, to help the earth station to perform coherent demodulation and to optimally sample the received bit stream. Typically, the burst continues with a unique word, possessing a good impulsive autocorrelation function, to provide a timing reference for the data frame. The reference burst concludes with a control and delay channel (CDC), which is used to give earth stations information about the timing delays which they should use, and to provide control over the initial acquisition and synchronisation process (see below).

The traffic bursts each commence in a similar way to the reference burst before the data (traffic) proper; the so-called preamble again providing carrier, clock and unique-word synchronisation.

Figure 8.9 illustrates an example of TDMA frame format (from INTELSAT and EUTELSAT), employing transmission at 120 Mbit/s with a frame duration of 2 ms. The guard band allocated between bursts here is about 1 μs. The frame is made up of traffic bursts preceded by two reference bursts, each of which is transmitted by a separate reference station. One is designated the primary

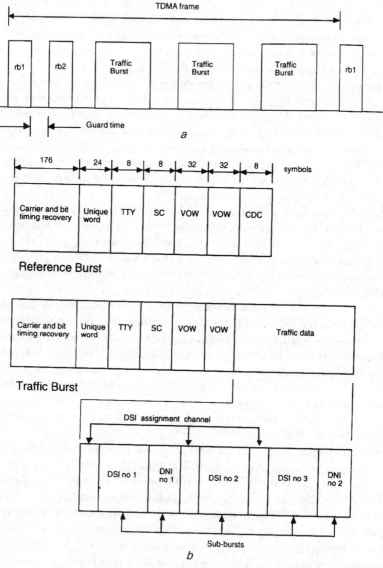

Figure 8.9   Illustration of TDMA frame format

    a Outline frame structure
    b Detailed structure

reference station, and the other the secondary; normally both operate, but terminals will respond to the secondary reference station if the primary should fail.

The reference burst here consists of a preamble and a CDC. The preamble comprises a carrier and bit-recovery sequence, a unique word, teletype order wire (TTY), a service channel (SC) for conveying management and control information and two voice-order-wire (VOW) channels (the order wires are used for exchanging instructions between engineers at earth stations). The CDC may address each earth station in turn, and effectively operates on a longer frame cycle of perhaps one second.

The traffic burst comprises a preamble, similar to that in the reference burst, and a traffic section. This includes both digital speech interpolated (DSI) and digital noninterpolated (DNI) traffic subbursts. DSI is a technique which takes advantage of the pauses naturally occurring within speech to improve the efficiency of digital voice-carrying channels. When there is no voice activity on a particular channel, that channel is assigned to another user which is commencing activity, thereby making more efficient use of the channel capacity. The assignment channels are used to signal the connection between the subbursts and the reconstituted traffic channels.

DSI has been well established in major network telephony systems, and can provide negligible degradation of speech quality, although clearly there can be problems under heavy demand or where usage is other than normal speech. DNI traffic refers to such other traffic, which may be facsimile, telex or continuous data. Nowadays, DSI is often employed on voice circuits in tandem with ADPCM speech coding; this is adaptive differential pulse-code modulation, and is a low-rate encoding (LRE) technique which recodes conventional 64 kbit/s PCM into 32 kbit/s, by effectively sending differences between successive samples rather than absolute values. The combination of DSI and LRE is often referred to as digital circuit multiplication equipment (DCME), and results in a typical capacity gain of 5 : 1 (around 2.5 times for DSI and twice for ADPCM).

8.6.2 *Synchronisation in TDMA*

TDMA synchronisation is important because of satellite motion and the different propagation ranges, which will affect the time at which earth stations must transmit so that their bursts do not overlap at the satellite. There are two types of synchronisation technique: open loop and closed loop. The essential difference between the two techniques is that in open-loop synchronisation the earth station simply estimates when it should transmit a burst, and allows adequate guard time either side to accommodate uncertainties, but in closed-loop schemes some feedback is applied to adjust the timing position of the bursts. Thus, a closed-loop scheme needs to directly or indirectly monitor the transponded downlink bursts, whereas this is not a requirement in open-loop synchronisation.

Synchronisation may be viewed as a two-stage process:

(i)   Acquisition — this is where the traffic station first acquires its position in the TDMA frame, and is generally open loop.
(ii)   Steady-state synchronisation — this is the means whereby a traffic station maintains its position within the frame, and is closed loop with feedback.

The precise mechanism depends upon the system architecture and the antenna beam configuration. Generally, a hub station will control the synchronisation of inbound (i.e. terminal-to-hub) traffic by means of timing instructions sent out to each terminal. If, however, a terminal is able to monitor the downlink frame pattern, it would be able in principle to adjust its own timing. This facility may, however, not be economic for low-cost terminals, and may not be feasible for scenarios where the uplink and downlink beams are geographically separated.

In many systems (e.g. INTELSAT telephony TDMA) open-loop acquisition is used initially by the reference station knowing the location of the acquiring terminal and the present location of the satellite. It then sends a D_n (delay) code within the CDC message, to enable the acquiring terminal to enter the frame appropriately, with sufficient guard time to avoid interference to other bursts. Subsequently, closed-loop feedback is used in the form of periodically updated timing information sent to enable the terminal to maintain accurate alignment of its burst within the frame.

8.6.3 Demand-assigned TDMA schemes

The use of fixed TDMA slots is inefficient for low-duty-cycle traffic. Demand-assigned TDMA permits slots to be dynamically allocated to earth stations requiring service, and the all-digital nature of TDMA coupled with software control permits a flexible response. Burst lengths may be varied, or slots allocated on a call-by-call basis, as requested through assignment control channels. Although reallocation could be performed on a frame-by-frame basis, it is also feasible to consider several consecutive TDMA frames as one super-frame, and allow reallocation of slots a few times per second.

Several DAMA/TDMA schemes are in existence, but the details and parameters are very system specific. With the development of commercial VSAT systems, a number of new and complex schemes compete against each other. Some systems use centralised control and others use distributed algorithms.

8.7 Satellite-switched TDMA and onboard processing

In trying to improve the capabilities of satellite communications networks, spot beams are used to cover individual regions rather than one beam covering all the regions. The benefits of using spot beams are as follows: first, a spot beam has a high gain which will increase the downlink EIRP to that region, permitting better performance (higher data rates and/or smaller antennas) and, secondly, the spot beams are isolated from one another allowing frequency reuse of the spectrum (provided that there is no overlap of coverage). This frequency reuse will increase the overall data-handling capacity of the system.

In a fixed TDMA system with spot beams, beam interconnections are normally static and only switched occasionally to accommodate major variations in traffic requirements. If full interconnectivity between the beams was required, one transponder would have to be allocated for each separate path between beams: this means that as the number of spot beams increases, the number of transponders required would increase at a far faster rate. (For example, to achieve full interconnectivity on a fixed TDMA satellite with ten beams, 90 transponders would be needed!) Full interconnectivity is clearly impractical if more than two or three spot beams are used.

This problem of full interconnectivity between spot beams may be solved by using satellite-switched TDMA (SS/TDMA); here, a transponder is allocated to each of the downlink spot beams. The uplink TDMA frames pass through the satellite switch which then routes traffic bursts to the transponder corresponding to the desired downlink beam. Switching takes place at various instants within the TDMA frame. Thus, SS/TDMA effectively allows full interconnectivity between uplink and downlink beams while employing a minimum number of transponders. The system requires time and space switching to be coordinated both on the ground and in the satellite. The detailed implementation and management is complex, and the subject of much development. Figure 8.10 illustrates the concept of SS/TDMA.

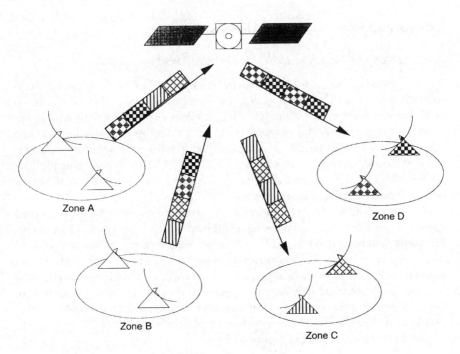

Figure 8.10   SS/TDMA concept

An example of a satellite using SS/TDMA is INTELSAT VI which has a six by six switching matrix (used for speech traffic). The uplink beam to downlink beam interconnectivity pattern is changed several (e.g. 20) times during each TDMA frame period, with the periods of switching set to suit traffic demands on the particular uplink to downlink beam pairs.

Switching can be performed at IF within a transponder, perhaps using PIN diodes. This is a simple form of onboard processing (OBP), which does not require demodulation to baseband on the satellite. More sophisticated satellites may employ OBP with signal regeneration at baseband, which can permit not only space switching but also time switching by means of buffering; very flexible and efficient usage of capacity can result. Such OBP may be especially advantageous, and provide necessary flexibility, for complex multiple-satellite networks, such as emerging low-earth-orbit mobile and personal satellite communication systems.

OBP can also provide onboard routing of packets for ATM (asynchronous transfer mode) systems.

(It is always arguable how far OBP is appropriate in practice for many satellite communication ventures. This is largely because any departure from a simple transparent transponder places constraints in terms of irrevocably linking the satellite to a particular ground-segment system architecture, and may remove the opportunity to purchase or apply transponder capacity where it may be found.)

8.8 Spread spectrum and CDMA

8.8.1 Spread spectrum for satellite communications

Spread spectrum (SS) is a form of modulation which offers the opportunity to partially reject interference. As such it has a long pedigree in military systems to facilitate communications under jamming, and more recently it has found application in some civil VSAT systems. It also offers the opportunity for multiple access, and when used for this purpose is called spread-spectrum multiple access (SSMA) or code-division multiple access (CDMA); CDMA is becoming increasingly popular for some terrestrial cellular mobile systems, and is also planned for some of the new LEO satellite systems.

We first consider SS, then examine its development into CDMA. Spread spectrum involves the signal being combined with a very wide bandwidth spreading function, determined by a unique code. There are two basic forms: direct sequence (DS) and frequency hopping (FH). In DS, the narrowband signal is multiplied directly by a pseudorandom code, and in FH the carrier hops in a pseudorandom pattern across a wide bandwidth, perhaps many tens of MHz. Although FH is widely used in military systems, civil satellite communication systems tend to use DS. The principle of operation of DS SS is shown in Figure 8.11, and that of FH in Figure 8.12.

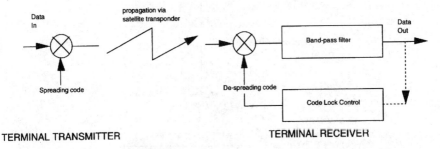

Figure 8.11   DS SS outline implementation

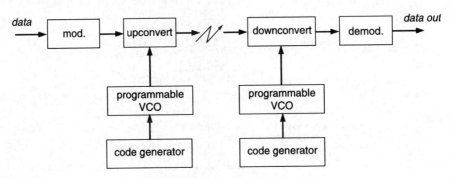

Figure 8.12   FH SS outline implementation

In DS SS the transmitter multiplies the data signal by a bipolar pseudo-random spreading code, at a rate typically between 1 Mchip/s and 10 Mchip/s (the term chip is used to represent a bit of the digital spreading code). Multiplication is again performed at the receiver with the identical spreading code, which has to be synchronised to the incoming signal to within a fraction of a chip width. This double multiplication restores the original unspread signal; it may be regarded as two phase reversals or, alternatively, as two modulo-2 additions. At the same time, uncorrelated interference is spread further at the receiver, and is then largely rejected by a bandpass filter. This is illustrated in Figure 8.13. The degree of rejection of uncorrelated signals is known as the processing gain, and is broadly the ratio of the spread BW (or the code-chip rate) to the data BW (or data rate); typical values are in the range ten to 30 dB. Not only other SS transmissions but any uncorrelated interfering signals are similarly largely rejected.

Spread spectrum in itself does not fundamentally provide any link budget advantage in terms of performance in additive thermal noise. The basic link bit-error ratio is determined by the link carrier-noise density, C/N_0, and it is of no first-order consequence that the carrier power, C, is spread. However, for a practical system, SS can offer benefits which may be very important and even necessary, and these will affect resultant system performance.

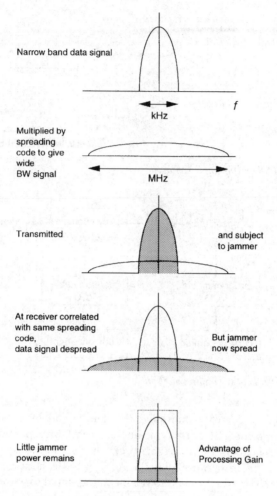

Figure 8.13   Spread-spectrum rejection of interference

SS may offer the following benefits for satellite communications:

(i)   Rejection of uncorrelated interference permits system operation where adjacent satellites (say at 1.5° spacing in the geostationary arc) may lie within the beamwidth of a small-terminal antenna. This may facilitate operation with small antennas at C-band, with its relatively longer wavelength and correspondingly large beamwidth.

(ii) Other link interference, multipath and adjacent channels may be similarly tolerated, thereby facilitating operation in shared bands.

(iii) The effect of SS transmissions upon other users is relatively benign, appearing simply as additional noise rather than as potentially destructive interference. Again, this allows operation in shared bands.

(iv)   Power flux-density values per unit bandwidth are reduced by virtue of the spreading (but the overall power flux density is unchanged). This may permit operation within the letter of regulatory limits for some high EIRP downlinks.

(v)   Spread spectrum generally provides good LPI (low probability of intercept), which may be highly significant in some military scenarios.

SS in itself makes inefficient use of bandwidth, but the bandwidth efficiency may not always be a real problem, especially in very-small-terminal systems. The real price paid for SS is the complexity and cost of the synchronisation circuitry in the receiver, with the difficulty in synchronising the despreading code. Code lock is generally maintained by means of a code-tracking loop, the behaviour of which is analogous to that of a phase-lock loop. Prior to such lock, however, it may be necessary to instigate a code-search phase, where all possible relative code states are examined to find the one producing a correlated product within a narrow BW. Such search and acquisition processes are usually performed sequentially, and can take an appreciable amount of time and represent a costly system overhead.

Spread spectrum (with or without a CDMA multiple-access capability, as described below) has been advocated for some VSAT systems, where the small ground terminal antenna apertures mean that interference will be received from (or radiated to) adjacent satellites. The interference-rejection capability of SS allows operation under such conditions. Essentially, the interfering power is uniformly spread so as to constitute an increased noise level, rather than represent potentially destructive narrowband interference against which large narrow beamwidth antennas would be needed. The effect of terrestrial interference (i.e. microwave links) is also mitigated.

8.8.2 CDMA

CDMA is a multiple-access technique based upon SS. Each user channel is spread to a wide bandwidth, raising the BW from typically a few kHz to a few MHz. The resulting SS signals are modulated onto a carrier frequency which may be the same for all users within the subnetwork, such that all transmissions are overlaid simultaneously within the common transmission bandwidth, with the powers from each transmitter adding.

At the receiver the input signal is correlated with the appropriate spreading function, suitably synchronised, to reproduce the originating data. Each transmit–receive user pair, or channel, is allocated a spreading code which is itself uncorrelated with codes in use by other channels. In this way transmissions from the other unwanted users in the band are largely rejected by the receiver-despreading process, as is any interference (including intermodulation products). The residual effect of such interference, known as the self-jamming noise, may be modelled essentially as additive gaussian noise.

The spreading codes themselves should be chosen as a set possessing low cross-correlation properties in order to minimise interference between user pairs.

Maximal-length sequences derived from shift registers (*m*-sequences) have tradi-tionally been used, but have a limited set size with useful low crosscorrelation properties, so other sequences may be preferred where more than a handful of users are involved. Software techniques allow a wide range of possible spreading codes, including Gold codes, Kasami sequences and Bent functions; for further reading see Reference 3. With large user populations, where some hundreds of user pairs may each require a unique code, limitations on the number of suitable codes may be a constraining factor.

Besides the spreading code, data information at perhaps a few kbit/s is modulated onto the spread signal, typically as BPSK. This data is recovered at the receiver output, following correlation with the local spreading code. As the number of users in the system increases, the aggregate level of self-jamming noise at the output of the despreader will increase and degrade the performance. The capacity limit to a CDMA system is not abruptly defined, but is a function of this signal-to-noise ratio (and corresponding bit-error ratio).

It can be shown that, in a simple model scenario, CDMA is fundamentally inferior to FDMA or TDMA in terms of capacity or spectral efficiency.[4] For a practical very-small-terminal scenario, however, the difference between CDMA and FDMA is not so great, especially where performance is power limited owing to small antenna size and where thermal noise predominates. As the total power in the system increases, the performance of CDMA becomes inferior to that of FDMA as the former is limited by the self-jamming noise. This leads to a graceful degradation of CDMA as the number of terminals in the network is increased. Figure 8.14 gives a rough illustration of the relative performance.

CDMA does, however, offer a number of advantages when other effects are taken into account, and particularly where interference from adjacent spot beams, or other systems, is a factor. This is akin to the terrestrial cellular situation with adjacent cell interference. Such factors can enhance CDMA per-formance to a level exceeding that which may be achieved by FDMA or TDMA in those scenarios.

Because satellite downlink EIRP may be shared between transmissions on a *pro rata* basis, and since other users contribute to the effective noise level, a

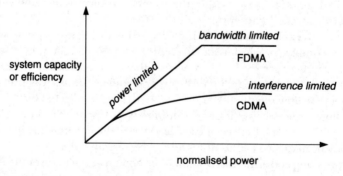

Figure 8.14   Simple system capacity comparison

CDMA system for speech traffic can benefit directly from voice activation[2,5]. If the carrier can be removed during speech pauses, the residual active channels can take advantage of the total EIRP at the same time as the average aggregate self-interference-noise contribution is reduced. Hence, the overall network capacity is optimised without the need for specific DSI techniques.

Note that in the satellite channel itself the CDMA transmission occupies a very wide bandwidth and, consequently, the channel signal-to-noise ratio (SNR) may be very small ($\ll 1$). The despreading process restores this to a decent value. The link C/N_0 requirements, however, which relate to the more relevant and fundamental noise spectral density, are unchanged to a first order. Thus it cannot be said that either spread spectrum or CDMA fundamentally implies low transmit power or a performance which is any different from FDMA etc., although it may be that other practical benefits can arise.

A feature of CDMA is that the spread signals in the satellite transponder give rise to noise-like IPs at low level and without peaks, having an effect smaller and more manageable than that of narrowband IPs (as might arise with SCPC/FDMA). As a result, a transponder may be operated closer to saturation than may an FDMA system, giving capacity benefits. Conversely, a CDMA system is relatively immune to narrowband IPs or interference. Another benefit of CDMA is resistance to multipath propagation, since once correlation lock has been achieved other multipath signals will represent simply uncorrelated interference. This is of value especially to mobile and VSAT systems. CDMA may additionally provide advantage where polarisation diversity is employed, through rejection of crosspolar components.

CDMA offers a further practical benefit in that the frequency agility of an FDMA transmitter/receiver is not required. In operation CDMA signals may be overlaid on the same carrier frequency, partially overlapping, or given non-overlapping frequencies; it is also possible to share a transponder with other signals on an FDMA basis.

The application of CDMA tends to be limited by the cost and complexity of the receiver, together with the time taken to achieve synchronisation. In simple theory terms its performance is inferior to that of FDMA or TDMA for a given power and bandwidth, but in practice the performance can be superior to FDMA allowing for the latter's limitations of guard bands and TWTA backoff. There is no need for network timing references as in TDMA, and speech duty cycles may be readily exploited. CDMA is invariably used in conjunction with forward-error-correcting (FEC) coding, and in practice may offer greater flexibility in this regard than either FDMA or TDMA.

Some military satellite communications use CDMA as a multiple-access scheme. These take advantage of the random-access nature of CDMA, where fixed or highly planned schemes are impractical, and where downlink EIRP needs to be maximised for operation to small terminals. (Although a degree of jamming and interference rejection is provided, other spread-spectrum schemes with much greater processing gain and code security tend to be used for specific antijam purposes.)

Spread-spectrum techniques (such as CDMA) became established in civil satellite communications with the early equatorial VSAT system operating at C-band in the US. The main reason for the adoption of CDMA in that system is understood to be the need for SS-derived rejection of adjacent satellite interference experienced with the relatively wide antenna beamwidths. It has also been suggested that the use of spread spectrum enabled operation within the letter of FCC flux-density regulation limits. It appears that few if any Ku-band VSAT systems currently employ CDMA, principally because the relatively narrow antenna beamwidths at the shorter wavelengths reduce potential adjacent satellite interference.

As a topic, CDMA has been given a boost by its application in terrestrial cellular systems (e.g. the US IS-95 scheme). A variant of this system is also employed in the Globalstar satellite personal communication network of 48 satellites (see Section 18.7). In satellite systems there has been some work under ESTEC sponsorship looking at synchronous CDMA for satellite communication application; here all codes are time aligned, and by choice of orthogonal sets the selfjamming noise may be reduced (see e.g. Reference 6); such schemes may, however, have certain difficulties and practical limitations. There is also renewed interest generally in SS, including CDMA, for new satellite services, as pressure on the radio spectrum increases, together with the density of traffic and band sharing.

The overall merits or otherwise of CDMA are highly scenario dependent, and the subject of considerable debate. In essence, it may be concluded that the main benefit to civil satellite communication systems is the ability to operate in a predictable and equitable manner where interference is likely, either from adjacent satellites (due to the wide antenna beamwidth) or from other coband operations. An operator wishing to introduce its own system into an already crowded environment might do well to use CDMA; it is not however a universal panacea, and if all operators used it little would be gained.

A general and readable reference to spread spectrum and CDMA may be found in Reference 7.

8.9 Packet-access techniques

8.9.1 Applications

Many small-terminal systems handle data in the form of occasional packets, giving bursty transmissions at a low duty cycle, and hence low average traffic rate. FDMA, TDMA or CDMA are generally more appropriate for real-time use with speech or continuous data traffic from a known set of earth stations, and may not be the most suitable multiple-access techniques in such cases.

Some small-terminal data systems are more concerned with EIRP and with terminal cost than with efficient bandwidth utilisation, especially as the overall data capacity required may not be large. This has led to the emergence of a variety of satellite access techniques having a pedigree based more upon local area networks (LANs) than upon telephony systems.

These packet-access techniques, or protocols, are generally based upon contention schemes, with a performance very much a function of traffic statistics. Their important parameters tend to be delay and throughput (i.e. channel-capacity utilisation). They may be used on their own, or as part of a hybrid access scheme, e.g. for a TDMA reservation signalling channel. They are also known as random-access protocols, and the term time-random multiple access (TRMA) is sometimes used. Random-access systems are used in mobile satellite systems for setting up connections (Chapter 19) and in VSAT systems (Chapter 22).

8.9.2 ALOHA schemes

The simplest random-access technique is the ALOHA contention protocol. Each earth station transmits its message whenever demand arises, as a short data packet on a common channel shared with other users. Some packets will experience collision with others at the satellite, and thus be corrupted or garbled. Any such collision is detected, and the packet is subsequently retransmitted, perhaps after a random delay, with the process being repeated until successful.

Although detection of the collision could be performed by the sender monitoring its own transponder channel, and retransmitting when a discrepancy from the packet sent is detected, it is unlikely that many satellite systems will be in a position to achieve such monitoring. More commonly, the destination station detects a packet error by means of built-in error-detection coding. Cyclic redundancy check (CRC) coding may be employed, using a number of code check bits appended to the data within the packet. The destination then arranges to send acknowledgement packets (ACKs) back to the sender for each correctly received packet; where an ACK is not received after a suitable delay, the sender retransmits the relevant packet. This is known as an ARQ (automatic repeat request) scheme.[8]

The merits of this ALOHA–ARQ access technique are simplicity and lack of need for central coordination or timing references. The problem is that useful throughput declines as more user traffic is demanded of the common channel, giving extensive collisions and large delays. Figure 8.15a illustrates packet collision, and Figure 8.15b shows how traffic on the channel is related to the demanded, i.e. useful, throughput. Under light-traffic loading conditions the system performs well, but it can saturate under heavy demand. It can be shown simply that the maximum throughput is 18 per cent ($1/2e$) of available channel capacity (which may be expressed in erlangs, a dimensionless unit of traffic).

8.9.3 Enhancements to ALOHA

The basic ALOHA scheme may be enhanced in several ways to improve performance. Variants are very well established in local area networks (LANs), for

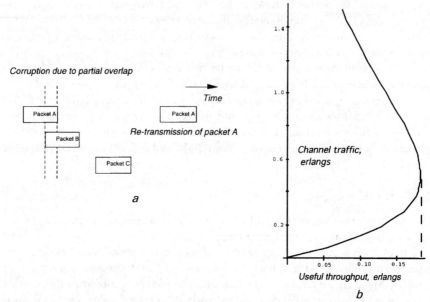

Figure 8.15   Simple ALOHA performance

   a Illustration of packet collision
   b Throughput characteristics of the ALOHA channel

example Ethernet. There are also a great variety of schemes under development for the small-terminal satellite communications market. Higher throughput and reduced delays are the goal, but this is invariably at the expense of complexity.

Selective reject ALOHA (SREJ ALOHA)

This is one of several strategies for retransmitting errored data with a minimum of overhead. A structure of small subpackets is formed from initial larger data packets, and only those subpackets which are corrupted are retransmitted. Some increase in throughput, and delay improvement, can be achieved with this technique. Several variants exist, e.g. the simpler go-back N protocol.

Slotted ALOHA

An improvement on the above technique, in slotted ALOHA packets are all the same size and are constrained to begin and end at fixed, regular time instants. Collisions caused by partial overlap of packets do not then occur, giving a doubling of throughput. In practical satellite networks the advantage over unslotted ALOHA may be less than at first sight, since difficulties arise owing to uncertainties in propagation delay, requiring a guard time to be introduced, and there is also the overhead of maintaining a time reference.

Reservation slotted ALOHA (R-ALOHA)

This technique employs a long frame structure comprising numbered slots. Once a station has successfully used a particular slot, that slot remains allocated to it until no longer required. This is beginning to approach a TDMA format, and can give good performance for both bursty and continuous data sources.

Reservation TDMA

A demand assigned packet-access scheme overcoming some of the problems of R-ALOHA. The TDMA frame consists of short reservation slots and a number of longer data slots. The hub controller assigns data slots to earth stations according to their requests; at termination of data traffic, slots are automatically assigned to other waiting earth stations in a structured order. It gives rise to potentially large delays, but allows efficient use of channel capacity.

Adaptive reservation slotted ALOHA

This is a technique whereby a network controller assigns a movable boundary within a TDMA-type frame structure, such that part is for pure slotted ALOHA and part for reservation slotted ALOHA. At low traffic demands the whole scheme is slotted ALOHA, and at high traffic demands it changes towards a TDMA-type structure.

Frequency hopped (FH) ALOHA

The channels in an FDMA structure are randomly hopped, in order to randomise traffic patterns and thereby even out the probability of collision. In some network configurations this may prevent excessive delays owing to dense clusters of traffic. (Such frequency hopping is not viewed as a spread-spectrum technique, but more a method of smoothing out the traffic distribution.)

CDMA-ALOHA

A scheme where a degree of spectrum spreading is applied. This means that packet collision will not necessarily result in fatal corruption and, provided the number of simultaneous users is small, several packets may be handled simultaneously over a single channel, yielding capacity benefits.

8.10 Hybrid access techniques, and comparisons

Many of the above multiple-access techniques may be used in combination to provide efficient schemes for satellite communication traffic demands. FDMA accesses at the satellite can themselves each represent a number of networks with different access techniques, and clearly great care has to be taken with power sharing and intermodulation products. Allocations may be requested and assigned through TDMA or ALOHA-type control channels, for which a specific frequency slot is provided.

TDMA is one of the most efficient schemes in terms of power and bandwidth utilisation, but in its fixed form it is unsuited to a large population of low duty cycle users. Demand-assigned TDMA has already been described, where the requests are made within an assignment control channel, which could again itself have a TDMA type of structure or an ALOHA structure. The state-of-the-art is represented by adaptive schemes where typically a frame structure has portions for TDMA bursts and portions for ALOHA access, with moveable boundaries between the two. Traffic statistics dictate which scheme is used at any time, and may also determine the boundary position. In general, speech would be handled by TDMA slots, and intermittent data by ALOHA slots.

There is usually no need for the access technique for each direction of transmission to be the same, and in any hub-based network there may be considerable asymmetry. A network with a central hub station might typically use MCPC-TDM for hub-to-terminal links, and SCPC FDMA or SCPC CDMA for terminal-to-hub links. (The INMARSAT system, for example, uses a combination of voice/FM/SCPC and data/TDM/PSK/SCPC for both shore-to-ship and ship-to-shore links, with a random-access channel for requests by ships).

The advent of spacecraft onboard processing allows for conversion of access technique in the payload. There is particular attraction in FDMA to TDM onboard conversion, where a number of SCPC uplinks are transmitted from the satellite as a composite TDM downlink stream; in this way the simplicity of SCPC FDMA for a large population of terminals is combined with the power efficiency of TDMA, at the expense of satellite cost and complexity.

All of the above satellite access schemes have their particular advantages, and may be used in combination with each other. Choice must be made for the particular network to make the most efficient use of satellite transponder capacity in terms of both bandwidth and power (which can be not only expensive but restricted in supply), coupled with high performance and low cost of terminal equipment. The choice of satellite access technique may also be affected by the need to work with the network-access protocols, and the traffic statistics.

Table 8.1 may represent a rough starting point in terms of the traffic model. It is almost impossible to declare which is the best technique, as it is 'horses for courses' and it is important to consider each network and application on its own merits. Clearly, the type of traffic and requirements are critical to choice of technique. Both the user and the service provider may need to optimise systems in terms of throughput and delay parameters. The efficiency of different schemes is also increasingly important, given the growing pressure on spectral capacity. The end user may, however, be simply interested in cost, or in other features of the overall network-management software (NMS) package.

Other general reading on satellite access techniques may be found in References 9, 10 and 11.

Table 8.1   Traffic models and multiple-access schemes

Traffic model	Appropriate multiple-access scheme
nonbursty traffic	fixed assigned TDMA, FDMA, or CDMA
bursty traffic, short messages	pure contention scheme (e.g. ALOHA)
bursty traffic, long messages, large number of users	reservation protocol with a contention reservation channel
bursty traffic, long messages, small number of users	reservation protocol with a fixed TDMA reservation channel

8.11 Conclusions

The standard multiple-access techniques in common use are FDMA, TDMA, CDMA and packet (or random) access. Practical schemes may be hybrids, combining the beneficial features of each, according to the traffic requirements and the operating scenario.

FDMA represents the simplest technique, suitable for a variety of system architectures and users, but suffers from potential intermodulation products. TDMA can provide more effective use of transponder power, but calls for sophisticated timing control; it is appropriate particularly for large relatively fixed telephony systems. (Satellite switched TDMA offers additional advantages where spot beams are employed.)

CDMA is an attractive technique where interference is a problem or where minimal overall control is available but is complex, particularly in terms of signal acquisition, and may not always represent optimum use of power and bandwidth.

Packet access techniques, such as ALOHA and its many variants, are appropriate for low-duty-cycle data-transmission systems with large user populations. They are particularly suitable for request channels in some reservation schemes, such as demand-assigned TDMA, and may provide part of a hybrid scheme.

8.12 Acknowledgment

Thanks are due to Des Prouse, of BT Goonhilly, for helpful advice, support and encouragement.

8.13 References

1 SPILKER, J.J.: 'Digital communications by satellite' (Prentice-Hall, 1977) ch. 9
2 JOHANSSEN, K.G.: 'CDMA versus FDMA channel capacity in mobile satellite communications', *Int. J. Satell. Commun.*, 1988, **6**, pp. 29–39

3 HA, T.T.: 'Digital satellite communications' (Macmillan Publishing Company, 1986)

4 VITERBI, A.J.: 'When not to spread a spectrum—a sequel', *IEEE Commun. Mag.*, April 1985, **23**, (4), pp. 12–17

5 JACOBS *et al*: 'Comparison of CDMA and FDMA for the Mobilestar system'. Proceedings of mobile satellite conference sponsored by JPL, Calif., USA, May 3–5, 1988

6 GAUDENZI, R. de, ELIA, C., and VIOLA, R.: 'Band-limited quasi-synchronous CDMA: a novel satellite access technique for mobile and personal communication systems', *IEEE JSAC*, 1992, **10**, (2), pp. 328–343.

7 DIXON, R.C.: 'Spread spectrum systems with commercial applications' (Wiley & Sons, 1994, 3rd edn.)

8 TANENBAUM, A.S.: 'Computer networks' (Prentice Hall International, Inc., 1996)

9 MARAL, G., and BOUSQUET, M.: 'Satellite communications systems' (John Wiley & Sons, 1998, 3rd edn.)

10 EVERETT, J.L. (ed.): 'VSATs—very small aperture terminals' (Peter Perigrinus, 1992)

11 MARAL, G.: 'VSAT networks' (John Wiley & Sons, 1995)

Chapter 9
Modulation and modems

T.G. Jeans

9.1 Introduction

If we have some message or information signal to transmit over a distance without using a cable, electrical or optical fibre, then we need to use a radio-frequency signal to radiate the information. Suppose we have a digital message to send, i.e. a message which is just a sequence of binary digits, 0s and 1s. We could connect this signal to a piece of wire, and some of the signal would radiate into space as an electromagnetic wave. In principle, that electromagnetic wave would induce a voltage on a corresponding piece of wire some distance away and, after (much) amplification, a replica of the signal, contaminated by noise and other interference, could be recovered.

Unfortunately, the efficiency of a piece of wire in radiating a signal is related to the length of the wire compared to the wavelength of the electromagnetic wave being radiated. Since the product of the frequency and wavelength of an EM wave in free space equates to 3×10^8 m/s, if we assume that we have a data rate of say 9600 bits (binary digits) per second, which could be filtered to a bandwidth of 4800 Hz, then this highest frequency component would have a wavelength of 62 500 m. Typical wire antennas radiate effectively when their dimensions are comparable with half the wavelength, which in this example would be 31 250 m, or 31 km. Now, although frequencies as low as 10 kHz or less are used to communicate to nuclear submarines, and consequently require antennas of length 15 km, most people are not able to pay for the festooning of the coastline with this amount of wire, and this is not a proposition for mobile receivers. Thus it is necessary to investigate how a high-frequency EM wave can be used to transport the information signal.

It is necessary to change some property of the transporting wave, or carrier wave, in sympathy with the information signal, or modulating signal. This operation is called modulation. The resulting modified carrier wave is called the modulated signal, and the information or message signal is called the modulating signal.

In general, we can modulate (or modify) the amplitude, the instantaneous frequency, the phase, or a combination of these parameters of a carrier wave denoted by $A \cos(\omega t)$, where A is the amplitude of the carrier, ω is the frequency and t is the time. The content of the bracket (ωt) is the total phase of the wave measured in radians, normally given the symbol θ, so that for a constant frequency carrier, the transmitted signal is:

$$S(t) = A \cos(\theta) = A \cos(\omega t) \tag{9.1}$$

9.2 Channel characteristics

We call the combination of the medium over which the signal must be transmitted, and its properties, the channel. Different channels have different properties and impairments, which will degrade the transmitted signal in different ways. For satellite communications the channel characteristics vary according to the satellite orbit, and the carrier frequency used. For satellites placed in geostationary orbit, the satellite appears stationary in the sky apart from a very small pointing wander, which puts a slow phase variation on the received signal. The main impairment on this channel is attenuation, since the satellite is so far away, and the addition of white gaussian noise, called AWGN in the textbooks.

9.2.1 AWGN

The noise is generated by the thermal motion of electrons in materials. It is called additive, because it just adds to the signal; white, by analogy with white light which contains equal amounts of all the colours and hence frequencies, so white noise has a constant power spectral density, often given as N_0 watts/Hz; and gaussian, because the noise amplitude probability density function is gaussian shaped.

Figure 9.1 shows the probability density function for white noise with an RMS voltage of σ volts, and gives the probability that the instantaneous noise voltage lies between v and $v + dv$ volts, plotted against v/σ.

$$\text{prob}(v < \text{noise voltage} < v + dv) = \frac{1}{\sigma\sqrt{2\pi}} e^{v^2/(2\sigma^2)} dv \tag{9.2}$$

The probability that the noise voltage is large is quite small, and the noise voltage generally has a small value.

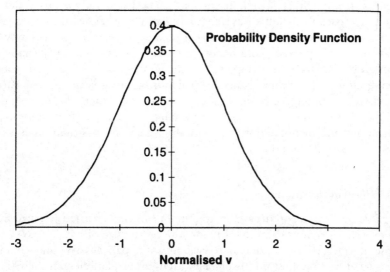

Figure 9.1   Probability density function of Gaussian noise

The noise power received by a demodulator is directly proportional to the bandwidth of the receiver. If the receiver bandwidth is B Hz, then the total noise power will be $B \times N_0$ watts, which is equivalent to σ^2 in communications calculations where one-ohm resistors tend to be assumed. It is clear that in order to reduce the effects of noise we want the narrowest possible receiver bandwidth, but this can lead to other problems, as discussed later.

9.2.2 Doppler effect

In a low-earth orbiting satellite, called a LEO satellite, the path loss is not such a problem, but Doppler shift and multipath problems become severe. When the satellite has a relative velocity of v m/s along the line of sight, then the received signal has a frequency shift on it given by v/λ where λ is the wavelength of the carrier signal. The frequency shift is positive as the satellite comes towards the receiver, and negative as it goes away. This Doppler shift can, according to the orbit and the carrier frequencies used, be many times the bandwidth of the receiver, for example for LEO satellites 50 kHz, and hence requires the use of frequency tracking in the receiver.

9.2.3 Multipath and shadowing

A more major problem is that the direct line of sight from the satellite is obscured or shadowed, so that the signal reaching the receiver, particularly at low satellite elevation angles, is made up of a number of signals reflected from buildings as well as a possible direct signal. At high elevation angles, the main problem is shadowing due to buildings and trees, but at low elevation angles the problem is a mixture of multipath and shadowing. In a multipath situation without a strong direct signal, if the path lengths of the signals differ by amounts comparable to half a wavelength at the carrier frequency, then the sum of the signals equates to zero! In practice, the satellite or the receiver is moving, so that the path difference, and hence signal strength, is constantly changing, giving fading, discussed in Chapter 19. The statistics of the fading are called Rayleigh fading where there is no direct line-of-sight signal path, and Rician where there is a direct line-of-sight signal in addition to other multipath signals. Since the phase of the resultant is also varying quickly, this does cause major problems for the demodulator, particularly where the information is being carried as phase modulation.

9.2.4 Other considerations

Power is a major problem in satellite and mobile communication systems. Each of these systems is power limited; the satellite by the size of its solar cell array and mobiles by the size of their battery. Thus, it is particularly vital for these systems to utilise the most power-efficient high-power amplifiers in their transmitters, where most of their power is used. Unfortunately, at the present time, such amplifiers become nonlinear at their most power-efficient operating points. A typical transfer characteristic for a travelling-wave-tube amplifier is shown in Figure 9.2. If an amplitude modulated waveform passes through such an amplifier, the peak values of the envelope will be flattened and the modulating signal information held as the envelope will be severely distorted. Hence, amplitude modulation is not suitable for satellite communication systems. In addition, any noise spikes or interference will add to the modulated signal, and when this is demodulated will appear in the recovered modulating signal. We need a modulation scheme which produces a modulated signal that has a

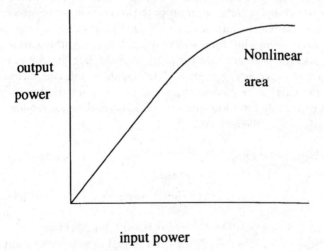

Figure 9.2   Characteristic of a TWT

constant envelope, which will be relatively unaffected by nonlinear amplifiers and, if possible, by additive interference and noise.

For any modulation scheme we want to get the greatest volume of data transmitted for a given RF bandwidth allocation, within the limitations of our transmitted power, which will affect the bit error rate (BER). We are interested in the bandwidth efficiency, also known as the spectral efficiency, which is defined as the data rate, measured in bits per second, divided by the total transmission bandwidth. Spectral efficiency is measured in bits/s/Hz. To lower the transmission bandwidth we can transmit fewer signal symbols per second by making each symbol represent several bits of information. Thus we could grab two bits at a time, which gives us four possible signal states to transmit during the time of two bits, or one symbol period. We could just transmit one of four possible phases of carrier shifted from the unmodulated carrier for this symbol period, as is done in QPSK modulation, for example. The number of signal states transmitted per second is called the baud rate. Because the symbol is transmitted for twice the bit period in this example, the bandwidth required is half what it was before.

In order to compare modulations we are interested in their error rate, which is related to the transmitter power. For constant envelope modulations, where the carrier amplitude is A, the carrier power, C, is $A^2/2$ W. If a bit period is given by T_b seconds, then the energy per bit E_b is CT_b or $A^2 T_b/2$. If the transmitted bandwidth is B Hz, and the noise spectral power density is N_0, then the total noise, N, is $N_0 B$ W. The ratio C/N is called the carrier-to-noise ratio and is related to E_b/N_0 by:

$$\frac{C}{N} = \frac{E_b}{N_0} \frac{1}{T_b B} = \frac{E_b}{N_0} \times (\text{bandwidth efficiency}) \tag{9.3}$$

The spectrum of the signal shows the frequency content of the signal, and hence the required transmission bandwidth. For many modulations the spectrum falls off quite slowly, either side of the carrier frequency. Adequate performance can still be obtained if much of the power in the outlying parts of the spectrum is removed by bandpass filtering, centred at the carrier frequency. This, then, means that other transmissions can be packed together, side by side in frequency. However, there will still be some spreading of signal power from the adjacent channel, and this is called adjacent-channel interference. Severe filtering causes the successive symbols to be smeared out in time, which will then lead to intersymbol interference, or ISI, which is discussed later.

9.3 Analogue amplitude modulation

If we choose to change the amplitude of the signal, then we are using amplitude modulation, which is the method used for medium and long-wave broadcasting in the UK. If we assume a message signal $m(t)$, then in ordinary amplitude modulation, the instantaneous amplitude of the carrier becomes

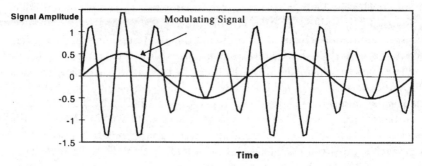

Figure 9.3   Amplitude modulation, showing the modulating and modulated signal

$(A + m(t))$ rather than A. Figure 9.3 shows the waveforms for a modulating signal which is a single frequency tone.

The envelope of the carrier wave contains the information that is being transmitted. As stated previously, because of power-amplifier nonlinearities, as well as its susceptibility to noise, this is not a suitable modulation scheme for satellite communication systems.

9.4 Analogue frequency modulation

Frequency modulation, where the instantaneous frequency of the carrier is varied about its nominal value in sympathy with the amplitude of the modulating signal, is used in satellite analogue television broadcast systems. If we assume a modulating signal $m(t)$ again, and carrier frequency ω_c, then the instantaneous frequency of the modulated carrier is given by:

$$\omega_i = \omega_c + k_f m(t) \tag{9.4}$$

where k_f is a constant. Since the instantaneous frequency is the rate of change of phase of the carrier, then the phase of the carrier is given by the integral of the instantaneous frequency. Hence, the expression describing an FM signal is:

$$S(t) = A \cos\left(\omega_c t + k_f \int_0^t m(t)dt\right) \tag{9.5}$$

The carrier deviates either side of its centre frequency by an amount equal to the maximum value of $k_f m(t)$ radians/s. This is called the peak frequency deviation. When it is much larger than the highest frequency contained in the modulating signal, then the modulated signal is said to be wideband FM. The ratio of the peak deviation to the modulating frequency is called the frequency modulation index, β. For the standard broadcast VHF FM radio system, the monophonic content of the speech is up to around 15 kHz, and the frequency deviation is 75 kHz, giving a β of five for this audio-frequency component. For satellite television the bandwidth of the video signal is 5.5 MHz and the deviation is around 13 MHz. Since it is difficult to calculate theoretically the total bandwidth of an FM signal, a rule of thumb called Carson's Rule is usually used. This states that for less than one per cent distortion, 99 per cent of the modulated signal is contained within a total RF bandwidth of 2 × (peak deviation + modulation frequency). For satellite systems this implies a required total bandwidth of 36 MHz.

Because the information in an FM signal is essentially contained within the time between zero crossings of the carrier waveform, that is the instantaneous frequency, the signal can be passed through limiting amplifiers without increased distortion, and so amplitude noise can effectively be greatly suppressed. An FM system exchanges increased output signal-to-noise power ratio for increased transmission bandwidth. In fact it can be shown that the output $S/\mathcal{N}$ ratio is $3\beta^2$ greater than the input $S/\mathcal{N}$ ratio for the equivalent AM signal.

Figure 9.4 shows the output $S/\mathcal{N}$ ratio as a function of the input $S/\mathcal{N}$ ratio of a typical FM system. Note that the output $S/\mathcal{N}$ ratio catastrophically degrades when the carrier-to-noise power ratio drops to around 10 dB, and this is called the threshold effect. This is not a good characteristic for a modulation scheme to be used on satellite systems, where transmitter power is at a premium. If the path losses increase owing to rain attenuation, or antenna pointing errors, the performance can change from being adequate to unusable for a few dBs change.

The onset of threshold can be delayed by five to seven dB by using an FMFB demodulator, shown in Figure 9.5, or by two to three dB using a phase-lock loop demodulator. The FMFB demodulator feeds the recovered modulating signal back to a voltage-controlled oscillator which mixes the incoming FM signal down to the IF and frequency demodulator. The feedback is such as to reduce the instantaneous deviation of the FM signal at IF. Consequently, the IF can

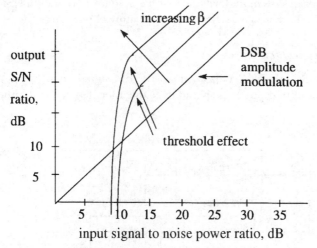

Figure 9.4   FM demodulator characteristics

have a narrower bandwidth and the frequency demodulator sees less input noise, which delays the onset of threshold.

9.4.1 Preemphasis and deemphasis

The discriminator gives an output proportional to the instantaneous frequency deviation of the FM signal from the carrier frequency, and the frequency is the differential of phase. The noise accompanying the FM signal causes the phase of the incoming signal to vary either side of its true value, by an amount linearly proportional to the noise amplitude. The action of the discriminator is to give an output noise related to the differential of the input noise which, since the power response of a differentiator is proportional to ω^2, means that the white power spectral density of the input noise is transformed into noise at the discriminator output with a power spectral density proportional to ω^2. This means that higher frequency components in the modulating signal are more contaminated with noise. Since the power spectra of typical modulating signals such as speech and video reduce at higher frequencies, a technique called preemphasis/deemphasis can be used to improve overall performance. An example used in the VHF FM band is as follows: the modulating signal is passed through a filter, the preemphasis filter, which boosts high frequencies from the boost frequency at 6 dB/octave. The output from this filter is used as the modulating signal. At the receiver the FM signal is demodulated as normal and then the recovered modulating signal is passed through the deemphasis filter with a characteristic which falls at 6 dB/octave from the boost frequency. The result should be a signal with the same frequency spectrum as that of the original. The output noise, with a power spectrum that rises at 12 dB/octave, will pass through the deemphasis filter. This will attenuate the noise from the boost frequency upwards, with its power response which falls at 12 dB/octave, so giving a flat power spectral density from the boost frequency upwards and reducing the output noise power; an improvement in system performance. Other systems may use different boost rates, which suit the characteristics of the signal more closely.

9.4.2 Spectral spreading

It is important that the power spectrum of the FM signal be as flat as possible so that the transmitted power is spread over the allocated channel bandwidth,

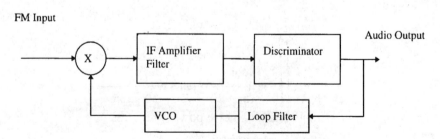

Figure 9.5   Block diagram of FMFB threshold extension

giving the lowest possible power peaks for a given total transmit power. This means that the lowest possible power will then spill over into the sidelobes of a receiving antenna that is aimed at a satellite next door in the geostationary orbit. In order to ensure that this spectral spreading occurs, a low-frequency triangular spreading waveform is added to the video signal prior to the modulator in order to smear out any spectral line components in the modulating signal. This waveform is below the frequency of any wanted message components and can be removed easily after demodulation.

9.5 Digital modulation methods

The main messages to be transmitted nowadays are in digital form, so the main part of this Chapter will be concerned with digital modulation schemes.

9.5.1 Filtering and bandwidth considerations

The bandwidth required for transmission is related to the bandwidth occupied by the original data signal, which we will assume is in the form of ± 1, representing logic 1 and 0, respectively. The spectrum is shown in Figure 9.6b.

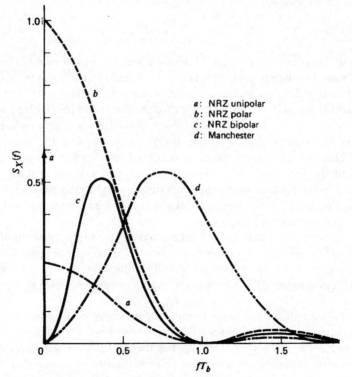

a: NRZ unipolar
b: NRZ polar
c: NRZ bipolar
d: Manchester

Figure 9.6   Power spectra of some baseband data waveforms

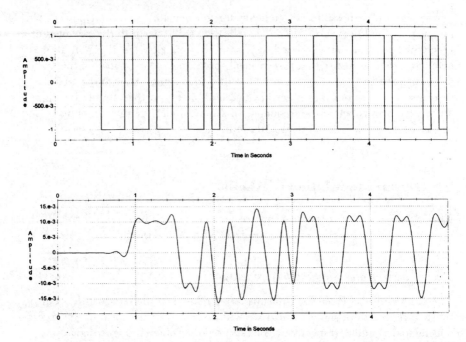

Figure 9.7   NRZ polar data filtered by an α = 0.35 RC filter
Upper trace = original data
Lower trace = filtered data (note filter group delay)

If this signal is passed through a lowpass filter, the waveforms of Figure 9.7 are obtained. The signal never reaches its maximum value if the filter has too narrow a bandwidth, but will pass too much noise if the bandwidth is too wide. There is an optimum bandwidth filter, called the matched filter, with an impulse response which can be shown to be the shape of the reversed signal pulse which maximises the output signal-to-noise power ratio. Passing the data stream through a lowpass filter gives rise to intersymbol interference, since the output response of the filter, even for a matched filter, will last for at least two bit periods. If we assume that we have perfect timing synchronisation then we need to look at the output of the filter only at the end of a bit period when the filter output reaches its maximum value.

Nyquist invented a class of filters which have zero intersymbol interference at the sampling times. The simplest is a perfect brick wall lowpass filter, the frequency response of which extends to half the bit rate, $R_b = 1/T_b$, shown in Figure 9.8 (the case α = 0). This has an impulse response shown in Figure 9.9 (also for α = 0).

It can be seen that if the data is input in the form of positive or negative impulses to this filter the output waveform will never have any intersymbol interference, as long as the output is sampled at the precise point corresponding to the end of a bit period. If the sampling point is slightly incorrect, then there is

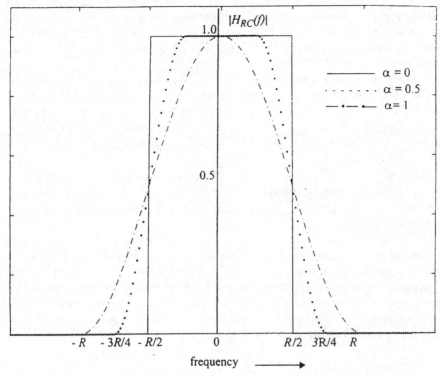

Figure 9.8   Frequency response of raised cosine filters

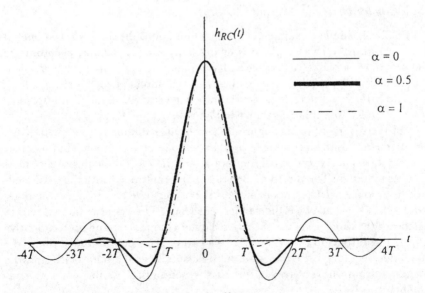

Figure 9.9   Impulse response of raised cosine filters

ISI. A filter class that has zero ISI at the sampling times is the raised cosine (RC) family. This has a frequency response that is flat from zero frequency to $(1 - \alpha)R_b/2$ where R_b is the bit rate, and α is the roll-off factor, which varies from 0 to 1 or is given as a percentage. The frequency response then rolls off in a cosinusoidal way to zero at a frequency of $(1 + \alpha)R_b/2$.

The frequency response and time response of raised cosine filters is shown in Figures 9.8 and 9.9 for various values of α. The bandwidth of a raised cosine filter is $\pm(1 + \alpha)R_b/2$, and the noise bandwidth of all the raised cosine filters is $\pm R_b/2$ Hz. Note that for values of α greater than zero the slope of the time response of the filter at the zero crossings is quite shallow, so that the effect of timing synchronisation errors is much reduced.

The effect of the intersymbol interference can be seen in the eye diagram, Figure 9.10, where segments of the time waveform of the filtered data stream of length equal to the bit period (or symbol period in the case of multilevel signals) are overlaid.

All possible data transitions are shown, and the opening of the eye shows the extent of the intersymbol interference. In practise, half of the filtering is performed in the transmitter and half in the receiver, so that each has a square-root raised-cosine filter in it known as an RRC filter. Where the data is filtered to band limit the transmitted signal, the filtering tends to be performed prior to upconversion to the final output frequency because filters which operate at high power levels at RF are difficult and expensive to manufacture.

9.5.2 Bipolar phase-shift keying (BPSK)

In BPSK, the simplest example of phase modulation, the data is taken one bit at a time, and the carrier phase is changed by 0 or π radians, according to whether the data is a 1 or a 0. Since a phase change of π radians is just equivalent to multiplying the carrier by -1, then a BPSK modulator consists of a level translator which changes logic levels to ±1 levels, which then multiply the carrier in an RF multiplier, as shown in Figure 9.11.

At the receiver, shown in Figure 9.12, a local oscillator is generated at the same frequency as the incoming carrier, and also in phase with it. This, together with timing recovery and synchronisation, are the most difficult parts of most receivers to design. The incoming BPSK signal is multiplied with the local oscillator to produce a DC component, positive or negative according to the data, and a component at twice the carrier frequency. The result is filtered with a matched filter which removes the twice-carrier component, and gives a positive or negative output that is sampled at the end of a bit period to determine the data value. Although the data rate may be specified, the data clock must be generated from the received data, and synchronised to the correct data-sampling epochs.

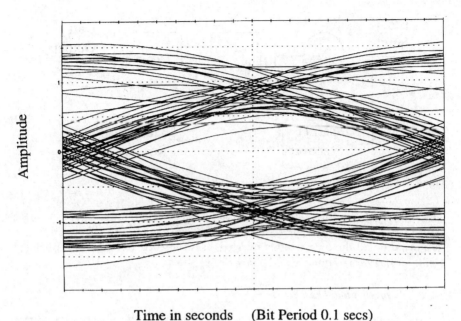

Time in seconds   (Bit Period 0.1 secs)

Figure 9.10   Eye diagram for RC filtered data in gaussian noise

If there is a phase error of θ radians between the local oscillator and the incoming carrier frequency, then the data amplitude from the filter is reduced by $\cos(\theta)$:

$$\text{input signal} = d_n \cos(\omega t)$$
$$\text{local oscillator signal} = \cos(\omega t + \theta) \tag{9.6}$$
$$\text{output from multiplier} = d_n \cos(\omega t) \cos(\omega t + \theta) = 0.5 d_n [\cos(\theta) + \cos(2\omega t + \theta)]$$

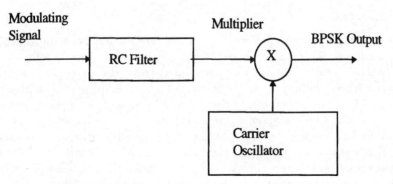

Figure 9.11   BPSK modulator

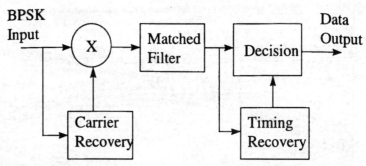

Figure 9.12　BPSK demodulator

The filter removes the component at twice the carrier frequency. If a phase error of $\pi/2$ occurs, then the data output will always be zero! If the error is π radians, then all the data will be inverted. It is therefore essential that the local oscillator has zero frequency and phase error.

9.5.2.1 Frequency and phase recovery

For BPSK-modulated signals there are two methods commonly used for carrier frequency and phase recovery: the squaring method with a phase-lock loop and the Costas loop.

9.5.2.2 The squaring method

If the BPSK signal is squared, that is, just multiplied by itself, then the signal $d_n^2 \cos 2(\omega_c t)$ will be produced, where d_n is the data, ± 1. By standard trigonometry, this can be split into a DC component and a component at double the carrier frequency. The effect of the data is eliminated by the squaring process.

The double frequency component can then either be filtered in a very high Q tuned circuit or passed to a phase-lock loop for further filtering. The frequency is divided by two to produce a signal at the carrier frequency which is then used to demodulate the BPSK signal as previously described.

9.5.2.3 The Costas loop

The Costas loop, shown in Figure 9.13, consists of a voltage-controlled oscillator nominally oscillating at the carrier frequency when its control input is at zero volts. The oscillator has two outputs in quadrature, that is outputs $\cos(\omega t)$ and $\sin(\omega t)$. If we assume that the VCO is at the correct frequency, but with a phase error of θ radians, then out of the top multiplier will come, after filtering, a component $d_n \cos(\theta)$, and out of the lower mixer $d_n \sin(\theta)$. These two signals are then multiplied together to produce $d_n^2 \sin(\theta) \cos(\theta)$, which eliminates the polarity of the data and gives a signal proportional to $\sin(2\theta)$ which, for small phase errors, will be proportional to θ. This signal is then filtered and applied to the VCO control so as to reduce the phase error to zero.

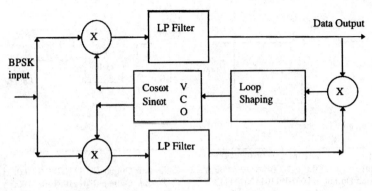

Figure 9.13   Block diagram of Costas loop

In practice, an integrator is required in the control loop to ensure zero phase error, since the VCO is never quite at the carrier frequency with zero volts on the control input and to maintain the correct frequency would have to operate with a constant phase error in order to generate the required DC offset at its input. The integrator allows a constant DC offset at its output for zero DC input, and hence zero phase error. When the loop locks, the phase error is zero and the filtered signal from the top mixer is the recovered data.

Unfortunately, both of these carrier-recovery methods suffer from a π-phase ambiguity. The squaring circuit produces a signal at twice the frequency which, according to which cycle is input to the frequency divider first, will give rise to a signal at the carrier frequency either in phase or π out of phase with the carrier. The Costas loop can lock with a π-phase error, since $\sin(\theta)$ equals $\sin(\pi - \theta)$. To overcome this problem a known sequence or preamble can be added to the beginning of all data sequences, so that if this preamble comes out of the receiver complemented then all the data can be complemented and thus be correct. An alternative solution is to differentially encode the data prior to modulation. A circuit for this is shown in Figure 9.14. Waveforms for data with and without a phase inversion are shown. The disadvantage of the scheme is that if a bit is in error from the demodulator then after the differential decoding process two bits will be in error.

The transmission bandwidth for RC-filtered BPSK will be $(1 + \alpha)R_b$ Hz, so that the bandwidth efficiency will be $1/(1 + \alpha)$ bits/s/Hz.

9.5.2.4 Error rate of BPSK

When a replica of the carrier is generated locally at the receiver, in phase with the incoming carrier, and used to demodulate the incoming signal, then the demodulation is said to be coherent demodulation. Under these circumstances, and in the case of AWGN, the error rate can be shown to be:

$$\text{BER} = Q\left(\sqrt{2\left(\frac{E_b}{N_0}\right)}\right) \tag{9.7}$$

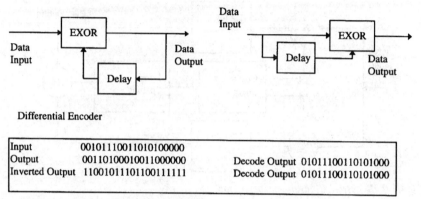

Differential Encoder

Input	00101110011010100000
Output	00110100010011000000
Inverted Output	11001011101100111111

Decode Output 01011100110101000

Decode Output 01011100110101000

Figure 9.14   Differential encoder and decoder

where the Q function is

$$Q(x) = \frac{1}{\sqrt{2\pi}} \int_x^\infty e^{t^2/2} \, dt \tag{9.8}$$

The Q function is tabulated in statistics tables and in telecommunication textbooks.

For differential demodulation, where the incoming signal is just multiplied by a one-bit delayed version of itself, hence negating the requirement for a carrier and phase recovery system but suffering the noise associated with the previous and current input, an increase of about 1 dB in the bit energy is required to achieve the same error rate as coherent demodulation at error rates in the region of 10^{-3}. Figure 9.15 shows characteristic error-rate curves for DPSK and BPSK.

9.5.3 Quadrature phase shift keying (QPSK)

Quadrature phase-shift keying, or QPSK, can be considered as the first upgrade of the MPSK modulation schemes, from 2PSK to 4PSK, or as the combination of two BPSK signals on carriers which are orthogonal, in this case phase shifted by $\pi/2$ radians, to each other. In its simplest form the input data is taken in groups of two called dibits. The even bits are doubled in length to a symbol period and used to BPSK modulate a carrier assumed to be $\cos(\omega_c t)$.

The odd bits are similarly time extended to twice the bit period and used to modulate a carrier $\sin(\omega_c t)$ also in a BPSK modulator. The two modulated signals are then summed to produce a constant envelope signal which has four possible phases relative to the unmodulated carrier, $\cos(\omega_c t)$, of ± 45 and ± 135 degrees. A block diagram of a QPSK modulator is shown in Figure 9.16.

Figure 9.17 shows the waveform of unfiltered QPSK and the constant envelope nature of the waveform. Since the bandwidth required for the individual BPSK signals is only that of the symbol rate, that is half the bit rate, the bandwidth efficiency of QPSK is twice that of BPSK.

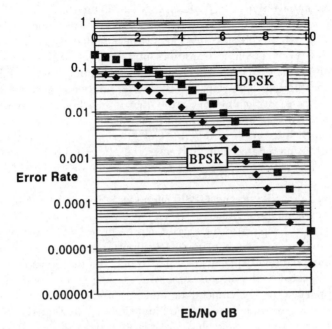

Figure 9.15   Error rates of BPSK and DPSK

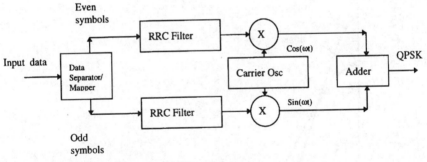

Figure 9.16   Block diagram of a QPSK modulator

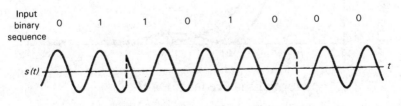

Figure 9.17   Waveform of unfiltered QPSK

The spectrum of unfiltered QPSK is shown in Figure 9.18. It is essential to filter QPSK to reduce its spectral occupancy. This gives an amplitude variation of the envelope at the phase transitions of the modulated carrier waveform, which is most severe at phase transitions of π in the QPSK waveform. This amplitude variation is of no consequence to the information content of the signal but, when this filtered signal is passed through a limiting or heavily nonlinear amplifier, the peaks of the envelope are flattened and the low amplitude parts are amplified. In the limit, the amplitude variations of the envelope are removed, which thus negates the effects of filtering, and hence the spectral occupancy of the modulated signal increases and causes severe adjacent channel interference. This is a major problem in QPSK modulation used in satellite communication systems.

In order to demodulate the QPSK signal it is passed to two quadrature mixers, as shown in Figure 9.19.

Assuming that a replica of the carrier can be generated at the receiver, in both phase and magnitude, then after RRC filtering at the mixer outputs the I-channel symbol data will appear in the I channel, and the Q symbols in the Q channel. If the local oscillator is in phase with the carrier none of the I-channel data will appear in the Q channel data, and *vice versa*. If there is a phase error, then there will be cross talk between the I and Q channels, related to the cosine of the phase error. In fact, if the phase error is 90 degrees, the I and Q channels will be interchanged.

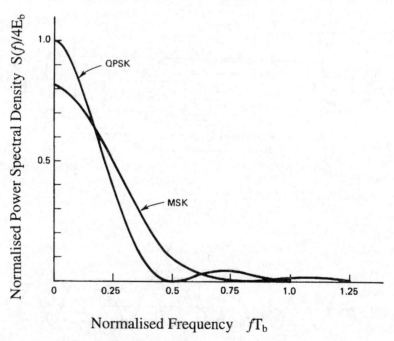

Figure 9.18   Spectrum of QPSK and MSK

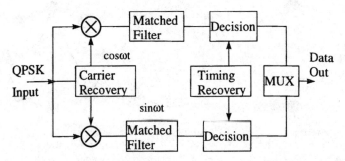

Figure 9.19 QPSK demodulator

Because each of the BPSK signals making up the QPSK signal is modulated on a quadrature carrier, if correct phase recovery is carried out there is no interaction between them. The error rate is then exactly the same as in BPSK. This is the great advantage of QPSK, that for the same E_b/N_0 as BPSK we can transmit double the bit rate for the same bandwidth.

Carrier recovery in QPSK is usually carried out in analogue receivers by raising the QPSK signal to the fourth power. Squaring a QPSK results in the production of a BPSK signal at double the carrier frequency. Squaring this signal results in a signal component at four times the carrier frequency which is passed to a phase-lock loop to reduce phase noise, and then divided in frequency by four and used to demodulate. Because of this division by four, correction must be made for a $\pi/4$ phase ambiguity in the carrier output.

9.5.4 Derivative PSK modulation schemes

9.5.4.1 Offset quadrature phase-shift keying

In order to try to mitigate the effect of the large-envelope variations of filtered QPSK, a variant called staggered QPSK (SQPSK) or offset QPSK (OQPSK) or offset keyed QPSK (OK QPSK) was developed. In the modulator, the quadrature data stream is doubled in duration to two bit periods, as in QPSK, but delayed by one bit period relative to the transitions in the inphase channel. This has the result that the phase of the modulated carrier changes every bit period, but only by $\pm\pi/2$ radians or 0. This results in reduced envelope variation when the modulated signal is filtered, and hence fewer problems when the signal is passed through limiting amplifiers. Although the timing synchronisation cannot be recovered as easily from this modulation as it can from QPSK, the error rate and bandwidth efficiency are the same.

9.5.4.2 $\pi/4$ QPSK

A further variant of QPSK, called $\pi/4$ QPSK, has become very popular for mobile applications because of its performance with limiting amplifiers and ease

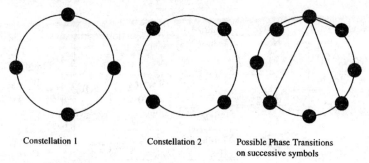

Constellation 1   Constellation 2   Possible Phase Transitions
                                                  on successive symbols

Figure 9.20   Constellation of π/4 QPSK

of demodulation. Essentially, there are two standard QPSK signal constellations that are rotated by π/4 as shown in Figure 9.20. The relevant signal constellation point is chosen from each set alternately. This means that there is always a phase change between successive transmitted symbols, even if the input data sequence is just 0.

The phase change will be ±π/4 or ±3π/4, so the phase change is between that of QPSK and SQPSK and its performance with limiting amplifiers is also intermediate. Because of the phase change per symbol it is very easy to synchronise to the symbols. A block diagram suitable for the generation of π/4 QPSK or π/4DQPSK is shown in Figure 9.21. The modulation is usually associated with differential modulation, so that the phase transmitted is a function of the current and previous input data dibits. It can be demodulated coherently, as in the QPSK receiver of Figure 9.19, but a major advantage is that it can be demodulated without much loss in performance using a limiting amplifier feeding a frequency discriminator. This is because the instantaneous frequency of a signal is equal to the differential of the phase with respect to time. Hence, turning this relationship around, the phase change over a symbol period is the integral over the symbol period of the instantaneous frequency. Since a frequency discriminator gives an output proportional to instantaneous frequency, integrating this over a symbol period results in four different values corresponding to the four possible phase changes available between successive symbols in π/4(D)QPSK. These are passed to a four-level decision circuit to recover the transmitted information. Because differential modulation is normally used, its performance in fading channels is better than that of QPSK and OQPSK.

9.5.4.3 Continuous phase-shift keying

In general, one can say that the smoother the phase trajectory of a modulated signal the faster the spectrum of a signal falls off and the closer such signals can be packed together. In continuous phase-shift keying, the phase of the carrier is allowed to be a function not only of the current data to be transmitted, but also

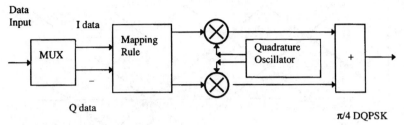

Data Input

I data

Q data

π/4 DQPSK

Figure 9.21   Schematic of π/4 DQPSK modulator

of previous data symbols. This means that, on demodulation, it is necessary to use knowledge of the previously demodulated data to recover the current data:

$$S(t) = \sqrt{\frac{2E_0}{T}} \cos(2\pi ft + \phi(t, \alpha))$$

$$\phi(t, \alpha) = 2\pi h \sum_i \alpha q(t - iT)$$

$$(9.8)$$

$q(t) = \int g(x)dx$ and is the phase pulse, $g(t)$ is the frequency pulse and extends from 0 up to LT seconds, h is the modulation index and has the value 0.5 for MSK, and α_i is the ith data symbol, value ± 1 for binary, ± 1, ± 3 for 4ary etc.

A sequence of data at the modulator gives rise to a particular phase trajectory, or a phase trellis, and the task of the demodulator is to discover the most likely set of data to give that phase trajectory, given that the phase estimates in the demodulator are subject to errors owing to noise and fading.

The advantage of this modulation is that it has a constant envelope and thus is suitable for satellite use; it has low inherent adjacent channel interference. However, for optimum performance it requires coherent demodulation and Viterbi decoding techniques, and has not found favour in satellite systems. When $q(t)$ is a linear ramp rising from 0 to 0.5 in one bit period, T, corresponding to a $g(t)$ of a pulse of width one bit period and amplitude $1/2T$, the phase trajectory is a series of ramps as shown in Figure 9.22, and for $h = 0.5$ corresponds to a phase change per bit of $\pi/2$ radians.

Since the instantaneous frequency relative to the carrier is the differential of the phase trajectory, it can be seen that the output for this particular form of CPSK is two frequencies, equally spaced around the nominal carrier frequency, and spaced by $1/4T$ Hz from the carrier frequency. This modulation is called MSK, minimum-shift keying, although some authors also use the term FFSK, or fast frequency-shift keying.

Other frequency pulse shapes used are the LRC, a raised cosine shape extending over L symbol periods and the gaussian pulse shape, which gives a gaussian cumulative distribution function shape for the phase pulse which for GMSK mobile phones extends over approximately three symbol periods. It should be noted that phase and frequency pulses which extend over more than one symbol period indicate that deliberate intersymbol interference has been introduced in order to reduce the bandwidth requirements of the modulated

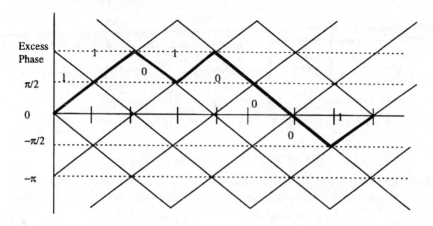

Figure 9.22   Phase trajectory of MSK

signal. One of the consequences of this is that in mobile conditions, where there is a Doppler shift on the incoming signal and hence an increasing phase shift in the signal, and where there is a multipath situation, there will be an irreducible error rate in the demodulation, even for very large received signal strengths. This is of little importance if the system uses, as in GSMK systems, speech algorithms which perform adequately at error rates of 1×10^{-4}, and the system maintains at least those error rates.

9.5.4.4 Mary phase-shift keying (MPSK)

In MPSK the input data is taken m bits at a time, where $M = 2^m$, to produce a symbol which is then transmitted as one of M possible carrier phases. For 8PSK, groups of three bits are taken at a time and one of eight possible carrier phases is transmitted for a symbol period which is three bit periods in length. The bandwidth required is thus reduced by a factor of three compared to BPSK for the same input data rate. It is obviously more difficult to distinguish between the received phases in the presence of noise as M increases, so that the E_b/N_0 required for a given error rate increases over BPSK and QPSK. For an error rate of 10^{-6} the E_b/N_0 required for QPSK is 10.5 dB, compared with 14 dB and 18.5 dB for 8PSK and 16PSK, respectively, assuming practical figures for raised-cosine filtering. Since satellite systems tend to be power limited, it is rare to find systems which do not use a variant of QPSK.

9.5.4.5 Quadrature amplitude modulation

In quadrature amplitude modulation both the phase and amplitude of the carrier are modulated. Consequently, QAM has a high spectral efficiency. The signal constellation of 16QAM is shown in Figure 9.23, where the data is taken four bits at a time and mapped onto one of 16 possible signals.

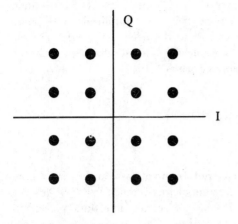

16 QAM Signal
Constellation
Note that there are 4
possible levels in either I and
Q channels, but only +/-3
levels in the output modulated
signal.

Figure 9.23   16QAM signal constellation

Each signal is naturally transmitted for a symbol period of four bit periods, so that the bandwidth required is only one quarter of the bandwidth of the equivalent BPSK system. Figure 9.24 shows a method of generating 16QAM, suitable for use with relatively nonlinear amplifiers, which uses two QPSK modulators. However, when a QAM signal is passed through a nonlinear amplifier, the signal is severely distorted and so this modulation scheme is not very suitable for use on satellite links. If the QAM signal were to be demodulated down to the bit level on the satellite, and then regenerated and remodulated on board the satellite, then QAM could be used. This requires much high-speed and power-hungry processing on the satellite, which used to be impossible, but with the rapid development of signal-processing technology in recent years there is a resurgence of interest in onboard processing satellites.

Typically for 64QAM, the spectral efficiency is around 3 bits/s/Hz, for an error rate of 10^{-6} for an E_b/N_0 of 18.5 dB, assuming practical values of raised cosine filtering. This should be compared with 64PSK, which has a spectral effi-

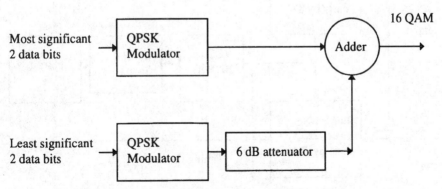

Figure 9.24   Method for generating 16QAM with two QPSK modulators

ciency of 3 bits/s/Hz for an error rate of 10^{-6} with an E_b/N_0 of 28.5 dB under the same filtering assumptions. However, being an amplitude and phase-modulated signal, QAM is more susceptible to the effects of interference and fading than MPSK, and so is more suitable for satellite channels with only gaussian impairments, such as geostationary orbits.

9.6 Practical satellite modems

9.6.1 Bit synchronisation

When the demodulator outputs the recovered stream of data the data is contaminated with noise and also suffers from symbol interference from the filtering processes which it has undergone. Instead of rectangular data pulses, the waveform is full of almost sinusoidal transitions. The receiver does not know when a bit begins, and it does not know the data rate exactly. Even if the data rate is known, the precision of typical crystal-controlled clocks is only one part in 10^6, so that the local bit clock would soon become misaligned with the data, and errors would be made. There are two problems: first, a data clock is required, synchronised to the actual received data and, secondly, the clock must be synchronised to the correct sampling point in the data waveform so that an optimum decision on the data polarity can be made. For analogue-based modems there are a number of common circuits; the early-late gate and the matched filter synchroniser are only two, and the early-late gate type is shown in Figure 9.25.

It should be noted that information to correct the frequency and sampling point of a synchroniser is only available when there are data transitions. A

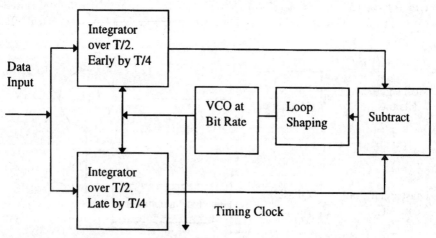

Figure 9.25   Early-late gate synchroniser

simple way to think of the operation of the system in Figure 9.25 is to think of the integrator as a correlator for the square input data pulses. The early integrator gives an output related to the time shift between the data and the VCO transition time, and the data polarity. The circuit then multiplies this signal by its sign so as always to obtain a positive result. Similarly, the late integrator gives an output corresponding to a delayed autocorrelation. If the early and late time shifts are equal relative to the data transitions, then the two outputs will be equal. If they are then subtracted a zero correction voltage will result. If the timing is early, the early integrator will output a lower voltage than the late integral, which on subtraction will lead to a negative timing-error voltage. This is then applied via loop-shaping filters to the VCO timing oscillator, which will alter its frequency and hence phase so as to reduce the timing error. In reality, the waveforms do not correspond to square pulses and compromises have to be made in the loop bandwidth. The lower the loop bandwidth the more noise is rejected, but the longer the loop takes to lock to the data rate.

In a coherent demodulator the recovered carrier has phase noise impressed on it from the carrier recovery loop. The higher the bandwidth of this loop, the faster the acquisition, but the larger the phase noise. If there is a residual Doppler shift on the incoming signal, then this leads to a phase shift during a symbol period. This may be large enough, together with the phase noise, to cause the phase-recovery circuit to cycle slip (because of the periodic nature of most phase-error detectors), or may cause the phase demodulator to make an incorrect decision as to the quadrant of the phase of the incoming phase-modulated signal.

With the advent of digital signal processing, more modulators and demodulators are being implemented as DSP systems, which use samples of the underlying data waveforms. Timing correction has to be performed rather differently, either by using a timing-error indication to change the frequency of the main sampling clock or by increasing the sampling rate to, say, 16 times the symbol rate and moving to the sample which is nearest to the correct sampling point prior to reducing the sampling rate to one sample per symbol again. Another alternative is interpolating the theoretical value of the sample at the correct sampling instant (from a timing-error detector) from the sample values available.

Similarly, rather different techniques have to be used for frequency and phase-error detectors. Since the sampling rate has to be at least twice the frequency of the information content of the signal, such techniques as raising signals to the fourth power as in QPSK carrier recovery are not used. Rather noisy algorithms for phase and frequency detectors are used, and joint estimation of timing, frequency and phase carried out. All-digital chips are now available for demodulating QPSK signals, with symbol rates up to 45 Msymbols/s, designed for digital television broadcasting.

In general, the satellite modem has the problems of very small amplitude signals, with E_b/N_0 of less than 6 dB on occasion, and interference from adjacent frequency channels, as well as adjacent satellites in orbit, thermal noise, Doppler shifts on received signals, nonlinear amplifiers, and fading and shadowing to contend with. Compare this to the relatively easy life of a modem used over a telephone line, with large signal margins, low attenuation, low interference, no Doppler, fading or shadowing. Hence the use of QAM for the telephone modem, and the overwhelming use of QPSK for satellite modems.

Chapter 10
Channel coding

P. Sweeney

10.1 Introduction

Channel coding is a signal-processing technique which makes the representation of information bits interdependent and introduces redundancy into the sequences so that noise averaging and error protection can be achieved. It originates from the development of information theory by Shannon[1,2] in the late 1940s which showed that any channel of known characteristics has a calculable capacity for information transfer. Provided that capacity is not being exceeded, it should be possible to achieve reliable communications with an error rate which can be reduced to any desired level by increasing the length of the codes used.

The passage of 50 years since Shannon's discoveries has not yielded practicable coding schemes which allow theoretical channel capacity and error rates to be achieved. There are certain recently discovered codes which appear to provide the closest approach to the Shannon performance, but mainstream practice is content to adopt more modest aims. These could include the reduction of error rates to be suitable for a particular application or to allow the extension of operations to difficult areas. A common objective would be to reduce the power requirements for acceptable error rates, since high power brings with it several expensive problems including effects induced by nonlinear power devices. In this cost-conscious age, it is unlikely that cost-effective solutions to digital communications can be achieved without coding.

10.2 Coded systems

Coding is usually introduced immediately prior to modulation, and decoding performed after demodulation, as shown in Figure 10.1. In principle, the functions of coding and modulation should be integrated. For binary channels the separation of the functions is acceptable, but a multibit interface between demodulator and decoder is desirable so that the demodulator can indicate confidence in its decisions on the binary symbols received over the channel. This soft-decision demodulation can virtually double the error-control power of the code compared with the having only hard-decision binary output from the demodulator. Not all decoders, however, can make effective use of soft decisions.

10.2.1 Types of code

An encoder takes a frame of k symbols at the input and computes a corresponding output frame containing a larger number (n) of symbols. The ratio of input to output symbols, k/n, is known as the code rate (R). Within this common approach there are many different algorithms for obtaining the output from the input, but two main classes of algorithm can be defined. For block codes the calculation of the output frame depends only on the current input frame, in other words the encoder has no memory of previous output frames. For tree codes the encoder retains memory of previous input frames and uses this in the encoding of the current input frame. If the tree code has finite memory and is linear (a property with a mathematical definition, but closely related to concepts of linearity in any system), then it is a convolutional code.

Block codes are commonly used for error-detection purposes, for protection of highly block structured data and for data-oriented applications requiring very low residual error rates. They are available in a range of lengths (output frame sizes, n) and dimensions (input frame sizes, k) and often called (n, k) codes. Examples include BCH codes, the Golay code and Reed–Solomon codes. The latter group together several bits into multilevel symbols before carrying out the encoding and decoding operations; this gives them an ability to correct errors

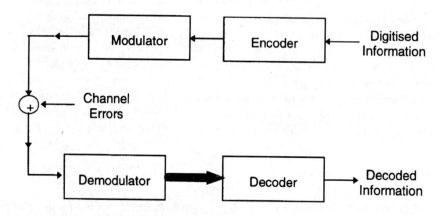

Figure 10.1   Channel coding in digital communication system

which occur in bursts on the channel. Block decoders cannot usually make use of soft decisions from the demodulator.

Convolutional codes are commonly used in applications requiring only moderate error protection, such as digitised speech. They are usually rate 1/2 having an input frame of one bit and an output frame of two bits. The performance of the codes improves with increasing memory, although this causes an exponential increase in decoding complexity. Higher-rate codes are available, although this, too, increases decoder complexity. Convolutional decoders can make use of soft decisions from the demodulator and many of the advantages of convolutional codes derive from this fact.

10.2.2 *Channel types*

The classic channel model against which codes are evaluated is the additive white gaussian noise channel. This assumes that noise is broadband with a gaussian (normal) amplitude distribution and that this noise is added to the received signal. The attractions of the model have more to do with ease of analysis than with its accurate representation of realistic channel conditions. Fortunately, conventional satellite communication channels, i.e. those using geostationary satellites communicating with fixed ground stations well within the antenna beam pattern and at moderate latitudes, conform reasonably to the AWGN model. The signal-to-noise ratios at the receiver will fluctuate slowly with atmospheric conditions, but over a reasonable period of time the AWGN model can be applied.

New developments in satellite communications are commonly concerned with communication to mobile terminals. Low-earth orbit (LEO) and other nongeostationary orbits are proposed to provide lower delays for such applications. Such applications bring problems of multipath interference and shadowing, resulting in a channel with behaviour which is quite unlike the AWGN model. The errors tend to occur in bursts and overall error rates are significantly higher than with AWGN for the same signal-to-noise ratio. The channel bit error rates fall more slowly as signal-to-noise ratio is increased and there may be a point where increasing signal-to-noise does not bring further reductions in error rates. Designing codes for such conditions is more difficult than for AWGN, but the potential gains are much greater.

10.3 **Error-detection strategies**

The simplest use of codes is to detect errors and use some other means to deal with those errors when they arrive. This can be done reasonably simply. For example, a block code may use c parity check bits on each frame. The probability that errors will produce a correct set of parity checks is 2^{-c}. We can easily set the value of c to make this probability as low as we like. For example, 20 bits of parity checks will reduce the probability of undetected errors below 10^{-6}, even in the worst reception conditions.

Having detected the presence of errors, it remains to decide how those errors will be handled. If the data has residual redundancy, it may be possible to apply some signal processing which will disguise the errors to make them insignificant. This is called error concealment. For example, corrupted pixels of an image or samples of a waveform could be reconstructed by interpolation from adjacent values or by extrapolation from previous values. If the received data is intended for human appraisal, the subjective quality may be little affected even when several errors occur. Error-concealment techniques obviously need to be designed with some knowledge of the application data, the inherent redundancies and the subjective effects of errors. They are of value in, and are being used for, many real-time applications.

The alternative to error concealment is to call for retransmission of data affected by errors. This requires the presence of a reverse channel and a proportion of channel capacity will be taken up by retransmissions. A particular problem is that the delays will be variable, depending on how many retransmissions are needed. As a result, retransmission techniques are difficult to use with real-time applications. Fortunately, these are often the applications for which error concealment can be used.

Retransmission-error control is often known as automatic repeat request (ARQ). Two important versions of ARQ are go-back-N (GBN ARQ) and selective repeat (SR ARQ), described in some detail by Lin and Costello.[3] In the former case, the transmitter returns to the point in the sequence where the errors occurred and repeats all frames from that point, regardless of whether they all contain errors. In the latter case, the transmitter repeats only frames that were in error; this is clearly more efficient, but the receiver must take responsibility for buffering the frames and putting them into the correct order. In both cases, achieving low frame error rates is important for efficiency. This can to some extent be achieved by making the frames relatively short, but fixed frame overheads such as parity checks and frame numbers eventually reduce the efficiency as frame sizes are reduced.

10.4 Forward error correction

A more powerful, although less reliable, use of codes is to attempt to recover the original information directly from the received data. This is known as forward error correction (FEC). To understand how a code may be used to correct errors, consider the example of a simple (5, 2) block code in Table 10.1.

The encoding of two-bit sequences can simply be done by looking up the appropriate sequence in the Table or by constructing logic which calculates the bit values in the code. In this case the code is systematic, having the information directly represented in the code. The other bits, known as parity checks, can be produced as the outputs of exclusive-OR gates with appropriate inputs — a char-

Table 10.1 *(5,2) block code*

Information	Code
00	00000
01	01011
10	10101
11	11110

acteristic of linear codes. Note that at least three bits would have to change in any code sequence in order to produce another code sequence; the code is said to have a minimum distance of three.

If a sequence is received with no errors, that sequence can be located in the Table or the validity confirmed by the correctness of the parity checks. In the case where errors are received, the decoder may compare what has been received with each code sequence in turn, counting the number of differences. The code sequence with the fewest differences from what was received is selected by the decoder as the most likely transmitted sequence. This method is called minimum-distance decoding or, sometimes, maximum-likelihood decoding, although it is truly maximum likelihood only if all information sequences are equally likely and the errors are independent from bit to bit.

As the code has minimum distance of three, we expect to be able to correct single bit errors since at least two more bits must change to reach any other code sequence. For example, if the code sequence 10101 is corrupted to 10001, it has two differences compared with 00000, three compared with 01011 and four compared with 11110. On the other hand, two errors may produce a sequence which is closer to a different code sequence. If 10101 is corrupted to 10011 then it is only one bit different from 01011, two different from the original sequence and three different from 00000 and 11110; it will therefore be miscorrected by the decoder. In some cases, however, the decoder will be able to recognise that more than one error has occurred and declare that there are uncorrectable errors. If the received sequence is 11001, for example, this differs by two bits from 10101 and 01011 and by three bits from 00000 and 11110. The detection of uncorrectable errors may be a valuable feature of codes, particularly block codes. Additional error detection may also be incorporated to improve reliability and used in a hybrid scheme in conjunction with error concealment or ARQ.

10.4.1 *Effects of forward error correction*

The performance of an FEC scheme is usually assessed in terms of the bit error rate achieved for a particular value of E_b/N_0. Here N_0 is the (single-sided) noise power spectral density and E_b is the energy per bit of information. The effect of this latter definition will be studied shortly.

When coding is added to a communication channel, there will be two effects on the channel's behaviour. The code will affect the error rates in a way which can be determined from the channel characteristics and the code properties. In the AWGN case it can be calculated, but for other channels it may be necessary to resort to simulation. The intention will be to reduce the error rates.

The other, unwanted, effect is to reduce the information bandwidth of the channel. We can now see the effect of our definition of E_b since more bits, and hence more energy, must go into transmitting the same amount of information. As a result, the value of E_b is increased by $10 \log_{10}(1/R)$ dB for a code of rate R. For example, a half-rate code will increase the value of E_b by 3 dB, corresponding to the doubling of the number of bits required for each message. This increase can be thought of as a penalty applied to the code performance because of the reduction in information throughput.

The overall effect of FEC is illustrated in Figure 10.2. For a given point on the uncoded performance curve, established either by calculation or simulation, there will be a corresponding point representing the coded performance. The former point will represent the value of E_b/N_0 experienced by the demodulator and the corresponding channel bit error rate. The latter point will represent the (higher) value of E_b/N_0 experienced by the decoder and its resulting, usually lower, output bit error rate.

If the final output bit error rate (BER) from the decoder is achieved with a lower value of E_b/N_0 than would be needed to achieve that same BER uncoded, then the reduction in E_b/N_0 is called the coding gain. It is a function of BER and can be negative (i.e. coding loss) at relatively high BER. Note, however, that the quoted BER and E_b/N_0 are the conditions relating to the decoder. The demodulator is experiencing a lower E_b/N_0 and a higher BER, which may be significant when considering the specification of the demodulator.

It should, perhaps, be pointed out that there are many applications in which BER is not the primary concern. In data messages, for example, the rate of

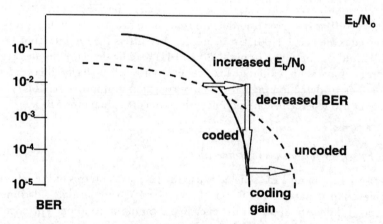

Figure 10.2   Overall effect of forward error correction

message errors will be far more important than the number of bit errors in any message error. It should also be realised that the characteristics of a coded and an uncoded channel will not be the same even when they provide the same bit error rate. In the case of the coded channel the bit errors will be clustered into bursts corresponding to decoder errors with long error-free intervals. The uncoded channel will tend to have its errors more evenly spread.

10.5 Convolutional codes

Convolutional encoders[3-6] incorporate memory of previous input frames of information and use those remembered frames along with the current information to determine the output frame. The input frame is almost always a single bit and the output frame is usually two bits (rate 1/2). The remembered frames are held in shift registers and encoding is carried out by fixed patterns of additions on current and remembered bits to produce the output bits. A typical schematic for a rate 1/2 encoder is shown in Figure 10.3.

The additions are by exclusive-OR gates (modulo-2 adders), required to produce a linear code. In this example, the encoding is carried out using the input bit and six previous input frames retained in memory. The code is said to have constraint length seven, indicating the number of input frames involved in the operation. The action of the encoder may be represented by a set of polynomials showing the contribution of the input and delayed inputs to each of the outputs. For this code, the polynomials would be:

$$g^{(1)}(D) = D^6 + D^5 + D^4 + D^3 + 1$$
$$g^{(2)}(D) = D^6 + D^4 + D^3 + D + 1$$

These may be represented by bit patterns in octal or hexadecimal form, e.g. 79, 5B in hexadecimal.

Encoding produces a continuous sequence of frames which could, in principle, be infinitely long. Nevertheless, analysis of this code will show that a finite number of bits can be changed in any code sequence in order to produce

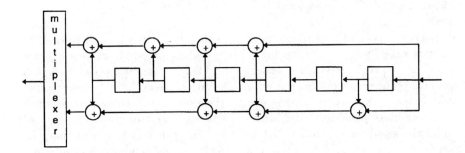

Figure 10.3   Convolutional encoder

another valid sequence. In this case, at least ten bits must change and the code is thus said to have a free distance (d_{free}) of ten.

Decoding of convolutional codes is usually by the Viterbi algorithm. This method models the encoder as a finite-state machine, the states being the encoder memory contents. The possible state transitions and associated outputs can then be determined. The incoming sequence is compared, frame by frame, with every possible code sequence; the method is thus a true implementation of maximum-likelihood decoding. The complexity is, however, contained by recognising that when two code sequences end at the same state the less good sequence can be discarded. Thus, the maximum number of sequences to be retained and updated depends on the number of encoder states, increasing exponentially with constraint length. This, in turn, limits the constraint lengths to relatively small values.

The Viterbi algorithm is usually implemented with soft-decision decoding. In principle, the entire sequence should be received before decoding, but in practice it is possible to decode after a delay of four to five constraint lengths with the decoder operating within a fixed window over the received sequence. Decoder errors produce bursts of output errors of around four to five constraint lengths.

Higher-rate convolutional codes can be produced from rate 1/2 codes by a process known as puncturing. Certain bits in the encoder output are omitted from the transmission to increase the effective code rate. For example, three bits into the encoder will produce six bits out; if two of those bits are not transmitted then the code will be effectively rate 3/4. The original rate half code (parent code) and the puncturing pattern must be carefully chosen to ensure that the final code is a good one, but good punctured codes are readily available. The decoder window must, however, be extended significantly compared with that for the parent code.

The convolutional code will produce a coding gain of up to $10 \cdot \log_{10}(R \cdot d_{free})$ dB in AWGN conditions, assuming unquantised soft decisions from the demodulator. Practical coding gains will be less because the operating BER is not low enough to reach this asymptotic value and the usual three-bit quantisation scheme for soft decisions will reduce coding gain by around 0.2 dB. The rate 1/2 code above, for example, gives coding gains of around 5 dB at an output BER of 10^{-5}. Using hard decisions would reduce coding gain by at least 2 dB.

Longer constraint lengths can be used in conjunction with another decoding method called sequential decoding. This is slightly suboptimum as it concentrates on the sequences which appear most likely, rather than considering all possible code sequences. The loss relative to Viterbi decoding is, however, small and is more than compensated for by the greater constraint lengths, and therefore more powerful codes, which can be used. Soft decisions can be used and may produce computational benefits, but extra coding gain may be produced more cheaply by using longer codes, so sequential decoding is not always associated with soft-decision decoding. The main drawback of sequential decoding is that the delays are variable and increase dramatically as the channel noise increases. This can make it difficult to implement for real-time applications. Sequential decoding is therefore more often a technique for offline decoding to produce low error rates.

10.6 Binary block codes

Block encoders operate only on the current input frame with no memory of previous frames. To compensate for this lack of memory, they often have large values of length n and dimension k. The codes are usually decoded with hard-decision demodulator outputs only. For a code of known minimum distance d_{min}, various combinations of error detection and correction capabilities are available, according to the formula:

$$d_{min} > t + s$$

where t is the maximum number of errors to be corrected and s ($\geq t$) is the number of errors that will be detected. For example, for a code with $d_{min} = 7$, the possibilities are:

$$t = 3, \quad s = 3$$
$$t = 2, \quad s = 4$$
$$t = 1, \quad s = 5$$
$$t = 0, \quad s = 6$$

The required number of errors to be corrected is set by the decoder, leaving the error detection determined by the formula, although it is possible that larger numbers of errors will result in detected uncorrectable errors rather than miscorrections.

The probability that a received sequence cannot be correctly decoded is easily calculated for random error channels. The output bit error rates are, however, more difficult to determine because of the possibility of detected uncorrectable errors. Asymptotic coding gains for a rate R, t-error-correcting code with hard-decision demodulation can, however, be calculated as $10 \cdot \log_{10}[R(t + 1)]$. Many block codes have higher asymptotic coding gains than those of commonly used convolutional codes, but the practically obtained values are less impressive unless low output BERs are required (below 10^{-8}).

Most of the useful block codes are of a type known as cyclic codes[4-7]. The encoding can be done using arrangements of shift registers with feedback to calculate the parity checks. Examples are the Golay code, a three-error-correcting (23, 12) code, and binary BCH codes, a family of codes of length $2^m - 1$ (integer m) requiring up to m parity checks for each error to be corrected. Decoding methods can be based on shift registers with feedback or can be algebraic, setting up and solving a set of simultaneous equations to locate the errors.

Block-code parameters can be modified in various ways to suit the application. One common modification is shortening, which reduces the number of information bits while maintaining the number of parity checks. Expurgation and expansion are two methods for increasing an odd value of d_{min} by one; the former converts an information bit to parity, keeping the same length, and the latter increases length by adding an overall parity check.

10.7 Coding for bursty channels

The codes encountered so far are suitable for random error channels. In many real situations there is some tendency for errors to occur in bursts. There are two approaches to coding for bursty channels. The first is to reorder the encoded bits before transmission so that bursts of errors on the channel are randomised when the bits are reordered before decoding. This approach is called interleaving. The second approach is to use special codes suitable for bursty conditions. Reed–Solomon codes provide the most important burst-error-correcting codes.

The concept of interleaving can most easily be understood by considering the block interleaving of block codes, as shown in Figure 10.4.

The rows of the array are filled with λ code words and the order of transmission is down the columns. A burst of errors spanning λ bits, as shown by the shading, affects only one bit in each code word. A t-error-correcting code thus corrects a burst spanning up to λt bits on the channel or a mixture of shorter bursts and random errors. The parameter λ is called the interleaving degree.

This type of interleaving can be carried out on convolutional codes, although it interferes with the continuous flow output of the encoders, imposing additional delays. An alternative is convolutional interleaving which can achieve similar interleaving degrees with lower delays. The information is demultiplexed into λ separate streams which are separately encoded. The output frames are remultiplexed, but with an offset applied to one bit of each output frame to offset them. The drawback is that λ separate decoders must be provided at the receiver. Convolutional interleaving can be applied to block codes as well as to convolutional codes.

Reed–Solomon (RS) codes are members of the BCH family. They are multi-level block codes, which means that bits are grouped together into symbols for encoding and decoding operations and the error-correction capabilities apply to

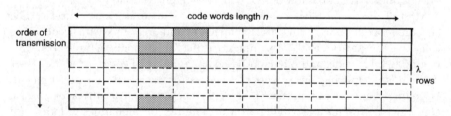

Figure 10.4   Block interleaved block code

symbols. The length of the code, measured in symbols, is one less than the number of values which any symbol can take. In bursty conditions many bits may have errors in a short interval but relatively few symbols may be affected. Although there are other multilevel codes, RS codes are of special interest because they need only two parity check symbols for each symbol error to be corrected.

As an example of an RS code, consider a code with eight-bit symbols. Since an eight-bit byte can take 256 values, the code is of length 255 bytes. To provide error correction for up to 16 byte errors, 32 parity checks are needed, resulting in a (255, 223) code, i.e. there are 233 bytes of information. A burst of errors spanning 15 bytes plus one bit (121 bits) could not possibly affect more than 16 bytes and hence will be correctable, although bursts spanning 122 to 128 bits might be correctable depending on their location relative to the symbol boundaries. In practice, the code could be used for mixtures of shorter bursts and random errors. The code is said to be phased-burst error correcting because of its sensitivity to error location relative to symbol boundaries.

Apart from being able to deal with a wide variety of different error characteristics, many RS codes have the property that uncorrectable errors are highly unlikely to be miscorrected, the probability of detected uncorrectable errors being much larger than the probability of miscorrection (often several orders of magnitude larger). This makes concepts of coding gain, measured on plots of BER, highly dubious. In terms of coding gain on the AWGN channel, the application must require very low error rates for RS codes to appear superior to convolutional codes, but their very low failure rate, ability to cope with a wide range of errors and high decoding speeds mean that they are well worth considering for many emerging applications.

10.8 Concatenation

The concept of applying two codes to a channel is well established. The more conventional way is to apply the second code after the first so that the decoder for the first corrects the errors of the second decoder. This is sometimes called serial concatenation to distinguish it from parallel concatenation where the two codes are applied in parallel and cooperate in the decoding.

Serial concatenation was first described by Forney[8] and is illustrated in Figure 10.5. The inner code is intended to correct most of the channel errors; it could be a convolutional code or a short block code and could be interleaved if the channel characteristics so dictate. The outer code is usually an RS code intended to correct the occasional bursts of errors from the inner decoder. With an inner block code with dimension which is the same as the symbol size of the outer code, each inner decoding error will be a one-symbol phased-burst error to the outer code. If an inner convolutional code is used, the inner decoder errors will be longer, requiring interleaving between the inner and outer codes, and they will not correspond with symbol boundaries.

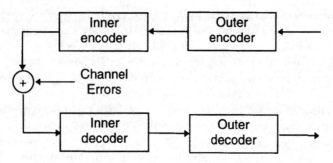

Figure 10.5   Serial concatenation

Since RS symbols are fairly short (usually no more than eight bits), the number of code words of an inner block code will be relatively small. This means that soft-decision decoding using full minimum-distance decoding will often be feasible for the inner code, even if it is a block code.

Parallel concatenation has been known for some time in the form of product codes where data is written into an array and different block codes applied to the rows and columns of the array. The decoders for the rows and the columns then cooperate to correct the errors. The subject has been given new impetus, however, by the use of tree codes known as recursive systematic convolutional (RSC) codes in conjunction with an iterative decoder (turbo decoder) to provide a coding scheme widely known as turbo codes.[9]

A typical parallel concatenation of recursive systematic convolutional codes is shown in Figure 10.6. Because of the feedback in the encoders, the effect of any information bit persists indefinitely providing infinite constraint length. The two encoders are the same in this example, although different encoders could be used. One encoder works on a data stream which is interleaved with respect to the original data stream, so the two streams of parity checks are essentially independent. Usually, the parity check streams are punctured to provide an overall code rate of 1/2 (all the data bits plus half the parity checks from each encoder).

The decoding process is an iterative one, with the data bits plus parity checks from one code being decoded to provide a new estimate of data bits to be interleaved and used with the parity checks from the second code for the next decoding. This process may continue for several iterations. Soft decision decoding is used which means that each decoder has to provide reliability estimates for its outputs. Two algorithms for decoding may be encountered, one called MAP (maximum *a posteriori*) and one called SOVA (soft-output Viterbi algorithm), an extension of the familiar Viterbi decoder.

The performance of this type of code is very dependent on the number of stages in the decoding, but BERs around 10^{-8} have been reported with E_b/N_0 around 2 dB. The interleaving means that bursty errors should be correctable. Overall this seems to represent an exciting development in coding, although

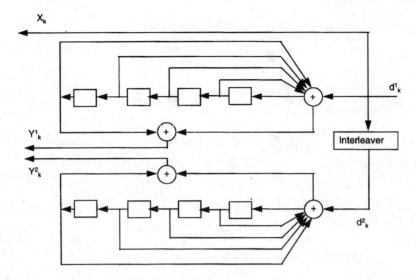

Figure 10.6   Parallel concatenation of RSC codes

there remain many theoretical and practical problems to solve, including complexity, delay, interleaver structures, channel estimation, quantisation and overcoming a BER floor which is often seen in the performance curves.

10.9 Coding for bandwidth-limited conditions

A problem of increasing concern to satellite communications is the availability of adequate bandwidth for a data channel. The approach to such a problem is to use a multilevel modulation which can carry several bits in each transmitted symbol. The difficulty is that this increases the error rate so that the need for coding is increased. The codes used need to be designed specifically for the modulation since errors are not all equally likely. Taking account of different error likelihoods is essentially the same problem as soft-decision decoding, so convolutional codes are the most likely choice.

The best-established method is that defined by Ungerboeck[10] who approached the problem of creating an overhead for coding by expanding the signal constellation and designing a code which would give the same information throughput as the original uncoded constellation, but with better error performance. Ungerboeck coding is therefore a performance/cost trade off rather than the more familiar performance/bandwidth trade off in coding.

Consider as an example the 8PSK constellation shown in Figure 10.7. The allocation of bit values to points has been done by a set-partitioning method so that adjacent points fall into different sets. One bit is left uncoded but the other bits, representing the sets, are encoded using the encoder of Figure 10.8. This arrangement is found to give 3 dB coding gain compared with uncoded QPSK.

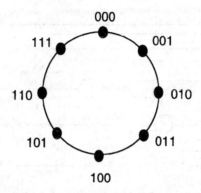

Figure 10.7　Ungerboeck mapped 8PSK constellation

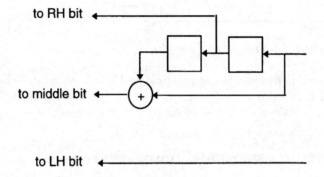

Figure 10.8　Rate 2/3 convolutional encoding for 8PSK

10.10 Application considerations

Choosing the best code for any application requires consideration of the data, the channel, the availability of bandwidth for coding and the development costs assessed against the projected market. The error-rate requirements must be judged realistically for the application since completely error-free communications cannot be achieved within a finite time and a finite budget.

Information intended for subjective human appreciation, e.g. digital voice, can often tolerate relatively high data rates. Digital speech, even when compressed using modern vocoders, can tolerate bit error rates of 10^{-3} or higher before the quality falls unacceptably. On the other hand, the transfer of large amounts of data for machine use (e.g. computer programs) requires extremely low undetected error rates. Fortunately, the applications which are the least tolerant of error often have the least delay constraints so that very long codes (e.g. from serial concatenation) or ARQ can be used. Real-time applications, particularly interactive ones, will have tight constraints on delay which mean that higher error rates are inevitable. If data rate is a problem, RS codecs will often be able to support higher bit rates, although Viterbi decoders are capable of several Mbit/s and the precise comparison will depend on the symbol size and error-correction capability of the RS code.

Although AWGN channels can be countered by convolutional codes if moderate BERs are needed, or by binary block codes for lower BER, bursty or fading channels will cause additional problems. Techniques such as interleaving are commonly used, but it is necessary to interleave over periods which are long compared with the fades or bursts, and this creates its own delays which may cause problems for the application. Convolutional interleaving can improve the interleaving achieved within a given delay, but at a cost of complexity.

If bandwidth is tight, the overhead of a rate 1/2 code may not be affordable and the higher complexity of punctured codes may be considered. Block codes generally correct fewer errors at higher rates and their decoder complexity falls accordingly; the rate/complexity trade off is therefore the opposite of convolutional codes. In case the bandwidth constraints are so tight that 8PSK or higher constellations must be used, convolutional codes again come into their own, although some block-coded solutions are available.

Since coding must be cost effective, significant development costs can be borne only if a mass market is envisaged. As a result, many applications adopt a solution which may not be ideal but has been chosen because of the ready availability of chip sets for the codecs. Constraint length seven, rate 1/2 convolutional code chips are readily available at low cost and provide soft-decision Viterbi decoding, sometimes with provision for puncturing to higher rates. Some RS codecs are also available, mostly using eight-bit symbols as convenient to byte-oriented data. Some, such as the example discussed earlier, are often employed for concatenation.

The conventional satellite communications emphasis on digital voice and other applications requiring low or moderate BER has led to an approach where a reasonable BER is provided for a specified proportion of the year, usually without coding. Users requiring coding then adopt a rate 1/2 convolutional code, perhaps with some interleaving, because of the near-gaussian conditions of the channel. At present, the penetration of other codes into satellite communications is limited. The scope for new solutions is, however, considerable given the prospect of data-oriented applications, high data-rate networks, mobile communications and the development of new coding concepts. Engineers providing the satellite-communication solutions for tomorrow's applications will have the need and the opportunity to consider a full range of techniques.

10.11 References

1 SHANNON, C.E.: 'A mathematical theory of information', *Bell Syst. Tech. J.*, 1948, **27**, pp. 379–423 & 623–656
2 SHANNON, C.E.: 'Communication in the presence of noise', *Proc. IRE*, 1949, **37**, p. 10

3 LIN, S., and COSTELLO, D.J. Jr.: 'Error control coding: fundamentals and applications' (Prentice Hall, Englewood Cliffs, New Jersey, 1983)

4 SWEENEY, P.: 'Error control coding: an introduction' (Prentice Hall, Hemel Hempstead, UK, 1991)

5 MICHELSON, A.M., and LEVESQUE, A.H.: 'Error control techniques for digital communication' (John Wiley, New York, 1985)

6 CLARK, G.C. Jr., and CAIN, J.B.: 'Error control coding for digital communications' (Plenum, New York, 1981)

7 BLAHUT, R.E.: 'Theory and practice of error control codes' (Addison Wesley, Reading, Massachusetts, 1983)

8 FORNEY, G.D.: 'Concatenated codes' (MIT Press, 1966)

9 BERROU, C., GLAVIEUX, A., and THITIMAJSHIMA, P.: 'Near Shannon limit error-correcting coding and decoding'. Proceedings of ICC '93, Geneva, Switzerland. pp. 1064–1070

10 UNGERBOECK, G.: 'Trellis-coded modulation with redundant signal sets', *IEEE Commun. Mag.*, 1987, **25**, (2), pp. 5–21

Chapter 11

Satellite systems planning

B.G. Evans

11.1 Introduction

The overall design of a complete satellite communications system involves many complex trade offs in order to obtain a cost-effective solution. Factors which dominate are the size, weight and thus DC power that can be generated from the satellite, earth-station size, complexity and frequency of operation and constraints upon interference which may apply via international bodies such as the ITU. We will concentrate here on a single satellite link which is applicable to both GEO (geostationary-earth orbit) and nonGEO systems, but with modifications to allow for the dynamics of the latter. Moreover, we will be concerned with the so-called link planning or link budgets. This is illustrated in Figure 11.1 in which a single link consists of:

- uplink (earth station to satellite);
- downlink (satellite to earth station);
- a satellite path.

Such a link, from earth station to earth station, does not form the total overall connection between customers or users. In the early days of fixed-service satellite communications each country had one or more large international gateway earth stations which interconnected to the terrestrial network (and hence the users) via the international exchange. Such earth stations (e.g. INTELSAT Standard A) were large (30 m (NB now 16 m) diameter dishes) and costly ($\approx$ US\$5 – 10 million). The user-to-user connection, which defines the overall path over which quality of services must be planned, thus involved terrestrial-network tails (Figure 11.2a). Modern satellite communication systems, on the

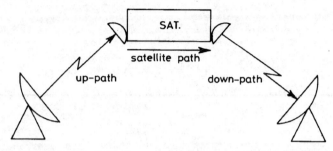

Figure 11.1   Single satellite link

other hand, can go direct, from user to user, and omit any switched path over a terrestrial network (Figure 11.2*b*). For a mobile system one connection will be direct to the satellite from the mobile and the other will be to the GES (gateway earth station) which again connects into a terrestrial network. In planning the end-to-end quality of a connection to international standards as laid down by the ITU, one must bear in mind the overall configuration into which the link fits, and make allocations accordingly.

Figure 11.2 assumes a single connection via the satellite which is not realistic, as most satellite systems operate in a multiple-access mode (e.g. FDMA, TDMA, CDMA etc.) with many earth stations accessing the satellite simultaneously. This situation is depicted in Figure 11.3, from which it will be seen that the satellite path itself, as well as the up- and downpaths, will introduce degradations to the transmission.

Now, with the above in mind, let us look at the individual sections of the link and calculate the power budgets for each. These will determine the sizes of satellite and earth stations needed, if the quality of service has been fixed and the

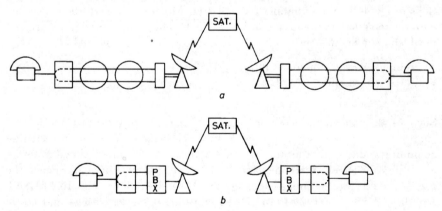

Figure 11.2   User-to-user connections

   a Via switched network
   b Point-to-point

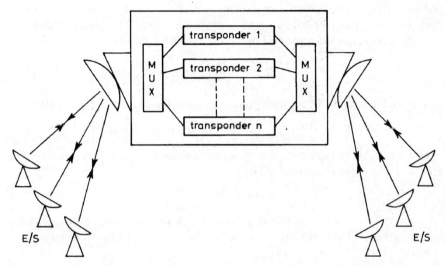

Figure 11.3   Multiple-access satellite system

traffic is known. Before embarking on this link-budget description we will introduce a few basic principles of transmission.

11.2 Basic transmission principles

11.2.1 Power levels

If one considers an isotropic transmitter in free space radiating total power P_T W, uniformly in all directions, then at a distance d metres from the source the power will be spread over the surface of a large sphere, on which the power flux density (PFD) is:

$$\text{PFD} = \frac{P_T}{4\pi d^2} \text{ W/m}^2 \tag{11.1}$$

In practical satellite systems we use directional antennas with gains $G(\theta, \phi)$, defined as the ratio of power per unit solid angle radiated in a given direction to the average power radiated per unit solid angle:

$$G(\theta, \phi) = \frac{P(\theta, \phi)}{P_0/4\pi} \tag{11.2}$$

The reference-angle condition $\phi = \theta = 0$ is called the boresight and is the maximum radiated power direction. The maximum gain, $G(0, 0)$, is thus a measure of the increase in power over that from an isotropic antenna emitting the same power. The PFD is thus:

$$\text{PFD} = \frac{P_T \cdot G_T}{4\pi d^2} \text{ W/m}^2 \tag{11.3}$$

(*Note:* From this point we assume that boresight conditions apply, $G(0, 0) = G_T$, the transmit antenna maximum gain.) The product, $P_T G_T$, is called the effective isotropic radiated power (EIRP). Thus:

$$\text{PFD} = \frac{\text{EIRP}}{4\pi d^2} \; \text{W/m}^2 \qquad (11.4)$$

(*Note:* In logarithmic terms with power being expressed as $10\log_{10}(P)$ dBW.)

$$\text{PFD} \cdot \text{dBW/m}^2 = \text{EIRP} \cdot (\text{dBW}) - 10\log(4\pi d^2) \qquad (11.5)$$

For an ideal receiving antenna with a physical aperture A m^2 we would collect power P_R W at the receiver given by:

$$P_R = \text{PFD} \times A \qquad (11.6)$$

A practical antenna with physical aperture A will not deliver this power as some energy will be reflected and some will be absorbed by lossy elements. Thus the actual power received will be:

$$P_R = (\eta A) \times \text{PFD} \qquad (11.7)$$

where η is the antenna efficiency and ηA is sometimes called the 'effective collecting area' of the antenna A_e. The antenna efficiency η accounts for all losses between the incident wavefront and the antenna output port (see Chapter 5 for details). For large parabolic reflector antennas the value of η is between 50 and 75 per cent, although for horn antennas it can be as high as 90 per cent. A fundamental relationship in antenna theory is:

$$G = \frac{4\pi Ae}{\lambda^2} \qquad (11.8)$$

thus the received power is:

$$P_R = P_T G_T G_R \left(\frac{\lambda}{4\pi d} \right)^2 \qquad (11.9)$$

This is the transmission equation in which the free-space path loss is $[\lambda/4\pi d]^2$. Whence again in logarithmic terms:

$$P_R = P_T + G_T + G_R - 20\log\left(\frac{4\pi d}{\lambda} \right) \qquad (11.10)$$

Note that in a practical system there will be additional loss due to absorption and scattering of the signal, which will almost certainly consist of a statistical parameter $L_e(\%)$ for precipitation, scintillation, multipath/shadowing etc. and a fixed component L_a to account for atmospheric absorption, antenna-pointing loss etc. The propagation loss $L_e(\%)$ on the slant path is explained in Chapter 6 and the multipath/shadowing in Chapter 19.

(*Note:* It is sometimes confusing that the free-space loss is a function of λ. This is merely due to the definition as the loss between two isotropic radiators and the fact that the gain of an isotropic radiator is by definition one giving $A_{ei} = \lambda^2/4\pi$.)

11.2.2 Noise levels

All radio receivers will have associated with them thermal noise of spectral density kT W/Hz. Thus the thermal noise in a bandwidth of B_n Hz is:

$$P = kTB_n \text{ W} \qquad (11.11)$$

where k is the Boltzmann constant in multiplicative form $(= 1.38 \times 10^{-23} \text{ J/K})$ and K is the Boltzmann constant in additive or logarithmic form $(= -228.6 \text{ dBW/K/Hz})$ and T is the noise temperature of the receiver.

In satellite communication systems the reduction of noise is all important owing to the low signal levels, hence the bandwidth is usually reduced to contain just the signal and immediate sidebands.

(*Note:* B_n above is the noise bandwidth of the receiver.)

Noise can emanate from purely physical components (anything that is lossy) and will be dependent on the physical temperature. However, it can also arise from pick up via an antenna in which case we define an effective noise and an effective noise temperature which is not related to a physical temperature but merely related in parametric form via eqn. 11.11.

A receiving system used for satellite communications comprises many different parts and we can define a system noise temperature of the whole receiving system (T_s), measured at its input, which when multiplied by kB gives the same noise power as the overall receiver, including all of the noise components. Thus with reference to Figure 11.4 the noise power at the input to the demodulator is:

$$P_n = kT_sBG \text{ W} \qquad (11.12)$$

where G is the total gain of the receiver to intermediate frequency (IF). Thus, if the antenna system (including feeders) delivers to the receiver input a power P_r W, the carrier-to-noise ratio at the demodulator input is given by:

$$\frac{C}{N} = \frac{P_rG}{kT_sBG} \qquad (11.13)$$

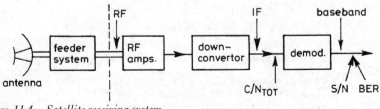

Figure 11.4   Satellite receiving system

It does not matter where the C/N is calculated but the defining point is at the input to the demodulator as the total noise here will determine the baseband quality. (*Note: B* is then the bandwidth of the filter prior to the demodulator.)

Now we can combine eqns. 11.13 with 11.10 to produce, in logarithmic terms:

$$\frac{C}{N}(\text{dB}) = \text{EIRP(dBW)} - 20\log\left(\frac{4\pi d}{\lambda}\right) + G_R - 10\log(kT_sB) \qquad (11.14)$$

Notice that in so doing we have equated P_R with P_r and thus assumed that the gain of the receiving antenna subsumes also the loss of the antenna feeding system, i.e. that G_R and T_s refer to the same physical point in the system (see Figure 11.5). P_R is the carrier power level denoted henceforth by C. For noise calculation the reference point is usually taken as the input to the first amplifier, e.g. the LNA.

Eqn. 11.14 gives us the basis of link planning, but prior to doing this let us look at the calculation of system noise temperature (T_s). In Figure 11.5 we show a typical communications receiver system with two RF amplifiers and one single down-conversion chain to IF. It can easily be shown that the cascaded amplifier

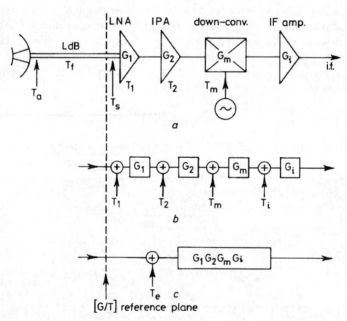

Figure 11.5   Noise equivalent circuits of receiver

    a Physical system
    b Equivalent ideal amplifier circuits
    c Overall equivalent circuit

system to the right-hand side (RHS) of the reference plane has an equivalent noise temperature (T_e) given by:

$$T_e = T_1 + \frac{T_2}{g_1} + \frac{T_m}{g_1 g_2} + \frac{T_i}{g_1 g_2 g_m} \tag{11.15}$$

where g is the numeric gain of the sections and not the dB gain e.g. G dB$= 10 \log_{10}(g)$. Note also that sections can have negative gain, e.g. loss L dB$= -G$ dB or $g = 1/l$.

Note: The relationship between noise temperature, T, and noise figure, N_F, is:

$$N_F = 10 \log \left(1 + \frac{T}{290}\right) dB \tag{11.16}$$

that is, succeeding stages of the receiver noise contribute less and less noise as the gain from each stage is accumulated. Certainly, the contribution from all but the first two stages is usually negligible and the importance of a low-noise front end is clearly seen.

Now, there is also a noise contribution from the left-hand side (LHS) of the reference plane from the antenna and from the lossy feeder system. Treating losses as gains of less than unity, then the contribution of the LHS is given by:

$$\alpha T_a + (1 - \alpha) T_f \tag{11.17}$$

where the feeder decibel loss $L = 10 \log(1/\alpha) dB$. The first term is the antenna effective noise or sky noise, T_a, referred through the loss of the feeder to the reference point and the second term is the output noise of the lossy feeder itself and is dependent on the physical temperature, T_f, of the feeder.

Thus the total system noise temperature is given by:

$$T_s = \alpha T_a + (1 - \alpha) T_f + T_1 + \frac{T_2}{g_1} + \frac{T_m}{g_1 g_2} + \frac{T_i}{g_1 g_2 g_m} \tag{11.18}$$

Here we see the importance of reducing the loss of the feeder system and of maintaining a low antenna noise temperature. The latter is an effective noise temperature that would produce the same thermal noise as actually measured at the antenna terminals due to the external pick up via the radiation pattern, i.e.:

$$T_a = \left(\frac{P_a}{kB}\right) K \tag{11.19}$$

The noise measured at the terminals of an antenna pointed towards the sky depends upon frequency of operation, elevation angle and the antenna sidelobe structure. In more formal terms the noise temperature will be derived from a complete solid-angle integration of the noise power received from all noise sources (e.g. galactic sources, stars, the earth's surface and surrounding objects):

$$T_1 = \frac{1}{4\pi} \int_{\Omega} G(\theta, \phi) T(\theta, \phi) d\theta \, d\phi \tag{11.20}$$

We see here that to produce a low-noise antenna its sidelobes must be minimised, especially in the direction of the earth's surface.

The earth-station system noise temperature described above is for clear weather conditions. When calculations of carrier-to-noise are performed for the worst-case weather conditions it is important to include the increase in system noise produced by the propagation fade, as well as the reduction in carrier power. For example, in a propagation fade:

$$C \downarrow \quad \text{and} \quad \mathcal{N} \uparrow, \qquad \text{thus } C/\mathcal{N} \downarrow \downarrow$$

(the arrows represent a reduction $\downarrow$ or increase $\uparrow$ in the value).

The increase in noise owing to a propagation fade, L_R dB, can be modelled by a lossy transmission line, the output noise temperature of which is given by:

$$(1 - \alpha_R) T_R \tag{11.21}$$

where:

$$L_R = 10 \log \left(\frac{1}{\alpha_R} \right) \text{dB} \tag{11.22}$$

and T_R is the equivalent temperature of the precipitation, which will vary from location to location. *Note:* In Europe, $T_R \approx 275$ K. The additional noise temperature given in eqn. 11.21 must be added to T_a before calculating T_s as in eqn. 11.18. The increase in noise due to rain can be similar in dB magnitude to the reduction in carrier due to the fade, and thus we have almost a double reduction of the $C/\mathcal{N}$.

11.3 Downlink budgets

Now referring to the downlink of the satellite connection as shown in Figure 11.6 we have:

$$\frac{C}{N_D} = \text{EIRP}_{\text{sat}} - (L_D + L_a + L_e(\%)) + G - 10 \log(k T_S B) \tag{11.23}$$

where N_D is the downpath thermal noise power, EIRP_{sat} is the available satellite effective isotropic radiated power, L_D is the downpath free space loss:

$$= 20 \log \left(\frac{4\pi d}{\lambda_d} \right) \text{dB} \tag{11.24}$$

For fixed nontracking antennas the value of G in eqn. 11.23 will be reduced by the pointing loss of the antenna. This represents the movement of the satellite

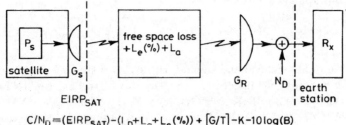

$$C/N_D = (EIRP_{SAT}) - (I_D + L_a + L_e (\%)) + [G/T] - K - 10 \log(B)$$

Figure 11.6 Downlink budget

over the main beam of the antenna pattern (which can be assumed to be parabolic) and is given by:

$$\Delta G = 10 \log \left(\frac{\alpha}{\theta_{3\,\mathrm{dB}}} \right)^2 \mathrm{dB} \tag{11.25}$$

where α is the off-axis pointing angle and $\theta_{3\,\mathrm{dB}}$ is the beamwidth of the antenna, both in degrees.

The value of d to be used in eqn. 11.24 is the path length from the earth station to the satellite. For a geostationary orbit satellite placed at ϕ_s longitude and an earth station at latitude θ_e and longitude ϕ_e the elevation, azimuth and path length to the satellite are:

$$\text{elevation angle} = \arctan \left(\frac{\cos \beta - \sigma}{\sin \beta} \right) \tag{11.26}$$

where

$$\begin{aligned}
\beta &= \mathrm{arcos}(\cos \theta_e \cos \phi_{es}) \\
\phi_{es} &= (\phi_e - \phi_s) \\
\sigma &= \frac{r}{r+h} \\
r &= \text{earth's radius} \\
h &= \text{satellite height above equator}
\end{aligned}$$

$$\text{azimuth angle} = \arctan \left(\frac{\tan \phi_{es}}{\sin \theta_e} \right) + 180° \text{ from true north} \tag{11.27}$$

Slant path $d = 35\,786(1 + 0.4199(1 - \cos \beta))^{1/2}$ km. For nonGEO satellites there will be a range of d from a minimum at zenith to a maximum at the minimum operational elevation angle of the system. Thus, the FSL variation needs to be considered.

It is convention in satellite communications to lump the receiving antenna gain and system noise temperature together to produce a figure of merit for the equipment called the G/T ratio, with units of dB/K.

Thus from eqn. 11.23 we have;

$$\frac{C}{N_D} = \text{EIRP}_{\text{sat}} - \left(L_D + L_a + L_e(\%)\right) + \left(\frac{G}{T}\right) - 10\log(kB) \qquad (11.28)$$

The G/T is usually defined and standardised by a satellite operator in order to ensure compatibility, homogeneity of service and tarriffing, e.g. INTELSAT standard-A, $G/T \geq 35$ dB/K at 4 GHz and 5° elevation angle, INMARSAT standard-A, $G/T \geq -4$ dB/K. Hence it can be seen that G/T can be positive or negative and this merely reflects the numeric value of G with respect to T.

For downlinks the ITU often specifies a maximum PFD on the surface of the earth in shared frequency bands and this needs to be checked with the appropriate ITU recommendation.

11.4 Uplink budgets

The uplink equation is very similar to that of the downlink and from Figure 11.7 is given by:

$$\frac{C}{N_U} = \text{EIRP}_E - [L_U + L_a + L_e(\%)] + \left(\frac{G}{T}\right)_s - K - 10\log(B) \qquad (11.29)$$

where N_U is the uppath thermal noise, EIRP_E is the earth-station effective isotropic power, L_U is the uppath free space loss

$$20\log\left(\frac{4\pi d}{\lambda_u}\right)\text{dB}$$

and $(G/T)_s$ is the G/T ratio of the satellite.

The G/T of the satellite is usually around 0 dB/K (slightly negative or positive) as the satellite antenna is small and the satellite noise is driven by the earth at temperature 290 K in the main beam for GEO satellites. There is thus not so much need for an ultra-low-noise amplifier on the satellite.

The EIRP_E of the earth station is usually large (≈ 80 to 90 dBW for the FSS) and must be carefully controlled since this will determine the carrier level

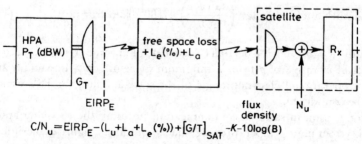

Figure 11.7   Uplink budget

through the satellite and hence the EIRP output from the satellite. There are ITU limitations on the transmitted EIRP from an earth station into the GEO and between orbits. These are designed to limit interference between orbit slots in GEO and between the GEO and other slots. For example in the FSS GEO situation there is a recommended antenna profile $(29 - 25\log(\theta)$ for off-axis angle $\theta)$ and limitation on off-axis power spectral density.

The C/N on the uppath will only be affected by rain in terms of the reduction of the carrier by the uppath rain fade as the noise seen by the satellite is dominated by the earth noise of 290 K which is similar to the rain noise, and thus this will to a first approximation not affect the background noise. So, unlike for the downlink, there is only a single effect on the uplink $C/N \downarrow$.

11.5 Satellite path

The satellite path degradation is dominated by the nonlinear final-stage power amplifier in the transponder. This can be either a TWT or solid-state amplifier. Typical nonlinear characteristics are shown in Figure 11.8. The received carrier levels on the uppath determine the output carrier levels, thus we can think of the amplifier as having transfer characteristics: PFD (dBW/m^2) in, to EIRP$_s$ (dBW) out. The input-power flux density may be related to the earth-station transmit EIRP via:

$$\text{PFD}_{\text{in}} = [\text{EIRP}_E - 10\log(4\pi d^2)]\text{dBW/m}^2 \qquad (11.30)$$

where again d is the slant range to the satellite.

Owing to the nonlinear power transfer characteristic, it is impossible to run the amplifier close to saturation for multicarrier operation. This will produce intermodulation products for FDMA and spectral spreading for TDMA operation. In Figure 11.8 we show diagrammatically the IMPs produced by several carriers occupying the same transponder. For multicarrier operation the spectral densities of individual IMPs will be approximately gaussian and overlap to produce a pseudointermodulation noise floor to the transponder (N_I). The characteristics of the latter are approximately gaussian as per the thermal noise. It is thus possible to trade off thermal and intermodulation noise as shown in Figure 11.8 in order to arrive at an optimum back off for the amplifier which maximises the total $C/(N_U + N_D + N_I)$. The transponder can then be assumed to operate linearly and the input PFD$_i$ will determine the output EIRP$_{\text{SAT}}$.

The back off for FDMA is usually in the range 3–10 dB and is determined by the uplink carrier power at the spacecraft. Accurate control of the carrier powers (± 0.25 dB) is needed to control the IMPs. In general, an operator will define a frequency plan for transponders to minimise intermodulation noise both within and between transponders. This can be done with a simple computer program knowing the nonlinear characteristics of the transponder and using a graphical interface; carriers can be moved around to determine optimum spacings. Note that the addition of new carriers will entail replanning. For

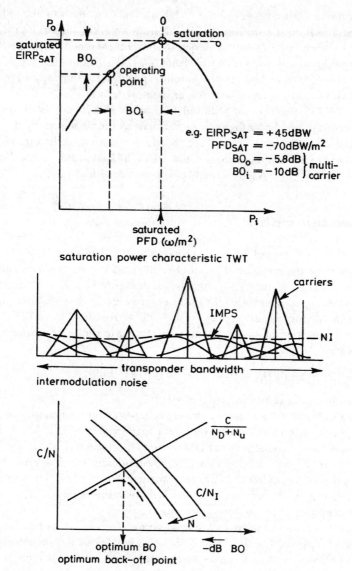

saturation power characteristic TWT

intermodulation noise

optimum back-off point

Figure 11.8   Intermodulation in satellite amplifier

TDMA the major consideration is spectral spreading and AM/PM conversion in the amplifier, and in general one can work much closer to saturation (1 – 2 dB).

It should be noted that the EIRP of the satellite is usually given in terms of an effective single carrier for link-budget calculations. In addition, this EIRP value is usually quoted as the on-axis (e.g. at the subsatellite point) value. Sometimes it is quoted as the beam edge EIRP (–3 dB down on the former value), but the earth-station aspect ratio with respect to these boundary values should be taken into account for accurate calculation, at a specific earth station.

The transponder will also introduce thermal noise which is amplified into the downlink. However, this is usually small compared with other components.

11.6 Overall link quality

As shown in Figure 11.9 the major consideration in planning an overall satellite link is the quality in the baseband. This is measured in terms of S/N for an analogue satellite system and in terms of BER for a digital system. In both cases the quality can be linked to the carrier to total noise ratio (C/N_{TOT}) at the input to the demodulator via the modulation equation, e.g.:

$$
\begin{array}{c} S/N \\ \text{or} \\ \text{BER} \end{array} = function \; of \; \left\{ \frac{C}{N_{TOT}} \right\} \tag{11.31}
$$

The exact details of the function will depend on the modulation scheme used, which is discussed in Chapter 9.

The total noise for a transparent transponder is a summation of all the link contributions, i.e.:

$$
N_{TOT} = \Sigma noise = N_U + N_D + N_I + (N_e + N_i) \tag{11.32}
$$

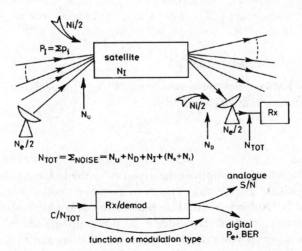

Figure 11.9   Overall noise and relationship between C/N_{TOT} and baseband quality

where N_e is the transmit earth station plus satellite equipment noise and N_i is the allocation made by the system planner for total interference noise. The overall carrier-to-noise is then obtained from:

$$C/N_{TOT} = \frac{C}{N_U + N_D + N_I + (N_e + N_i)}$$

$$= \left\{ \left[\frac{C}{N_U} \right]^{-1} + \left[\frac{C}{N_D} \right]^{-1} + \left[\frac{C}{N_I} \right]^{-1} + \left[\frac{C}{(N_e + N_i)} \right]^{-1} \right\}^{-1} \quad (11.33)$$

Note: when performing the calculations in eqn. 11.33 the C/N values must be numeric and not in decibels. It is sometimes easier to evaluate these equations in terms of noise density (N_0) rather than the noise power (N). Thus C/N_0 replaces C/N in the above:

$$C/N_0 \text{ dBHz} = C/N + 10\log(B) \quad (11.34)$$

11.7 Satellite-link design for specified quality

As previously mentioned, a complete connection from user-to-user will have associated with it an ITU-T specified quality of service. Following normal transmission system design practice ITU defines a hypothetical reference circuit for a satellite link which encompasses earth station to earth station equipment. It must be remembered that the hypothetical reference connection may involve some terrestrial tails and therefore some of the noise or errors must be associated with these. Provided that this is accounted for, then the remaining problem is to determine the link between baseband quality and demodulator C/N_{TOT}. The latter will be modulation and service dependent. Normally, analogue satellite links operate in frequency modulation (FM) and digital satellite links operate in phase-shift keying (*M*ary PSK) or derivative modulation schemes (see Chapter 9).

11.7.1 Analogue frequency modulation

The familiar FM equation relating input C/N and output baseband S/N for a linear demodulator may be expressed as:

$$\frac{S}{N} = \frac{C}{N} \frac{3B \cdot \Delta f^2}{(f_U^3 - f_L^3)} \quad (11.35)$$

where B is the noise bandwidth at the input, Δf is the frequency deviation and f_U, f_L the baseband signal limits. The above equation applies provided that the demodulator is working above threshold. Normally one would arrange a working point which was $L_e(\%)$dB, (i.e. the propagation margin) above threshold for a given per cent of time as specified by the ITU availability requirements for the service.

Now for an FDM block of multichannel telephony the above in logarithmic form may be reduced to:

$$\frac{S}{N} = \frac{C}{N_{TOT}} + 20 \log \left[\frac{\Delta f_{rms}}{f_m} \right] + 10 \log \left[\frac{B_{if}}{b} \right] + W + P \qquad (11.36)$$

where N is the baseband noise in the top telephone channel of width b Hz and f_m is the top modulating frequency. Δf_{rms} is the so-called test-tone frequency deviation and is the deviation produced by the 1 mW test-tone injected at the zero reference point of the system, i.e. S is a 1 mW signal, hence the quality can be expressed simply in terms of the noise N, usually expressed in pWOp.(NB.O refers to the georeference point and p the fact that the noise is psophometrically weighted.) The latter assumes that the noise is weighted (W), using standard ITU weighting circuits $(W = 2.5$ dB) and that preemphasis is applied yielding an advantage in the top channel of $P = 4$ dB.

Some domestic satellite systems use single-channel-per-carrier (SCPC) companded FM modulation schemes and the equivalent signal to baseband noise for these is given by:

$$\frac{S}{N_b} = \frac{C}{N_{TOT}} + 20 \log \left[\sqrt{\frac{3}{2}} \left[\frac{\Delta f_r}{f_m} \right] \right] + 10 \log \left[\frac{B_{if}}{f_m} \right] + CA + W + P' \qquad (11.37)$$

where Δf_r is the RMS frequency deviation produced by the single channel, CA is the companding advantage due to the action of the syllabic compander employed $(15-19$ dB), and P' is the preemphasis characteristic in the SCPC system.

Broadcast quality television is still mainly transmitted via FM (1998), although digital transmission is now being introduced. The performance of TV/FM is given by:

$$\frac{S}{N} = \frac{C}{N_{TOT}} + 20 \log \left[\sqrt{3} \left[r \frac{\Delta F_{pp}}{f_m} \right] \right] + 10 \log \left[\frac{B_{if}}{f_m} \right] + W_T + P_T \qquad (11.38)$$

where S is the peak-to-peak amplitude of the picture luminance, N is the RMS weighted noise power, ΔF_{pp} is the peak-to-peak frequency deviation produced by a sinusoidal baseband signal at f_r, f_m is the top frequency of the unified weighting filter used, e.g. 5 MHz, and r is the ratio of peak-to-peak amplitude of the luminance signal to peak-to-peak amplitude of the video signal including synchronising pulses $(r = 0.7$ for 625/50 systems).

The sinusoidal signal, f_r, is usually taken as 15 kHz (representing a test signal of half white plus half black lines) and for this frequency the preemphasis and weighting advantages $(W_T$ and $P_T)$ for different TV systems are given in Table 11.1.

The IF bandwidth, B_{if}, is the noise bandwidth and is related to the FM carrier bandwidth by:

$$B_{if} = (\Delta F_{pp} + 2f_m')k' \qquad (11.39)$$

Table 11.1   Improvement on signal/noise ratio due to deemphasis and weighting

Number of lines	Highest video frequency f_m (MHz)	Name of system	Effect of deemphasis network on noise (dB)	Effect of preemphasis network on signal at 15 kHz (dB)	Effect of preemphasis/ deemphasis on SNR at 15 kHz P_T (dB)	Effect of the weighting network on deemphasised noise W_T (dB)	Total improvement on SNR (dB)
525	4.2	M	12.9	−10	2.9	9.9	12.8
525	5.0	unified*	13.1	−10	3.1	11.7	14.8
625	5.0	B,C,G,H	13.0	−11	2.0	14.3	16.3
625	5.5	I	13.0	−11	2.0	10.9	12.9
625	6.0	D,K,L	13.3	−11	2.3	15.8	18.1
625	5.0	unified*	13.0	−11	2.0	11.2	13.2

*SNR is measured within the 0.01–5 MHz band with the unified noise weighting network (CCIR, rec. 568, 1982)

where f'_m is the top frequency in the baseband video signal (6–8 MHz) and k' is a constant which relates the signal-energy spectrum to the noise bandwidth; $k' \approx 1.1$.

11.7.2 Digital modulation

A digital baseband bit stream modulates an RF carrier in either amplitude, phase or frequency or in some hybrid combinations of these. The choice of digital modulation type is governed by the reduced power available from the satellite (hence the need to operate at low C/N) and the spectral efficiency and energy distribution of the modulated waveform. One problem with the satellite channel is that the nonlinear power amplifiers tend to restore sidebands in the modulated RF spectrum. Hence, the choice is a trade off of the above and will depend greatly on the systems application.

For conventional satellite systems where spectral efficiency is not so important but power conservation is, phase-shift keying (PSK) is the usual choice. Here, one codes the phase to $M = 2^m$ symbols at T second intervals, and in the presence of thermal noise at the demodulator one may derive the bit error rate (BER) performance against the energy per bit to noise density ratio (E_b/N_0) (see Chapter 9). The two most commonly used modulation schemes are binary (BPSK) and quaternary (QPSK) and both have the same BER performance for Gray encoded phases with coherent demodulation when expressed in terms of E_b/N_0:

$$\text{BER} \begin{array}{c} \text{(BPSK)} \\ \text{(QPSK)} \end{array} = \frac{1}{2} \text{erfc} \left[\sqrt{\frac{E_b}{N_0}} \right] \tag{11.40}$$

However, note that when considering Mary PSK ($M > 2$) the error rate performance should be specified in terms of the symbol-error rate (SER), and a good approximate relationship is:

$$\text{BER} = \frac{\text{SER}}{\log_2(M)} \tag{11.41}$$

For BPSK, $M = 2$, symbol rate $R_s = R_b$, $E_s = E_b$, BER = SER.

But for QPSK, $M = 4$, symbol rate $R_s = R_b/2$, $E_s = 2E_b$, $\text{BER}_4 = \frac{1}{2} \text{SER}_4$.

It should be noted that modems work on incoming symbols, but performance in the baseband is still required in terms of BER.

Now the relationship back to C/N_{TOT} is given by

$$\frac{C}{N_{TOT}} = \left[\frac{E_b}{N_0} \times \frac{R_b}{B} \right] = \left[\frac{E_s}{N_0} \times \frac{R_s}{B} \right] \tag{11.42}$$

The spectral impurity of the carrier when transmitted through a nonlinear channel can be improved by using variants of PSK. In particular offset-keyed PSK and MSK have this characteristic but require more complex demodulators. OK QPSK and MSK have the same BER performance as QPSK, as seen from Figure 11.10. The latter Figure also shows differential modulated PSK (DPSK and DQPSK) and differentially encoded but coherently demodulated versions of the modulation schemes (for more details see Chapter 9).

Differential encoding is sometimes used to avoid phase ambiguity of the demodulator, and differential modulation is adopted for its simplicity. Both schemes operate for the same BER at higher E_b/N_0 and hence will require more satellite power. For greater spectral efficiency (R_b/B transmitted bit rate/bandwidth, bit/s/Hz) one needs to increase the number of phase states, M. Current work on trellis-coded modulation and QAM shows great promise in providing good spectral efficiency as well as good spectral purity.

Although large values of E_s/N_0 reduce the errors caused by thermal noise, the error rate in a practical system is also influenced by intersymbol interference, and this increases with R_s/B. An ideal Nyquist filter with minimum bandwidth would allow R_s/B of unity. Theoretical analysis tends to use a unity value, but practical satellite links operate with (R_s/B) values in the range 1.2 to 2. For a raised-cosine performance channel the bandwidth is $R_s(1 + \alpha')$ where α' is the roll-off factor of the filter e.g. 0.5 or 50 per cent roll-off factor. A full analysis of the intersymbol-interference performance incorporates the complete link and involves all filters in transmitters and receivers as well as nonlinear performance

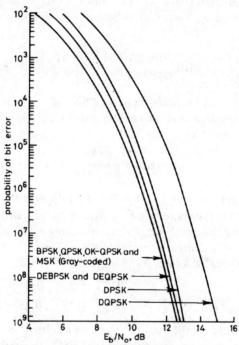

Figure 11.10   Power efficient modulation techniques for satellite channels. Theoretical $P_C = f(E_b/N_0)$ performance of coherent MPSK, DEBPSK, coherent QPSK and DQPSK modems (Gray encoded). Additive white gaussian noise and intersymbol-interference-free model

in the satellite transponder. Filter characteristics are then selected to minimise the overall ISI.

11.7.3 Channel coding

Channel coding involves taking η source-information bits and in the encoder adding r redundancy bits in either a block or a convolutional manner for purposes of error control (detection, or detection and correction (FEC) see Chapter 10 for details). The encoded output bit rate is thus increased to $R_c(R_c > R_b)$ by the reciprocal of the code rate, ρ, as shown in Figure 11.11.

$$R_c = \frac{R_b}{\rho} \tag{11.43}$$

where

$$\rho = \frac{\eta}{\eta + r} \tag{11.44}$$

typically

$$\rho = \frac{7}{8}, \frac{3}{4}, \frac{1}{2}, \text{ etc.}$$

The mechanism of adding the redundancy bits and detection and correction of errors is the property of the code type. Various FEC code types (e.g. Hamming, BCH, Reed–Solomon) are selected according to the types of errors needing correction, which depends on the channel characteristics and on the permissible complexity of the decoder (e.g. syndrome, Viterbi etc.). As far as the transmission is concerned this allows an improved BER for the same transmitted power (e.g. E_b/N_0). This improvement is measured in terms of the coding gain, see Figure 11.12, which can be between three and six dB. Note that this operation relates to the baseband and not the RF section of the link; coding may be added to a system, without change of its modulation, to enhance its performance. The enhancement in dBs of the coding gain can be traded off via the link budget, in terms of reduced satellite power, smaller antennas, improvement in interference protection etc. which, with the falling cost of electronics, may be a very

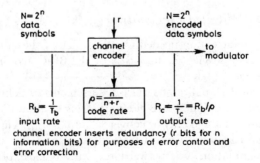

Figure 11.11   Channel coding

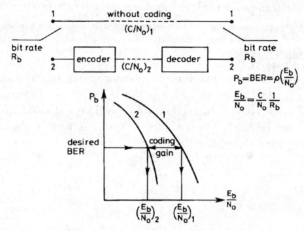

Figure 11.12   Coding gain (after Maral and Bousquet, Wiley, 1986)

attractive systems option for buying dBs. Alternatively, coding can be chosen to optimise the power and bandwidth available from the satellite.

11.7.4 Interference degradation

It has already been mentioned in eqn. 11.33 that interference (N_i) can reduce the overall quality of the satellite link; for analogue systems it increases the noise and for digital systems the BER. As an example, if we add an interfering carrier to the thermal noise for digital demodulation we will degrade the BER according to the C/I ratio as shown in Figure 11.13. Usually, satellite systems design assumes a certain value of C/N_i or C/I in the link budget calculations, e.g. eqn. 11.33. A detailed analysis of the interference between all types of carrier, e.g. FDM/FM, SCPC CFM and PSK, TV/FM and PSK, is carried out in the procedures of Appendices 57 and 58 of the radio regulations when determining internetwork interference of earth-station coordination (see Chapter 4).

11.7.5 Service quality and availability

The remaining information needed for the design of the satellite system is the type of service, the service quality (S/N or the BER) required and the availability of the service.

A summary of the quality-of-service requirements as defined by the ITU is given in Section 11.10. These are given in terms of allowable baseband noise, signal-to-noise or BER and may be converted into C/N_{TOT} via appropriate modulation equations given in this text. Satellite systems design is then a matter of sharing the degradations out between the various parts of the link and the equipment, interference etc., so as to obtain an economic solution in terms of transmitter power, antenna sizes, satellite size and interference minimisation.

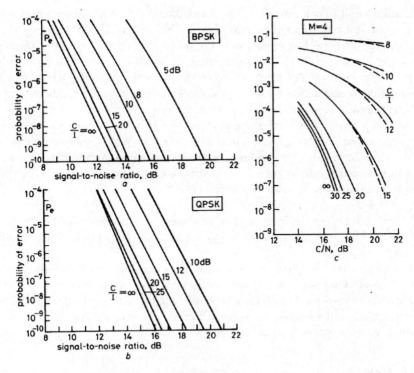

Figure 11.13   *BER = f(C/N, C/I) for BPSK (a), QPSK (b) both for gaussian noise interference and for a single unmodulated tone interference. (After Prabhu, 1969). (Reprinted with permission from the AT&T Technical Journal, © 1969 AT&T) (Figure 11.13c is reproduced with special permission of the International Telecommunication Union (ITU), Geneva, Switzerland, from the latter's publication 'Volume IV/XI,2 of the Recommendation and Report of the CCIR, Report 388-5 figure 6b page 165' (1986))*

The availability of the service is a function of the propagation-link availability figure (e.g. 99.9%) and the earth-station and satellite availabilities. The latter are usually high compared with the propagation-link availability and thus this is usually assumed to approximate to the service availability. However, one must check, via the earth-station-availability calculations, e.g. MTBF etc., and the satellite operator's published reliability figures that this is a reasonable assumption. Overall availability figures may range from 99.99% for a high-quality international satellite link to 95 per cent for a handheld mobile system or a television broadcast system.

11.8 Link budgets

Finally, to the link budget and the bringing together of all the components in the Chapter. The basic idea is to start from the service requirements of QoS and availability and to calculate from these and the modulation/coding schemes to be employed a required C/N_{TOT} to provide the service. The second step is to perform the physical link equations — up, down and satellite — and to sum up the C/Ns so calculated to determine what the satellite system can provide in terms of power, noise etc. Comparison of the latter with the requirements will leave a link margin which should be positive but not excessively so, otherwise the link has been overengineered. A positive margin of between 1–3 dB is quite usual and represents all those components which it has not been possible to calculate in the budgets but which may affect the link quality (examples may be phase noise, a.m./p.m. conversion in amplifiers, nongaussian interference etc.). The link budget is usually performed in dBs and set out in tabular form so that simple additions can be made of the component parts. An example is shown in Figure 11.14.

In the above all of the losses and gains should have been included once. It is easy to double count and thus over engineer the link. The margin should really be a true margin of those incalculable items.

The theory in this Chapter has tended to concentrate on FSS links where rain is the predominant cause of fade margins. However, exactly the same techniques can be used for MSS links where the fade margin will be dominated by the multipath and shadowing loss (see Chapter 19) statistics and for the lower frequency bands, below 3 GHz, rain will be negligible.

11.9 Conclusion

The above theory should enable the reader to design satellite links for prescribed quality of service and availability and to dimension the earth stations in terms of dish sizes, transmitter powers, receiver noise, performance etc. Familiarity with the concepts and technologies will only be acquired by performing sample calculations. Some examples of such calculations are to be found in Appendix A at the end of this book.

11.10 Appendix: service specifications*

11.10.1 Analogue

(i) Voice circuits (CCIR Rec. 353-4, 1982)

Noise power at a point of zero relative level in a telephone channel should not exceed:

- 10 000 pWOp, 1 minute mean power, for more than 20% of any month;
- 50 000 pWOp, 1 minute mean power, for more than 0.3% of any month;
- 1 000 000 pWOp, unweighted (with an integrating time of 5 ms) for more than 0.01% of any year.

---

*CCIR is now ITU-R; CCITT is now ITU-T

Uplink from hub		
Transmit power	0	dBW
Hub antenna gain	+50	dBi
Hub EIRP	+50	dBW
Free space loss (14 GHz)	−207	dB
Atmospheric loss	−1	dB
Rain loss	−3	dB
Pointing loss	−0.6	dB
Satellite G/T	+3	dB/K
Boltzmann constant	−228.6	dBW/Hz/K
Uplink C/N_{OU}	70	dB-Hz
Satellite		
Saturated EIRP	+45	dBW
Back-off output	−15	dB
Downlink EIRP/carrier	+30	dBW
C/N_{OIM}	70	dB-Hz
$C/I_{OU} = C/I_{OD}$	73	dB-Hz
Downlink to VSAT		
Free space loss (12 GHz)	−205.7	dB
Atmospheric loss	−0.5	dB
Rain loss	−6	dB
Pointing loss	−1	dB
VSAT antenna gain ($D = 1$ m, $\eta = 0.6$)		
* *including 1 dB feeder loss*	+38.8	dBi
VSAT noise temp	+25.5	dB-K
VSAT G/T worst case	+13.3	dB/K
Boltzmann constant	−228.6	dBW/Hz/K
Downlink C/N_{OD}	+58.7	dB-Hz
Overall C/N_{OTOT}	+58	dB-Hz
Requirements of service		
Service bit rate (100 kb/s)	50	dB-bit
E_b/N_0 (inc. 3 dB coding gain)	6	dB
Required C/N_{OReq}	56	dB-Hz
Margin	+2	dB

Figure 11.14   An example of a link budget from a hub earth station to a small VSAT terminal via a Ku-band satellite

(ii)   TV broadcast quality (CCIR Rec. 567-1, 1982)

The quality objective corresponds to a minimum signal-to-weighted noise power ratio:

$$\frac{S}{N}\left[= 20\log\left[\frac{V_{LPP}}{V_{Nrms}} \right]\right] \begin{array}{l} \geq 53 \text{ dB for more than } 1\% \text{ of any month.} \\ \geq 45 \text{ dB for more than } 0.1\% \text{ of any month.} \end{array}$$

11.10.2 Digital

(i) Voice circuits (CCIR Rec. 522-1, 1982)

This deals with PCM telephony applications:

- 10^{-6} for more than 20% of any month (10 min mean value);
- 10^{-4} for more than 0.3% of any month (1 min mean value);
- 10^{-3} for more than 0.01% of any year (1 s mean value).

(ii) Voice and data within the ISDN (CCIR Rec. 614)

(See also CCITT Rec. 982 and CCIR Rec. 521 for the hypothetical reference digital path HRDP.)

This applies to FSS satellite systems operating below 15 GHz, and stipulates that the BER of the 64 kbit/s connection at the instant of the HRDP should not exceed, during the available time, the following objectives:

- 10^{-7} for more than 10% of any month;
- 10^{-6} for more than 2% of any month;
- 10^{-3} for more than 0.03% of any month.

Chapter 12

Earth-station engineering

J. Miller

12.1 Introduction

It is important to understand why satellite communications have become popular in recent years and to what uses they are put. Satellite communications can be made available from almost any point on the earth's surface to any other point, in both an international and domestic environment. They are high in quality and reliability, and it is these attributes which have led to the popularity of satellite communications.

We are familiar with the televising of great international events via satellites as well as the convenience of worldwide telephone and data communications. We have also become more familiar, over the last few years, with an increasing number of data services and the development of low-bit-rate television for video-conferences and visual information transfer.

As system planners, it is our task to ensure that any earth-station design introduced into a satellite communications systems has the necessary virtues of giving reliable, high-quality communications. This Chapter will examine the ways of achieving these ends.

There may well be instances of marginal services where the quality and reliability may not be as important as the capital cost of supplying such a service. In these cases we may have to turn our ideas upside down in search of a cost-effective solution to this problem.

12.2 What is an earth station?

Let us first look at the types of communication covered by satellite, so that we can build up a picture of the services required from our earth station. These fall

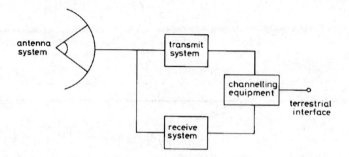

Figure 12.1   Simplified earth-station block diagram

into a number of categories, for example, services such as telephony, data communications and videoconferencing require earth stations which can transmit and receive simultaneously. Other techniques, such as telemetry, data collection and distribution, television broadcasting, electronic-document transmission etc. may require one central station and large numbers of remote stations with either receive-only or transmit-only capabilities.

We can thus build up the simplified earth-station block diagram of Figure 12.1, which shows our station consisting of an antenna system, a receiving system, a transmitting system and channelling equipment; the latter processes whatever kind of signal we are dealing with into a suitable form for onward transmission by satellites or for use terrestrially. This block diagram covers all types of earth station although not all will employ every block shown.

12.3 Typical system configuration

The next step is to look inside the blocks to discover the kinds of process which are required to complete an earth-station design and how these are accomplished.

Figure 12.2 shows a typical earth-station system block diagram, excluding the power system.

For most earth stations the largest item, in terms of size and cost, is the antenna system — including antenna, drives and tracking system. This should provide sufficient low-noise gain to help establish an economical, efficient communications system and, in cases with narrow antenna beamwidths operating to inclined orbit satellites, an antenna-pointing capability. This is not considered in detail in this chapter as the antenna subsystem is covered in Chapters 5 and 15.

The low-noise receiver system, in combination with the antenna, must provide sufficient amplification to the extremely weak satellite signals, without the addition of much thermal noise, to enable the following receiving stages to perform their functions with an adequate carrier-to-noise ratio. It must cover the complete band of interest (e.g. 3.625 to 4.2 GHz for C-band on most INTELSAT satellites) and may be used to supply reference signals to the

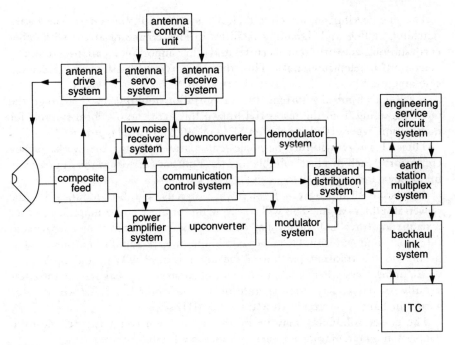

Figure 12.2   Typical earth-station block diagram

antenna tracking system, as well as providing the main receiver communication signals. The amplifier gain of the low-noise receiver system is an important factor in reducing the contributions of subsequent stages to the overall receiver system noise temperature, T, which is part of the figure of merit or G/T described in Chapter 11. For this reason the low-noise receiver system is normally located as near as possible to the antenna feed.

The microwave receiver system or down convertor must accept the broadband (multicarrier) signal from the low-noise receiver and split it into several narrowband chains, each of which will be dedicated to one receive band. This will require a power division of the receive signal followed by a down conversion of the wanted signal to a common intermediate frequency (IF) which may be at 70 or 140 MHz. At this stage, by the use of suitable filtering, the single carrier or band of interest can be selected. After this selection, the IF-based signal undergoes a demodulation process appropriate to the modulation scheme applied to the transmitted signal.

In some arrangements, the down convertor may well be located remote from the antenna structure, and consequently a so-called interfacility link, normally consisting of lengths of semiflexible elliptical waveguide or coaxial cable, may be employed. For some earth stations 4 GHz optical-fibre systems are used to provide the interfacility link.

The engineering service circuit (ESC) system will contain the necessary patching, calling and signalling facilities to give communications with other earth stations, network control centres and other important locations, usually by voice and/or teleprinter links. This, then, is a typical earth-station receiving system.

For the transmitting system, the earth-station multiplex will arrange the outgoing signal from the terrestrial link, if this exists, into a form suitable for modulation, where IF carriers at 70 MHz or 140 MHz will be produced.

These IF signals will then be processed separately, or together, in frequency convertors which produce suitable low-level output carriers in the appropriate frequency band (e.g. 5.850 to 6.425 GHz for C-band).

The resultant upconverted microwave signals will be amplified by the power-amplifier system. This will consist of one or more power amplifiers, either broadband (travelling-wave tube, solid-state power amplifier) or narrowband (klystron), and with suitable power combining to give a unified output for connection to the transmit portion of the antenna feed. TWTs and SSPAs are wideband devices where the entire band of interest (e.g. C-band) is instantaneously available. Klystrons operate over a narrower band and are tunable across the band of interest in 40 MHz or 80 MHz steps.

The power combining may be by the use of fixed couplers, variable ratio couplers, filter circulator combiners or other low-loss techniques.

Figure 12.3 shows a typical earth-station power-distribution system with commercial power available. In this Figure, the incoming high-voltage feeder is terminated at the unit substation, where all voltage transformation and metering functions (for costing purposes) are performed. This substation will normally form part of a power building which will also contain the standby engine generators and the switching-control cubicle. The engine generators will supply the station complex in the event of a failure of the commercial power. This is normally detected in the control cubicle in terms of overvoltage, undervoltage, overfrequency and underfrequency.

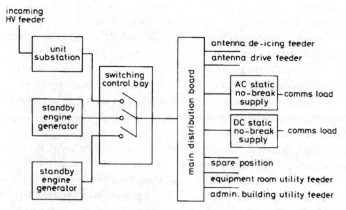

Figure 12.3   Typical earth-station power-distribution system

Any such failure will initiate the starting sequence for one of the engine generators which, after a run-up period of a few seconds, will supply the station load. If the first engine generator fails to start, or fails while in service, the starting sequence for the second engine generator will be initiated and again, after a run-up period of a few seconds, this latter generator will supply the station load. On the return of the commercial power to its normal condition, after a period of time during which the power is checked, the station load will revert to the commercial supply and the engine generator will be run down automatically.

From the switching control bay, a feeder runs to the main distribution board from which the main facilities feeds are run via fused switches or nonfused circuit breakers. Two of these feeders are of interest as they supply the AC to AC no-break power and the AC to DC no-break power. In order to ensure the no-break nature of these supplies, each unit is equipped with standby batteries which are normally being trickle charged. In the event of failure of the commercial power, and during the time that it takes for the engine generators to run up and take over the load, the batteries supply the necessary power to those portions of the equipment which require an uninterrupted supply of power.

For earth stations without a source of reliable commercial supply, it is usually to supply three engine generators, any one of which is able to take the station load, a second one would assume a standby role and a third could be subject to regular routine maintenance.

12.4 Major subsystems

12.4.1 Power amplifiers

Power amplifiers can range from solid-state amplifiers (SSPAs) of a few watts of output power to large travelling-wave (TWT) tube amplifiers with output powers of up to 3 kW. At the upper range, the costs of prime power and space required become dominant, whereas at the lower range, the adequacy of the device to perform the required task and whether or not there is any margin for expansion may well dominate. Smaller earth stations may utilise SSPAs for a small number of channels. The robust design of the smaller SSPAs and their reliable nature makes them ideal for remote sites. Typical examples would be a 20 W SSPA for a small remote terminal transmitting, say, ten SCPC channels, and a 3 kW TWTA for a large earth station operating in a multicarrier mode transmitting, say, a number of high-capacity digital carriers.

The maximum output power from SSPAs has increased steadily, with 400 W now available at C-band. The improved linearity of SSPAs compared with tube amplifiers increases their useful capacity for multicarrier operation, owing to improved intermodulation performance. Their useful capacity for multicarrier operation is therefore comparable to that of a considerably larger TWTA. However, large SSPAs can have high-power consumption and cooling requirements. For this reason in some applications a small TWTA will sometimes be used when power consumption is critical. For example a single 400 W SSPA unit may require 3 kW of prime power and therefore need to be cooled to maintain a stable equipment-room temperature. The costs of airconditioning plant and power for a 1:2 redundant system can be prohibitive unless cooling air for the SSPA can be sourced from and exhausted to the outside environment. This cooling arrangement is standard practice for large klystron and TWT amplifiers.

From the above examples of amplifier design a number of choices arise. For stations having a requirement to transmit more than one carrier, a choice between wide and narrowband transmission has to be made. Under these circumstances, multicarrier intermodulation considerations have to be balanced against narrowband combining losses, and in both instances output power requirements have to be balanced against prime power costs. For stations transmitting only a single carrier, it is necessary to have some idea of whether expansion of the station is likely and in what form it will occur (i.e. more carriers or expansion of existing carrier). In either case, provision should be made in the original system design. For stations operating with single-channel-per-carrier systems it is likely that narrowband multicarrier operation will occur, again leading to intermodulation limitations.

In Chapter 11, the satellite link has been analysed and noise budgets produced. An amount of noise can be allocated to that produced by so-called third-order intermodulation products from multicarrier transmissions. From this noise level an intermodulation EIRP density limit can be deduced, which can then be used as a basis for the selection of amplifier size.

12.4.2 Low-noise amplifiers

Low-noise amplifiers have changed dramatically in construction over the past twenty years. Initially they were cryogenically-cooled parametric amplifying devices, giving noise temperatures of around 20 K at 4 GHz, which needed large amounts of external equipment such as helium compressors, stainless-steel gas lines, charcoal absorber traps etc. The costs and frequency of maintenance were high and often the amplifiers were rather inaccessible.

Improvements in antenna efficiency and feed techniques, as well as increased satellite capability, have led to a need for less stringent noise-temperature requirements for low-noise amplifiers. The amplifiers have also developed to such a state than now uncooled gallium arsenide field-effect transistor (GaAsFET) and high electron mobility transistor (HEMT) amplifiers are available with noise temperatures as low as 30 K at 4 GHz. Such devices are more compact than parametric amplifiers, with a far higher reliability and almost no maintenance requirements, as well as being far cheaper. These amplifiers are also inherently wideband devices.

For a number of small-dish applications, such as television receive-only and business services, the low-noise amplifier has been further refined into the low-noise converter. This consists of several stages of FET amplifiers followed by, usually, a single down-conversion process to a standard IF signal of around 1 GHz. The unit is normally attached directly to the antenna feed on the antenna structure itself.

12.4.3 Ground communications equipment

Generally, this area of equipment covers frequency conversion, modulation and demodulation. In particular cases, the methods of achieving these may vary in detail, but the same overall concepts are used.

12.4.3.1 Frequency converters

The upconverters translate the incoming IF signals of 70 MHz (or 140 MHz) from the modulator subsystem into RF signals of 6 GHz (C-band) or 14 GHz (Ku-band), using double frequency conversion. Conversely, the down converters translate the incoming RF signals into IF signals for application to the demodulator subsystem. The group delay equaliser (GDE) preequalises the delay caused by the reactive components in the system and satellite transponder. The IF filter in the transmit IF module filters out any unwanted frequencies of the transmit signal to ensure that it is within the bandwidth allocated by the satellite operator. A typical simplified upconverter system is shown in Figure 12.4. Frequency synthesisers are most commonly used in the up- and downconverter modules to provide greater ability to change the RF carrier frequencies. The converters are required to have good linearity to prevent unwanted intermodulation products from forming in the system. For digital systems the phase-noise performance of the converters is also critical, especially for low-bit-rate digital applications, to ensure a good bit-error-rate performance.

12.4.3.2 Modulation/demodulation

The function of modulators in an earth-station system is to superimpose analogue or digital information onto an IF carrier. The demodulators extract the analogue or digital information from the incoming IF carrier.

In digital transmission, phase modulation is used and is most commonly quadrature phase-shift keying (QPSK). In the INTELSAT system, intermediate data rate

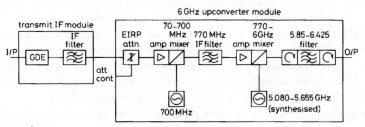

Figure 12.4   Typical simplified upconverter system

(IDR) and INTELSAT business services (IBS) modes of digital transmission have almost replaced older analogue FDM/FM and SCPC techniques (see Chapter 8).

Modern digital modem designs generally fall into two distinct classes. Closed network modems tend to be simple in design and only operate to other identical modems. Open network modems generally have increased flexibility of programmable settings and conform to recognised international standards (such as INTELSAT or EUTELSAT) to ensure interoperability between manufacturers' equipment.

Many designs include the capability to operate at a wide number of data rates, usually at set multiples of 64 kbit/s. Some designs allow operation at custom data rates and sub-64 kbit/s. Common data rates used in international telephony include 512 kbit/s, 1 Mbit/s, 2 Mbit/s, 6 Mbit/s and 8 Mbit/s. Earth stations used to restore high capacity cable networks often operate at 45 Mbit/s, with a 36 MHz transponder, and as high as 140 Mbit/s and 155 Mbit/s where 72 MHz satellite transponders are available. Forward error correction (FEC) is applied at rate 3/4 or 1/2, or occasionally rate 7/8, which reduces the bit error rate (BER) of the data carrier. Sometimes outer coding such as Reed–Solomon coding is applied to further improve the link BER. Carriers over 45 Mbit/s may have particular modulation, FEC and outer coding schemes applied to ensure optimal performance.

Modems may have a variety of terrestrial data interfaces depending on customer/network requirements. At lower data rates V35 and RS449 interfaces are used. Telephony earth stations operating sub 2 Mbit/s carriers often use standard G703 interfaces. These interfaces operate at 2 Mbit/s, with internal drop and insert multiplex in the modem selecting the appropriate part of the 2 Mbit/s data stream for transmission on and reception from the satellite carrier. Often each one of these interfaces will be available on the modem in order for the manufacturer to standardise manufacturing processes. This also allows the use of a common standby modem, capable of supporting any specification of satellite carrier or terrestrial interface.

An overhead framing structure (additional 96 kbit/s) has been defined to allow transmission of engineering service channels (ESCs) and maintenance alarms. A typical configuration of the modem system is shown in Figure 12.5. The required configuration should consider the degree of redundancy required and the number of upchains/downchains for which the station is being equipped. Allowance must be made for the differences in clock frequency between distant earth stations, and for the effects of a constantly moving satellite (the Doppler effect). Many modem designs include a Doppler/plesiochronous buffer as a standard feature. The required buffer size is dependent on the data rate, satellite Doppler shift and whether the distant earth station derives its timing from a receive IDR carrier or national clock source. Typical designs for 2.048 Mbit/s carriers allow selection of buffer sizes up to 32 ms in 4 ms increments.

In analogue systems, frequency modulation is the normal modulation process, and the most commonly used analogue modes of operation are

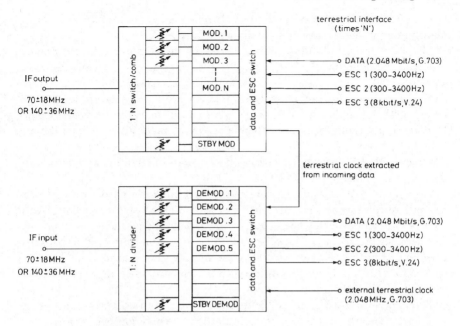

Figure 12.5   1:N redundant IDR modem arrangement

FDM/FM and TV/FM for television carriers. In the FDM/FM mode of operation, where the audio channels are frequency-division multiplexed, the terminal equipment modulates and demodulates the carrier as well as providing preemphasis, deemphasis and energy dispersal. For the TV/FM mode of operation, the most common method with INTELSAT is to use an FM carrier utilising the FM subcarrier technique for sound, which combines one of two sound subcarriers above the video baseband. The transmit/receive systems then deal with this composite carrier, ensuring that filtering is adequate for both the video and audio channels. The use of digital transmission for television is increasing, typically using QPSK carriers of various bandwidths, depending on the quality of video required.

12.4.4 Multiplex equipment

The multiplex equipment is generally located at the earth-station site, although some of the equipment may also be located at the international switching centre.

Digital multiplex can take individual digital channels and assemble them, using time-division multiplex techniques, into a single digital bit stream for application to the modulator subsystem. The reverse process takes place in the receive direction. It can also be used to extract certain designated data streams from larger bit-rate streams. We must therefore make a careful choice of the multiplex equipment required to fulfil the traffic plans.

The use of circuit multiplication is increasing in order to reduce the amount of satellite bandwidth required for a given number of voice circuits. Low-rate encoding (LRE) provides a gain of 2:1 for both voice and data by using four-bit ADPCM encoding instead of eight-bit PCM encoding. If digital speech interpolation (DSI) is employed, gains of up to 3:1 can be achieved on voice circuits. Digital circuit multiplication equipment (DCME) incorporates both LRE and DSI and theoretical gains of 5:1 are possible. In practice, however, the gain rarely exceeds 4:1 owing to the increasing amounts of facsimile and data in the public-switched telephone network.

With increasing amounts of data traffic, especially on corporate networks, packet multiplexers are increasingly being used. These systems efficiently utilise the available bandwidth where a mix of voice, data and perhaps videoconferencing services are required. Low-rate encoding algorithms are used to packetise voice onto a frame-relay-compatible data stream. Data services such as LAN connections can share the same transmission path, with their apparent data rate varying with the available bandwidth.

The choice of DCME must be considered against the background of the equipment cost, the space segment cost and the size and number of the carriers to be transmitted.

12.4.5 Control and monitor equipment

Equipment for control and monitoring can currently take many forms, from the simplest local panel, with warning lights and some control switches, to a full blown PC or UNIX system with VDU and colour graphics with the capability of system reconfiguration by keyboard operation. In general, the kinds of function which may be available are full antenna-movement control, power-amplifier control including output power level change capability, switch-over of redundant equipment, indications of faults and alarm conditions which may be on an area-by-area basis rather than an individual equipment basis etc.

A well designed control and monitor system could support all of the equipment shown in Figure 12.6, a typical INTELSAT standard-A earth station. Such a system allows more reliable operation of the station and increased maintenance efficiency.

A typical computerised control and monitor system will have a number of mimic screens, organised in a hierarchy of screens for the main systems and subsystems. The main screen will show a summary status of all of the systems at the earth station and may include ancillary equipment such as power, airconditioning and security. Generally, this will show the system status in a sequence of colours for normal operation, failure, in maintenance or not equipped (future expansion). More detailed screens will show equipment status, system configuration and key performance settings. For example, transmit frequencies and EIRP

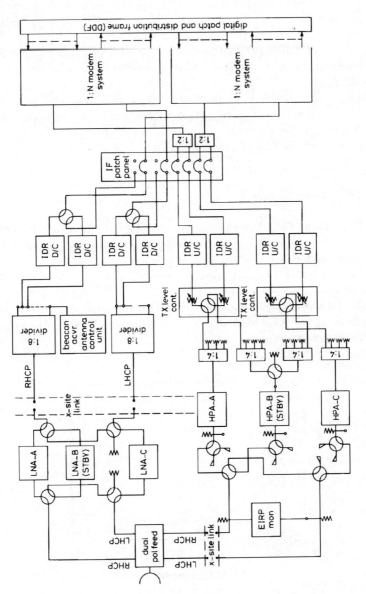

Figure 12.6   Typical standard-A earth station IDR communications system

for each transmitted carrier may be displayed on the HPA transmit subsystem screens.

Statistical functions may include recording key performance variables over time, for example, beacon level or carrier EIRP. Data on equipment faults and outage times may also be available for calculating system availability either for the customer's service or for specific types of equipment at the earth station.

Most computerised systems are Windows based and fully reconfigurable by the operator, for example when more equipment is to be added to the earth station in an expansion project. Some systems have a database of many types of manufacturers' equipment, which can be used to control the equipment when it is installed.

An alarm and equipment status history log can be valuable when fault finding. Some systems can also assist in scheduling and recording routine maintenance procedures. Some may also be capable of supporting a LAN network so that a variety of users can have access to the system.

12.5 Equipment costs

Figures 12.7, 12.8 and 12.9 have been produced from an extensive equipment-cost database which is used for producing budgetary pricing information for particular projects as and when they arise.

The costs have been averaged and the curves represent cost trends rather than actual prices.

Figure 12.7 shows average cost against amplifier subsystem noise temperature for redundant pairs of low-noise amplifiers at C-band. There has been a large drop in LNA costs between 1992 and 1996, largely owing to the introduction of

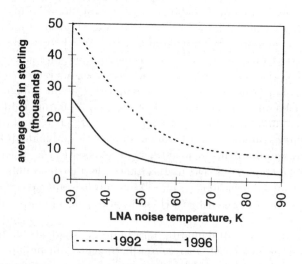

Figure 12.7   Average C-band LNA cost against noise

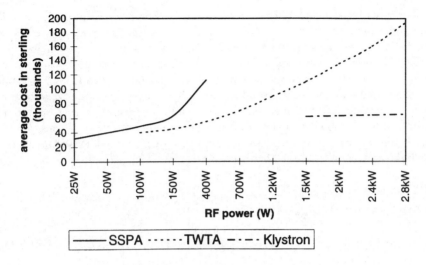

Figure 12.8   Average C-band HPA costs against RF power

HEMT (high-electron-mobility transistor) LNAs. Costs tend to fall as the noise temperature increases, but this trend is flattening out.

Figure 12.8 similarly shows the average cost of redundant pairs of high-power amplifiers against nominal RF output power for C-band operation. The Figure illustrates the comparative costs of solid-state power amplifiers, travelling-wave-tube amplifiers and klystron amplifiers. Although different types of amplifier are sometimes similarly priced, it must be borne in mind that the capabilities of each type vary considerably. The suitability of a certain device type needs to be carefully evaluated for a particular type of application.

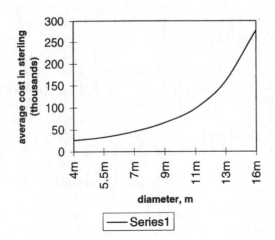

Figure 12.9   Average cost against antenna diameter: C-band

Figure 12.9 has been added for completeness. The price rises steadily as the antenna diameter increases, owing to the increased amount of steel and aluminium. As well as the diameter increasing the amount of structural material required, the design windspeed also has a key impact on cost. The graph shows costs for standard windspeed antennas (e.g. 125 mph survival). An antenna designed to operate in locations subject to higher windspeeds or hurricanes must have a significantly stronger and hence more expensive structure. Some antennas are designed to survive during high-wind periods in a stow position, to lessen the effect of the wind. In addition, the larger antennas will probably include more complex tracking arrangements.

12.6 System design

How do system designers use such curves as Figures 12.7, 12.8 and 12.9 for planning purposes?

12.6.1 Standard earth-station design

We first look at the service requirement for the earth station in question and at the quality of service required. If an operational standard exists for such a service, then we will have to meet that.

Let us assume that we are looking at a particular value of figure of merit (G/T) to give us that service quality. Tables 12.1, 12.2 and 12.3 show a summary of the earth-station characteristics required for operation to INTELSAT, INMARSAT and EUTELSAT satellites, respectively. There are usually a number of possible combinations of antenna diameter and low-noise amplifier temperatures which will give us the required result, and it is obviously necessary to perform cost trade-offs to give an acceptable combination. However, we must also look at our transmit requirements. It is no use selecting a small-size antenna which gives the correct G/T with a suitable low-noise amplifier, if we must then use expensive high-power amplifiers to give us the required EIRP to the satellite. We now have, therefore, a three-way balance to strike between antenna size (and cost) against low-noise amplifier noise temperature (and cost) and high-power amplifier output power (and cost). If this were not enough to consider, we also have to look at the site area required to house our combination of equipments, and the building size requirements. In many countries in the world the cost of electricity over the ten to 15 year life of an earth station can be such as to make the additional cost of a larger than necessary antenna to meet G/T requirements economical at the initial stages of a project. Savings would then be made on the cost of prime power of the HPAs, which is a major contribution to overall running costs.

It should also be borne in mind that, although the curves given are continuous, the equipments come in discrete sizes, be it output power (HPAs), diameter (antennas) or noise temperature (LNAs), although for the latter the

Table 12.1 INTELSAT *earth station characteristics*

No.	Station type	Freq. band polarisation (GHz)	Approx. antenna diameter (m)	Typical EIRP* (dBW)	G/T (dB/K)	Approx. receive antenna gain (dB)	Modulation type	Access method
1	Standard A	6/4 circular	15	88	35	>54	FDM-FM CFDM-FM SCPC-QPSK TV-FM IDR IBS TDM-QPSK	FDMA TDMA
2	Standard B	6/4 circular	11–14	85	31.7	>51	CFDM-FM SCPC-QPSK TV-FM IDR/QPSK IBS/QPSK	FDMA
3	Standard C	14/12 linear	11	75–87	37	>59	FDM-FM CFDM-FM IDR IBS	FDMA
4	Standard D	6/4 circular	4.5 (D-1) 11 (D-2)	53–57	23–32	>43 >51	SCPC-CFM	FDMA
5	Standard E	14/12 linear	3.5–10	49–86	25–34	>50	IBS/QPSK IDR/QPSK	FDMA
6	Standard F	6/4 circular	4.5–10	46–76	23–29	>29	IBS/QPSK IDR/QPSK	FDMA
7	Standard G	6/4,14/12 circular, linear		Parameters decided by users, subject to approval of INTELSAT					
8	Standard Z	6/4,14/12 circular, linear		Parameters decided by users, subject to approval of INTELSAT					

*EIRP's *shown are typical and depend upon the* G/T *of the receive earth station*

Table 12.2   INMARSAT earth station characteristics (second generation)

No.	Station type	Freq. band polarisation (GHz)	Approx. antenna diameter (m)	Typical EIRP* (dBW)	G/T (dB/K)	Approx. receive antenna gain (dB)	Modulation type	Access method
1	Fixed satellite	6.4, RHCP	10	75	N/A	N/A	FM & Digital	SCPC & TDMA
2	Fixed satellite	3.6, LHCP	10	N/A	30.7	49.2	FM & Digital	SCPC & TDMA

*EIRP's shown are typical and depend upon the G/T of the receive earth station

Table 12.3   EUTELSAT earth station characteristics

No.	Station type	Freq. band polarisation (GHz)	Approx. antenna diameter (m)	Typical EIRP* (dBW)	G/T (dB/K)	Approx. receive antenna gain (dB)	Modulation type	Access method
1	Standard	Pol: linear X, Y Freq: 14/11	11–13	82.5–85.5	37	59.8–61.3	PSK	TDMA
							TV-FM	FDMA
2	TV-only	14/11	5.5	82.5	30.5	54.4	TV-FM	FDMA
3	Business	14/12	5.5	57.52	29.9	55		
			3.5	57.70	26.9	51	SCPC-PSK	FDMA
			2.4	56	22.9	47.2		

*EIRP's shown are typical and depend upon the G/T of the receive earth station

step sizes are small. This can sometimes mean that complete design optimisation is not possible and that the final arrangement chosen may be for other reasons, such as commonality of spare parts holdings or familiarity of staff with one particular type of equipment, or some strategic reasons.

12.6.2   Customer's premises earth-station design

If we consider an application requiring a small earth station to be located at the plant or office of someone whose main business is not satellite communications, then a rather different set of parameters will, necessarily, come into the design equations.

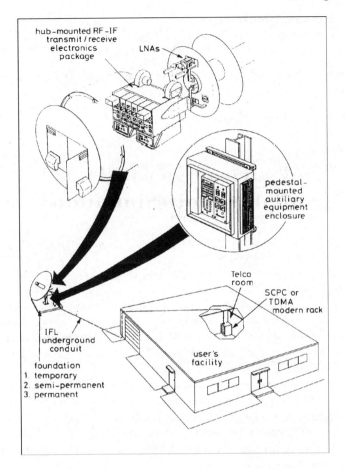

Figure 12.10   Typical site layout: customer premise earth station

The equipment will need to be as compact and unobtrusive as possible, exhibit high reliability and permit quick and easy fault-finding and repair on a modular basis.

Figure 12.10 shows a typical arrangement for such an application. With this concept as much of the electronics as possible is mounted in the antenna hub, with only the need for IF signal connections plus remote control and monitor connections to a second equipment location, shown here as the telecommunications room of the company concerned.

Let us put ourselves in the position of the system designer of this arrangement and see how the many variables can be reconciled to give a cost-effective solution.

First, we have to consider possible system applications, say:

- digital network (star or mesh);
- video uplinks;

- videoconferencing (with compressed video);
- audio/video broadcast uplinks;
- international links (e.g. INTELSAT IBS, IDR etc.).

This will give us a range of transmit effective isotropic radiated powers (EIRP) for the various carrier applications, say:

50 to 80 dBW	C-band
50 to 85 dBW	Ku-band

and also a range of receive G/T requirements, say:

20 to 30 dB/K	C-band
25 to 35 dB/K	Ku-band

Of course, for a full range of services there would undoubtedly be a much wider range of both output powers and receive figures of merit required, but in order to achieve a modular and cost-effective solution it is necessary to put realistic bounds upon a design concept.

The modular approach adopted in this design is such that the hub-mounted package contains redundant low-noise receivers, redundant power amplifiers and their associated power supplies, redundant upconverters and redundant downconverters together with all necessary forced air cooling, control, monitor and alarm equipment.

A significant cost in all forms of engineering is associated with new editions of products, each being slightly different from the next and each requiring drawing changes. For this modular approach to be cost effective it must exist as a self-contained package and be equally applicable to the smallest and largest antenna in the proposed range, as well as being capable of housing the range of power amplifiers proposed.

To cope with these differing requirements we will need a range of antenna diameters of, say:

3.5 to 9.0 m

and a typical power amplifier range of, say:

10 to 400 W	C-band
1 to 300 W	Ku-band

With the use of typically available 35 K C-band uncooled LNAs or 150 K Ku-band uncooled LNAs, the required range of G/Ts can also be met.

Antennas mounted on customer premises tend to be smaller, typically a 3.7 m antenna with a 20 W SSPA will support a variety of digital services. Larger bandwidth services, perhaps supporting the transmission of full motion video or wideband Internet service links, generally require a larger antenna and high-power amplifier.

When it comes to choosing the right combination for a particular application, the designer has to look at the balance between some or all of the following:

- space available to site the station;
- cost-effective solution for initial capital investment;
- long-term running costs, e.g. space segment charges, electricity charges, spares etc.;
- need for system expansion, enhancement;
- cost and provision of maintenance services/watchkeeping staff.

12.7 Environmental and site considerations

When a decision has to be made as to a suitable location for an earth station, a number of physical considerations have to be taken into account. Typical of these are the external temperature and humidity, rainfall, snow etc., likely wind conditions, particularly corrosive atmospheric conditions, and the likelihood of earthquakes etc. The effects of some of these can be minimised by careful site selection, but other considerations may well outweigh these possible actions.

Other items which must be taken into account are the possibility of radio-frequency interference (RFI) and electromagnetic interference (EMI) into the earth station from other sources and from the earth station into other RF installations.

A clear line of sight to satellites of interest is paramount, and this can be critical in the case of the small roof-top type of station. However, this can also conflict with the requirement to avoid external interference. Availability of sufficient space to locate the equipment, transportation to the site (e.g. good road access) and reliable electric power at the site must also be considered.

Owing to the possible disastrous effects of RFI upon the earth-station operation, it may well be necessary to carry out a radio-frequency survey at a number of possible sites before a final selection is made.

As an operator, it is essential to specify all the possible site constraints and environmental factors to potential manufacturers, and as a manufacturer it is essential to plan into the design of all equipment the necessary systems to permit operation under the specified environmental and RFI conditions.

12.8 Testing and acceptance

Having identified the requirement, selected the site, prepared the specification, selected the equipment and placed contracts, it is necessary to be satisfied that the equipment and systems meet the requirements of the specification. In addition, it may well be necessary to prove to a third party (e.g. the satellite operator) that this earth station will not cause any problems to other users of the satellite or to any adjacent satellites. This results in several levels of testing.

12.8.1 Unit testing

Unit testing normally takes place in the factory of the unit manufacturer. The test data will be sent to the system integrator and in general the tests will only be witnessed if the unit is of a new design.

12.8.2 Equipment testing

The equipment (or subsystem) will be extensively tested in the factory. If the equipment is of a new design, type-approval testing will also be conducted including environmental and mechanical performance evaluation. The full equipment test data will be sent to the customer prior to witness testing, who will generally attend factory testing for selected tests. These customer witness tests are usually repeated when the equipment is installed on site.

Depending on the subsystem, the customer witness tests will include the measurement of such parameters as input/output levels and frequency, gain, noise performance and amplitude/frequency response, stability etc.

12.8.3 System testing

Where a complete system has been ordered from the contractor, as much of the system as possible will be assembled in the factory and the performance of the system verified. Full system testing will be performed on site and, since this is generally where the system is accepted from the contractor, it is also termed acceptance testing.

The system tests are designed to prove that the system performance meets all of the mandatory requirements of whichever satellite system is used. These include antenna verification tests where transmit/receive gain is measured, and such tests as off-axis EIRP to ensure that the emissions meet ITU-R limits and do not interfere with adjacent satellites. The receive-system figure of merit (G/T) will also be measured to verify that the quality of service will be achieved. The system tests will verify that the customer's specification is met, as well as fulfilling the mandatory requirements of the satellite operator.

The transmit and receive test will generally include as a minimum:

- antenna tests for gain, G/T, using the satellite or a radio star as appropriate;
- antenna transmit patterns, copolarisation and crosspolarisation;
- level setting and EIRP capability;
- EIRP and frequency stability;
- overall transmit amplitude and group delay response;
- noise performance and spurious emissions;
- intermodulation product levels;
- RF radiation leakage;
- receive amplitude/frequency response.

After satisfactory completion of transmit and receive system tests, the overall system will be configured for loop testing. For this, the transmit system is looped back via a frequency translator through the receive system, and carrier performance is monitored for a soak test typically lasting for 24 hours.

12.8.4 Line-up testing

When the earth-station commissioning has been completed, line-up tests will be scheduled between the new earth station, the earth stations to which it will operate and the monitor/control earth station of the satellite operator.

During these line-up tests, the carrier EIRPs will be set to provide the required receive carrier-to-noise ratios. The monitor/control earth station of the satellite operator will generally assist during the line-up tests.

On successful completion of the line-up tests, traffic can be carried over the satellite link.

12.9 Maintenance

Now that the earth station is ready to go into operation it is imperative that a correctly formulated plan is in existence for keeping it in operation. To this end, supplies of spare parts must be available as well as sufficient test equipment for routine maintenance and fault finding. Maintenance instructions must be prepared for all equipment and systems as well as charts showing recommended maintenance intervals. Maintenance records, when well kept, can be an excellent indication of fault trends leading to early diagnosis of problem areas and timely corrective action. Staff must be available with the necessary skills to carry out this maintenance either on a round-the-clock attendance basis or a call-out basis for unattended sites. Full training will be required to ensure that maintenance staff are familiar with the equipment. The standby/redundancy philosophy for the earth station is a key factor here.

12.10 Conclusions

This Chapter has identified the major subsystems of a satellite communications earth station, has described some of the factors leading to the selection of particular items of equipment, discussed siting matters and interference problems, and looked at testing, acceptance and ongoing maintenance aspects.

Although, out of necessity, the approach has been brief, it is hoped that this Chapter has given the reader an insight into some of the likely design challenges associated with earth-station engineering.

Chapter 13

Satellite engineering for communications satellites

P. Harris and J.J. Pocha

13.1 Introduction

The potential of satellites for communications was first realised by A.C. Clarke, who published a paper in 1945 proposing a three-satellite system to provide worldwide relay communications from the geostationary orbit. The geostationary orbit is unique since its orbit rate of 24 hours precisely matches the Earth's rotation allowing one satellite to provide fixed coverage over a large region. With the development of launch vehicles in the 1950s permitting accurate injection of satellites into orbit, communications from this natural resource could finally be achieved.

Satellites have been used for communications since the early 1960s when relatively simple systems were used to provide telephony communications. These were spinner satellites with only limited payload capability. During the 1970s and 1980s three-axis-stabilised satellite designs were introduced which were capable of carrying much larger payloads although still using the geostationary orbit. In the 1990s the satellite industry has shown rapid growth, especially with the development of constellations of satellites using the low-earth orbit for mobile communications. Satellites for geostationary orbit have continued to grow in size, matching the increase in launch-vehicle capability and demand for high-power broadcast satellites.

Figure 13.1 shows how the EUROSTAR satellite platform has grown to meet the demand for larger satellites. EUROSTAR 1000 was used for the INMARSAT 2 contract with the first launch in 1990 with a launch mass of just under 1400 kg. The latest variant, the EUROSTAR 3000 is designed for a launch mass > 4000 kg.

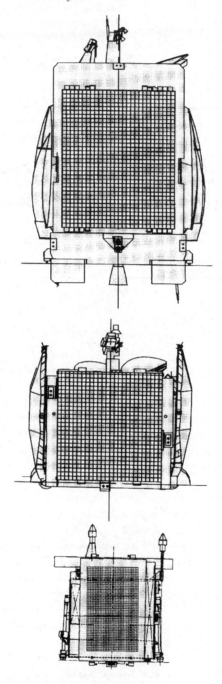

Figure 13.1   Growth in EUROSTAR platform size
       Top: EUROSTAR 3000; Middle: EUROSTAR 2000+; Bottom:
       EUROSTAR 1000

This Chapter outlines the design drivers of a geostationary satellite before describing the individual subsystems of a three-axis-stabilised satellite. Orbit fundamentals and a brief overview of launch vehicles are also covered.

13.2 Satellite design drivers

There are four main design drivers for satellites:

- the communications requirements (the payload);
- the launch vehicle;
- the space environment;
- orbit characteristics.

13.2.1 Communications payload

The main purpose of a satellite is to carry the communications payload which provides revenue-earning services to the customers in a reliable and timely fashion. This traditionally utilises what are referred to as bent-pipe transponders, which in simple terms function by: receiving an uplink signal, downconverting the signal frequency, channelising and amplifying the signal before retransmitting it to the coverage area. This was first performed by the TELSTAR satellite in 1962 and most satellites still use this classical communications architecture.

The main communications requirements of a satellite system are defined as follows:

- EIRP, the combination of the high-power amplifier (HPA) RF output power and antenna gain, measured in dBW;
- number of transponders and overall reliability;
- frequency band;
- coverage area;
- G/T for uplink and downlink, measured in dB/K;
- reliability of payload.

A payload design must be established to meet these communications requirements, also taking into account the constraints imposed by the satellite design (and launcher). The main impacts of the payload on the satellite are:

- payload mass;
- payload power and dissipation;
- size and number of antennas;
- payload layout.

These are explained below.

Present-day satellites have payload mass of typically up to 400 kg and payload powers up to 6000 W and the trend is for increased size payloads. This leads to a satellite mass over 3000 kg at launch. For an Ariane shared launch, the cost quoted is approximately US $30 000 per kg leading to a total cost of around US $90 million for this launch mass, simply to inject the satellite into a geostationary transfer orbit (GTO). With such a high cost per kg, one of the fundamental design drivers is to minimise the mass of the satellite and thus satellite designs come in a number of sizes to allow optimal matching of the payload with the satellite platform. The satellite size selected depends on the payload mass as well as the payload power, which itself is a strong mass driver affecting the size of the satellite power and thermal subsystems.

Although 6000 W of power does not appear to be a lot when compared to terrestrial uses, in space this power is generated by the satellite solar arrays which convert solar radiation into a steady source of power. At the earth's distance, the solar-flux power density averages around 1370 W/m^2 which, after allowing for solar cell efficiencies of around ten per cent for silicon (at end of life, EOL), requires a total solar array area of over 40 m^2. Owing to inefficiencies within the satellite payload, some of the generated power is wasted as heat, which must be radiated into space. This is performed by large radiator panels which form part of the satellite structure.

Conventional antennas are limited in diameter to around two to three m owing to the constraints imposed by the launch-vehicle fairing size. Larger antennas are being developed which use unfurlable mesh and deploy in orbit to overcome the fairing constraints, although at an increase in complexity.

Payload requirements are a strong driver on the spacecraft configuration; the payload is split into antennas and repeater equipment. The repeater consists of electrical equipment which includes a number of units that dissipate large amounts of heat. All of this equipment must be kept within its operating-temperature limits. This leads to repeater equipment being mounted on the satellite radiator walls which are not directly illuminated by the sun, except obliquely during the solstice season. The antennas must be pointed at the earth and, since they do not dissipate heat, they can be mounted on parts of the satellite which are directly exposed to solar radiation with thermal insulation being used to avoid excessive heat loads. They must, however, point towards the earth which may require that they are deployed. A more detailed commentary on payload-design impacts is provided in Chapter 14.

13.2.2 Orbit impacts

The selection of the orbit for the satellite has two main impacts on the satellite.

First, it affects whether the satellite has to raise its orbit from that in which the launch vehicle has placed it. This mainly affects the sizing of the propulsion subsystem, and for a geostationary satellite almost doubles the satellite mass at launch.

Secondly, different orbits have different periods, ground coverages and solar-illumination conditions. The orbital height and inclination affects the ground coverage provided to the users as well as affecting whether continuous or

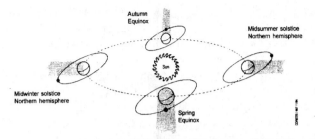

		Relative distance (AU)	Declination (degrees)	Relative Flux
Equinox	Vernal	0.996	0	1.008
	Autumnal	1.0034	0	0.993
Solstice	Summer	1.016	23.4	0.889
	Winter	0.984	-23.4	0.948

Figure 13.2   Seasonal variation in geostationary orbit

intermittent communications are provided. The solar illumination of the satellite varies with the orbit, which affects how long the eclipses are and how frequently they occur. This affects the sizing of the power and thermal subsystems.

Figure 13.2 shows the orbital geometry in a geostationary orbit. During the course of a year the satellite experiences seasonal effects, so that during the solstices the satellite is either below or above the ecliptic by a maximum of 23.5°. This is referred to as the declination, and reduces the solar-array power since the array is no longer normal to the sun. It also results in illumination of one of the two radiator walls, the affected radiator changing between the winter and summer solstice. During the equinoxes, the satellite passes through the earth's shadow and experiences an eclipse. There is also a slight variation in the distance between the earth and the sun during the year, being closest during the winter solstice and furthest during the summer solstice.

More detailed discussions on the impact of orbit are presented in Section 13.3.

13.2.3 Space environment

The environment experienced by a satellite in orbit is very different from that on earth, requiring different technical solutions compared to, say, the aircraft industry.

The radiation environment in orbit varies with the satellite's orbital height, inclination and longitude. The magnetic fields around the earth trap electrons and protons in the Van Allen belts, so that certain areas effectively cannot be used, owing to their very high radiation levels which cause problems with radiation-sensitive electronic components.

Even in the relatively benign geostationary orbit, the radiation environment is several orders of magnitude above that experienced on earth, so measures must be taken to protect sensitive electronics using shielding from both the satellite structure as well as from the units' construction. In order to be considered for use on board a satellite, electronic components must be able to withstand a total dose of > 50 krads since it is difficult to shield below this level.

Cosmic rays can cause software changes due to single-event upsets (SEU), since their high energies are impossible to shield against. Therefore, error-correction and detection software is flown on board the satellite, continuously monitoring and correcting any spurious bit changes.

Surrounding the satellite is a plasma sheath which would naturally charge the satellite to a high negative potential unless protective measures were taken. The external surfaces of the satellite are made conductive and the spacecraft structure is grounded to avoid different potentials within the satellite. Recently, the effects of deep dielectric discharges within the satellite have become a concern, where charge deposited within a dielectric material such as coaxial cable can result in discharges internal to the satellite, damaging electronics.

The satellite experiences zero gravity in orbit and its liquids therefore have no fixed shape or form and there is no natural convection to remove heat. For a liquid propulsion system the liquid must be fed to the thrusters, and either capillary action or a pressurised membrane must be used to force the liquid out of the desired outlet.

Heat must be removed from the satellite by radiation, since conduction will only transfer the heat around to other parts of the system. Payload equipment which dissipates a lot of heat is mounted directly on dedicated satellite radiators to maximise heat removal.

The lack of an atmosphere removes the possibility of heat convection but also causes certain materials to outgas when exposed to a vacuum. In extreme cases, the outgassing can cause structural weakness with the outgassed materials damaging delicate sensors, so care must be taken when selecting materials for use on a satellite.

13.2.4 Launch vehicles

The launch vehicle is used to inject the satellite into either a geostationary transfer orbit, GTO, or directly into a geostationary orbit. Launch vehicles are characterised by their payload mass into orbit, available volume for the satellite (fairing diameter and height) as well as their cost and reliability. Each launcher will also have a different launch environment with which the satellite design must be compatible. Launch vehicles are described in more detail in Section 13.5.

13.3 Satellite orbits

13.3.1 Introduction

Satellites remain in orbit owing to a delicate balance of forces. A satellite is said to be in pure orbital motion when the two principal forces governing its motion are exactly in balance. These two forces are the gravitational attraction

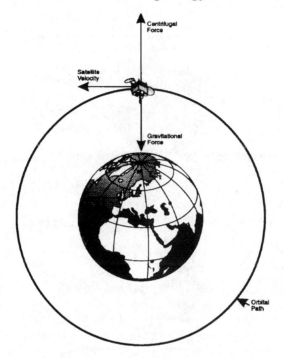

Figure 13.3   Balance of forces in earth-centred orbits

of the earth and the centrifugal acceleration of the satellite's circular motion around it. There are other forces, but these are generally much smaller and cause only small perturbations. This is illustrated in Figure 13.3.

The application of a satellite largely dictates its orbit. There are several different types of earth-centred orbit (shown in Figure 13.4) which can be classified as follows:

- low-earth orbits (LEO) have altitudes between 200 and 1000 km and low or moderately low inclination to the equatorial plane, with moderate eccentricity;
- medium-altitude-earth orbits (MEO) with attitudes of around 10 000 km, and inclinations in the range 45° to 60°;
- polar orbits (PO) are LEOs with high inclinations and therefore take satellites over the earth's poles;
- highly elliptical orbits (HEO) come near to the earth at closest approach (perigee) and move very far away on the opposite side of the orbit (apogee);
- Molniya orbits are a special type of HEO used for some communications satellites, with the apogee located at maximum latitude;
- geostationary-earth orbit (GEO) has zero inclination and a rotational period equal to that of the earth;

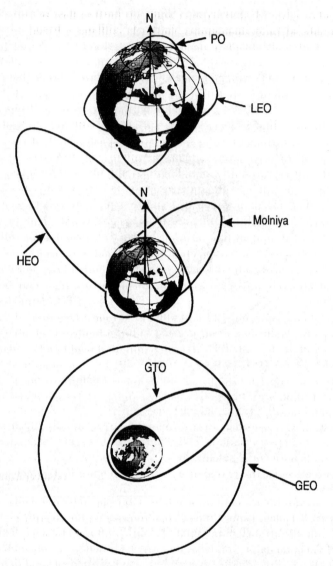

Figure 13.4   Different orbit types

- geostationary transfer orbit (GTO) is an eccentric orbit designed to position a satellite so that it may be placed into GEO efficiently.

To ensure continuous communications between two points on the ground, a communication satellite should be continuously visible from both points. Satellites in GEO fulfil this requirement because their period (the time taken to go once round the earth) is exactly the same as the rotational period of the earth.

Such satellites remain stationary with respect to the ground so that the ground stations can have fixed antennas which do not need to track the satellite.

The characteristics of GEO are as follows:

- period — 24 hours;
- shape — circular with earth at the centre;
- plane — coincident with the earth's equatorial plane;
- height — 35 786 km or 22 236 miles (5.61 earth radii) above the earth's surface.

Satellites in GEO can see almost one third of the earth's surface, and three of them arranged at 120° around the orbit can provide communications between any two points, except those near the poles.

LEO and MEO orbits are being increasingly used for satellite-borne mobile communications (see Chapter 20). The use of such orbits requires a large number of satellites to provide global coverage, with the number of satellites depending on the height of the orbit and coverage provided by the antennas. For example, the Iridium program will use 66 satellites in low-earth circular orbits in six orbital planes, providing complete coverage of the globe. The ICO and the Odyssey programmes will use 12 satellites in medium-earth circular orbits in two and three orbital planes, respectively, also providing worldwide coverage. The advantages and disadvantages of these orbits compared to GEO are outlined in Table 13.1.

13.3.2 Principles of orbital motion

In 1609 orbital mechanics stepped from the world of superstition into that of science, with the publication by Johannes Kepler of his 'Astronomia Nova'. In it he announced his first two laws of planetary motion; many years and much painstaking work later, he published his third law. For years the work lay neglected and little known. In 1685 Edmund Halley (of comet fame) was visiting Newton at Cambridge and mentioned Kepler's ideas. He asked Newton: 'If the sun pulled the planets with a force inversely proportional to the square of their distances, in what path ought they to go?' To Halley's astonishment Newton replied without hesitation: 'Why in ellipses, of course. I have already calculated it, and have the proof among my papers somewhere.' He was referring to the work he had done 20 years ago as a 23-year old graduate taking two years off because the plague had shut down the University. Only in this casual way was Newton's greatest discovery made known to the world.

When Halley recovered from his shock, he advised his reticent friend to develop completely, and publish, his theory of planetary motion. Two years later, in 1687, Newton's 'Principia Mathematica' was published.

Kepler's laws

First law — the orbit of each planet is an ellipse, with the sun at a focus.

Second law — the line joining the planet to the sun sweeps out equal areas in equal times.

Table 13.1   Comparison between LEO / MEO and GEO

GEO	LEO and MEO
provides continuous coverage	provides intermittent coverage from each satellite
cannot provide polar coverage	can provide continuous global coverage
system requires few satellites	system requires a constellation of satellites
satellite is stationary relative to an earth-fixed user, hence fixed high-gain antennas can be used; ideal for fixed satellite, and direct broadcast services (FSS and DBS)	satellite moves relative to an earth-fixed user, hence either widebeam or tracking narrowbeam antennas are required by the user
elevation angle to satellite can be low, increasing signal loss and requiring high power	for high-orbit inclination, elevation angle is high, reducing blockage; ideal for mobile satellite services without fixed antenna
signal propagation loss is high and Doppler effect is negligible; satellite power must be high	signal propagation loss is low, but Doppler effect must be considered; satellite power can be lower
signal propagation time delay is high (≈ 0.5 s)	signal propagation time delay is much lower for users in the same coverage
energy requirements to reach GEO are high, especially for high-latitude launch sites	energy requirements to reach operational orbit are lower
easy to provide on-orbit spare satellite	each orbit plane must be provided with an on-orbit spare satellite
satellite power generation is more efficient as eclipse seasons are short	satellite power generation must cope with extensive eclipse seasons, although individual eclipses are shorter
single ground station required for control and monitoring	requires ground network to coordinate handover between satellites and traffic routing
standard graveyard strategies are available	graveyarding the satellite in LEO and debris from a collision could be a problem

Third law — the square of the period of a planet around the sun is proportional to the cube of its mean distance from the sun.

The force on an earth satellite due to gravity is given by:

$$F_g = \frac{GMm}{r^2}$$

where G = universal gravitational constant = 6.6720×10^{-11} m^3/kg·s^2

M — mass of the earth — 5.9742×10^{24} kg
m = mass of the satellite
r = distance between the satellite and the centre of the earth

The centrifugal acceleration due to orbital motion is given by:

$$Fc = \frac{mV_t^2}{r}$$

where V_t is the satellite's velocity normal to the radius vector.

Equating the two we can obtain the orbital velocity of a satellite in circular orbit of radius r:

$$V_t = \sqrt{\frac{GM}{r}}$$

Typically, for a LEO satellite 200 km above the surface of the earth, the circular orbit velocity would be 7.78 km/s. For a geostationary satellite, the velocity would be 3.08 km/s.

13.3.3 Orbital elements

Five independent parameters are required to define an orbit's size, shape and orientation in space, and a further parameter defines the satellite's position in the orbit. These six parameters are called the orbital elements. They are:

a = semimajor axis of orbit ellipse (= radius of circular orbit)
e = eccentricity of orbit (eccentricity = 0 for circular orbit)
i = inclination of orbit to equatorial plane
Ω = right ascension of ascending node (RAAN)
ω = argument of perigee
v = true anomaly of satellite

and are illustrated in Figure 13.5. In this Figure, the axis set IJK is earth centred and inertially fixed, with K pointing upward through the earth's north pole, I in the earth's equatorial plane and pointing in the direction of the vernal equinox (i.e. where the sun crosses the equatorial plane in spring) and J completes the orthogonal triad.

Some other useful terms are:

- apogee: the furthest point in the orbit from the earth;
- perigee: the closest point in the orbit to the earth;

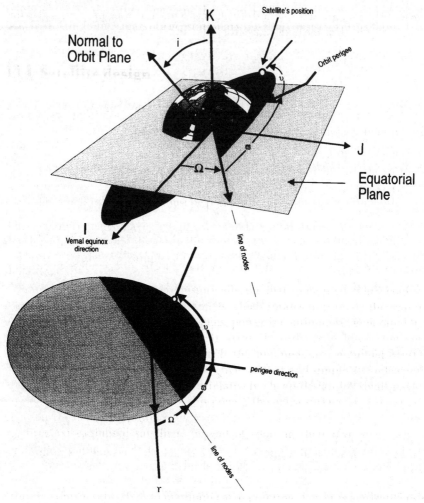

Figure 13.5   Orbital elements

- apse line: the line joining apogee to perigee;
- node: the point at which an orbit crosses the equatorial plane; ascending node, if the satellite is travelling northwards, and descending node if the satellite is travelling southwards;
- argument: angle made with the node line (i.e. the line joining the ascending and descending nodes);
- right ascension: angle made with the I axis, in the equatorial plane.

13.3.4 Orbital perturbations

If the force of gravity due to a spherical earth was the only external force acting on the satellite, its orbit would be fixed forever in the IJK axis frame previously defined. Various factors, however, conspire against such simplicity. These are:

- the influence of other attracting bodies, e.g. sun, moon;
- the nonspherical nature of the earth;
- atmospheric drag;
- solar radiation pressure;
- the satellite's own manoeuvres.

We shall consider the effect of each in turn.

Influence of other attracting bodies

For most earth-centred orbits, the influence of the sun, moon and other planets is very small. For orbits close to the earth (LEOs) their influence can, in most cases, be ignored, as other perturbations dominate (see Table 13.2).

For geostationary and highly-eccentric orbits, the sun and moon impose important perturbations which need to be understood and accounted for.

For highly eccentric orbits the sun and moon affect the orbit's inclination, argument of perigee and the perigee height; the effect depends critically upon the orientation of the orbit with respect to the sun and moon. The perigee height, in particular, can be strongly affected and, if this perturbation is not allowed for, could be reduced to such an extent that aerodynamic drag becomes important and the satellite's lifetime can be seriously affected.

For geostationary orbits, the sun and moon affect the inclination and ascending node of the orbit. If uncorrected this perturbation would increase an initially zero inclination to around 15° before again reducing to zero, over a

Table 13.2   Comparison of relative acceleration (in G) for a 370 km earth satellite

	Acceleration in G on 370 km earth satellite
Earth	0.89
Sun	6×10^{-4}
Mercury	2.6×10^{-10}
Venus	1.9×10^{-8}
Mars	7.1×10^{-10}
Jupiter	3.2×10^{-8}
Saturn	2.3×10^{-9}
Uranus	8×10^{-11}
Neptune	3.6×10^{-11}
Pluto	10^{-12}
Moon	3.3×10^{-6}
Earth oblateness	10^{-3}

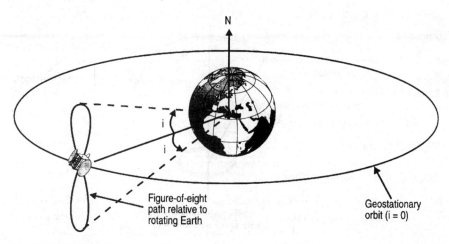

Figure 13.6   Satellite in geosynchronous orbit of nonzero inclination

period of 54 years. A geostationary orbit with a nonzero inclination describes a figure-of-eight path with respect to the rotating earth, as shown in Figure 13.6. The satellite reaches northerly and southerly latitudes equal to the inclination of the orbit plane. Such behaviour would require a large ground-based antenna to track the satellite continuously, making communications awkward and expensive. The year-on-year inclination change owing to lunisolar perturbations is between 0.75° and 0.95°, depending upon the orientation of the moon's orbital plane. To preserve good communications, most satellites are maintained to inclinations of 0.1° or less, through periodic manoeuvres to correct the inclination. Such manoeuvres are known as north-south station keeping.

The nonspherical nature of the earth

The earth is not spherical in shape. It closely approximates to an oblate spheroid which, relative to a sphere of radius equal to the equatorial radius (6378.14 km), is flattened at the poles by 21.4 km. Furthermore, the equatorial cross-section is an ellipse rather than a circle, with the major axis aligned to longitudes of 11.5° W and 161.9° E and the minor axis aligned to longitudes of 75.1° E and 105.3° W. Beyond this there are local deviations from the spheroid of several tens of meters.

Of greatest importance to satellites is the earth's oblateness and, in addition, for geostationary satellites, the elliptic nature of the equatorial crosssection.

The earth's oblateness particularly affects the orientation of near-earth orbits. For circular orbits the effect is a constant rotation of the orbit normal about the earth's polar axis, for inclinations less than 90° this rotation is in a westerly direction, for inclinations greater than 90°, the rotation is in an easterly direction. For exactly polar and equatorial orbits, there is no effect at all (see Figure 13.7).

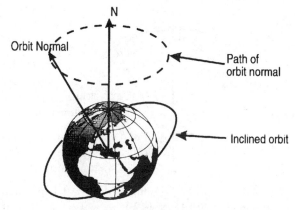

Figure 13.7   Orbit precession due to earth's oblateness

In addition, for near-earth elliptic orbits, the argument of perigee is also affected. For inclinations less than 63.4°, the argument of perigee increases, and for inclinations greater than 63.4° the argument of perigee decreases; at an inclination of 63.4° there is no effect. Molniya orbits are at this particular inclination, in order to maintain apogee position.

For geostationary orbits, the elliptic equatorial crosssection of the earth has a small but persistent effect. It causes the satellite to drift in longitude away from its nominal position. If left uncorrected the satellite would execute long-period (several years) oscillations about the nearest stable longitude, i.e. about the minor axis of the ellipse (see Figure 13.8). Since this means that the satellite would no longer be stationary manoeuvres, known as east-west station keeping, are executed periodically to counteract the effect. Typically, a geostationary communication satellite's longitude is maintained within ±0.1° of its operational longitude.

Atmospheric drag

Atmospheric drag only affects satellites of low altitude or those with low perigees. A satellite flying through the atmosphere loses orbital energy and, therefore, orbital altitude. If the effect goes unchecked the satellite will eventually reenter the denser regions of the atmosphere and be destroyed by kinetic heating. Typically, altitudes above about 1000 km are essentially drag free and satellite lifetime is not affected. At 500 km, however, drag compensation would be necessary for long life.

The effect of atmospheric drag is influenced strongly by the level of solar activity. When this is high, the earth's atmosphere heats up and expands. The air density at very high altitudes can increase by more than an order of magnitude, and the effect of drag is correspondingly increased. Geostationary and high-altitude satellites are not affected by aerodynamic drag, except during their transfer orbit phases (i.e. for a few days).

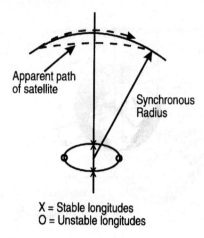

Figure 13.8   The satellite's apparent motion due to elliptic equatorial crosssection

Solar radiation pressure

The influence of solar radiation pressure is only significant for satellites which carry large sun-facing solar arrays. Communications satellites come within this category, and hence this effect needs to be evaluated and, usually, compensated for.

For geostationary satellites, the influence of solar radiation pressure is to set up an eccentricity in an initially circular orbit or to change the eccentricity of an already eccentric orbit. To a ground-based observer, orbit eccentricity manifests itself as a daily oscillation of the satellite about its mean longitude. This motion is also referred to as libration.

The acceleration on the satellite due to solar pressure is given by:

$$\frac{S(A)}{(m)}\frac{(1+R)}{2}$$

where: S = solar radiation pressure for a perfectly flat reflecting surface
  $= 9.1 \times 10^{-6}$ N/m^2
 A = satellite area normal to the sun direction
 M = satellite mass
 R = surface reflectance; ≈ 1 for perfect reflector; ≈ 0 for perfect absorber.

It can be seen that solar radiation pressure particularly affects satellites with large area-to-mass ratios. Furthermore, the effect will vary through the lifetime as the satellite's propellant is consumed, and as reflectance properties change with age.

This effect adds to the longitude drift of the satellite due to the elliptic nature of the equatorial crosssection, and both effects are countered by the satellite's station-keeping strategy. For satellites with large area-to-mass ratios, the solar radiation pressure dominates the east-west station-keeping strategy.

Satellite manoeuvres

When a satellite activates its thrusters, translational and/or rotational motions result. The main manoeuvres that satellites undertake are associated with:

- transition from initial to operational orbit;
- attitude control;
- maintenance of operational orbit;
- deorbiting at end of life.

For GEO the majority of the propellant carried by a satellite is used to transition from the orbit achieved by the launch vehicle, to the satellite's operational orbit. This generally involves changes to the orbit's semimajor axis and inclination.

For geostationary satellites, orbit-maintenance manoeuvres are also very demanding on propellant, specifically those manoeuvres required for inclination control (north-south station keeping).

The manoeuvres required for attitude control and end-of-life deorbiting take up only a very small proportion of the total propellant load. The significance of a manoeuvre is well described by the velocity change or delta-v achieved by the manoeuvre. Table 13.3 shows typical delta-v for a geostationary satellite launched by Ariane 4 and having a ten-year operational lifetime.

In general, the propellant used for a manoeuvre, Mp, is given by:

$$Mp = Mo[1 - \exp(-\Delta v/(\text{Isp}.\eta.g))]$$

where: Mp = propellant mass used for manoeuvre
Mo = satellite mass at start of manoeuvre (including mass of propellant)
Δv = delta-v produced by the manoeuvre
Isp = specific impulse of the propulsion system
η = manoeuvre efficiency
g = acceleration due to gravity (9.81 m/s^2).

For example, a 2000 kg satellite would use 788 kg of propellant to achieve a delta-v of 1509 m/s with a propulsion-system-specific impulse of 310 s and a manoeuvre efficiency of 0.99. This is typical of an orbit-transition manoeuvre for a geostationary satellite launched by Ariane 4. If to this is added the propellant mass

Table 13.3 Manoeuvres for a typical geostationary satellite

Manoeuvre type	Delta-v (m/s)
Orbit transition	1509
North-south station keeping	470
East-west station keeping	35
Deorbiting	8

Note: attitude manoeuvres cannot be characterised by delta-v and are not included in this Table.

required for all the other functions shown in Table 13.3, it is not surprising that over half of the launch mass of a geostationary satellite consists of propellant.

13.4 Satellite design

13.4.1 Types of satellite

For a satellite in geostationary orbit, there are two fundamental pointing requirements:

(i)   The satellite antennas or instruments must be pointed towards the earth continuously.
(ii) The satellite solar array must be pointed towards the sun continuously.

These requirements have led to two fundamental types of satellites being developed: the spin-stabilised satellite and the three-axis-stabilised satellite.

Figure 13.9 shows examples of both spin-stabilised and three-axis-stabilised designs.

Spin-stabilised satellites were historically the first to be used. They consist of a cylindrical drum covered in solar cells which rotates to provide the required gyroscopic stability. As the satellite spins, power is continuously generated from the illuminated part of the drum solar array. In order to achieve continuous earth coverage, the platform on which the antennas are mounted must be despun. The relatively benign thermal environment and simple attitude control and propulsion subsystems lead to a relatively cheap design for small payload powers of up to 1–1.5 kW. Above this power, the extra surface area required to generate the necessary power leads to a mass and cost penalty.

Three-axis-stabilised satellites, instead of rotating their own structure, use internal spinning wheels to compensate for any disturbance torques. Either momentum wheels or reaction wheels are used depending on the design of the satellite. The satellite body is rotated in pitch at one revolution per day by the attitude-control subsystem to maintain earth pointing. Power is generated using a deployable solar array made up of a number of panels. Additional panels can be added to increase the power generated while minimising the impact on the satellite design. As the satellite rotates, the solar array tracks the sun using a solar-array drive mechanism (SADM) so that all the cells are in sunlight all of the time.

Since most communication satellites being built today utilise three-axis-stabilised designs, the rest of this section focuses on their subsystem design. Figure 3.10 shows an exploded view of a three-axis communication satellite.

13.4.2 Attitude determination and control subsystem, ADCS

The ADCS performs all the necessary operations to maintain the satellite accurately pointing towards the earth and on station, as well as controlling any transfer orbit phase.

Despun antenna locked
onto Earth beacon

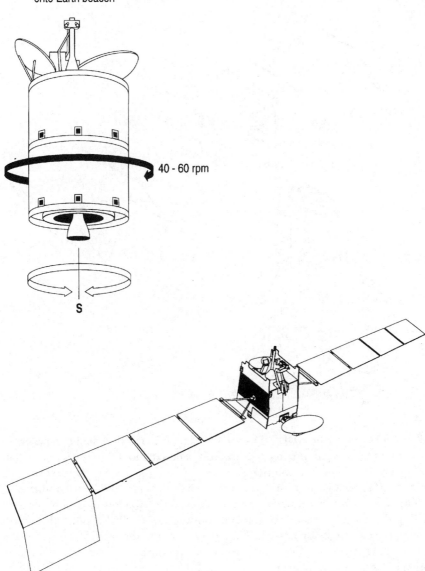

40 - 60 rpm

S

Figure 13.9   Spin-stabilised and three-axis-satellite designs

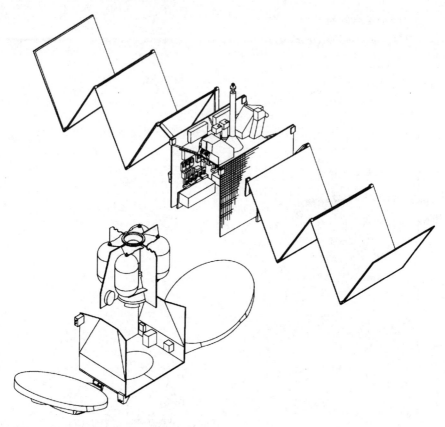

Figure 13.10   Three-axis-stabilised satellite design

Before describing the ADCS subsystem design, it is useful to define the coordinate system for the satellite in its earth orbit. The satellite's $+z$ face is pointed at the earth with the $+y$ face pointing due south. The right-hand triad is completed by the $+x$ face pointing in the direction of the orbit path. Roll, pitch and yaw axes are also defined with pitch rotating the satellite around the satellite's y axis moving the coverage east west and roll rotating the satellite around the x axis moving the coverage north south. Finally, yaw rotates the satellite around the z axis resulting in the coverage moving both north south and east west. These are shown in Figure 13.11.

Satellites experience varying disturbance torques as they circle the earth, mainly from solar radiation pressure, which tend to depoint the spacecraft. To compensate for this the satellite uses either a momentum-biased or reaction wheel system.

A momentum-biased system uses a momentum wheel spun up at high speed within the satellite to provide a net angular momentum. This provides gyroscopic stability against any external disturbance torques. The wheel operates at

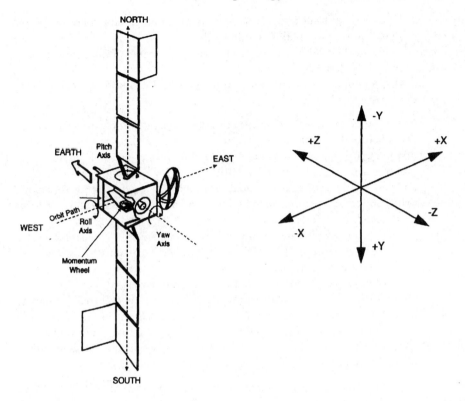

Figure 13.11   Coordinate system of a satellite

a nominal speed, with any disturbance torques spinning the wheel up or down to maintain the satellite pointing. Periodic momentum off loading is required to maintain the wheel within its operating limits. The momentum wheel also provides the one revolution per day rotation rate to keep the satellite pointing towards earth.

A zero degrees of freedom (DOF) momentum-biased system provides control around only one axis, usually the pitch axis. Although pitch control is provided using this method, roll and yaw control must also be provided to control the satellite. One elegant solution is to locate solar sails on the outer panels of the solar array which, by adjusting the solar-array wing position, provides roll and by extension yaw control. However, for a large roll repointing of the satellite, a one DOF momentum-biased system is better. This can be implemented by using two wheels with their axes in the roll plane, but inclined in the pitch axis, and varying each wheel speed to achieve the desired pointing.

A reaction wheel system, by contrast, has no net angular momentum but instead relies on wheels spinning in one direction or the reverse to react to any disturbance torques, ensuring that the satellite maintains an earth-pointing direction. Four wheels are usually used; one each in the *x, y* and *z* axes of the satellite and the fourth inclined with components in all three planes for redundancy. This configuration allows full pitch and roll repointing, within the constraints of the ADCS design, but is less inherently stable when compared to a momentum wheel design.

The ADCS consists of three main components and a simplified schematic is shown in Figure 13.12:

(i)   Sensors, to determine the pointing attitude in roll, pitch and yaw as well as any attitude rate. These include infra-red earth sensor, IRES, which provides roll and pitch attitude data to control the satellite on station, and roll and pitch pointing accuracies of better than 0.03°. Gyros are used to provide roll, pitch and yaw reference data in any satellite orientation, e.g. during apogee engine firing. Sun sensors are used to maintain the array's sun pointing and used as an attitude point of reference.

(ii) Actuators, to compensate for any disturbance torques and control the solar arrays and satellite pointing. These include momentum wheels, reaction wheels, solar-array drive mechanisms (SADM) and thrusters.

(iii) Control electronics and databus, to process the sensor data and send commands to the actuators. These are usually microprocessor controlled, allowing flexibility to reprogram the control loops as well as permitting automatic routines to be run on board the satellite to simplify operations on the ground. The ADCS electronics also contain emergency sun reacquisition loops in case of satellite depointing.

Typically, on-station pointing accuracies are of the order of 0.1° in roll and pitch and 0.2° in yaw.

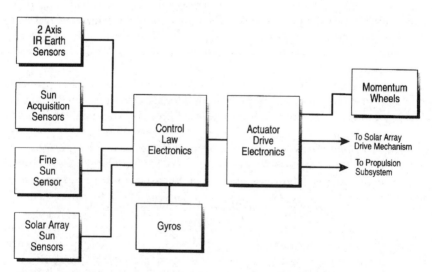

Figure 13.12   ADCS functional schematic

13.4.3 Power subsystem

Of all the elements on a satellite, the power subsystem is the most dependent on the customer requirements, specifically on the number of transponders required and their power. It supplies power to the payload during both sunlight and eclipse conditions, and consists of three main elements:

- solar array;
- battery;
- conditioning electronics and harness.

Solar array

The solar array provides power to the satellite during sunlight conditions by using photovoltaic cells to convert solar radiation into electric power.

For a three-axis-stabilised satellite, the solar array is maintained pointing at the sun using a solar-array drive mechanism (SADM) which rotates the array in pitch only, at a constant rate. Owing to the variation in the position of the sun during solstice and equinox, there is a slight seasonal variation in power output, being higher in equinox than at the solstices. The radiation environment in geostationary orbit leads to a degradation in power output of the solar cells, since they are shielded by only a thin coverglass. This leads to a reduction in the mean power output throughout the array's life, the array being sized for the worst-case requirement at end of life (EOL) after allowing for one string failure. The solar-array output power variation with life is shown in Figure 13.13.

Solar arrays are designed to be modular, allowing additional panels to be added or subtracted to meet the payload power requirements. Typically, silicon cells are used which have EOL efficiencies of nine to ten per cent.

In order to meet the demand of future satellite missions with increased payload powers, various advanced solar cells are under development and some are already being used. Their typically power densities after 12 years in orbit are shown in Table 13.4.

Gallium arsenide cells have been flown in space and have a higher conversion efficiency than silicon, which compensates for their heavier solar cell. Recently, solar-array designs have been proposed which use array edge-mounted concentrators to increase the effective solar radiation density illuminating the solar cells, and thus increase the power output from the solar array.

Batteries

Batteries supply all electrical power during the eclipse period, with most missions now requiring full payload operation during eclipse. The two most common batteries used are nickel hydrogen (NiH_2) for payload powers above 1.5 kW and nickel cadmium (NiCd) for smaller powers.

In geostationary orbit, there are two eclipse periods per year lasting 45 days each with the eclipse length varying, but reaching a maximum of 72 mins at the

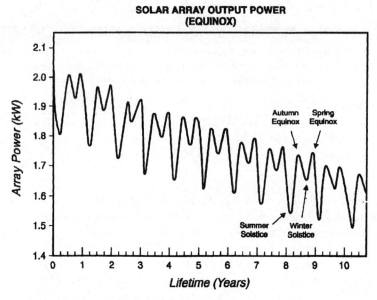

Figure 13.13   Solar-array output power variation

equinoxes. The batteries are sized for the maximum eclipse and allow for one cell failure per battery. NiH_2 batteries utilise pressure vessels which are more mass efficient at high payload powers but are considerably larger in size.

Batteries also provide power during any satellite emergency where the arrays may temporarily lose sun pointing.

Future developments are aimed at increasing the energy density of the batteries to reduce their overall mass. Batteries represent a considerable mass penalty considering the time for which they support the payload, and weigh between 15–20% of the dry mass of the satellite for high-power missions. Lithium carbon (LiC) batteries represent a promising development for space having been extensively utilised for terrestrial purposes, e.g. camcorders, and are expected to be flight qualified in the next few years.

Table 13.4   Advanced solar cells

Cell type	Power density, W/m²	Mass density, kg/m²
Silicon	114	0.94
High efficiency Si	129	0.79
Gallium arsenide, GaAs	160	1.28
Dual junction GaAs	195	1.28

Table 13.5 Energy densities of different battery technologies

Battery	Energy Density Wh/kg	Comments
NiCd	15	old technology
NiH$_2$	44	large pressure vessels
LiC	100	used on earth but not demonstrated in space

Power conditioning electronics

The conditioning electronics and harness provide a number of functions including:

- battery monitoring;
- battery charging and discharging, trickle and main charge;
- regulation of bus in sunlight and eclipse.

The power bus can either be semi- or fully regulated at voltages usually between 30–100 V. A fully regulated bus provides a constant voltage on the power bus utilising one bus to connect all sources and loads. Battery-discharge regulators are used during eclipse to increase the battery voltage to that of the regulated bus.

A semiregulated bus maintains a fixed voltage during sunlight, but follows the battery voltage during eclipse. This utilises separate power buses and for a two-bus design each is supplied by one array wing and one battery, with the loads split equally between the buses.

The power harness is used to distribute the electrical power around the satellite to the various electrical units.

A typical power subsystem schematic is shown in Figure 13.14. This is for a dual bus semiregulated design.

Power subsystem sizing

Solar array calculations: The solar array must be sized to provide sufficient power to the payload and platform during both solstice and equinox. Although the payload load remains essentially the same, the platform power loads and the available array power vary between the two conditions.

During solstice the power available from the solar array is reduced, since the arrays are inclined to the plane of the sun by up to 23.5°. This reduces the array power to about 0.91 of that available in equinox. During equinox the heater power must be increased to compensate for the more efficient radiators and power must be provided to recharge the batteries, increasing the total load compared to solstice.

The solar array is sized to provide between five and ten per cent power margin over the satellite loads, after allowing for various loss and degradation factors including one string failure.

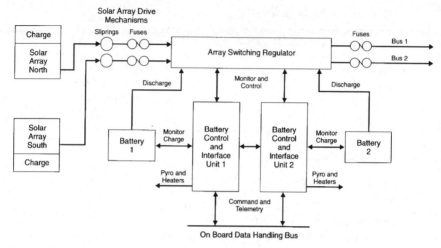

Figure 13.14   Power subsystem schematic

The solar array area required for the equinox case can be found using the calculation below:

$$SA_{\text{area}} = \frac{P_{\text{eq}} \times 1.05}{(No_{\text{pan}} \times Area_{\text{pan}} \times 2 \times Solar_{\text{flux}} \times Eff_{\text{cell}})}$$

where SA_{area} is the total solar array area in m^2, P_{eq} is the satellite power required at equinox in W including battery charge, No_{pan} is the number of panels per wing, $Area_{\text{pan}}$ is the area of one panel in m^2 allowing for layout efficiency of typically 0.95, Eff_{cell} is the solar cell efficiency at EOL and $Solar_{\text{flux}}$ is the solar-flux power density with the average at earth's distance $= 1370$ W/m^2.

Battery calculations: Batteries consist of a number of cells, each of the same capacity and connected together in series to give the required voltage. During eclipse the battery voltage decreases as the cells discharge. To calculate the battery capacity required, an allowable depth of discharge (**DOD**) appropriate to the battery type is used, with 55% for NiCd and 80% for NiH$_2$ being typical for geostationary orbits. Higher **DOD**s than these can lead to long-term damage to the cells and are thus avoided.

To allow for the possibility of battery-cell failure, each cell is bypassed by a diode which must be taken into account in the sizing calculations.

The battery capacity required is given by the following formula:

$$C_{\text{bat}} = \frac{P_{\text{ecl}} \times h}{N_{\text{bat}} \times V_{\text{disc}} \times DOD}$$

where C_{bat} is the capacity of the battery in A hr, P_{ecl} is the satellite power required during eclipse, h is the maximum eclipse length in hours and is 1.2 for geostationary orbit, N_{bat} is the number of batteries and V_{disc} is the worst case discharge voltage of the battery in volts assuming one cell has failed and equals (number of cells per battery − 1) × the average cell voltage.

13.4.4 Telemetry, command and ranging subsystem, TCR

The TCR subsystem provides telecommand (TC) and telemetry (TM) capability between the satellite and ground-control station during both the transfer orbit and onstation phases. During the transfer orbit, wide antenna coverage is required to maximise the chances of receiving commands and/or sending telemetry to the ground station. On the station, more directive antennas or horns are used to improve the TC/TM link margin and permit the use of smaller ground stations. The TM/TC frequencies used tend to be at the edge of the communication frequency bands for coordination purposes. Figure 13.15 shows a schematic of the TCR subsystem.

Telemetry is used to monitor the satellite, providing detailed information on the state of all equipment via the telemetry format. Each format consists of a number of frames with each frame containing a number of datawords. Typically, the telemetry format repeats every 15−25 s and is continuously transmitted to the ground using dedicated TCR transponders.

Critical functions are sampled more often than once per format, e.g. IRES pitch and roll data, and less critical functions are sampled less often, e.g. NiH_2 battery cell pressure and voltage. It is also possible to provide dwell TM channels which allow telemetry parameters to be sampled continuously, if some anomalous behaviour has been observed.

Telecommands are used to change the state of the units on board the satellite, for example to reconfigure the payload. Redundant receivers demodulate the uplink signal and pass the command to the central interface unit. This decodes and verifies the command before sending it to the relevant unit via a serial databus. In case there is an onstation emergency involving satellite depointing, automatic switching to wide antenna coverage is provided.

Telecommands are also used to initiate an operation, e.g. battery charging. As seasonal changes occur on board the satellite, it is necessary to reconfigure

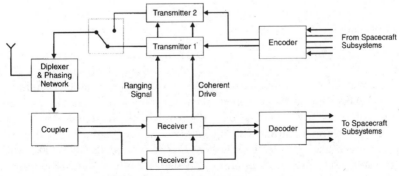

Figure 13.15   Simplified schematic of TCR subsystem

the platform to take account of the different thermal conditions and/or to prepare for the eclipse season.

Recent advances have focused on reducing the ground operators' workload, either by using the increase in the capability of the onboard microprocessors to automate operations or by using expert systems to monitor the satellite from the ground.

The central interface unit is microprocessor controlled and more recently there has been a trend to centralise all critical platform functions within one avionics unit (with the necessary redundancy). This reduces the number of units, and therefore mass and cost, as well as simplifying the testing.

Ranging is also performed using this subsystem, being required to accurately locate the spacecraft during the critical transfer-orbit phase as well as on station. During the transfer orbit, apogee engine firing (AEF) must be performed accurately to avoid wastage of propellant and so ranging is performed before and after each firing. This also allows evaluation of the AEF performance after the first burn, which in turn allows accurate positioning of the satellite into the final drift orbit after the final burn. On station, ranging is performed after significant manoeuvres to ensure that the satellite remains within its station-keeping box. This is typically ± 0.05 to $0.1°$ in both longitude and latitude from the allocated orbital slot for a full station-keeping mission. Ranging is performed using either tone or pseudo random noise code ranging (PRN) measuring the phase shift or time delay, respectively, and then calculating the distance.

13.4.5 Combined propulsion subsystem, CPS

The propulsion subsystem is initially used to provide the impulse to transfer the satellite from a geostationary transfer orbit (GTO) to the final geostationary orbit. On station, the propulsion subsystem is used to maintain the satellite's position within the on-station box.

A combined propulsion system is shown in Figure 13.16. This utilises common propellant tankage for both transfer orbit and onstation phases, allowing any excess propellant from the orbit transfer burns to be used on station.

Transfer orbit phase

The satellite is injected by the launcher into a geostationary transfer orbit (GTO) with the satellite itself providing the impulse to achieve the final geostationary orbit. Most commercial satellites use a high-performance liquid-apogee engine (LAE) utilising the bipropellants nitrogen tetraoxide (N_2O_4) and monomethyl hydrazine (MMH) to provide the required thrust. These propellant are hypergolic, i.e. they spontaneously ignite when they come into contact with each other, producing a high specific impulse (or Isp) of typically $310-320$ s. During the transfer orbit burns, the liquid bipropellant system is pressure regulated to ensure constant pressure to the LAE and optimum performance.

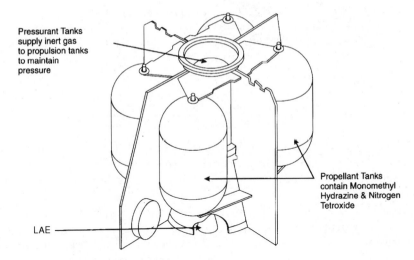

Pressurant Tanks
supply inert gas
to propulsion tanks
to maintain
pressure

Propellant Tanks
contain Monomethyl
Hydrazine & Nitrogen
Tetroxide

LAE

Figure 13.16   Bipropellant CPS tanks plus LAE

Solid-propellant apogee boost motors (ABM) are also used but have lower performance (Isp $\approx$ 290 s) and consequently are only appropriate for smaller satellites or older designs.

On station

Most present-day satellites use bipropellant thrusters to provide the required onstation manoeuvres including east-west station keeping, north-south station keeping (if required), any satellite repositioning and the final graveyard burn at EOL. After the LAE burns, the pressurisation system is isolated from the tanks to increase the reliability of the subsystem and the thrusters are operated in blow-down mode. This results in a slight reduction in Isp of the thrusters over life as the pressure drops. Two separate thruster branches are provided for redundancy, which allows for a thruster or its valve to fail without affecting performance.

Recent developments in thruster technology allow the use of electric propulsion to be considered for onstation operations, which reduces the amount of propellant mass required for station keeping by raising the specific impulse above that of existing bipropellant systems.

There are a number of electric-propulsion thrusters currently under development, with one of the most promising utilising stationary plasma thruster (SPT) technology developed by the Russians and flying successfully on board a large number of Russian satellites.

Ion thrusters are also under development, offering a doubling of specific impulse performance compared to the SPT technology. These are being developed by European, American and Japanese companies. A comparison of the different types of thruster is shown in Table 13.6.

Table 13.6   Performance of different propulsion technologies

Performance parameter	Bipropellant	Ion thruster	SPT-100
Isp, s	290	3000	1500
Thrust, N	0.5–10	0.025	0.08
Power, W	30	700	1500

When calculating the propellant mass required for a manoeuvre, the effective specific impulse of the thruster is used which takes into account any inefficiencies of the thruster, e.g. canting of thruster to avoid plume effects on the antenna or solar array.

Electric thrusters have a much lower thrust which results in a much longer manoeuvre length. They also require considerable amounts of power which can be provided from the available solar-array margin at the beginning of life, together with cycling of the battery at a low DOD as the array margin reduces. Electric propulsion is expected to replace chemical propulsion for north-south station keeping as customer confidence in it grows and satellite masses increase.

13.4.6 Structure

The satellite structure must provide sufficient mounting area to accommodate all the units and be capable of withstanding the most severe conditions during launch. It is split into a communications module (CM) and a service module (SM), allowing parallel build and integration to shorten the satellite time to delivery.

Figure 13.17 shows a typical structure for a communication satellite. This comprises a central thrust cylinder which carries the loads to/from the launch vehicle. The cylinder is mounted on the launch vehicle adapter via a cone interface using a clampband to attach the satellite to the launch vehicle. After launch this clampband is released and the satellite is pushed away from the adaptor by a number of powerful springs. Surrounding the central cylinder are four shear walls which provide support to the box-like walls as well as the top floor. The top floor $(+z)$ and the east and west walls $(\pm x)$ are used to mount payload antennas and sensors and any low-heat dissipating equipment, and the north and south radiator $(\pm y)$ walls are used to mount any equipment with high heat dissipation.

The central cylinder is made from an aluminium honeycomb core wound with carbon-fibre-reinforced plastic (CFRP). The walls and floors are made of aluminium honeycomb of varying thickness to provide high stiffness for low mass and allow easy fixing of units to the structure using inserts.

The structure is designed to cope with the launch environment including acceleration, compression and vibration loads, as well as acoustic noise, all of which vary with launch vehicle. The worst-case loads are defined in terms of the first lateral and axial frequencies of each launcher plus the quasistatic loads

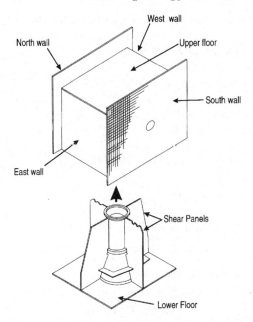

Figure 13.17   Typical structure design of a satellite

(QSL), which defines the simultaneous axial and lateral loads' envelope. The structure is designed and tested for worst-case loads of all the launchers, with a safety factor of 1.3–1.5. During the satellite programme, a coupled-load analysis is performed for each satellite/launcher combination to ensure launch-vehicle and satellite safety.

13.4.7 Thermal subsystem

The thermal subsystem must maintain all equipment within acceptable operating limits during all phases of the mission. To do this, the heat flow in and out of each unit is controlled, with its location dependant on the amount of dissipation that it produces. The thermal design must also cater for the variation in solar illumination during the year.

In geostationary orbit, a satellite is orientated with its solar arrays north-south and the payload antennas pointing towards the earth. Effectively, the sun appears to rotate around the satellite's x and z faces throughout the day, and thus the y walls are not directly illuminated, except at the solstices where one or other of the y radiators is obliquely illuminated.

Therefore, all equipment with high heat dissipation is mounted on the y radiator walls, with low-heat-dissipating equipment mounted on either the x or z faces, depending on the layout requirements. To reduce the heat input from the sun onto the y walls even further, second surface mirrors (SSM) are mounted on the y radiator walls to reduce sun absorbance and increase their emittance.

Payload units that are mounted on the *y* radiator walls include high-power amplifiers (TWTA + SSPA) and output multiplexers (OMUX). LNA that are required to be kept at low temperature are also mounted on the *y* radiator walls in a separate cool area.

In order to take full advantage of the available radiator area, thermal doublers and heatpipes are used to spread localised heat sources over a larger area. Heatpipes can either be surface mounted or embedded within the structure walls. They function by evaporating a fluid within the heatpipe, removing the heat which is forced by pressure along grooved ducts. On reaching the cooler part of the pipe, the vapour condenses and the heat is rejected. Ammonia is commonly used as the fluid.

To isolate the spacecraft from the external, widely varying, thermal environment extensive use is made of multilayer insulation (MLI), particularly on the *x* and *z* faces of the satellite, and on the *y* walls where excess radiator area is not utilised.

Heaters are used in conjunction with thermostats to maintain the temperature of equipment above a minimum threshold. They maintain the temperature of the thrusters mounted outside the structure, as well as the tanks and pipework within the satellite, to above that of the propellants' freezing point. Also within the satellite box, the walls and unit faces are painted black to equalise the internal temperature. Figure 13.18 shows the thermal design of a satellite.

Promising new developments include capillary-pumped loops (CPL) which use capillary action to remove heat from one location to another via flexible

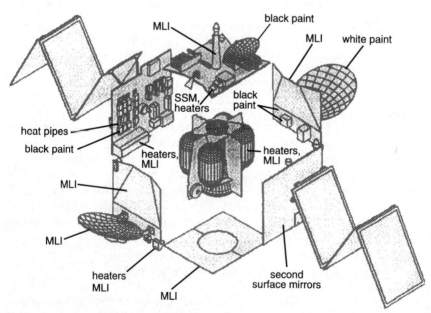

Figure 13.18   Thermal design of a satellite

pipes and can be used to precisely control the temperature of a unit, regardless of location, allowing more flexibility in unit positioning. Radiators which deploy in orbit are also being developed to increase the available radiator area for higher power missions without having to oversize the overall structure.

13.5 Future large geostationary mobile communication satellites

Although considerable attention was focused on constellations of mobile satellites to provide global coverage in the early 1990s, e.g. Iridium, the satellite industry has more recently started to look again at providing handheld mobile communications from geostationary satellites.

This is of particular interest to large countries or regions which want true handheld mobile communications services provided by satellites under their own control but which do not want, or cannot afford, to operate a global constellation.

Several key developments have made it more feasible to provide mobile communications from a geostationary satellite. These are:

- development of large unfurlable mesh antennas with diameters of 12 m and above (mainly in the USA);
- development of larger launchers including: Boeing Sealaunch, Proton M and Ariane V with increased mass capability into GTO (five tonnes and over);
- development of larger satellites to take advantage of improved launcher performance including Matra Marconi Space's EUROSTAR 3000, Hughes' HS 702 and Lockheed Martin's AS2100;
- maturity of geostationary satellite design for mobile communications e.g. INMARSAT 3 and AMSC;
- development of wideband and narrowband processors (DSP);
- development of compensation techniques to counteract echo in geostationary satellite mobile links.

Providing mobile communications from a geostationary satellite requires designs capable of supporting payload masses of up to 1000 kg and payload powers up to 8 kW. This leads to satellites with masses in the range of four to five tonnes and electric propulsion for station keeping, making them the largest satellites yet seen in the commercial industry. The designs have high heat dissipations mainly from the SSPAs and payload processors which require new thermal concepts, utilising capillary-pumped loops as well as deployable radiators.

Large unfurlable antennas of the order of $12-15$ m diameter are required to achieve the link budget for the small mobile handsets with limited EIRP and G/T. The limited frequency spectrum allocated for mobile systems requires high frequency reuse with a large number of spot beams. Designs with both one combined or two separate transmit and receive antennas are being considered,

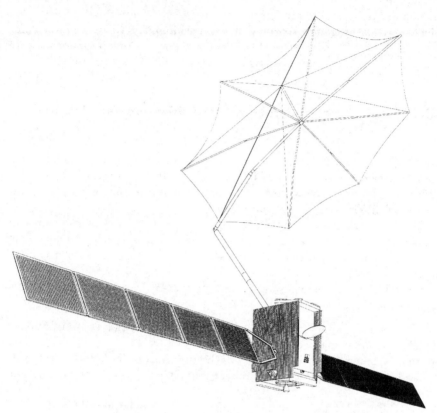

Figure 13.19   Example geostationary mobile communication satellite

although there are concerns that passive intermodulation (PIM) could cause problems with a single antenna solution at L-band. Figure 13.19 shows one example of a geostationary mobile communication satellite design.

The large complex processors required for channelisation, routing and shaping of the spot beams have not been flown on commercial satellites before. Since their power and, to a lesser extent, their masses are a critical function of the application-specific integrated circuit (ASIC) technology used, the satellite design must incorporate large margins in both mass and power to avoid excessive satellite growth during the programme.

13.6 Launch vehicles

13.6.1 General

A wide range of launch vehicles is available to satellite users and manufacturers. Table 13.7 shows a selection of current launchers which are illustrated in Figure

Table 13.7 Launch vehicles

Vehicle	Country of origin	Launch site	Multiple payloads	GTO capability (kg)	Operational status
Ariane 4	ESA	Kourou	yes (up to 2)		operational
40				2050	
42P				2840	
44P				3320	
42L				3380	
44LP				4060	
44L				4520	
Ariane 5	ESA	Kourou	yes (up to 3)		operational
Single payload				6800	
Dual payload				5900	
Triple payload				5500	
Delta II 7925	USA	Cape Canaveral	no	1870	operational
Delta III	USA	Cape Canaveral	no	3810	operational
Atlas 2	USA	Cape Canaveral	no		operational
2A Block 1				3066	
2AS Block 1				3697	
2AR				3810	
Proton D-1-e	Russia	Baikonur	no	2100 (into GEO)	operational
Long March 3	China	Xichang	no		operational
3A				2500	
3B				4500	
3C				3500	
Sea Launch	Russia/US	on equator	no	5250	Zenit3 vehicle operational, launch system in development
H-2	Japan	Tanegashima	yes	4000	operational

13.20. It is the function of the launch vehicle to either inject the satellite into its operational orbit, or into an intermediate or transfer orbit from where the satellite's own propulsion system can efficiently achieve the final orbit.

Owing to the limitations of chemical propulsion systems and the high energy requirements for orbits suited to communications satellites, launching is a costly business. Typically, a launch into geostationary orbit can cost as much as the

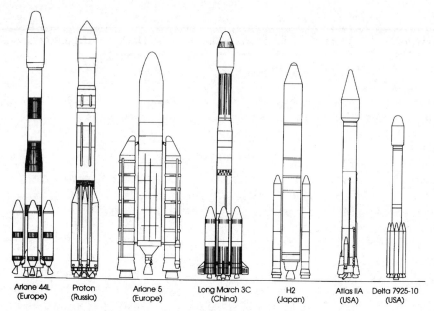

Ariane 44L Proton   Ariane 5   Long March 3C   H2   Atlas IIA   Delta 7925-10
(Europe)   (Russia) (Europe)   (China)   (Japan) (USA)   (USA)

Figure 13.20   Selection of currently available launch vehicles

Table 13.8   Launch-site characteristics

Launch site	Country	Minimum achievable inclination for GTO (°)	Easterly velocity (m/s)	Launch-site latitude (°)
Korou	French Guyana	7.0	436	5.2
On equator	Pacific Ocean	0.0	465	0.0
Cape Canaveral	US	28.5	409	28.7
Baikonur	Kazakhstan	48.0	323	46.0
Tanegashima	Japan	28.5	403	30.0
Xichang	China	28.5	410	28.2

satellite itself. Currently, launch costs into orbits of interest to communication satellites are in the region of $20 000 to $30 000 per kg of satellite mass. All ways are therefore sought to minimise launch mass.

An important consideration to this end is the selection of the launch site. The geographic location of the launch site largely determines the efficiency of the launch process. For low-earth and geostationary transfer orbits, the launcher seeks to maximise use of the earth's spin rate to provide the necessary easterly orbital velocity. For this reason a launch due east, i.e. at a launch azimuth of 90°, is most efficient. Since the launch site must be in the resulting orbit plane, the

Table 13.9   Useable fairing diameter

Vehicle	Useable fairing diameter (m)
Delta II	2.74
Long March 3A	3.00
Atlas 2	3.65
Ariane 4	3.65
Long March 3B/C	3.65
Delta III	3.75
Sea Launch	3.75
Proton D-l-e	3.82
Ariane 5	4.57

launch latitude defines the minimum inclination which can be achieved without the use of inefficient dog-leg manoeuvres in the ascent trajectory. Since geostationary orbits are nominally at zero inclination, the closer the launch site is to the equator the more efficient is the launch process. Furthermore, the closer the launch site is to the equator the greater the easterly velocity due to the earth's spin rate.

Table 13.8 shows the easterly velocity due to the earth's spin rate and the minimum inclination possible from a selection of the world's launch sites. Note that the minimum GTO inclination is sometimes slightly different from launch-site latitude. This is due to trajectory constraints, range safety constraints and multiple manoeuvres of the upper stage of the launch vehicle.

One way of reducing launch cost is to launch more than one satellite on the same launcher. Currently, Arianespace is the only commercial launch agency which handles multiple manifesting of geostationary satellites. However, multiple payloads are becoming the standard for launching constellations of low-earth orbit communication satellites like Iridium, Globalstar etc. For this purpose special dispensers are used.

The disadvantage with multiple payloads is that with only a single launch window, transfer orbit optimisation for more than one mission is not possible.

The increasing demands made by communication service users mean ever larger and more powerful satellites, and thus heavier and more capable launch vehicles. Not only is greater payload capacity required, but greater payload volume is also necessary to accommodate the large antennas which must now be carried. Usable fairing diameter is gradually increasing; Table 13.9 shows the trend.

13.6.2 Launch-vehicle interfaces

Most vehicles offer the satellite builder a small range of standard adapter structures which provide the interface between the satellite and the vehicle and carry the separation system and umbilical harnesses. These are commonly known as the 937, 1194 and 1666 adapters, in ascending order of size; the number relates to the diameter (in mm) of the interface ring. They all rely on the Marmon clamp separation system in which the satellite and adaptor interface rings are clamped together by steel blocks that are held in place by a tensioned metal hoop band. When the band is severed, the blocks fly outwards and release the interface rings. Positive separation velocity is achieved by the expansion of four or more compressed springs. Typical separation velocities are in the range 0.5 to 1 m/s.

All launch vehicles impose certain environmental conditions upon the satellite which depend upon the characteristics of the vehicle and its ascent trajectory. The satellite must, of course, be designed to withstand this environment. By design, most vehicle environments are very similar, which makes it possible for a satellite to be compatible with several launch vehicles with very little penalty. The main elements of the environment are:

- axial and lateral loads — quasistatic and random;
- thermal;
- maximum aerodynamic flux at fairing jettison;
- acoustic;
- limitations on minimum satellite axial and lateral natural frequencies.

The task of matching a satellite to a launch vehicle in all respects is carried out through a well ordered process of definition of requirements, analyses, compatibility checks and reviews. This process runs in parallel with the satellite design and build processes, and culminates in the flight readiness review that finally releases the satellite for launch.

13.6.3 Reliability

Of paramount interest to the user is the reliability of the launch vehicle. Not only is a high reliability vital to mission success, it also has a direct impact on mission cost both through the cost of insuring the launch and the satellite's operational lifetime.

Most manufacturers aim at a target reliability of around 95 per cent. What they actually achieve is more variable. For geostationary orbit missions over 20 years to February 1995 the trend is remarkably consistent and is shown in Table 13.10 for each vehicle.

13.7 Commercial satellite programmes

13.7.1 Satellite procurement

This Section outlines the lifecycle of a satellite from its initial procurement, through the design and development phases and onto its operations before finally being in the graveyard at the end of its useful life.

Table 13.10   Launch-vehicle reliability

Vehicle	Reliability	Comments
Ariane 4	0.96	over all missions up to flight no. 90
Proton D-1-e	0.96	over last 50 launches
Atlas 2	1.00	over last 16 launches
Delta 2	0.98	over last 50 launches
Zenit (Sea Launch)	0.88	over all missions, and over ten years
H-2	1.00	three launches only

There are two main approaches to satellite procurement:

- delivery on the ground, where the satellite is supplied to the customer and the customer arranges for the satellite to be placed into geostationary orbit;
- delivery in orbit (DIO) where the satellite is delivered fully tested in orbit, with the supplier arranging for the launch-vehicle procurement, the insurance of the launcher and satellite, the delivery of the satellite to its onstation location and its testing and commissioning.

In addition to the procurement of the satellite, the customer may require supply of ground-segment equipment to control and monitor the satellite including the TCR control station(s), communication subsystem monitoring equipment (CSM) and any specific in-orbit test (IOT) equipment.

The customer usually procures a satellite through the issuing of a request for proposal (RFP) to the main satellite suppliers, which defines key factors such as: the communication performance, life of the satellite, launch date and the orbital slot from which the performance criteria must be met. In addition, this defines the deliverables including documentation, the satellite testing requirement and quality-assurance programme to be followed. The proposals are submitted to the customer and, after a technical and commercial evaluation, the selection of the preferred satellite supplier is made.

Figure 13.21 shows a typical lifecycle from contract award to end of the satellite's operational life.

13.7.2 Satellite development

Once the contract between the customer and supplier has been issued, the satellite design and development phase can begin, also known as phase C/D. A preliminary design review (PDR) is usually held as soon as possible to agree the requirements for the satellite design, sweeping up any changes requested by the customer. After the PDR the satellite build can begin, high-reliability parts are procured and orders placed for any subcontracted equipment not built within the company.

As the satellite structure is manufactured, the panels and cone cylinder are assembled into a communications module (CM) and a service module (SM).

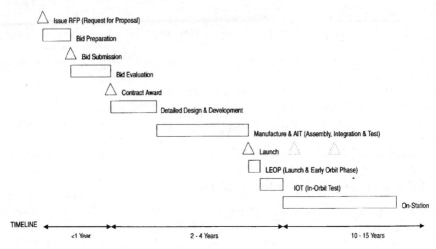

Figure 13.21   Typical satellite lifecycle

Thermal finishes are applied as appropriate, which may include fitting thermal doubler and heatpipes to the structure panel(s) underneath where the units will be attached. As platform units become available they are attached to the SM, similarly with the repeater equipments being attached to the CM. Where substantial development of units is required, an engineering qualification model (EQM) is built first which proves the design, although it may not include any redundancy or high-reliability parts to reduce costs. Once the EQM has been built a flight or prototype flight model is built and used for the flight or prototype flight satellite. The amount of new development and complexity of the payload affects the length of a programme which varies from about 24 months for a satellite requiring little new development to over 36 months where substantial development is required.

A number of other major programme reviews are held including: the satellite critical design review (CDR) where the specified requirements are confirmed by analysis and the satellite design is frozen, the test readiness review (TRR) which occurs before environment testing is begun and confirms that all the flight units are fitted to the satellite and the flight readiness review (FRR) which occurs before the satellite is launched, confirming that both the satellite and the launcher are ready for launch.

The installation of equipment onto the satellite is performed by assembly, integration and test engineers (AIT) who, working in a clean-room environment, are responsible for the build quality of the satellite. In addition to attaching the units to the structure, they help to install the harness to power the units as well as the databus to control the units. Once the SM and CM are fully populated, the two modules are integrated and environment testing of the complete satellite can begin. This involves taking the satellite to special facilities which can simulate the environmental conditions both during launch as well as in orbit.

For example, thermal vacuum cycling is performed to simulate the worst hot and cold cases predicted, after which electrical performance is tested to ensure there has been no change in performance. Other tests include: vibration tests,

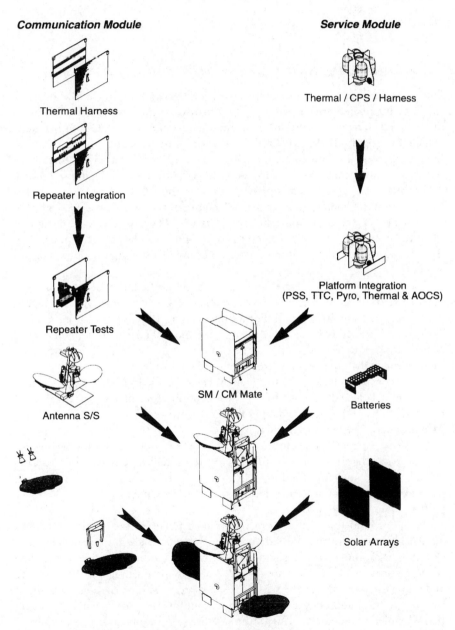

Figure 13.22   The AIT phase of a typical satellite programme

electromagnetic compatibility tests, RF performance tests and deployment tests.

Once this testing has been performed the satellite is delivered in its flight-ready state to either the customer on the ground or sent to the launch site for delivery in orbit.

13.7.3 Launch and early-orbit operations and in-orbit test

During this phase the satellite is injected into a geostationary transfer orbit by the launch vehicle with the satellite itself providing the necessary orbit-raising manoeuvres to reach geostationary orbit under ground control. Once the launch vehicle has lifted off, the satellite LEOP team must wait until the satellite has separated from the launcher and the automatically initialised routines have been run before they can be sure that the satellite has survived the launch. Automatic routines are performed to initialise the TCR subsystem so that it can transmit telemetry and receive telecommands; the other electrical sub-systems are powered so that the array panels, once deployed, can supplement the batteries, which have been the only source of power since launch, and the orientation of the satellite can be determined using the attitude-control sensors.

Once the initial safety of the satellite has been verified, the satellite's perigee must be raised from its initial altitude of about 200 km to geostationary height while removing any inclination. For a bipropellant propulsion subsystem this is performed by a number of LAE burns with the final burn parameters being chosen to place the satellite in a near geostationary orbit or drift orbit to allow the satellite to drift towards the final onstation longitude.

To support this phase, a LEOP control centre is used which is networked to a series of TTC earth stations to permit reception of telemetry/transmission of telecommands to the satellite during LEOP. Control staff work in shifts to allow 24-hour operations support during this phase.

To ensure correct operations before the satellite's operational life begins, an in-orbit test programme is used to fully test the satellite's payload and also the correct operation of the satellite subsystems. This, as a minimum, tests each transponder path through the satellite (prime and redundant) before handover to the customer is accepted.

Special ground test facilities are required to test each transponder's performance by uplinking the individual channel communication frequencies and verifying that the received signal is within the required performance. The coverage provided by the antenna is also tested by performing cuts through the antenna-gain patterns and verifying the coverage. After this phase has been completed, the equipment can be used for communication subsystem monitoring (CSM) which allows monitoring of the satellite's communications performance, ensuring that only authorised users are accessing the satellite transponders.

13.7.4 Satellite operations

Typically 24 to 28 months after the satellite contract is signed, the satellite is ready to start earning revenue for its investors. Satellite operational lifetimes of 12 years are commonplace, and this figure may include part of the life being spent in an inclined drift orbit where the satellite payload continues to function but the satellite propellant is finished, and the satellite is no longer able to correct the inclination drift which it experiences. Although satellite lifetimes of 15 and even 20 years will become common, there is a limit to the usefulness of increased lifetimes owing to the improvements in satellite payload technology as well as the impact which increased lifetimes have on the satellite's design (increased redundancy, increased radiation dose, thermal cycling etc.).

Major operations during the life of a satellite include:

- north-south and east-west station keeping where the satellite's natural drift is compensated for by using onboard propellant;
- eclipse operations where the battery must supply power during the equinoxes with each eclipse season lasting for 45 days and occurring twice per year;
- thermal operations; during the year the satellite experiences seasons as the sun position varies, requiring heaters to be switched on or off, also, throughout the satellite's life the performance of the second surface mirrors degrades leading to a general warming of the satellite;
- payload switching, owing to either switching of traffic between coverages or to a payload failure and switching over to a redundant unit.

In the past telecommands were sent by ground controllers to a satellite one at a time, sent back to the ground in telemetry for verification, before being executed on board the satellite, this process being a very time and labour-intensive operation. Nowadays the purchaser may own a fleet of satellites, so methods of reducing the operational complexity are being investigated in order to maintain the same number of ground staff while increasing the number of satellites under their control. Therefore, operations are having to become more automatic, either using expert systems on the ground or via limited autonomy on board the satellite, e.g. use of time tagged commands. The ground operator must now perform more value-added tasks, e.g. scheduling of payload traffic switching, while still actively monitoring critical operations on board the satellite and reacting to any anomalous situations.

Finally, in order to make way for a replacement satellite in the increasingly crowded geostationary arc, once a satellite has served out its useful life it is switched off and put into a graveyard orbit, at a slightly higher altitude. In order to do this, accurate calculation of the onboard propellant is required throughout the satellite's life, to permit graveyarding while there is still a sufficient quantity.

Chapter 14

Payload engineering

D.R. O'Connor

14.1 Payload definition

A communication satellite payload is the system on board the satellite which
provides the link for the communications signal path. Traditionally, this link
was between two ground stations but present-day payloads now provide not only
for this link but also can provide interconnectivity for a large number of mobile
users directly to each other or via the ground stations.

Payloads are providing this extra functionality to meet the needs of more
mobile populations, rapidly changing traffic demands and variable operational
scenarios. To do so they have become more complex and more powerful, which
has become possible through advances in technology, reducing the mass and
power consumption of the electronics equipment for a given level of functionality
or performance.

14.2 Payload function

The primary and most fundamental functions of the payload are to receive and
filter the uplink signals and provide frequency conversion and amplification of
the signals for retransmission, thus satisfying the requirements for the uplink
G/T, the downlink EIRP and the system frequency planning. In addition, the
payload must provide minimum C/N_0 degradation, variability of the high-gain
amplification, channelisation for efficient, linear amplification and processing
and high reliability in functionality. With modern payloads flexibility is
provided using digital and/or analogue processing and connectivity for routing
traffic between beams.

14.2.1 Amplification

Consider the link budget for an end-to-end satellite link. Figure 14.1 shows such a link graphically. The vertical scale is signal power level (relative to isotropic where appropriate). This particular graph is a hypothetical 30/20 GHz system, but the overall shape is the same for all satellite links. It can be seen that the high ground-station transmit EIRP is reduced by a large uplink path loss. The effective gain of the satellite receive antenna increases the effective signal level to around −100 dBW at the input to the repeater. Working through the downlink budget in a similar way shows that the power level at the output of the repeater may have to be around +20 dBW.

It is worth noting that the downlink path-loss is dependent only on frequency and orbital height, neither of which is usually a parameter that can be chosen by the payload designer. Similarly, the earth-station size and perfor-mance are usually predefined. The payload designer may be able to adjust the antenna gain to some extent, but will be limited by the maximum size which the satellite or launcher can handle. Whatever the details of the budget, these con-straints mean that the repeater will be required to provide around 110 to 130 dB of amplification. Furthermore, this amplification must produce acceptably low levels of distortion.

14.2.2 Frequency translation

The limited size of the satellite bus means that the transmitting and receiving antennas will be close to each other. In fact, in many systems these items of equipment are one and the same piece of hardware. In either instance there will be some degree of electromagnetic coupling between transmit and receive. The

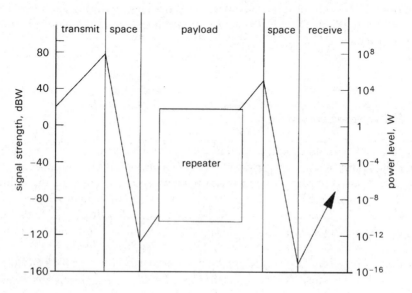

Figure 14.1   Example 30/20 GHz satellite system

best isolation that can be expected at a given frequency is about 100 dB (i.e. one part in 10^{10} of the transmit signal will leak back into the receive antenna). We have said that the repeater must have a gain well in excess of 100 dB, thus this transmit leakage will be larger than the desired uplink signal. This is clearly an unstable situation and the repeater would oscillate uncontrollably if this feedback were not prevented.

The solution adopted is to use different frequencies for transmitting and receiving. The unwanted transmitted signal can then be reduced to an acceptable level by frequency-selective filtering, leaving the desired uplink signal untouched. Thus the satellite repeater must incorporate a frequency translation somewhere in its amplification chain.

14.2.3 Channelisation and on-board processing

In fixed satellite service (FSS) and direct broadcast service (DBS) payloads channelisation is required to limit the number of signals through any one amplifier and so prevent intermodulation products from degenerating the quality of the signals. Channelisation provides a means for efficient amplification of the signals limiting the effects of the nonlinear properties of the higher power amplifiers which generate the intermodulation products. In payloads where there are multiple-beams antennas, as in mobile satellite service (MSS) payloads, the channelisation scheme allows routing of the signal through to the correct beam.

Channelisation is provided by highly selective filters. In FSS and DBS payloads these are physically large items having a demanding specification with respect to channel isolation. In multibeam operation it is advantageous to provide switching of channels to increase bandwidth to meet the needs of increased traffic patterns. Some limited onboard processing is provided in analogue signal processors as used in the INMARSAT 3 MSS payload. In this payload SAW filters are used in the processors for channelisation with the switching between channels provided by GaAs FET monolithic integrated circuits (MMICs).

Complete connectivity can only be provided by complex digital processors. These processors recover or regenerate the baseband information from the microwave signal by demodulation. The digital baseband signal is then switched, processed and routed to the appropriate output where it is modulated onto a microwave carrier at the downlink frequency. These payloads, therefore, are sometimes called regenerative compared with the conventional, transparent payloads which operate irrespective of the modulation schemes used.

14.3 Payload constraints

The payload designer must fulfil the functions outlined in Section 14.2 within the constraints imposed by the space environment and by the limitations of RF and baseband hardware. The most important of the constraints are as follows.

14.3.1 Mass

Payload mass minimisation is perhaps the most all-embracing design criterion. Any increase in payload mass may result in a significantly heavier spacecraft. This will have a direct consequence on manufacture and launch cost and hence service viability. For Ariane 5, a launch cost coefficient of around $30 000 per kg is quoted. A typical modern communications payload may weigh in excess of 300 kg with figures increasing to between 700 and 900 kg in the near future.

14.3.2 Power consumption

The satellite bus (structure) is capable of providing a certain amount of electrical power from its solar arrays (or batteries during eclipse). The payload will be allocated the greater amount of this power, usually 300–600 W. Each piece of equipment in the payload is designed to minimise its power consumption, but the most important items are the high-power amplifiers which consume the bulk of the allocation. The efficiency (RF output power divided by DC consumption) of the chosen high-power amplifier will critically determine the number of channels which a payload can provide. Also, it must be remembered that any power not emitted as RF will manifest itself as heat, complicating thermal design.

14.3.3 Thermal control

Many electrical performance parameters of a payload are temperature dependent. Thus, in order to keep performance fluctuations as small as possible, payload temperature must be controlled. The payload, usually mounted in the interior of the satellite, does not see the extremes encountered by external equipment but is nonetheless required to operate over a wide temperature range (typically 50 °C). Internally-generated heat must be carried away to prevent any undue temperature rise, as this may have a prejudicial effect on long-term reliability and a system noise-figure performance. For this reason, low-noise equipment is mounted as far as possible from major heat sources, such as high-power amplifiers.

14.3.4 Transmission requirements

Up to now it has tacitly been assumed that the required 120 dB repeater gain is applied uniformly to all parts of the required bandwidth. In fact, microwave equipment is seldom completely flat across any appreciable bandwidth. Active equipment (such as amplifiers) often intrinsically has more gain at one end of the operating bandwidth than at the other. Passive equipment (such as filters) relies on resonance or electrical path-length effects which are both frequency dependent. Further, imperfect matching between microwave elements causes interfering reflections, resulting in a gain ripple across the bandwidth. Careful design can go a long way towards mitigating these effects. Gain slopes in some equipment can be cancelled by equal and opposite slopes in others. If a significant residual slope remains, special gain-slope equalisers can be introduced. The payload specification usually defines the required level of gain flatness.

The delay time encountered by a microwave wavefront when propagating through a payload (known as group delay) would also ideally be frequency independent, and again practical equipment does not perform ideally. Payloads typically require the maximum group-delay variation to be a few nanoseconds across the required bandwidth. Group-delay variations in different parts of the payload can sometimes be cancelled out. However, a group-delay variation across bandwidth which is then followed by a nonlinear process (e.g. saturated high-power amplification) cannot be cancelled out by equal and opposite variations subsequent to the nonlinear process.

The transmission characteristics of the power amplifier itself are also important. One byproduct of nonlinear amplification is the phenomenon of AM/PM conversion whereby any amplitude modulation present at the amplifier input is converted to phase modulation at the output. The extent to which this happens is dependent on the detailed design of the power amplifier and on its input signal level. A typical payload specification will allow a few degrees of spurious phase modulation to result from one dB (peak-to-peak) amplitude modulation.

14.3.5 Noise

Noise sources in payloads are generally divided into phase-noise and thermal-noise contributions, although both types are, in fact, thermal in origin. The phenomenon of multipaction is also mentioned here as it results in a high level of broadband noise. Payload noise will directly add into the overall satellite link noise and is usually rigorously specified.

Noise-temperature calculations for thermal noise are dealt with in Chapter 11. A good payload design will have equipment with a low noise figure near the front end. This means front-end losses must be minimised. Further into the repeater (after about 30 dB of gain), the equipment noise figure has very little effect on overall link noise and design is dominated by other considerations. The first amplifier in a payload chain is termed the low-noise amplifier (LNA) and its performance will define the overall payload noise figure. It must be remembered that a satellite antenna is generally looking at the warm earth at a temperature of approximately 300 K, and so receives an appreciable amount of input thermal noise. Thus, there is little point in demanding satellite LNA noise figures below a certain point, as input noise dominates.

Phase noise is effectively unwanted phase modulation on a signal and can be viewed as spurious sidebands on a wanted carrier. An ideal local oscillator is a fixed single-frequency source, but in reality there is always some rapid fluctuation around the nominal frequency. This fluctuation will be passed, via the frequency-conversion process, into the signal path. A saturated amplifier will also increase the phase noise of a signal passing through it. The overall level of phase noise will affect the signal-to-noise ratio of the satellite link. The exact mechanism is dependent on the modulation scheme, but a rule of thumb that phase noise should contribute about one per cent of the overall effective link noise is often used. Once the overall maximum repeater phase noise level has been specified it must be apportioned between the local oscillators and any saturated amplifiers. Higher-frequency oscillators generally have a higher phase noise associated with them. A typical phase-noise specification will define maximum sideband levels, relative to the carrier, at a certain frequency displacement from the carrier. This specification is frequently the key design driver for the payload local oscillator.

One final source of payload noise that should be mentioned is noise resulting from multipaction. Unlike thermal and phase noise, multipaction is a breakdown mechanism and can be avoided entirely by careful design. It occurs in high-power microwave equipment operating in a vacuum. Free electrons present in an evacuated cavity are accelerated by an electric field. If this field is sufficiently intense, each electron will liberate more than one secondary electron when it collides with the cavity wall. A reverse in the electric field will then accelerate these secondary electrons towards the opposite side of the cavity where more electrons are liberated. If the frequency of the electric field is such that synchronism is maintained between electron transit and field reversal, then an avalanche breakdown occurs. The resulting broadband noise may render the equipment unworkable and sustained multipaction leads to permanent physical damage. Multipaction can be avoided by ensuring unwanted cavities (e.g. voids in solder connections) do not occur and necessary cavities (e.g. in microwave filters) have dimensions such that multipaction does not occur at the operating frequency.

14.3.6 Spurious signals

The ideal payload would simply amplify and translate the frequency of the uplink signal. In reality, as we have seen, there is always the addition of some amount of thermal and phase noise. As well as this broadband noise, there are also a number of mechanisms whereby discrete spurious (unwanted) signals are introduced into the transmission path by the payload.

Perhaps the most important of these is intermodulation distortion. Linear equipment may be characterised by the equation:

$$S_{out} = aS_{in} \qquad (14.1)$$

where

$$S_{out} = \text{output signal}$$
$$S_{in} = \text{input signal} = A \sin(\omega t)$$
$$a = \text{transfer characteristic (gain or loss)}$$

In reality, equipment is rarely completely linear. A more general expression, allowing for nonlinear behaviour, is:

$$S_{out} = a_1 S_{in} + a_2 S_{in}^2 + a_3 S_{in}^3 + a_4 S_{in}^4 + \cdots \quad (14.2)$$

where

$$a_n = \text{transfer coefficients}$$

It can be seen that if S_{in} is small, the first term dominates and the expression reverts to the linear case. This is a valuable general result. In small-signal cases (generally near the input of the repeater) equipment can be treated as linear, but as the signal level rises, nonlinear effects must be considered.

Intermodulation distortion results when two or more carriers pass through nonlinear equipment. In a two-carrier case:

$$S_{in} = A \sin \omega_1 t = B \sin \omega_2 t \quad (14.3)$$

where $\omega = $ angular frequency.

Substituting in eqn. 14.2 and expanding the sinusoidal terms shows that an infinite range of frequencies will be generated. These will comprise the two original frequencies, ω_1 and ω_2, plus intermodulation products of the form $n\omega_1 \pm m\omega_2$ where n and m are integers. High-order products $(m + n > 3)$ are, however, relatively small in amplitude and can often be ignored. Using the nonlinear transfer expression only as far as the third-order term yields output frequencies ω_1, ω_2, $2\omega_1$, $2\omega_2$, $3\omega_2$, $3\omega_1$, $2\omega_1 - \omega_2$, and $2\omega_2 - \omega_1$. The first two of these frequencies are the desired outputs, the next four are well outside the operating bandwidth of the equipment and can be ignored and the last two terms are the third-order intermodulation products and are usually within the operating bandwidth. Not only are these the principal spurious frequencies encountered in a satellite repeater, but their level (I_3) compared to the desired output level (C) is used as a figure of merit to define the linearity of a piece of equipment.

Unlike the noise figure, C/I_3 ratio depends on the output power of the device or equipment. As C is proportional to the wanted signal level and I_3 proportional to the cube of this, when the output power of an item rises by 1 dB the C/I_3 value drops by 2 dB. Thus by knowing the C/I_3 at any given power, the C/I_3 at any other power can be evaluated.

When a series of items are put together, the total C/I_3 is calculated using the following expression:

$$\left(\frac{C}{I_3}\right)_{Total}^{-1} + \left(\frac{C}{I_3}\right)_1^{-1} + \left(\frac{C}{I_3}\right)_2^{-1} + \left(\frac{C}{I_3}\right)_3^{-1} + \cdots$$

All C/I_3 must be expressed in linear terms.

When converting between C/I_3 in dB and linear terms, the expression:

$$C/I_3 \text{ dB} = 20 \log\{[C/I_3]\text{lin}\}$$

is used.

This is different from the usual 10 log expression since intermodulation products often add coherently.

Intermodulation products are usually associated with active equipment, but passive devices such as connectors and filters can generate them, particularly when high powers are involved. Passive intermodulation products can be caused by any nonlinear process, but oxide layers on metal-to-metal interfaces or saturated ferromagnetic materials are frequent causes.

Another major source of spurious signals in a payload is the local oscillator. The frequency-conversion process will produce the input frequencies plus or minus the local oscillator frequency. Only one of these two products is required, the other is filtered out, but some residual spurious signal usually remains. Also, the local oscillator fundamental and its harmonics may break through into the RF chain. Finally, the frequency-conversion process can be viewed as not being entirely effective, as some amount of signal may pass through unconverted and become a spurious component. The level of all these signals should be assessed by analysis during payload design and appropriate levels of filtering introduced.

14.3.7 Reliability

Until recently satellite payloads were required to operate to specification in orbit for seven-and-a-half years. A thirteen or fifteen-year lifetime requirement is, however, now the norm; future systems may have to last even longer. The reliability requirement is usually specified as a probability of survival over the full mission lifetime. A 90 per cent probability is typical for a payload, and meeting this requirement is fundamental to payload design.

The reliability of an electronic component may be characterised by its failure rate λ, often expressed in FITS (failure instances in 10^9 hours). Typical failure rates for space-qualified parts are given in Table 14.1, although actual figures vary widely. Failure rates are related to reliability by the expression:

$$R = e^{-\lambda T}$$

where R is reliability and T is time (in appropriate units).

Table 14.1   Typical failure rates for space-qualified parts

Part	FITS
Resistor	0.08
Chip capacitor	0.02
Field-effect transistor	1.00
Solder connection	0.10

Active payload equipment may contain hundreds of electrical components and the single equipment reliability over a ten-year lifetime is usually inadequate. In these instances, some form of redundancy is used.

If a single piece of equipment has a reliability R over a given time period, the chance of a failure is $(1 - R)$. If two identical redundant pieces of equipment are used instead, then the chance of both failing is $(1 - R)^2$ and the chance of at least one not failing is $1 - (1 - R)^2 = 2R - R^2$. Thus, for example, if the single equipment reliability is 0.90 then the one from two redundant reliability is 0.99. Similar expressions can be derived for other levels of redundancy.   •

In fact, the reliability performance improvement in redundant systems can be even greater than this if the unit which is not being used is turned off. The failure rates of electronic components are often considered to decrease by a factor of ten when not powered. This standby redundancy can only be used in equipment where turn on, in the event of a failure, can be rapid enough not to significantly interrupt payload operation.

14.3.8 Electromagnetic compatibility

The satellite payload is a very closely integrated system. Electromagnetic emissions from one piece of equipment may adversely affect another. Maximum emission levels are therefore specified for each piece of equipment to ensure the whole system operates to specification when integrated. There are two types of emission: radiated emission where free-space propagation can couple together physically unconnected equipment, and conducted emissions where signals travel through paths designed for other purposes (e.g. power supply, telemetry and telecomand lines). Similarly, the susceptibility of equipment to these emissions is specified.

14.3.9 Ionising radiation

The background natural radiation level in geostationary orbit is more intense than at the earth's surface. Low-earth orbit is a still harsher environment. Electronic components can be susceptible to damage by ionising radiation. Analogue semiconductor devices show a loss in gain and an increase in noise figure after sustained exposure to radiation. Digital memory chips can spontaneously introduce errors in a radiation environment. Some components are, however, much less susceptible than others. In general, the payload must be constructed using only robust components and any software used must be fault tolerant.

A radiation sensitivity analysis is performed during the payload design phase. If the aggregated radiation dose which the components will receive over the mission lifetime would cause an unacceptable performance degradation, then some localised shielding can be introduced.

14.4 Payload specification

The constraints outlined in Section 14.3 are detailed for any one satellite in a payload specification document. This may be hundreds of pages long, but a summary example is presented in Figure 14.2 for a simple four-channel Ku-band payload.

Maximum mass	55 kg
Maximum DC power consumption	500 W
Maximum thermal dissipation	400 W
Number of channels	4
Input power level (per channel)	−100 dBW
Output power level (per channel)	+14 dBW

Operating frequencies, MHz

	Input	Output
Channel 1	14000−14036	12000−12036
Channel 2	14040−14076	12040−12076
Channel 3	14080−14116	12080−12116
Channel 4	14120−14156	12120−12156

Thermal noise temperature	260 K	
Phase noise level		− 40 dBc at 100 Hz offset
		− 70 dBc at 1 kHz offset
		− 100 dBc at 10 kHz offset

Transmission requirements	
Gain variation (with life, temperature)	1.5 dB
Gain variation over any 36 MHz	0.5 dB
Group delay variation (with life, temperature)	3 ns
Group delay variation over any 36 MHz	1 ns
AM/PM conversion	5°/dB
Linearity C/I_3 with two nominal carriers	10 dB
Reliability over ten years	0.9

Figure 14.2   Example payload summaries specification

14.5 Payload configurations

We have seen the functions that the satellite payload must perform and what constraints the unique environment of space imposes on the payload designer. This Section shows how these considerations together drive the payload design.

Figure 14.3 shows the trivial case, which we will use as the starting point, where the payload comprises a single transponder which is a simple connection between the transmit and receive antennas. This does not meet the gain requirement of 110 to 130 dB.

Figure 14.4 introduces the necessary gain in one block. If we assume that this amplifier is relatively broadband it will amplify the full range as received signals—wanted and unwanted uplinks as well as the full spectrum of noise. Thus, very little of the satellite power will go into amplifying the desired signal. This is not an efficient situation.

Figure 14.5 introduces an input filter. This only permits a relatively narrow spectrum of signals to enter the amplifier, allowing it to function primarily on the wanted bandwidth. We must now consider the question of frequency translation. In Figure 14.5 we have a forward gain of, say, 120 dB and a feedback isolation of only 100 dB (see Section 14.2.2). Thus, the overall loop gain is +20 dB. This configuration would clearly oscillate.

Figure 14.6 introduces a frequency translation. The required 120 dB of gain is now provided for signals received at f_u and transmitted at f_d. Transmitter leakage at f_d may enter the receive antenna but will be rejected by the input filter which is tuned to pass only signals at frequency f_u.

Figure 14.3 *Single transponder providing connection between transmit and receive antennas*

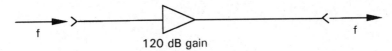

Figure 14.4 *Gain introduced in one block*

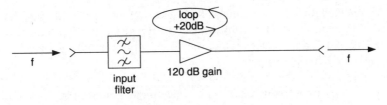

Figure 14.5 *Introduction of input filter*

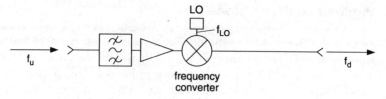

Figure 14.6   Introduction of frequency translation

The frequency-translation effect is implemented by feeding a microwave mixer with the uplink frequency and an onboard generated CW signal (at frequency f_{lo}) produced by a local oscillator unit.

Thus, the configuration of Figure 14.6 provides the basic transponder functions of amplification and frequency translation. The next refinement to be considered is the post-frequency translation filtering requirement. The output on the mixer will contain the products $f_u + f_{lo}$ and $f_u - f_{lo}$. Only one of these signals is wanted (usually $f_u - f_{lo}$, as downconversion is more common on satellites than upconversion); the other must be filtered out. Thus, a new filter must be introduced after the mixer. At this point in the transponder there has already been an appreciable amount of signal gain and hence a small amount of loss is not noise critical. This new filter can therefore have a significant loss without affecting the overall transponder noise figure. Because this constraint is relaxed, the post-mixer filter can be used to provide other filtering requirements as well as rejection of unwanted mixing products. Group-delay equalisation and the rejection of breakthrough frequencies are often also performed at this point in the transponder.

There is one more important filtering operation to be performed—that of rejection of the image frequency.

We have said that the uplink, downlink and local-oscillator frequencies are related, in a downconverting transponder, by the expression:

$$f_d = f_u - f_{lo}$$

but there also exists a frequency f_i such that:

$$f_i = f_u + f_{lo}$$

Signals which enter the transponder at frequency f_i (spurious signals and noise) will be translated in the mixer into the products $f_i + f_{lo}$ and $f_i - f_{lo}$. The latter can be rejected by the post-mixer filter but the former is in the required passband at frequency f_d. Thus, only one of the image frequency products can be rejected after the mixer. Rejection of the other must be performed by the input filter rejecting f_i in the first place.

Figure 14.7 shows the transponder, now with two distinct filters. The first is the input filter with a passband around f_u rejecting the image frequency and protecting the input amplifiers, and the second is the post-mixer filter with a

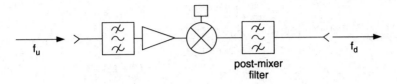

post-mixer
filter

Figure 14.7   Transponder with two distinct filters

passband around f_d, rejecting the unwanted mixing products and also perhaps performing group-delay equalisation.

At this point it is necessary to consider the distribution of the gain in the transponder. Signals passing through the mixer and the succeeding filter will suffer a loss of about 10 dB. Thus, the amplifier must provide at least 10 dB more power than is necessary for the downlink. Clearly, it would be sensible to have gain after the mixer where all the power available could be transmitted. The entire gain requirement cannot, however, be moved to after the mixer as an appreciable amount of gain is necessary prior to any losses in order to minimise transponder noise.

Figure 14.8 shows a solution whereby the gain is split into two segments, each side of the mixer. Details of the gain distribution between the two segments vary greatly from satellite to satellite. However, since gain at high frequency is generally more difficult to achieve than at a lower frequency, the bulk of the amplification is usually performed after the mixer $(f_u > f_d)$ with only about 30–40 dB before the mixer in a low-noise amplifier (LNA).

Looking now in detail at the post-mixer gain it can be seen that about 80 dB is necessary. This is a very wide power range and is best handled by two distinct amplifiers. These are usually termed the channel amplifier and high-power amplifier (HPA).

Figure 14.9 shows this split. The high-power amplifier must provide the required transponder output power and usually has a gain of around 40 dB. The overriding importance of efficiency in the HPA means that the other functions cannot usually be implemented in this equipment. The channel amplifier, however, operates at a much lower power level and can afford to have a poorer efficiency. Similarly, the channel amplifier does not require a low noise figure as the transponder noise performance is determined by the LNA. Thus, this equipment is often used to provide other functions, such as gain control.

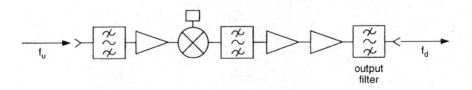

output
filter

Figure 14.8   Gain is split into two segments, each side of mixer

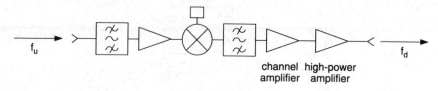

Figure 14.9   Channel amplifier and high-power amplifier split

HPAs operate at maximum efficiency near saturation where their linearity is poor. Most transponders operate in a nonlinear mode to some extent to improve efficiency and this results in spectral spreading (i.e. the output signal bandwidth is greater than the input signal bandwidth). This can cause cochannel interference if some form of filtering after nonlinear amplification is not introduced.

Figure 14.10 shows this output filter. This payload configuration now performs all the required major tasks and would provide a single-carrier satellite link.

In practice, the payload will need to operate on many carriers. Intermodulation distortion often prevents more than one carrier being amplified in single high-power equipment. Low-power equipment, however, can operate effectively in multicarrier mode. Thus, at some point in the payload, channelisation is required. This splitting is achieved by separating carriers by frequency onto different paths, a bank of filters known as a demultiplexer. The demultiplexer also defines the exact channel bandwidths; individual demultiplexer filters are often known as channel filters. Each channel is then amplified separately prior to recombination after the HPA in another bank of filters known as the output multiplexer. The output multiplexer filters often also perform the filtering functions of the output filter. Figure 14.11 shows such a multichannel payload.

The reliability of the system must now be considered. Each piece of equipment in Figure 14.11 is a single point failure, as any malfunction would prevent the entire transponder channel from operating. This is not a satisfactory situation for a system which is required to operate for many years without maintenance. Reliability is improved by introducing redundancy.

Figure 14.12 shows the system with standby units introduced. The input filter and demultiplexer are reliable units and are not redundant.

The LNA is one from two redundant. The input redundancy connection is achieved with a switch, as low loss is important. The output connection is

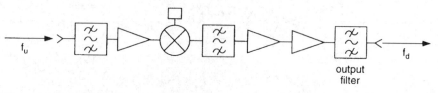

Figure 14.10   Introduction of output filter

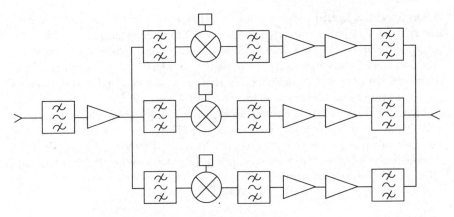

Figure 14.11   Multichannel payload

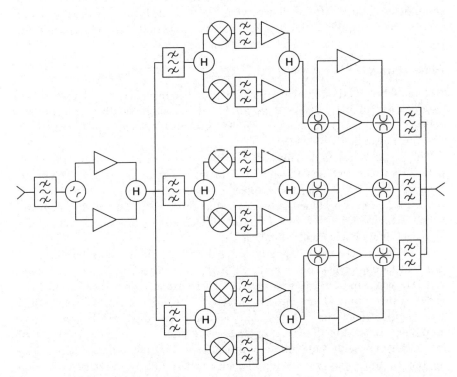

Figure 14.12   System with standby units

achieved with a hybrid power combiner because the resulting 3 dB loss here is not critical. The downconverter, post-mixer filter and channel amplifier are again one from two redundant with hybrid splitters and combiners at each end. The local oscillators may have their own redundancy. If the local-oscillator frequency is the same for each channel, reliability can be further improved by allowing interconnection between local oscillators on different channels.

The HPA is frequently the single most unreliable item of payload equipment and redundancy is essential. Unfortunately, HPAs are also very heavy and numbers must be minimised. Figure 14.12 shows a three from five redundancy scheme with switches at input and output. (Note that any two HPA failures still allow three channels to operate.)

14.6 Typical configurations

Figure 14.12 shows a generalised satellite transponder configuration, suitable for many applications. There are, of course, many variations on this design suitable for different applications. For example, the positions of the downconverter, demultiplexer and channel amplifier are often interchanged. In particular, because the frequency conversion has a large overhead in terms of equipment requirements, the downconverter is usually located before the demultiplexer. Some specific variations on Figure 14.12 are outlined in this Section.

14.6.1 FSS and DBS payloads

FSS and DBS functionalities are typically provided by conventional, transparent payloads and in many cases will be combined on a single satellite with common (shared) antennas. An example is the KOREASAT payload which is shown in simplified form in Figure 14.13.

This payload has an antenna beamforming network which discriminates, by polarisation, the DBS and FSS signals. The input signals are fed from the beamforming network into separate receivers for each dedicated FSS and DBS repeater. The receivers for the two repeaters are similar, having an integrated low-noise amplifier, downconverter and local oscillator in each receiver chain, with each chain being dual redundant.

From the receivers the signals are fed to the input multiplexers which provide the channelisation. For the DBS repeater there are six channels each 27 MHz wide and for the FSS repeater there are 12 channels each 36 MHz wide. At the outputs of the input multiplexers are input switch networks which provide the redundancy for the channel amplifiers and travelling-wave-tube amplifiers (which are the HPAs in this system). The requirement for the DBS repeater is to provide a maximum of three channels at any time, so there is a six for three redundancy of the amplifier chain. The FSS repeater has a flexible redundancy arrangement where 16 amplifier chains are utilised to provide the

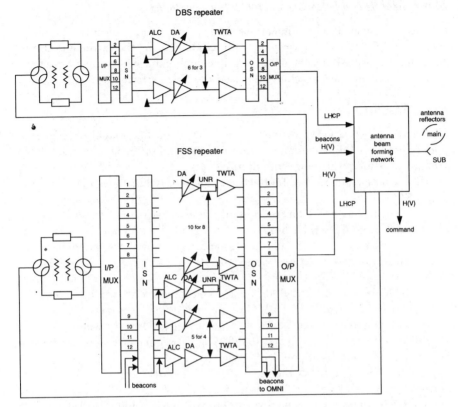

Figure 14.13   FSS and DBS payload

total requirements for 12 operating. The output switch network provides the
necessary routing to the output multiplexer which recombines the channels onto
a single signal path to feed back to the beamforming network.

Typical characteristics of the payload are as follows:

mass = 13.6 kg
power consumption = 1205 W
frequency bands = 14.2 GHz receive/12.5 GHz transmit (FSS)
= 14.5 GHz receive/11.6 GHz transmit (DBS)
TWTA RF power = 14 W (FSS)
= 120 W (DBS)
EIRP = 51.5 dBW (FSS)
= 61 dBW (DBS)
G/T = 16 dB/K (FSS)
= 15 dB/K (DBS)
design lifetime = 12 years

14.6.2 Mobile and personal communications payloads

Mobile and personal communications systems are characterised by the use of very low G/T and EIRP terminals, with handsets being the extreme example. This, in turn, necessitates the use of multiple, high-gain, spot-beam coverage on the mobile link in order to achieve the necessary link performance.

INMARSAT 3

INMARSAT 3 is a GEO regional system based on four satellites in geostationary orbit providing communication for sea, land or air mobile users. Each satellite is of identical design providing a total of seven spot beams and one global beam per satellite. A simplified view of the satellite payload is shown in Figure 14.14.

The payload has two information paths. The forward path receives the C-band signals from the service provider's fixed ground station and retransmits to the mobile user over an L-band link. L-band signals from the mobile user are converted to C-band and retransmitted to the service provider by the return path.

To double the uplink bandwidth, the C-band signals to the satellite are circularly polarised in two directions. The antenna system separates the left and right-hand components and passes the signals to the receiver where they are amplified and downconverted to L-band. The C-band receiver has two for one redundancy for each of the polarisation signals.

From the C-band receiver the signals are fed to an IF processor which downconverts the L-band signal to 160 MHz where SAW filters split out 40 discrete channels. The outputs from the SAW filters are converted back to L-band and fed to a cross bar switch which routes each channel to the selected antenna beam. The processor has a modular construction and contains a high degree of redundancy.

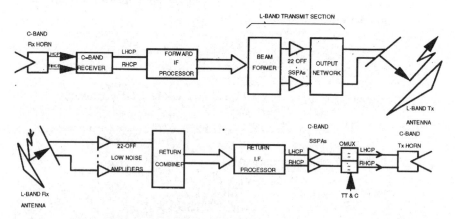

Figure 14.14   INMARSAT 3 payload

The antenna beams are formed by a passive beamforming network which takes the eight channelled beams (seven spot beams and one global beam) from the one processor and creates the drive signals for the 22 antenna feed elements. Each element drive signal is amplified by solid-state power amplifiers operating in a three for two redundancy, before being routed through an output network to the antenna feed elements. This output network allows all 22 amplifiers that are operating at any one time to equally share the total RF power to the antenna.

The return link from the mobile user comprises a similar 22 element-fed antenna operating at L-band with each element having an associated dual-redundant low-noise amplifier. A passive combiner takes the 22 separate signals and reconstructs the return beams. The return IF processor reconstructs the return signal to be transmitted to the service provider via dual-redundant upconverter/power amplifiers and a dual polarised C-band horn.

Future systems

New systems are being designed and developed to provide for a large number of mobile handset users. These systems may typically have 50 to 300 beams. They may also provide a total global coverage via a number of satellites in the lower ICO (intermediate circular orbit) or LEO (low-earth orbit).

To meet the requirements of these new systems, miniaturisation of all equipment and advanced digital signal-processing techniques will need to be utilised. These have only become feasible recently owing to the availability of chip and wire technology, GaAs MMICs, high-integration, radiation-hard, application-specific integrated circuits (ASICs) and advanced software algorithms to give sufficiently low mass and power consumption.

There are two approaches to the design of digital onboard processors determined by the bandwidth of the mobile link beamforming: the narrowband approach uses digital beamforming to route capacity by beamsteering on a channel-by-channel basis and the wideband approach uses spatial switching to route capacity to beam parts and beamforming is performed by analogue means across the full bandwidth of each beam.

The trade off between narrowband and wideband processors depends strongly on the mission requirements. For the narrowband case, each mobile link chain operates on the full mobile link bandwidth, irrespective of the beam bandwidth. In the wideband case, however, each mobile chain can potentially be minimised according to individual beam bandwidth. The choice, therefore, depends on the number of elements, number of beams, the beam bandwidth and the mobile bandwidth.

This can be explained further by considering in detail the two approaches. Figure 14.15 shows the generic architecture associated with the narrowband digital processor of the forward link of a mobile communications payload. The incoming signal is firstly amplified and downconverted with antialiasing filtering (AAF) to near baseband for A/D conversion. Digital demultiplexing of

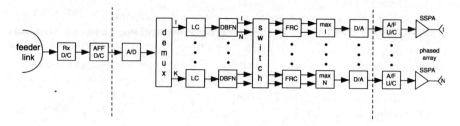

Figure 14.15 *Payload showing generic narrowband processor architecture*

the feeder link band then divides the wideband signal into K separate signals corresponding to granularity blocks which are the smallest units of capacity routable between beams. These are characterised by a bandwidth containing single or multiple carriers each of which may typically carry a small number of time-multiplexed user channels.

The signals from the digital demultiplexer are in digitally sampled form, sampled at a rate consistent with the individual granularity block bandwidth. These may then have independent level control (LC) applied as a simple scaling of the digital samples.

Each feeder link block signal has a separate digital beamforming network (DBFN) function, the purpose of which is to determine the mobile link beam properties for that specific block. The block signal is split N ways (where N is the number of elements for the phase array) and a complex weight multiplies samples for each of the element paths to provide amplitude and phase control. To create a spot beam in a given direction for a given block, the weights would be such as to create the required linear phase gradient across the phased array. Thus, the beamforming is implemented at the decimated sample-rate character-istic of the narrow bandwidth of the individual block, hence the term narrowband beamforming.

The switch provides flexible mapping to the mobile link frequency slots. The same phase array is used to provide all the beams and therefore signals must be combined for each antenna element. For blocks that are mapped to the same frequency slot on the mobile link, samples are combined by simple addition in a frequency reuse concentrator (FRC). Samples for all mobile frequency slots are frequency multiplexed in a digital multiplexer. For a given phase-array element multiplexer the number of the inputs is the number of slots in the mobile link (not all necessarily occupied) and is less than K because of the frequency reuse. The output is a sampled form if the full element signal is at a rate consistent with the full mobile bandwidth.

Each element signal is D/A converted, antiimage filtered (AIF) and is upcon-verted to RF and amplified prior to going to the phase-array antenna element.

The return link essentially involves the inverse functions. Within the return digital beamforming function there is the requirement to add samples for a given block following complex weighting. If direct mobile to mobile channels are required, a linkage is provided from the return link after receive beamforming into the forward link path prior to transmit beamforming.

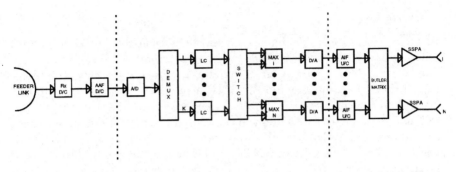

Figure 14.16   Payload showing generic wideband processor architecture

Figure 14.16 shows the generic architecture associated with the wideband digital processor of the forward link of a mobile communications payload. In this case the beamforming is implemented using analogue techniques. This offers the potential to significantly decrease the complexity of the digital processing relative to the narrowband approach if it is acceptable to limit the occupied contiguous bandwidth of a given beam to be a relatively small fraction of the overall mobile bandwidth. Such a limitation may be acceptable within certain practical systems where the upper limit on beam traffic can be defined.

The signal flow is the same as for the narrowband case up to and including the digital demultiplexing and level control of the feeder link blocks. Beam routing in this case is performed by explicit spatial switching of blocks to beams. Flexible frequency mapping is included within the switch function using RAMs to map between input and output time frames in the same way as for the narrowband architecture.

Digital multiplexing is on a beam-port basis. The dimension of the beam digital multiplexer is determined by the maximum contiguous bandwidth of the beam and may be significantly lower than the dimension of the element-based digital multiplexer of the narrowband approach which must cover the full mobile bandwidth.

Beam-port signals are D/A converted and upconverted to RF: upconversions will, in general, be different between beams to place the beam band at the required location within the overall mobile linkband.

Beamforming is performed using analogue technology using, for example, a Butler matrix which can efficiently generate the necessary phase gradients to generate a uniform grid of beams.

14.7 Payload equipment

The functions of each item of payload equipment have been described in Sections 14.5 and 14.6. In this Section some of the current technologies are described that are used to implement this equipment.

14.7.1 Antennas

Antenna systems for communication satellites generally produce shaped radiation patterns to cover a single or group of countries from geostationary orbit. This can be achieved by direct radiating arrays of horns, helices or microstrip patches, where the excitation of the elements can be synthesised to produce the desired patterns, or more typically by a reflector antenna between 500 and 3000 mm in diameter fed from one or more feed horns. The radiation pattern may either be formed by the position and excitation of a group of up to 200 feeds reflecting off a 2–3 m reflector (INTELSAT 7 and 8 series), or by distorting the traditional offset parabolic reflector to form the pattern shaping (or even a combination of both techniques) e.g. (EUTELSAT/HOT BIRD/ASTRA/ST1).

Payloads for communication satellites have to be able to transmit and receive signals over a wide frequency range (typically 10.7 to 18 GHz for a Ku-band system, for example) and often using the same antenna for transmit and receive functions. In some applications, where complex multimatrix antenna designs are employed, Tx/Rx isolation is very important and separate transmit and receive antennas are implemented on the payload. An example of this is the INMARSAT 3 payload which is shown in Figure 14.17. Operating at L-band it comprises two separate reflectors about 2.4 m in diameter. The 22-element feed array is shown in Figure 14.18.

Polarisation of the antennas can be circular (LHCP or RHCP) for C- and X-band or linear (vertical or horizontal) for Ku-band. A typical Ku-band payload may also require operation in dual polarisation, where dual-shell gridded reflectors are employed in prime focus systems with separate feeds for the vertical and horizontal polarisations. The reflector shells share a common geometry with one reflector radiating through the surface of the other producing similar radiation patterns for the satellite coverage. This is possible because the two shells are polarised by incorporating orthogonally-orientated reflecting strips on the surface of essentially transparent reflectors (aluminium strips on Kevlar shells). These reflector systems offer extremely high crosspolar discrimination (33–37 dB) in the radiation patterns, and can produce shaped ground-coverage patterns by profile-shaping reflector surfaces. The reflectors are fed from wideband low crosspolarised corrugated feeds working in the HE_{11} mode. Isolation between feed horns can reach -70 dB depending on the reflector geometry.

Alternative designs of reflector antenna use Gregorian or Cassegrain geometry, where the feed-horn energy is reflected from a subreflector onto the main reflector to form the radiation pattern. A single corrugated feed is used with an orthomode transducer (OMT) to provide inputs for dual-polarised operation. Isolation at the inputs is limited by the OMT performance to around -50 dB. Both the reflectors in the design can be surface-profile shaped to optimise the radiation pattern for a given ground coverage. However, the theoretical crosspolar and isolation performance is not as good as a dual-gridded reflector system (-34 dB), but offers a compact, lightweight solution for a

Figure 14.17   INMARSAT 3 payload showing the antenna system in the anachoic chamber during antenna testing

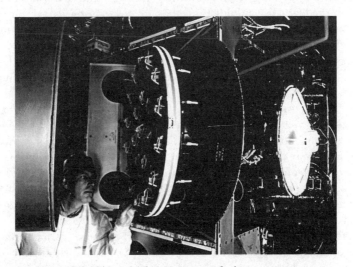

Figure 14.18   INMARSAT 3 multielement antenna feed array

typical Ku-band antenna. Reflectors can be formed from carbon fibre/aluminium honeycomb sandwiches or exotic combinations of Kevlar, Kapton and carbon-carbon matrices to allow good thermal performance and in-orbit surface accuracy.

Larger multibeam systems are now being developed based on multimatrix and digital beamforming techniques. These systems need reflectors to 15 m in diameter, demanding deployment or unfurling in orbit, and can have feed systems of ~100 horns or helices producing ~250 shaped spots over the surface of the earth.

14.7.2 Input filter

At most current satellite frequencies the input filter consists of a microwave cavity filter, although lumped element circuits can be used at lower frequencies. The rejection requirement usually necessitates a series of cascaded cavities. This, however, increases loss, and one solution is to excite a number of resonant modes within a single cavity. These double- and triple-mode filters show reduced size and mass as well as loss for a given rejection requirement. The most advanced current designs mix dual- and triple-mode operation in a single filter.

Construction is usually of plated aluminium or the alloy invar. Invar has a very stable thermal characteristic but has the drawback of being a high-density material.

In payloads which have a single shared transmit and receive antenna, isolation by frequency is achieved in a two-channel multiplexer (diplexer). The input filter then forms one part of this diplexer.

14.7.3 Receivers

Input filters can be integrated with other functions, and it is usual to combine them with the low-noise amplifier, local oscillator and downconverter functions into a single piece of equipment called the receiver. The receiver will utilise the latest high-electron mobility transistor (HEMT) technology in order to achieve the lowest possible noise figures. Since noise figure is very critical, discrete devices are used, often in bare chip form, rather than GaAs monolithic microwave integrated circuits (MMICs) which tend to sacrifice optimum performance during the design process in order to ensure the maximum manufacturing yield. MMICs can be used as gain stages after the downconversion where their noise figure does not affect the overall receiver performance. The circuits are realised in substrates made from alumina. At frequencies above 20 GHz quartz can be used as its lower dielectric constant means that manufacturing tolerances are not so tight. Conversely, at low frequencies (<1 GHz), a high dielectric material such as barium tetratitanate may be used to reduce circuit size. Figure 14.19 shows an LNA module at S-band.

The primary element of the downconverter is the mixer, which consists of a number of diodes or transistors that are switched rapidly on and off by the local oscillator signal. There are a number of configurations in which these diodes can

Figure 14.19   S-band LNA module

be arranged. The double-balanced approach elegantly produces some natural cancellation of unwanted mixing products. Filtering needs to be included in the downconverter to remove the unwanted sidebands and local-oscillator signal components. An example of a downconverter module is given in Figure 14.20.

Figure 14.20   Ku-band downconverter module

The local oscillator is a microwave frequency generator. The frequency accuracy and low phase-noise requirements are provided by a low-frequency reference oscillator which then has its output frequency (typically 5–50 MHz) increased to the required value.

The reference is usually a quartz crystal oscillator, although rubidium and caesium clocks are sometimes required for specialised applications. The output frequency of the reference is invariably temperature dependent and this can result in unacceptably large frequency drifts. For this reason, the reference is often placed in a temperature-compensated oven to maintain it at a constant, elevated temperature. Because this often takes some time to reach operating temperatures, a redundant local oscillator cannot be turned off when not in use, unlike other equipment.

The reference frequency can be increased to the required local-oscillator frequency using transistor multipliers and filters. An alternative approach is to employ a microwave voltage-controlled oscillator, the output of which is phase-compared with the reference. This phase-locked loop approach can offer significant mass and power savings over the frequency multiplier technique. An example of a phase-locked local oscillator is shown in Figure 14.21.

In MSS payloads the requirements are for a large number of local oscillators to satisfy the various frequency conversions within the receivers, upconverters

Figure 14.21   Local oscillator module for a Ku-band receiver

and processors. In this case a single frequency generator is used in which all the local oscillators are integrated into a single piece of equipment and use a single pair of redundant reference oscillators and a single pair of redundant DC/DC converters.

An example is given in Figure 14.22 which shows a frequency generator of the type used in the INMARSAT 3 and ARTEMIS payloads. A single local-oscillator module is shown in Figure 14.23 and in this hybrid and surface-mount technology is used to minimise mass and connectivity requirements.

Figure 14.22   Integrated frequency generator equipment

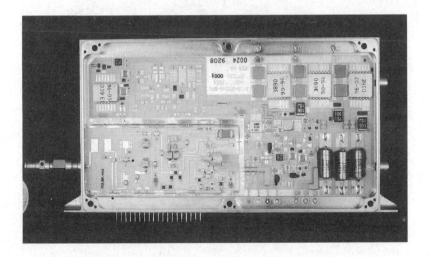

Figure 14.23   Single local-oscillator module used in a frequency generator

14.7.4 Channelisation filters

In Ku-band payloads for FSS and DBS services or in X-band payloads for military services, the channelisation is achieved by demultiplexers or input multiplexers. These consist of a number of channel filters fed, via some form of splitting network, from a common input port. The splitting network can be a manifold or, as loss is not critical at this point in the payload, a circulator tree arrangement where the input is presented to each channel in turn. The channel filters comprise microwave cavity resonators. Latest input multiplexers use a dielectric material to increase the Q and hence allow a reduction in size of the cavity. An input multiplexer built using this technique is shown in Figure 14.24.

The output multiplexer cannot use the dielectric-resonator technology because loss is more critical at that part of the payload owing to the high power levels. These power levels can also cause thermal problems and risk of multipaction.

The cavities of the output multiplexer are made from invar to provide stability over the temperature range. Dielectric inserts can be introduced across gaps if the design is considered to be vulnerable to multipaction. Other techniques can also be used such as partial pressurisation of the cavity or application of a constant DC electric field to prevent electron oscillation. An example of an output multiplexer is given in Figure 14.25 as used on the HOT BIRD payload.

In MSS payloads the channelisation occurs at lower frequencies allowing the use of surface-acoustic-wave (SAW) filters. These filters are integrated with the microwave/RF functions, as in the processors for INMARSAT 3, using fairly

Figure 14.24   Dielectric resonator input multiplexer

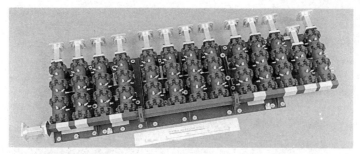

Figure 14.25   Output multiplexer

conventional microwave-integrated-circuit (MIC) or surface-mount technology (as described in Section 14.7.3).

14.7.5 Channel amplifiers

Generally, the channel amplifier uses the same technology and fabrication techniques as the receiver. As noise figure is not critical and the quantity of equipment is much greater, the channel amplifiers extensively utilise MMICs to achieve a lower-mass and lower-cost product.

Many payloads require the full saturated output power to be available for a range of input powers. This can be because of different types of ground terminals accessing the satellite, or for varying propagation conditions. The variable gain requirement in FSS, DBS and military payloads is met by the channel amplifier. Also, temperature and lifetime-induced gain changes can be compensated for in the channel amplifier. Gain alignment is provided by PIN diode attenuators which can be realised in MMIC form. The amplifiers and attenuators are biased and controlled by ASICs, also in chip form. An example is given in Figure 14.26.

14.7.6 High-power amplifier (HPA)

For conventional FSS, DBS and military payloads where single-beam antennas are used, the only efficient means of providing the high power required is by a

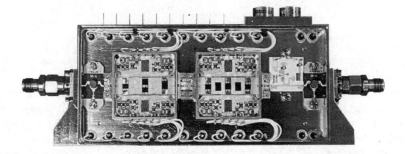

Figure 14.26   Ku-band channel amplifier

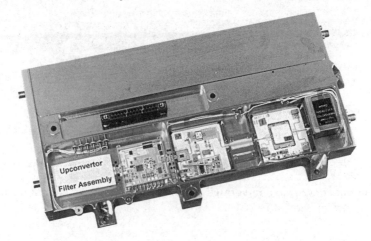

Figure 14.27   S-band dual-channel SSPA

travelling-wave-tube amplifier (TWTA). These amplifiers use a modulated electron beam to amplify the microwave signal to achieve the high power and wide bandwidth required at reasonable efficiencies. The TWTA is a relatively unreliable piece of equipment and the high-voltage supplies needed make it large and heavy.

As more payloads utilise multibeam antennas, the amount of equipment increases and lower powers are required. This favours the adoption of solid-state power amplifiers (SSPAs) which use bipolar transistors and FETs (depending on their frequency of operation) using technology similar to the other active equipment described previously. An example of a dual-channel SSPA is given in Figure 14.27. The large discrete-power device can be seen at the output driving an isolator which protects the device from poor, undefined terminations of the equipment. These units are much lighter than the TWTAs and are far more reliable. They are more efficient for the lower output powers and offer an improved linear performance at these powers.

14.7.7 Onboard processors

Onboard processors for satellite payloads can be realised in analogue or digital technology. INMARSAT 3 uses analogue processors with SAW filters for channelisation and MMIC switches for the connectivity. As systems become more demanding with respect to the amount of traffic handled and increased beam connectivity, digital technology needs to be used. This functionality is realised using submicron, radiation-hard ASICs which are mounted in multichip modules (MCMs) as shown in Figure 14.28. MCMs comprise several ASICs each resulting in a complex, highly integrated packaging approach enabling significant reductions in size and mass.

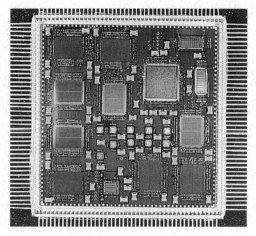

Figure 14.28   Multichip module for digital onboard processor

14.8 Future systems

As miniaturisation progresses, more complex subsystems can be envisaged in place of the many individual items of equipment in conventional payloads. This is depicted in Figure 14.29, which shows a future payload comprising complex antenna systems and a large, onboard digital processor.

The antenna systems are highly integrated arrays with a large number of LNAs and SSPAs embedded in the directly radiating antenna. The LNAs and SSPAs will have integral downconverters and upconverters, respectively, to provide the interface to the complex switching networks which connect to the digital processor.

The digital processor is reprogrammable in flight providing total flexibility in the utilisation of the payload to meet any variation in the traffic demands or beam patterns required.

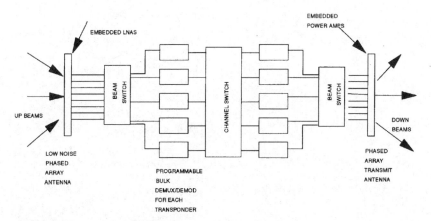

Figure 14.29   Future payload concept

14.9 Acknowledgments

The author would like to acknowledge Mr. D. J. Flint who was the author of the original chapter published in previous editions as well as Matra Marconi Space for making available and permitting information and photographs to be used in this text.

Chapter 15

Earth-station and satellite antennas

B. Claydon

15.1 Introduction

The dramatic increase in satellite communications capacity, capability and flexibility over the last two decades has been made possible by improvements in techniques and in the design of components both in the space and ground segments. A key factor in achieving these improvements has been advances made in antenna technology. The antenna provides the vital link between the ground and the satellite, performing such complex operations as:

- the simultaneous reception and transmission of communication signals;
- the rejection of interference from neighbouring systems, both space and terrestrial;
- the maintaining of accurate pointing between earth station and satellite.

As overall system requirements become more stringent and natural resources such as the geostationary orbit and frequency spectrum more saturated, the antenna rapidly becomes a critical subsystem demanding more sophistication in design and the development of new synthesis techniques. Specifications on both terrestrial and satellite antenna performance, especially as regards copolar isolation, crosspolar discrimination and sidelobe radiation characteristics, reflect these advances and provide a significant challenge to designers.

In this Chapter both earth-station and satellite antenna technology is reviewed. Obviously, it is impossible to cover every aspect of such wide subject matter and the reader is referred to authoritative texts such as References 1 and 2.

15.2 Earth-station antennas

Advances in satellite design have led to a reduction in the size of earth-station antennas as used in the fixed satellite service to provide telephony, television uplinking and data dissemination. The original INTELSAT Standard-A antenna of 30 m diameter has been replaced by an 18 m requirement and very small aperture terminals (VSATs) with diameters as small as 1 m have become prevalent. High-power broadcasting satellites such as ASTRA and EUTELSAT HOT BIRD providing direct-to-home television have led to a proliferation of very small receive-only antennas in the range of 40 to 80 cm diameter. Most earth-station antennas in the fixed satellite service employ reflector configurations. However, as the antenna size reduces, the interest moves towards flat-plate antennas initially with fixed beams but, especially for future nongeostationary satellites, with electronic scan capability.

In the mobile maritime, land and aeronautical communications field, the use of the lower frequency L-band demands the implementation of printed and patch antenna designs which are either portable or can be located conveniently on the structure of the mobile vehicle. Within the next few years, mobile satellite communications will be extended to handheld telephones with simple small antennas produced in large volumes.

The proposed use of Ka-band satellites, such as SPACEWAY and Teledesic, for interactive multimedia applications will herald a new era in earth-station antenna technology with uplinking capabilities for domestic systems.

15.2.1 Typical RF performance specifications

Antennas for earth-station applications in the fixed satellite service are required to radiate narrow beams with high gain in both the receive and transmit frequency bands (mostly C- and Ku-band). For most systems, reflector configurations are employed in which special attention is given to achieving high antenna efficiency while minimising potential interference characteristics associated with off-axis copolar sidelobe and crosspolar radiation.

The important factor in determining the efficiency of geostationary orbit utilisation is the off-axis radiation pattern associated with the earth-station antenna, particularly in the angular region of $0-50°$ from boresight. Based on antenna data available in the 1960s, the CCIR adopted a reference radiation pattern for use in large antenna interference calculations of the form:

$$E(\theta) = 32 - 25 \log \theta \text{ dBi} \tag{15.1}$$

where θ is the angle in degrees from the boresight axis.

This results in a minimum satellite spacing of $2.9°$,[3] and was adopted by space-segment operators such as INTELSAT as a mandatory specification in the earth-station uplink band. With the growth in the number of satellites populating the geostationary arc, especially in prime longitudinal locations, the minimum satellite spacing has been reduced to $2°$ and the transmit sidelobe specification tightened by 3 dB. The current transmit sidelobe specification adopted by INTELSAT, based on ITU Recommendation S.580.5, is:

$$
\begin{aligned}
29 - 25 \log \theta \text{ dBi}, & \qquad 100\lambda/D \leq \theta \leq 20° \\
-3.5 \text{ dBi}, & \qquad 20° < \theta \leq 26.3° \\
32 - 25 \log \theta \text{ dBi}, & \qquad 26.3° < \theta \leq 48° \\
-10 \text{ dBi}, & \qquad \theta > 48°
\end{aligned}
\tag{15.2}
$$

where D/λ is the antenna diameter in wavelengths.

This specification should be met for 90 per cent of all sidelobe peaks within 3° of the geostationary arc. When D is greater than 100λ the minimum value of θ is taken as 1°. It should be noted that the only concession given to small antennas, such as used for VSATs or satellite newsgathering terminals, which have a wider beamwidth and much slower fall-off of sidelobe level with absolute angle compared to larger antennas, is the initial angle at which the specification

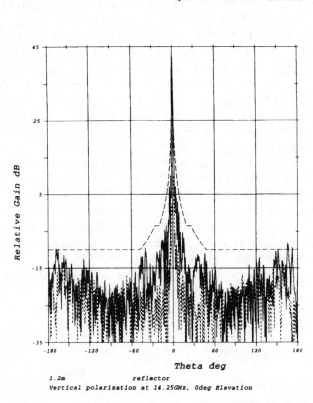

Figure 15.1   Radiation pattern of a 1.2 m VSAT antenna at 14.25 GHz compared to INTELSAT specification

commences. Hence, a critical parameter in the design of small antennas is the sidelobe radiation pattern. This has led to the use of the offset reflector config-uration and, in some cases, modification of the normal circular shape of the reflector periphery to reduce off-axis radiation in the plane corresponding to the geostationary arc. As an example, Figure 15.1 shows the measured radiation pattern associated with a 1.2 m VSAT antenna indicating full compliance with the INTELSAT specification.

At present, specifications on earth-station antenna sidelobes are only mandatory in the transmit mode since this can cause interference to other systems. In the receive mode, the envelope is only a recommendation to afford protection from unwanted signals.

Current INTELSAT and EUTELSAT satellites, along with other domestic and regional systems, reuse frequency bands on orthogonal hands of polarisation in order to increase the channel capacity of the system. A fundamental require-ment of such a system is to maintain a tolerable level of cochannel interference between signals on the two hands of polarisation. This is termed crosspolar inter-ference (XPI) and both earth and space-segment antennas contribute. Dual-polarised satellite links are designed to operate with an XPI of better than 27 dB under clear-sky conditions,[4] and a mandatory polarisation purity standard has been set by INTELSAT for new C-band earth-station antennas, equivalent to an XPI of 30 dB. The INTELSAT specification, with circular polarisation, is for an earth-station antenna voltage axial ratio of 1.06 in the direction of the satellite. This translates into a requirement of 30.7 dB crosspolar discrimination (XPD) within the tracking sensitivity of the antenna. In the case of EUTELSAT, which employs orthogonal linear polarisations at Ku-band, the polarisation discrimination of the earth-station antenna is required to be at least 35 dB within a cone defined by the antenna pointing angle, generally accepted to be the −1 dB copolar beamwidth.

Although requirements for polarisation discrimination involve improvements in all aspects of the antenna design, the impact is greatest on the feed-system complexity. Additional components are required for the separation of the ortho-gonal channels and more stringent specifications are placed on existing components such as the feed horn and polariser. Polarisation tracking facilities may also be provided at an earth station to overcome depolarisation effects caused by propagation through the atmosphere.

The above specifications on the RF performance of earth-station antennas often have to be achieved under operational environmental conditions of high wind speeds, very low temperatures at which icing may occur and heavy rainfall.

15.2.2 Earth-station antenna configurations

15.2.2.1 Axisymmetric reflector antennas

Conceptually, axisymmetric reflector antennas represent the simplest configura-tions potentially capable of meeting the RF specifications for earth-station applications. The major advantages of such configurations are that they are mechanically relatively simple, reasonably compact and, in general, fairly inexpensive. The circular symmetry of the main reflector leads to considerable cost savings in the manufacture of the reflecting surface, backing structure and antenna mount.

The simplest form of axisymmetrical configuration is the single paraboloidal reflector, with the phase centre of the primary feed horn located at the focus. For antenna diameters greater than 3 m, the necessary long waveguide run from the feed to a conveniently located electronics box, generally at the rear of the reflector, is undesirable. A more compact axisymmetrical configuration, especially for larger antenna diameters, can be realised by the introduction of a smaller subreflector. The feed horn is now located near the rear of the main reflector, eliminating the need for long and lossy waveguide runs. Two types of subreflector configuration are possible:

(i) The Cassegrain, where the subreflector is a section of a hyperboloid situated within the focus of the paraboloidal main reflector.
(ii) The Gregorian, where the subreflector is a section of an ellipsoid located outside the focus of the paraboloidal main reflector.

Both systems result in similar RF performance characteristics, although the Cassegrain is more commonly employed in earth-station applications.

The RF performance, in terms of antenna efficiency and sidelobe envelope, of axisymmetric antennas is inhibited by the aperture blockage due to the primary feed or the subreflector, together with the associated support structure. In a dual-reflector Cassegrain system, subreflector blockage can lead to an increase of 5 dB to the peak sidelobe levels attainable by suitable choice of the aperture illumination distribution and without aperture blockage. Moreover, a reduction in peak gain of between 0.1 and 0.5 dB is typical. Direct radiation from the primary feed horn which is not intercepted by the subreflector also leads to a degradation of the antenna sidelobe performance in the forward direction. This degradation can be minimised by first ensuring that the radiation associated with the feed horn itself has low sidelobes and, secondly, by increasing the percentage of energy intercepted by the subreflector. Although the first requirement can be well satisfied by using a corrugated or dual-mode feed horn, the second can lead to a reduction in antenna efficiency owing to the increased tapered illumination distribution in the antenna aperture. The optimum antenna efficiency of approximately 65 per cent is achieved with a subreflector edge illumination taper of 13 dB; however, this leads to high feed spillover lobes in the antenna secondary radiation pattern. By increasing this edge taper to 25 dB, a significant improvement in the antenna radiation pattern is achieved, but the antenna efficiency is reduced to around 55 per cent. A partial compromise can be achieved by introducing blinkers to the subreflector in the azimuth planes which correspond to the geostationary orbit only.

An improvement in antenna efficiency can be realised by shaping the reflector profiles of an axisymmetric antenna and, thereby, controlling the aperture amplitude distribution while still maintaining a uniform phase distribution. With this technique it is feasible to achieve a reasonable wide-angle sidelobe performance and high antenna efficiency simultaneously by employing a high-illumination taper at the subreflector edge and then shaping the reflector profiles to achieve near-uniform illumination in the main reflector aperture. The near-uniform aperture distribution will, however, lead to higher sidelobes close to the boresight direction.

For antennas with aperture diameters of at least 200λ, reflector profile shaping can be undertaken successfully using a geometrical ray optics technique.[5] Most large earth-station antennas employ such shaped reflector profiles to improve their efficiency factor. Smaller antennas can, of course, be designed using the same technique, but their performance would be limited by the effects of diffraction, especially by the small subreflector. For such antennas, diffraction optimisation techniques have been derived.[6,7]

Figure 15.2   Cassegrain antenna (courtesy of Vertex Communications Corporation)

Axisymmetric dual-reflector antennas are commonly used in applications in the fixed satellite service, as hubs for VSAT networks and to provide broadcast uplink facilities. Figure 15.2 shows a typical Cassegrain antenna.

15.2.2.2 Asymmetric (or offset) reflector antennas

For many years, axisymmetrical reflector antennas have been employed for earth-station applications because of their straightforward geometry and mechanical simplicity. However, when improved electrical performance in terms of efficiency and sidelobe radiation is required, the axisymmetrical antenna is limited by the effects of aperture blockage. Although the geometry of an asymmetric, or offset, reflector is less straightforward, the removal of all blockage effects brings about major improvements in performance.

Similar to its axisymmetric counterpart, the offset paraboloidal reflector can be utilised as a single reflector fed from the vicinity of its prime focus, or in a dual-reflector configuration. Again, the latter can be arranged in either a Cassegrain or Gregorian configuration. In this case, the compactness of the Gregorian geometry is generally preferred.

As with the axisymmetric system, an offset reflector can be realised with shaped reflector profiles. However, even with the conventional conic arrangement, antenna efficiencies of at least 70 per cent and sidelobe envelopes better than $29 - 25 \log \theta$ dBi can be readily achieved. With shaped reflectors, efficiencies of 84 per cent have been reported.

The current trend towards improved sidelobe radiation, especially as regards smaller earth-station antennas, has led to the adoption of the offset configuration. Such designs can equally be applied to larger earth-station antennas, although the mechanical and alignment problems together with the requirement for steerability make the configuration slightly unattractive from the manufacturer's point of view. The greatest use of the offset geometry has been for small and medium-sized antennas for the professional VSAT market and the domestic reception of direct-broadcast satellite television, where the volumes required have justified investment in specialised tooling. Heavy presses traditionally used for stamping out car body parts have been used to manufacture offset reflectors of up to 1 m in diameter. Techniques such as the use of compression moulding of composite materials like sheet moulding compound (SMC) have been used to provide low-cost, high-quality reflector surfaces. Figure 15.3 shows a typical single-offset reflector antenna used in VSAT applications.

A performance limitation of single-offset reflector antennas is the inherent high crosspolarisation which maximises in the azimuth plane when linear polarisation is used. For dual-polarisation systems this can be minimised by increasing the reflector focal length, introducing a compensating multimode feed or by using a dual-reflector configuration. Figure 15.4 shows a multimode feed which uses the higher-order TE_{21} waveguide modes to effectively cancel the inherent crosspolar lobes associated with the offset geometry. A difficulty with this solution is that it is restricted to a bandwidth of approximately eight per cent.

Figure 15.3   Single-offset VSAT reflector antenna (courtesy of Channel Master (UK))

Figure 15.4   Crosspolar cancelling multimode feed for single-offset reflector antennas (courtesy of ERA Technology)

The use of the dual-reflector configuration, as illustrated in Figure 15.5, essentially increases the nominal focal length of the system and, consequently, is not restricted by bandwidth considerations. However, the positioning of the subreflector relative to the main reflector is critical.

In general, the projected aperture periphery has been circular. However, new designs involving either a square or diamond periphery have been introduced which provide higher gain than the conventional design together with very low sidelobe radiation in the plane corresponding to the geostationary orbit.

Coupling this with the advantage of being easily folded in a clam-shell arrangement has led to the extensive use of such antennas with satellite newsgathering vehicles. A typical configuration with a 1.5 m single-offset diamond antenna mounted on the roof of an SNG vehicle is shown in Figure 15.6. Segmented reflector designs which can be transported in suitable flight cases have also been used for flyaway newsgathering systems.

Figure 15.5   Dual-offset VSAT antenna (courtesy of Vertex Communications Corp.)

Figure 15.6   Satellite newsgathering vehicle using a 1.5 m single-offset Diamond antenna (courtesy of BAF Communications Corporation and ERA Technology)

15.2.2.3   Primary feed system

The complexity of a primary feed system for an earth-station antenna depends on the number of functions it has to provide. This can range from a simple horn and circular-to-rectangular transition for a receive-only application to a horn, polariser, orthomode transducer and possibly diplexers for a four-port receive and transmit circularly polarised system.

A critical element in the feed chain is the primary feed horn which, in a high-capacity communications system, must provide efficient illumination of the antenna and good crosspolar response over both the receive and transmit frequency bands. Extension of the currently-used Ku-bands up to 18 GHz leads to even greater demands on the performance characteristics of the feed horn.

The prime characteristics which must be exhibited by the feed horn include:

- axially-symmetric radiation patterns for good antenna efficiency;
- low crosspolarisation, especially in dual-polarised systems;
- low return loss;
- a well defined phase centre which is independent of frequency;
- wide bandwidth properties.

Hybrid-mode corrugated waveguide feeds radiating the HE_{11}° mode are now extensively used in earth station applications. The HE_{11}° mode may be regarded as the sum of a balanced mix proportion of the TM_{11} mode and the TE_{11} mode of smooth circular waveguide. Horns radiating this hybrid mode are theoretically capable of satisfying the design requirements for both equal *E*- and *H*-plane copolar beamwidths, to provide efficient rotationally symmetric illumination of the reflectors, and low crosspolarisation. In principle, the cross-polarisation may be completely cancelled if the balanced hybrid mode condition is satisfied, which can only occur at a single frequency. The crosspolarisation performance can be seriously degraded also by the excitation of unwanted higher-order modes both in the throat of the horn and along its length.

In cases where bandwidth is not so important, for example a receive-only station, a dual-mode ($TE_{11} + TM_{11}$) Potter horn may be applicable. These primary feed horns are discussed in more detail in Chapter 5.

15.2.2.4 Array antennas

As antenna sizes become smaller, the interest in array antennas increases. To date the main usage of array antennas has been at L-band with the INMARSAT mobile communications satellite system. For these applications, the array antenna must be compatible with the mobile, eg. truck, aircraft, and also be capable of scanning the beam either mechanically or electronically. Typical of such an antenna is the blade phased array for aeronautical communications as discussed in Chapter 19.

Array antennas have also been used for receive-only applications such as direct-to-home television. In most cases, these have utilised printed elements and feed networks leading to only average efficiency compared to the reflector configuration. More recently, less lossy waveguide technology has been used, which has been proved to be readily fabricated, cost effectively in volume, using injection-moulded plastic techniques. Such an array antenna is illustrated in Figure 15.7 showing its modular form of construction.

At present, the use of array antennas, especially those employing beam-scanning techniques, is prohibited by cost compared to the relatively simple reflector antenna. However, only a few years ago the same was thought about offset reflectors, but the volume requirements of small antennas for VSATs and direct-to-home television changed the market. The possible move to nongeostationary Ka-band multimedia satellites in the coming years may well see the same opportunities for phased-array antennas.

15.2.3 Antenna tracking consideration

For large earth-station antennas, such as those employed in the INTELSAT system, loss of signal owing to imperfect pointing of the antenna boresight towards the satellite must be held to a few tenths of a decibel. In practice, this is generally achieved by automatic tracking of a beacon signal. As the antenna diameter decreases the beamwidth increases and the need for tracking is less obvious. In most applications at Ku-band tracking is not considered necessary for antennas with diameters of less than 4 m.

The most accurate, and most expensive, form of tracking is a multimode monopulse system. This is generally reserved for antennas used for in-orbit

Figure 15.7   Flat-plate waveguide aperture array antenna (courtesy of ERA Technology Ltd)

testing (IOT) or telemetry, tracking and command (TT&C) of satellites. Higher-order asymmetric waveguide modes are excited in the feed aperture if the antenna points away from the designated beacon source. By extracting these modes from the normal communications signal, the resultant error signals may be used to drive a tracking receiver and control the antenna pointing mechanisms. This RF sensing arrangement offers a high tracking sensitivity without unduly compromising the communication channel performance. For output in polar form, the system relies on the accurate measurement of amplitude and phase detected from a single higher-order circular waveguide mode[9]. For Cartesian tracking, the configuration is essentially that of an amplitude monopulse comparator and, although involving greater complexity, exhibits important operational advantages over the polar form. In particular, the normalisation process of the error signals can render the tracking system insensitive to the effects of depolarisation for certain mode configurations.

For most commercial earth-station antennas a less-expensive step-track system is widely used. In such a system, the antenna is made to turn a small, predetermined distance in one direction; if the signal received from the satellite increases in power as a result of this movement, the equipment deduces that the antenna has turned in the correct direction and makes a further similar move.

If, however, the signal decreases, an incorrect move is deduced and the antenna is moved in the opposite direction. Certain levels of intelligence can be integrated into such a system[10].

A new technique for RF beam scanning has been described[11] which takes the form of generating proportions of higher-order waveguide modes within the antenna feed horn in order to electronically squint the antenna secondary pattern in azimuth and elevation, respectively. A mode generator is introduced within the primary feed system to selectively produce the wanted higher-order mode using PIN diodes controlled by a microprocessor which is compatible with existing control systems. This system leads to a pointing accuracy superior to existing conical scan or step-track systems and approaching that of a traditional monopulse system without the requirement for a separate and expensive tracking receiver. This system has particular advantages for applications such as ships and oil platforms where it can be used instead of a mechanically-dependent stabilised platform.

15.3 Satellite antennas

Satellite antenna subsystems have increased significantly in number, size and complexity over the years. Early INTELSAT spacecraft carried only a single communications antenna which provided global coverage. The first European communications satellite, EUTELSAT 1, carried six reflector antennas for transmission and reception to and from coverage areas as illustrated in Figure 15.8. Current generations of both operators' spacecraft involve sophisticated antenna technology to provide system enhancement.

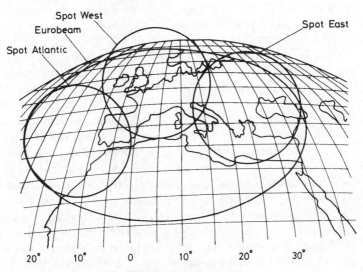

Figure 15.8   Coverage areas for EUTELSAT 1 Satellite

As in the case of the earth station, most satellite antennas utilise high-performance reflector configurations to generate highly directional focused beams towards the earth. Space within a satellite launch vehicle is invariably restricted and, in general, antenna diameters of no more than 3 m can be accommodated and even that requires the antenna to be partially stowed during launch and then deployed in orbit. Larger antenna diameters can be envisaged but require folding or even inflation of the reflector surface[12].

Even for small satellite antenna dimensions, the mass can be significant and considerable effort is expended in minimising it as far as possible. For example, current satellite-borne reflectors are generally manufactured as a sandwich construction, consisting of an aluminium honeycomb core with face sheets on either side. Carbon fibre is often employed for these sheets because of its high strength/mass ratio and excellent thermal stability, resulting in a strong, lightweight stable reflector surface which obviates the need to provide an additional microwave-reflecting layer[13]. To reduce thermal extremes, Kapton blankets are usually employed to wrap the rear of the reflector, the antenna support structure and the primary feeds.

15.3.1 Circular and elliptical beam coverage

Initial requirements for satellite antennas centred on the use of circular or elliptical beams which would illuminate the required coverage area within, approximately, their half-power beamwidth. The global coverage beam was, and in several applications still is, a prime requirement with a half-power beamwidth of approximately 17°, being the angle subtended by the earth relative to a satellite in the geostationary orbit. For most missions, the global coverage beam can be realised using either a smooth-walled fundamental-mode conical horn or a corrugated conical horn. The edge of coverage gain specification is assured with either type of antenna and both have their relative attractions. The smooth-walled conical horn is lighter and cheaper to manufacture but exhibits a high crosspolar component which may restrict its usage in a dual-polarisation application; in this context, the hybrid-mode corrugated horn may be preferred.

When smaller beamwidths are required, the mass and dimensions of a horn on its own become prohibitive and a reflector antenna is employed. In its front-fed axisymmetric configuration, gain reduction, scattering and depolarisation caused by aperture blockage from the primary feed horn, waveguide run and support struts can be a fundamental limit, especially when the reflector diameter is small in terms of the wavelength. For this reason, the offset reflector, which removes all aperture blockage effects, has become favoured for spacecraft application. In addition to the advantages gained in performance, the use of offset reflectors can lead to overall antenna geometries better suited to the spacecraft environment where either the feed horns associated with several reflectors are placed on a single support tower protruding from the spacecraft platform, or the feed horn is actually located on the platform and close to the associated electronics, thereby reducing losses associated with waveguide runs.

A major disadvantage of the single-offset reflector, especially when frequency reuse by polarisation discrimination is implemented, is the depolarisation associated with the use of linear polarisations which can present a severe limitation. For satellite applications, a favoured approach for achieving a high degree of polarisation purity in this case is by means of a gridded reflector. The reflector comprises wires or gratings which act as the reflecting surface for the desired polarisation. This approach allows inline stacking of two reflector systems, with orthogonal grid rotations, to produce orthogonally polarised beams from the same aperture area; a typical arrangement is illustrated in Figure 15.9. The front reflector is formed by a grid which acts both as a polarisation-selective reflecting surface for the horizontally polarised signal and as a polarisation-selective transmission filter for the vertically polarised back reflector system. The back antenna can be solid or a polarisation-selective surface.

An alternative approach to the dual-gridded reflector is the use of a dual-offset reflector configuration which, again, is not limited by high crosspolarisation. Recently, this latter approach has become popular since the manufacturing time is considerably less than for the dual-gridded system. Both configurations are applicable to shaped beam coverage requirements, described in the next Section.

The realisation of an antenna beam with an elliptical crosssection can be achieved using several techniques, including:

- a single reflector having a small asymmetric profile distortion but a circular aperture;

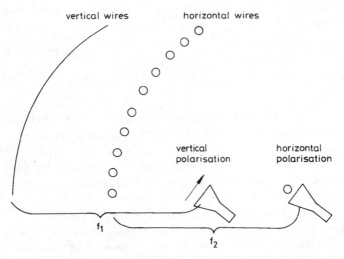

Figure 15.9   Dual-gridded offset-fed reflector

- a single or dual-offset reflector with an elliptically-contoured main reflector aperture;
- a reflector with circular aperture illuminated by an array of primary feed elements.

In those cases where a reflector with an elliptical aperture is proposed, efficient illumination of this aperture coupled with high polarisation purity presents several difficulties. The use of a rectangular corrugated primary feed horn[14] and an elliptical corrugated horn[15] have been considered, although both have disadvantages which result in degraded overall performance.

An alternative elliptical-beam dual-offset antenna configuration has been proposed[16] in which the elliptical-beam characteristics are achieved by shaping the two reflector surfaces. This approach offers the advantage that the subreflector possesses a circular boundary so that high-performance axisymmetrical primary feed horns with a high degree of polarisation purity, such as the corrugated conical horn, may be used. A similar approach but using only a single-offset reflector has been described by Wood and Boswell[17]. The reflector has a circular projected aperture and the beam crosssection is shaped by a small asymmetric distortion of the reflector profile.

15.3.2 Shaped and contoured beam coverage

In most present applications, the use of a circular or elliptical beam to service a particular coverage zone is inefficient, resulting in high levels of power being directed towards unwanted geographical locations, such as areas of the sea. A more effective solution is achieved by tailoring the satellite antenna radiation pattern to the desired coverage area, thereby concentrating a high proportion of the radiated power as uniformly as possible within that area, while minimising the radiation elsewhere.

Two approaches exist for contouring the radiation pattern of a satellite reflector antenna:

- use of a conventional offset parabolic reflector system combined with a multi-element feed array, rather than a single element;
- retaining the single feed element but modifying, or shaping, the profile of the reflector (or reflectors).

The former technique has seen application in several satellite antenna systems, including those in the INTELSAT series. The INTELSAT VI spacecraft utilises two such antennas at C-band (one for the downlink and one for uplink) each providing two fixed beams for hemispheric coverage and four isolated beams for zone coverage, as illustrated in Figure 15.10. Each antenna comprises an offset parabolic reflector and 144 dual-polarised, dual-mode feed horns. The two feed arrays mounted on the spacecraft platform are shown in Figure 15.11.

The contoured coverage is achieved by suitably controlling the amplitude and phase of the signals which are fed to the various primary feed elements by means of a beamforming network. The necessary amplitude and phase

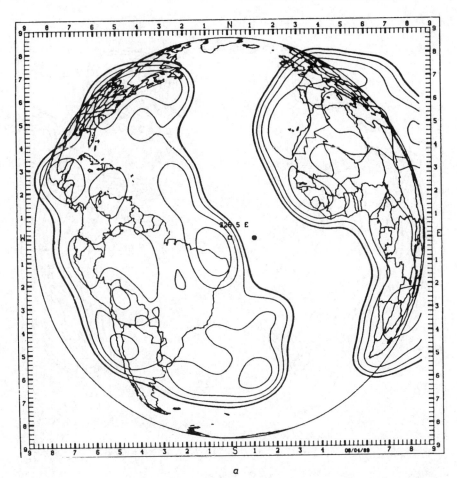

Figure 15.10a   Typical shaped beam coverage at C-band for INTELSAT VI spacecraft: Hemibeams

excitation coefficients are determined by synthesis techniques which employ mathematical optimisation procedures.

By increasing the number of elements in the primary-feed array, the flexibility of this approach is enhanced. For example, sidelobe suppression can be incorporated thereby reducing potential interference into neighbouring systems. By incorporating some method of varying amplitude and phase to each feed element, the radiation characteristics associated with a satellite antenna in orbit could be reconfigured to accommodate changing traffic demands or to provide discrimination against unwanted interference.

One of the critical components for contoured-beam spacecraft antennas is the beamforming network. Several technologies are available for realising this network including waveguides, coaxial cable, stripline and microstrip. For example, in the case of the INTELSAT VI C-band antennas, the network is

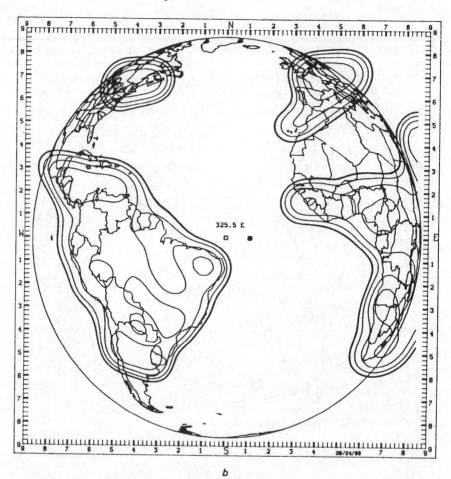

325.5 E

Figure 15.10b   Typical shaped beam coverage at C-band for INTELSAT VI spacecraft: Zone beams

designed as an air-supported transmission line utilising printed circuit centre conductors and honeycomb sandwich ground planes. This technology can also be used at higher frequencies (11/14 GHz) for low-power applications; however, for high-power requirements, waveguide networks are required. The latter can be designed using couplers and frequency-independent phase shifters.

As the degree of beam contouring and reconfigurability becomes higher, so the dimensions of both the reflector and the feed array increase. This has led, in some cases, to the use of direct radiating feed arrays to provide contoured-beam coverage[18].

The alternative approach to achieving contoured-beam coverage of a given geographical area is to use reflector profile shaping with a single primary feed horn. A simple illustration of this is the generation of an elliptical beam using a parabolic reflector with asymmetric profile distortions, as described in the

*Figure 15.11   C-band receive and transmit feed arrays for INTELSAT VI spacecraft
(courtesy of Hughes Space and Communications)*

previous Section. A further example is the use of a single shaped reflector to provide a shaped beam with enhanced edge of earth coverage gain which results in uniform flux density across the earth's surface. Such an antenna was employed on the MARECS satellite for maritime communications [19].

The number of degrees of freedom can be enhanced by employing a dual, rather than a single, reflector system and modifying the profiles of both the main and subreflectors.

Using reflector-shaping techniques to achieve shaped beam coverage has become normal practice with many spacecraft manufacturers, mainly since it removes the need for the large, heavy and lossy waveguide feed network. The coverage areas for the EUTELSAT HOT BIRD range of direct-to-home broadcast satellites, Figure 15.12, are typical of those realised using shaped-reflector technology.

The antennas used to generate these beams use the dual-gridded approach, discussed in the previous Section, to achieve dual-polarisation capability. Figure 15.13 illustrates one of the antennas utilised on the HOT BIRD 2 spacecraft

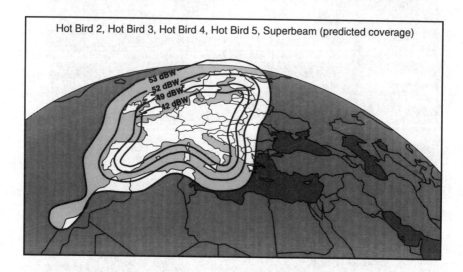

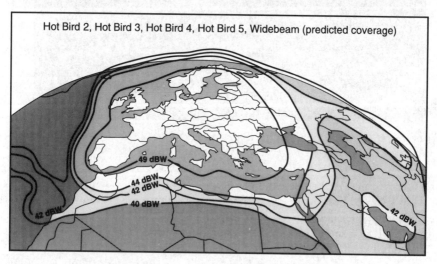

Figure 15.12   Predicted coverage areas for the HOT BIRD satellites (courtesy of EUTELSAT)
  a   Superbeam
  b   Widebeam

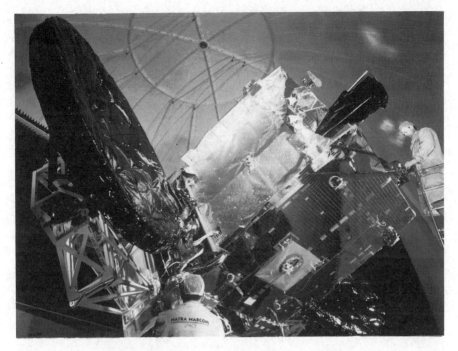

Figure 15.13   *HOT BIRD 2 dual-polarised shaped reflector antenna (courtesy of Matra Marconi Space)*

launched in 1996. The photograph shows the two gridded reflectors, covered by a thermal blanket, with the two corrugated feeds at the relevant focal points to the right.

New synthesis algorithms allow the design of shaped reflector antennas radiating more than one contoured beam from a single reflector surface. For example, Reference 20 considers the design of a single shaped reflector with two separate feeds to generate the two hemispherical beams for the INTELSAT VI spacecraft with sufficient spatial isolation to enable frequency reuse.

15.3.3 Multibeam antennas

To meet requirements for sufficient usable bandwidth with flexible high EIRP levels operating to small earth-station terminals implies the use of multiple spot-beam satellite antennas. Such antennas also offer the advantage of increasing the level of frequency reuse by spatial discrimination, whereby more than one nonadjacent beam can utilise a common frequency band. Polarisation discrimination can be used in conjunction with spatial discrimination to realise an even greater level of frequency reuse. The coverage scenario also provides flexibility in the traffic-to-beam allocation.

Typical of the current state-of-the-art for multibeam antenna design is the concept applied to INMARSAT 3 to provide a complex set of spot-beam

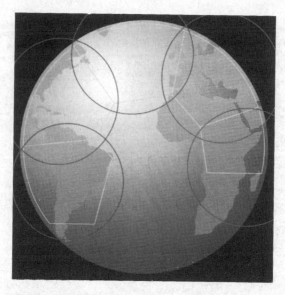

Figure 15.14   INMARSAT 3 spot-beam coverage from 15.5° W (courtesy of Matra Marconi Space)

coverage with the necessary isolations for land, maritime and aeronautical mobile services. Here, the world is divided into regions which are each covered by both global beams and up to six spot beams from one of four satellite locations in the geostationary arc; Figure 15.14 illustrates one such coverage from the satellite at 15.5 ° W.

The antenna design incorporates the principle of a multimatrix semiactive reflector antenna patented by ESA[21] and also used for the ESA Artemis LLM to provide overlapping spot beams across Europe[22]. For INMARSAT 3, the beams are generated from feed clusters comprising five helix feeds from a total of twenty-two with adjacent beams sharing some feed elements[23]. Each feed cluster is fed via a Butler-like matrix from equally excited power amplifiers. Beam selection is provided by low-level phase-only control at the input to the amplifiers. The complete antenna under test in an anechoic chamber is shown in the photograph of Figure 15.15.

Multibeam antennas figure strongly in the new batch of satellite systems to provide communications to handheld telephones for operation commencing in 1998 (see Chapter 20). This new commercialism of satellite communications has also led to a new sensitivity among the leading satellite manufacturers in describing their current technology. Consequently, details on antenna design are scarce at the present time.

The new breed of mobile satellite systems will operate in either a low-earth orbit (LEO) or a medium-earth orbit (MEO) rather than the classical geostationary orbit (GEO) for communications satellites. By employing constellations

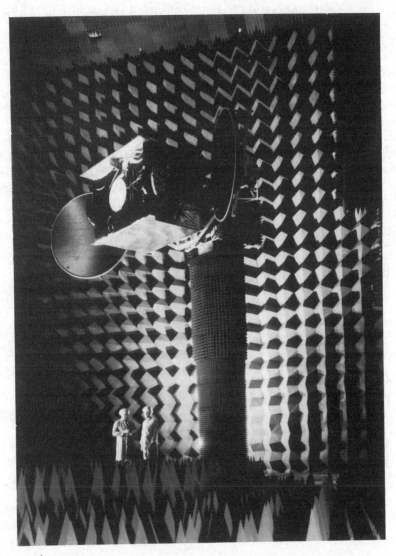

Figure 15.15   INMARSAT 3 antennas under test (courtesy of Matra Marconi Space)

of satellites in various orbital planes, dependent on the actual system, continuous coverage of the entire earth's surface can be provided.

One of the new mobile satellite systems is that of ICO Global Communications (see Chapter 20). This system comprises a constellation of ten satellites in medium-earth orbit arranged in two planes of five satellites each[24]. The communications payload uses a high degree of digital technology for functions such as channelisation and beam generation which have traditionally been performed using analogue techniques. Separate receive and transmit S-band phased-array antennas, with aperture dimensions of approximately two metres, provide 163 service link beams. In the transmit antenna, output signals are routed to the electronics section which comprises 127 upconverter/solid-state power amplifiers each of which is connected to one of the 127 radiating elements and bandpass filters of the 2 GHz transmit array. The array and the beam-forming coefficients supplied by the digital processor generate the 163 contiguous spot beams, each having approximately a 4° half-power beamwidth. Feeder link antennas at C-band connect each satellite to a series of ground-based satellite access nodes for coordinating and routing traffic.

Multibeam antennas and digital onboard processing techniques will also figure strongly in the proposed new satellite systems, offering broad bandwidth at Ka-band for interactive multimedia applications. These will lead to even more sophistication in the design of spacecraft antennas.

15.4 References

1 'Satellite communication antenna technology' (Elsevier Science Publishers BV, 1983)
2 RUDGE, A.W. *et al.*: 'Handbook of antenna design vol. 1' (IEE Electromagnetic Waves Series 15, Peter Peregrinus Ltd, 1964)
3 WITHERS, D.J.: 'Effective utilisation of the geostationary orbit for satellite communication', *Proc. IEEE*, 1977, **65**, pp. 308–317
4 MARSH, A.L.: 'Dual polarisation technology—Part 2—The introduction of dual-polarisation operation at UK earth stations', *Post Office Elect. Eng. J.*, 1977, **70**, pp. 178–184
5 GALINDO, V.: 'Design of dual-reflector antennas', *IEEE Trans.*, 1964, **AP-12**, pp. 403–408
6 WOOD, P.J.: 'Reflector profiles for the pencil-beam Cassegrain antenna', *Marconi Rev.*, 1972, **35**, pp. 121–138
7 POULTON, G.T.: 'The design of Cassegrain antennas for high efficiency and low VSWR'. Proc. fifth European *Microwave* conference, 1975, pp. 61–65
8 WATSON, B.K., DANG, N.D., and GHOSH, S.: 'A mode extraction network for RF sensing in satellite reflector antenna'. IEE second int. conf. on *Antennas and propagation*, York, 1981, pp. 323–327
9 BRAIN, J.R.: 'A simple multimode tracking system', *Marconi Rev.*, 1979, **87**, pp. 164–180
10 EDWARDS, D.J., and TERRAL, P.M.: 'The smoothed step-track antenna controller', *Int. J. Satell. Commun.*, 1983, **1**, pp. 133–139
11 DANG, R., WATSON, B., DAVIS, I., and EDWARDS, D.: 'Electronic tracking systems for satellite ground station'. Fifteenth European *Microwave* conf., Paris, France, 1985
12 RIEBALDI, G.: 'Antenna mechanical technologies within ESA'. ESA workshop on *Mechanical technology for antennas*, ESA SP-225, 1984
13 BRAIN, D.J., and RUDGE, A.W.: 'Efficient satellite antennas', *Electron. Power*, 1984, pp. 51–56
14 ADATIA, N.A. *et al.*: 'Study of an antenna system for an experimental TV satellite' Final report on ESA Contract 3285/77/NL/AK, 1979
15 VOKURKA, V.J.: 'Elliptical corrugated horns for frequency re-use applications'. IEEE international symposium on *Antennas and propagation*, Quebec, Canada, 1980
16 PONTOPPIDAN, K., and NIELSEN, P.H.: 'Dual reflector system shaped for elliptical beam with good efficiency and sidelobe performance'. IEE *Antennas and propagation* conf., ICAP83, Pub. 219, Norwich, 1983
17 WOOD, P.J. and BOSWELL, A.G.P.: 'Elliptical beam antenna for satellite applications' IEE conf. on *Antennas for aircraft and spacecraft*, Conf. Pub. 128, 1975

18 BORNEMANN, W., BALLING, P., and ENGLISH, W.J.: 'Synthesis of spacecraft array antennas for INTELSAT frequency re-use multiple contoured beam', *IEEE Trans.*, 1985, **AP-33**
19 WOOD, P.J.: 'Shaped-beam reflector design for earth coverage applications'. IEE conf. on *Antennas for aircraft and spacecraft*, London, 1975
20 PEARSON, R.A. *et al.*: 'Application of contoured beam shaped reflector antennas to mission requirements'. *ICAP93*, Herriott Watt University, Edinburgh, 1993
21 ROEDERER, A.G., and CRONE, G.A.E.: 'Recent European antenna technology for space systems'. Proc. *ISAP'96*, Chiba, Japan, 1996
22 ROEDERER, A.G. *et al.*: 'Some European satellite-antenna developments', *IEEE Antennas Propag. Mag.*, April 1996, **38**, (2)
23 GREENWOOD, D. *et al.*: 'Multimatrix beamforming for semi-active antennas at L-band'. IEE colloquium digest 1991/175, 1991
24 POSKETT, P.: 'Satellite system architectures'. Second European workshop on *Mobile/personal satcoms*, 9–11 October, 1996, Rome, Italy

Chapter 16

Satellite networking

M. A. Kent

16.1 Introduction

The majority of this book concentrates on satellite communication systems from a satellite communication engineering perspective. This Chapter addresses the issue from an overall network perspective in order to clarify what the network (and its services) expects from satellites and the implications which satellites have for network design.

The Chapter will be primarily concerned with main network services (such as telephony) where satellites have to compete/integrate with terrestrial facilities, rather than independent/specialised networks (such as broadcast television using satellites) where the characteristics of satellites are built into the service offering.

Analogue, PDH and SDH transmission, leased and switched analogue and ISDN services, and ATM transport will be considered. Certain network-specific equipment is also covered (for example, DCME).

16.2 Services

There are two broad categories of telecommunication service, namely main network services and custom networks.

16.2.1 Main network services

Those services carried on the main network, i.e. the network composed of a complex mix of switching and transmission to carry a variety of services (both switched and leased), are known as the main network services. Services in this category include:

- voice;
- voiceband data;
- 64 kbit/s digital data;
- broadband.

16.2.1.1 Voice

Voice services are characterised from a transmission perspective by a bandwidth of 300 Hz to 3.4 kHz and by parameters such as noise, distortion, echo and delay.

16.2.1.2 Voiceband data (facsimile, datel etc.)

Voiceband data enters the network in the same way as voice, i.e. via a 3.1 kHz channel. Various types of modem are used to interface the data terminals with the network, and speeds up to 9.6 kbit/s are readily achievable. Higher bit rates are possible, but the sensitivity to channel impairments increases and throughput problems become more likely as the speed rises. Current modem developments have achieved 28.8 kbit/s (V.34) for data and 14.4 kbit/s (V.17) for facsimile, with some work exploring proprietary methods for higher bit rates.

16.2.1.3 64 kbit/s digital data (ISDN and leased network)

It is also possible to access the digital network directly, rather than via a voiceband modem. Initially, this access was to point-to-point leased circuits, often submultiplexed, to carry multiples of the lower-rate services originally carried as voiceband data on the analogue network, but services are emerging to make use of this facility in both switched (in particular the integrated services digital network — ISDN) and leased formats. ISDN is described by the I.xxx series of ITU recommendations.

16.2.1.4 Broadband

2 Mbit/s videoconferencing is a good example of a broadband service, and it is relatively straightforward to offer such services on leased circuits through the main network, provided that they are compatible with the existing transmission hierarchy. The majority of automatic switches in the current network (i.e. those carrying dial-up services) are limited to 64 kbit/s crossconnection, but a lot of interest is being shown throughout the world in developing techniques to offer broadband services at a variety of bit rates across the switched network.

The recent developments in SDH crossconnection equipment now permit automatic switching of bit rates up to the VC-4 path rate (140 Mbit/s), and also switching of section rates (STM-1 and STM-4), which enables wide bandwidth leased services to be readily and rapidly offered to customers.

The network requirements of these services are covered in Section 16.3.

For all these services, it is not just the end-to-end customer requirements which are important. Satellite usage must also take into account the signalling/routing constraints of a particular network configuration.

The requirements of these services may also differ depending on whether they are carried on a dedicated (leased) circuit within the main network or a switched connection.

16.2.2 Custom networks

Some services are carried on custom-designed networks independent of the main network. The network to carry these specialised services can be optimised to the specific service in question.

16.2.2.1 Broadcast television

Satellites play a major role throughout the world in providing television services direct to the home, and the satellites used are often custom designed for this purpose.

16.2.2.2 Television distribution

Television operators use satellite channels extensively to transfer material around the world between studios prior to transmission through the terrestrial facilities used for the majority of television reception in the UK. The satellites used for this transfer are often the same as the ones used to provide the main network services described in Section 16.2.1, but they are totally separate in networking terms and subject to different constraints.

16.2.2.3 Small dish/VSAT-type data networks

These networks use dedicated earth stations, usually less than 5 m in diameter, to provide a range of services to business customers. Many of the services offered are identical to those carried via dedicated leased circuits on the main network, particularly when the main network penetration to a particular geographical location is limited and the only access possible is via a dedicated satellite network of this kind. In addition, there is a range of unidirectional services offered by this type of network which cannot be provided by the main network. They use a hub earth station to transmit to a large number of small receive-only earth stations at customer premises.

16.3 Network description

This Section describes the elements of the main network and some of its features which impact on its ability to carry the services described in Section 16.2. It should be stressed that this is very much an introduction to what is a very wideranging subject, and so the concepts have been simplified to concentrate on those features necessary to enable the reader to understand the main principles from a satellite-system designer's point of view. Furthermore, it does not deal with custom networks, as these are dealt with elsewhere in this book.

16.3.1 Generalised network description

Figure 16.1 illustrates the generalised concept of the main network. It is composed of a series of nodes (where circuit interconnection takes place) and transmission links between these nodes. The nodes can be either manual/automatic crossconnection pieces of equipment used on the leased network to establish semipermanent connections between two customers, or fully automatic switches which respond to customer dialling carried in signalling messages to set up a connection which only lasts for the duration of the call. The transmission links employ a variety of transmission media (optical fibre/metallic cable, terrestrial microwave radio and satellite) to carry groups of circuits at varying capacities depending on the particular link (from single circuit local lines through to tens of thousands of circuits grouped together between main network nodes).

Both the nodes and the transmission links can be either analogue or digital at the moment, but the majority of the global network will be digital by the end of the decade. Therefore, the rest of this Section concentrates on the digital

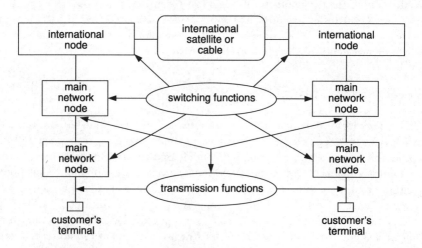

Figure 16.1   Main network concept

network structure with reference to some of the features of the analogue network, where appropriate.

16.3.2 Leased networks

Figure 16.2 shows the typical structure of a network for carrying both analogue and digital leased lines. It reflects the situation where the majority of the network is digital and therefore the major difference between the two service offerings is in the local-access portion of the network.

Satellites can, in principle, be used on any section (or combination of sections) of this network. However, apart from the specialist applications described in Section 16.2.2 of this Chapter, the main usage by European countries is on the international section of the network between international gateways worldwide. In the case of large countries, where the distance between main network nodes is large, they are also used as part of the national network infrastructure.

16.3.3 Switched networks

Figure 16.3 illustrates the diverse nature of the circuit-switched main network and the increasingly complex local-access options. Packet-switched networks are essentially of the same structure as far as satellite-system applications are concerned.

The use of satellites within circuit-switched networks is similar to their use for leased networks, described above. An added complication, which restricts their use, is the need to carefully control circuit routing through a switched

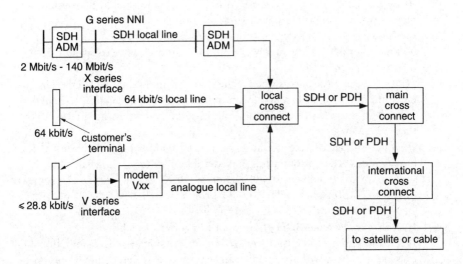

Figure 16.2   Leased-network structure

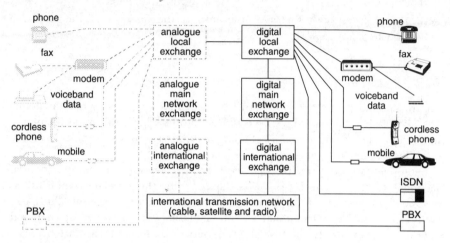

Figure 16.3   Circuit-switched main network

network to minimise the chance of picking up two satellites hops on a particular call—see Sections 16.7.2 and 16.7.1.

16.3.4 Main-network service rates

16.3.4.1 Local access

The local lines shown in Figure 16.2 can employ a variety of transmission techniques to access the network, depending on the services required and the size of the customer involved, but the following are the main methods of access:

(i)   *Analogue*: the standard two wire, 3.1 kHz bandwidth local line used for the majority of customers on the network.

(ii)   *n × 64 kbit/s*: leased circuit customers have been able to access the network at 64 kbit/s (or $n \times 64$ kbit/s, $n = 1$ to 30) for a number of years using ITU-TX series interfaces.

(iii)   *144 kbit/s*: the basic access rate for the international standard ISDN. The signal comprises two 64 kbit/s information channels, which can be switched independently through the network, plus a 16 kbit/s signalling link to control these channels.

(iv)   *2 Mbit/s*: this access rate can either be used for wideband leased circuit access, in which case the customer has considerable freedom in how the channel is split up, or to connect a private exchange (PBX) at the customer's premises to the main network with 30×64 kbit/s information channels plus a 64 kbit/s signalling channel to control them.

(v)   *Wideband leased lines*: SDH equipment now offers wideband services (up to 140 Mbit/s) direct to customers. SDH transmission systems using optical fibres are installed, with the terminating equipment at the customer's premises. G.703 terminating cards are available to offer 2 Mbit/s, 34 Mbit/s, 45 Mbit/s and 140 Mbit/s.

(vi) *ATM*: ATM services are currently under trial with some customers. These use SDH local access with ATM adaption equipment also at the customer's premises.

16.4 Main-network transmission technologies

16.4.1 Analogue transmission

The analogue transmission hierarchy is rapidly being replaced in the network by digital hierarchies. Analogue satellite channels using a variety of access/modulation techniques are still used internationally to support this hierarchy.

16.4.2 Plesiochronous transmission — PDH

Figure 16.4 illustrates the structure of the plesiochronous digital transmission hierarchy in Europe. There is another standard based on 1.5 Mbit/s (shown in Figure 16.5) used in countries such as the USA and Japan. Interworking between these two hierarchies is achieved by means of the hierarchy shown in Figure 16.6.

Higher bit rate line systems are also used extensively throughout the network (e.g. 565 Mbit/s), but these are not internationally standardised rates.

16.4.3 PDH and satellites

Satellites offer a variety of means to accommodate these main network transmission rates; the two major ones being 120 Mbit/s TDMA and 2 Mbit/s IDR (intermediate data rate). Both TDME and IDR are defined in detail by INTELSAT specifications. In addition, higher-bit-rate IDRs are available, for

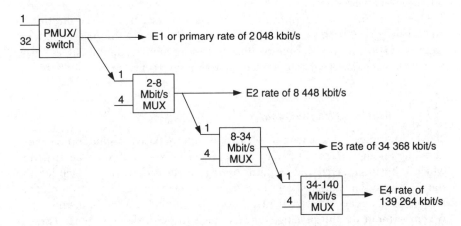

Figure 16.4   Plesiochronous digital hierarchy (2048 kbit/s primary rate)

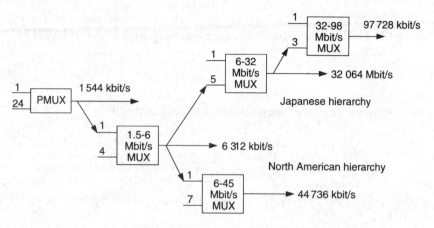

Figure 16.5   Plesiochronous digital hierarchy (1544 kbit/s primary rate)

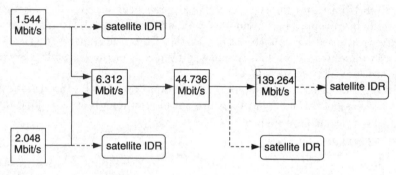

Figure 16.6   Interworking hierarchy

example 140 Mbit/s and 45 Mbit/s, which, currently, are used primarily for submarine-cable restoration purposes.

Therefore, there is a significant difference between the interconnection of nodes using satellite links, which have a preponderance of thin routes between pairs of earth stations, and interconnection using modern digital cables where each line system has a capacity measured in thousands of circuits.

16.4.4 Synchronous transmission, SDH

The current digital transmission network is mainly based around a plesiochronous digital hierarchy (PDH) composed of a very complex frame structure which is bit interleaved and contains timing justification between each level in the hierarchy. The net result of this complex frame structure is that it is difficult to, for instance, access a particular 2 Mbit/s channel from a 140 Mbit/s line system without demultiplexing stage by stage through the hierarchy. There is also a very limited amount of overhead capacity within the frame structure to facilitate management and performance monitoring of the traffic channels.

SDH is now being deployed as an overlay network by most operators, and uses byte interleaving instead of bit interleaving to permit simple access to payload within the frame structure. The bit rates of SDH are much higher than those used for the PDH, reflecting the developments in transmission technology since PDH was first introduced.

Note that in North American countries, the SONET (synchronous optical network) is being deployed. This is similar to the international SDH approach, and interconnection between the systems is relatively simple.

16.4.4.1 Overview of SDH hierarchical levels

The SDH primary rate is called STM-1 (synchronous transport module-1), and has a bit rate of 155520 kbit/s. Each frame consists of payload space capable of carrying a PDH 140 Mbit/s signal completely, with extra capacity for error-checking and management channels.

The currently defined higher SDH levels are STM-4, STM-16 and STM-64 (see Table 16.1). These are made up of four STM-1s and 16 STM-1s (etc.), respectively, through a simple byte-interleaving process of the original STM-1s. This means that it is a technically simple task to identify the bytes belonging to individual STM-1s, since the system is fully synchronous at these levels. The consequence of this is that an individual STM-1 can be extracted from the STM-4/16 signal and an alternative STM-1 put in its place, without demultiplexing the whole signal, as would be required in the PDH. This process is known as drop and insert, and is a major advantage for network operators.

16.4.4.2 Transporting PDH signals within SDH

The SDH primary-rate signal, the STM-1, is capable of carrying most PDH signals; Table 16.2 illustrates how.

Note that the unsupported PDH rates shown in the Table can be transported in multiplexed form at the next highest supported rate, however, the advantages of ease of identification of payload are lost.

Table 16.1

SDH level	Bit rate (kbit/s)
STM-0/R	55840
STM-1	155520
STM-4	622080
STM-16	2488320
STM-64	9953280

Table 16.2

PDH level	PDH bit-rate	SDH container
1	1.5 Mbit/s	VC12 (VC11 USA)
1	2 Mbit/s	VC12
2	6 Mbit/s	not supported
2	8 Mbit/s	not supported
3	34 Mbit/s	VC31
3	45 Mbit/s	VC32
4	97 Mbit/s	not supported
4	140 Mbit/s	VC4

The term VCn represents virtual container and the number corresponds directly to the PDH hierarchical level which it supports. It can be seen that the second hierarchical level of the PDH is not directly supported by the SDH. These signals (6 and 8 Mbit/s) can still be carried in the 34/45 Mbit/s and 140 Mbit/s signals within the SDH.

The virtual container contains overhead space in addition to the PDH signal. This overhead includes a pointer, which shows where the PDH framed signal starts within the VC. Since the SDH signals are unlikely to be synchronised to the incoming PDH signals, the pointer is necessary in order to recover the PDH signal at the distant end of the VC path. Also included in the overhead is an error-checking mechanism (called bit-interleaved parity, BIP), a path trace indicator, which labels the VC, and other information to enable automatic protection of VC paths in the event of network failures.

16.4.5 Transporting ATM signals in SDH

ATM is discussed in more detail later in this Chapter, however, it is possible to map ATM cells directly into the STM-1 payload space using the AU-4 pointer to indicate the first cell. Since the cells are all of fixed length, they can be read directly from the bit stream once the pointer has identified the start of the first cell in the frame.

16.4.6 SDH frame structures

The following subsections describe the synchronous digital hierarchy in a tutorial form. (They are not intended to replace the relevant ITU-T recommendations!)

16.4.6.1 Virtual containers (VCs)/tributary units (TUs)

When a plesiochronous signal is presented to an SDH multiplexer, it is first mapped into the appropriate container (as defined in Rec. G.709). Containers exist in four types:

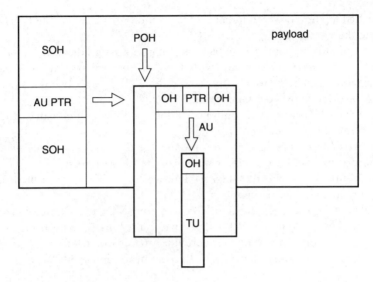

Figure 16.7   SDH frame construction

(i)   C-1: 1544/2048 kbit/s (PDH primary rate).
(ii)   C-2: 6312/8448 kbit/s (not supported).
(iii) C-3: 32 064/44 736/34 368 kbit/s.
(iv)   C-4: 97 728/139 264 kbit/s.

After this operation, a virtual container is created by adding the appropriate path overhead (POH) to the container. The path-overhead structure depends upon the bit rate of the signal. It provides in-service performance monitoring using bit-interleaved parity (BIP), which is an errored block-detection technique. Space is allocated within the POH to allow for monitoring of distant end-path failures and errors to assist in maintenance and performance monitoring. Bits are also reserved for automatic path restoration and path labelling.

16.4.6.2 Virtual containers

VCs exist in two categories and four types:

(i)   VC-1: basic or lower order; carries a C-1.
(ii)   VC-2: basic or lower order; carries a C-2 (not used).
(iii) VC-3: higher order/lower order; carries a C-3 and/or VC-1s.
(iv)   VC-4: higher order; carries a C-4 or VC-3s and/or VC-1s.

The higher-order VCs have more POH than their basic counterparts.

In order to combine a basic VC into a higher-order VC, the basic VCs are combined into tributary units (TU), and subsequently into tributary unit groups (TUG). These TUs are no more than the original basic VC plus a group of bytes, called a pointer, which indicates the position (or phase) of the POH with respect to the system clock.

(*Note:* this use of pointers enables plesiochronous signals, which may not have an exact phase relationship with the synchronous system clock, to be combined into synchronous transport modules (STM) (see below), since the phase of the original plesiochronous signal is held by the pointer.)

Two of these TUs are then combined to form a TUG, which can in turn be combined into a higher-order VC.

Higher-order VCs are then given a pointer (or pointers) and combined into an administrative unit (AU). This is similar to the TU described above. With the addition of further overhead, called section overhead (SOH), the synchronous transport module (STM) is complete.

TUs also provide a mechanism for concatenating VCs, to accommodate non-standard and/or wide bandwidth tributary signals. Finally, both negative and positive justification opportunities are provided in this structure.

For further information refer to ITU-T recommendation G.707.

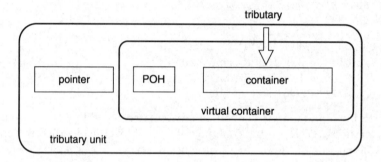

Figure 16.8   TU frame format

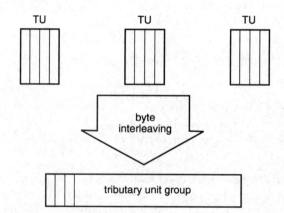

Figure 16.9   TUG frame format

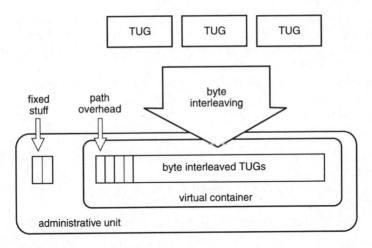

Figure 16.10   AU frame format

16.4.6.3 Synchronous transport module (STM) 1, N and R

The synchronous transport module-1, or STM-1, is the primary synchronous digital hierarchy rate, which is 155 520 kbit/s. The STM-1 can be four-byte interleaved to form an STM-4 (622 080 kbit/s) or 16-byte interleaved to form an STM-16, etc. Synchronous transport modules can be referred to as STM-N generically.

STM-R, the reduced bit rate STM-1, is a special STM with a bit rate of 51.84 Mbit/s, and could not be byte interleaved to form an STM-1.

The satellite community should note that all levels of the synchronous digital hierarchy contain a considerable percentage of overhead, much of which is at present undefined. Carriage of this overhead may represent a significant

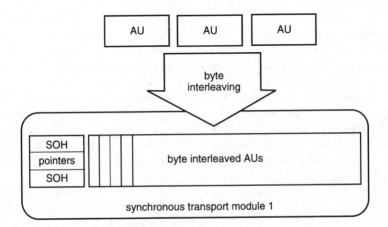

Figure 16.11   STM-1 frame format

penalty for satellite systems. For example, for the synchronous transport module, which contains 9×9 bytes of section overhead and 9×261 bytes of payload (an administrative unit), the section overhead represents $81/(2349+81) = 3.33\%$ of overhead.

For further information refer to ITU-T recommendations G.707.

16.4.6.4 SDH pointers (TU, AU)

The TU pointers indicate the phase of a VC-1, 2 or 3 with respect to the start of the STM-N, and the AU pointer indicates the phase of a VC-4 with respect to the start of the STM-N. These pointers allow signals of slightly differing bit rate and phase to be combined into the synchronous bit stream made up of STM-Ns in any synchronous digital hierarchy transmission system.

For further information refer to ITU-T recommendation G.707.

16.4.6.5 SDH protocols, e.g. the DCCr and DCCm seven-layer protocol stacks

The data communications channel (DCC) is a 192 kbit/s $+ 576$ kbit/s $= 768$ kbit/s communications channel embedded within the STM-N SOH. It is expressed as a sum of two bit rates because the STM-N SOH is split into two parts, called the regenerator section overhead (RSOH) and the multiplexer section overhead (MSOH). Thus, the two data communications channels are called the DCCr and the DCCm and are 192 and 576 kbit/s, respectively.

These channels are intended for network management use and the information contained within them would normally be presented at a Q interface point and F interface point to the management network. For more information on management structures and interfaces see recommendations G.773 and G.782. M.3400 describes the TMN architecture in detail, where these interfaces can be used.

The regenerator section channel would normally be available at the regenerator section termination (RST) point, and the multiplexer section channel would be available at the multiplex section termination (MST) point. The implication is that a regenerator would be able to access the DCCr, and multiplexing equipment would be able to access the DCCr and the DCCm.

Finally, the DCC is defined using a seven-layer protocol stack, similar in structure to the seven-layer OSI stack. Each layer employs its own protocol suite, and the technicalities and interactions of these layers are not well understood. The effects of satellite delay and error distributions on this protocol stack should be cause for concern within the satellite community.

For further information refer to ITU-T recommendation G.707, which details the bit assignments, and recommendations G.782, G.783 and G.784 which detail the DCC protocols and usage.

16.4.6.6 SDH management

Each of the virtual containers and the synchronous transport modules has some overhead, called path or section overhead. This overhead information is designed to permit network management, such as comprehensive in-service monitoring, signal labelling/identification, protection switching and tributary synchronisation/justification (with the pointers).

In addition to this information, the data communications channel is provided to allow other management information to be transferred across national and international networks.

At present, the path overhead is well defined for all levels, the section overhead is only partly defined, and the data communications channel lacks any agreed defined purpose. It is not doubted that a purpose for the DCC will be found, but the messages have not yet been defined. In view of the inherent complexity of the DCC protocol stack, and the lack of knowledge concerning plans for the information to be transported on the DCC, the satellite community should be addressing this issue.

Other issues which the satellite community should be considering include the uses for the protection switching bytes (for example, see G.781/2/3/4). The k bytes are reserved for protection switching of paths at all levels. The satellite community should be aware of the implications of automated restoration switching where SDH paths use satellite links, especially which criteria are selected for switching, acceptable delays in switching etc.

16.4.6.7 *Synchronisation and pointers*

The use of pointers and justification opportunities to allow frequency/phase justification of plesiochronous tributaries has been considered above. There are, however, considerable questions concerning synchronisation across borders of the STM-N signals.

It should be noted that the synchronous digital hierarchy does not require globally-synchronised clocks.

At present, these issues are not well understood. The satellite community should consider them in terms of satellite links.

16.4.6.8 *New equipment*

Network operators around the world are introducing equipment based on this new standard. Although the current PDH will continue to account for the

Table 16.3   ITU-T recommendations

G.707	definitions of the SDH frame structures
G.773	protocol suites for Q interfaces for management of transmission systems
G.781/2/3/4	muxes and management of SDH networks; the source of much new terminology
G.957/8	definitions of the optical interfaces for SDH
G.803 and G.805	definitions of architecture of SDH based networks; much new terminology

majority of the digital network for many years to come (there are still significant amounts of analogue equipment around the world today despite many years of digitalisation), SDH is the standard for new systems in the 1990s. The implications of this change need careful consideration by the satellite community. In particular, the minimum line system rate is 155 Mbit/s, although there are network interfaces at rates broadly compatible with the existing levels of the PDH hierarchy. This level of capacity is too large for the majority of satellite routes currently in operation in the main network and the ITU-R is currently addressing this problem.

16.4.7 SDH and satellites

16.4.7.1 INTELSAT scenarios

INTELSAT, in conjunction with its signatories and the ITU-T and ITU-R standards bodies, is to develop a series of SDH-compatible network configurations with satellites forming part of the transmission link.

A full description of these network configurations, referred to by INTELSAT as scenarios, is outside the scope of this Chapter. Recent chairman's reports of the ITU-4 SG4 contain fuller descriptions of these scenarios.

In summary, the options are given below.

16.4.7.2 Full STM-1 transmission point to point

This has required the development of an STM-1 modem capable of converting the STM-1 digital signal into an analogue format which can be transmitted through a standard 70 MHz transponder. This development is now completed, and working prototypes available.

In addition, there is (as yet) no recognised need for this amount of capacity via an SDH satellite link. Current high bit rate PDH IDR links are generally used for submarine-cable restoration (although there are some exceptions), but for SDH cables, the capacity of submarine cables is such that a complete current generation INTELSAT satellite would have to be held in reserve for SDH cable restoration. This is clearly not a cost-effective use for telecommunications satellites.

16.4.7.3 STM-R uplink with STM-1 downlink—point to multipoint

This scenario suggests a multidestination system and requires considerable onboard processing of the SDH signals, however, the advantage is flexible transponder usage for the network operator(s) using this system. The approach is not generally favoured by most network operators for reliability and future proofing reasons; it may prevent alternative usage of the satellite transponders in the future, and the additional complexity is likely to reduce the reliability/lifetime of the satellite and increase its initial expense.

16.4.7.4 Extended TU-12 IDR

This approach is favoured by a large number of signatories, since it retains the inherent flexibility of the satellite (regarded as a major advantage over cable systems), and would require the minimum of alterations to satellite and earth-station design. Additionally, some of the management advantages of SDH are retained, including end-to-end path performance monitoring, signal labelling and other parts of the overhead. Current development work is centred around determining which aspects of the data communications channels could also be carried with the TU-12.

Since the bit rate of the TU-12 is not much greater than an existing 2 Mbit/s PDH signal (about 2.33 Mbit/s), it is likely that minimal rearrangement of the transponder band plans would be required, with the possibility of mixing PDH- and SDH-compatible IDR carriers. Additionally, development work is currently taking place to modify existing IDR modems to carry the TU-12 signal, rather than the more expensive options of developing new modems (for example, for the STM-1 and STM-R options).

16.4.7.5 PDH IDR link with SDH to PDH conversion at the earth station

This is the simplest option of all, but it does not provide the operator with true SDH compatibility. All of the advantages of SDH are lost, with additional costs incurred in the SDH to PDH conversion equipment. In the early days of SDH implementation, it may be the only available method, however.

16.4.7.6 Variations

Other variations have been proposed, usually point-to-multipoint variations of the above.

16.5 Asynchronous transfer mode

16.5.1 General

ATM is a protocol designed to be carried by both the PDH and the SDH hierarchies. It is not, as is commonly misunderstood, directly linked with SDH. ATM is more closely related to packetised data protocols such as the ITU-T X.75 system, but it differs from standard packetised protocols in that it is specifically intended to provide both constant and variable bit-rate transmission (as opposed to constant bit rate transmission from the PDH or SDH).

16.5.2 ATM and B-ISDN

ATM has been selected by the ITU-T as the preferred switching system for broadband ISDN (i.e., it performs both transmission and switching functions).

16.5.3 ATM cell structure

ATM consists of cells which are 53 bytes in length. Of these 53 bytes, 48 bytes are reserved for payload and five bytes are used as overhead. The payload area has no error checking built in, it is up to users to provide their own. The overhead of the cell has error protection; additionally, it contains the routing information (which is updated at each switching node), and a flag which indicates to the ATM switching equipment the importance of the particular cell. Cells may be labelled as unimportant if, for example, the user exceeds their agreed nominal bit rate. If the network then suffers congestion, the unimportant cells may be discarded by the next adaption function (ATM switch). This technique is called policing.

16.5.4 Adaption of ATM cells into SDH and PDH transmission networks

ATM cells are presented to the PDH or SDH network node by adaption equipment, and are routed around the constant bit rate network by ATM switches. A virtual path is set up between two switching nodes; a virtual-path connection is a series of concatenated virtual paths and is intended to eliminate variable delay between the possible different cell routings for delay-sensitive payloads (e.g. speech), by maintaining a constant path length for the duration of the transmission.

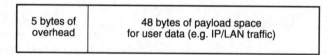

Figure 16.12   Simple sketch of ATM cell

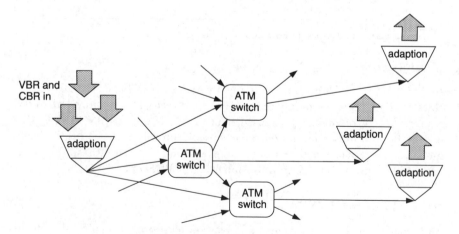

Figure 16.13   Principles of ATM cell adaption

Since ATM will be supported by SDH or PDH it should have no direct impact on satellite system design as the satellite link will be designed to carry the SDH or PDH signal. However, the QoS requirements given in I.356 suggest that the performance requirements of G.826 may not be adequate for ATM transmission in all cases. This is under consideration in both the ITU-T and the ITU-R.

ATM cells can be mapped into an STM-1 payload space or a 34 Mbit/s PDH signal (ETSI countries only), the mappings are given in ITU-T recommendation G.832.

16.5.5 Current ATM trials

Current ATM trials in Europe are using ATM switches to function as crossconnects between European nodes, with PDH transmission of cells between the switches. Project James, which is an example of this, interconnects European academic institutions in a similar way to the UK Janet network.

16.5.6 ATM performance parameters

Key performance parameters for ATM are cell-delay variation (CDV) and cell-loss ratio (CLR). CDV is a result of the dynamic routing possibilities of ATM, and has a direct impact on the size of the buffers required in the receiving equipment to be able to reconstruct the original signal. The overall instantaneous delay for a given virtual path will be determined by the longest individual-cell delay which the receive buffer can treat.

Cell-misinsertion rate (CMR) is another key parameter, but is likely to be a secondary result of error activity. If a cell header is damaged by error activity, it could be inserted into another stream, producing an additional cell for that stream. Other parameters are cell-error ratio (something which can only be estimated in service), severely-errored cell block (based on a variable block length which is bit-rate dependent, again, only estimatable in service) and SESatm, which is derived from SECB.

These parameters are given in ITU-T recommendation I.356 against the various quality-of-service classes for ATM. I.610 provides maintenance principles and functionality, including defining maintenance cell flows. M.2201 (just begun) will provide bringing-into-service and maintenance limits for ATM virtual paths, including those supported by satellite links.

16.5.6.1 Delay and echo cancellation

The possibility of using echo cancellation associated with ATM is being considered at present. Because of the wide range of possible CDVs, even in national networks for smaller countries such as the UK, echo cancellation may be considered for trunk calls.

16.5.6.2 *ATM and error performance*

Cell-loss ratio is the main error-performance measure of ATM. Since the payload space within an ATM cell is not performance monitored by the transmission network, it is not possible to determine whether the payload of an individual cell has been errored during transmission. The loss of complete cells (i.e., unrecognisable headers) is measurable, however, and is therefore used instead. (*Note:* the transmission network, SDH or PDH, provides some measure of error performance of its payload space, which will include all of the ATM cells transported. However, it is not possible to relate errors detected by the transmission network to individual ATM cells transported.)

16.6 Major network features which impact on service-carrying ability

This Section deals with three network features which have a major impact on the service-carrying ability of the network, namely: digital circuit multiplication equipment (DCME), synchronisation and signalling. They are all features which are of relevance to satellite integration into the network, although many of the problems associated with them are common to both terrestrial and satellite-based networks.

16.6.1 *Digital circuit multiplication equipment (DCME)*

DCME enables network operators to make maximum use of their circuits by concentrating a number of calls from a set number of circuits onto a smaller number of circuits. This is achieved by exploiting the considerable redundancy which exists in any particular telephone call, e.g. people spending a large amount of time listening rather than talking, speech bursts having pauses between words/phrases/sentences and 64 kbit/s PCM coding being very inefficient. Efficient bandwidth utilisation is crucial to the economic viability of satellites in the main network, and so this technique is used extensively, particularly on IDRs in the INTELSAT network. It is also used on long-distance submarine cable systems such as the transatlantic optical-fibre cable TAT 8.

Figure 16.14 is a conceptual diagram of DCME showing the two functions responsible for the circuit gains which are achievable (typically 4:1), namely: digital speech interpolation (DSI) and low-rate encoding (normally adaptive differential pulse-code modulation, ADPCM), although LD-CELP, which can theoretically achieve 4:1 compression gain, is being trialled for use in the network. (*Note:* circuit-multiplication equipment is also used in some parts of the network which makes use of one or other of these techniques independently.)

DSI exploits the gaps between speech bursts (including periods when a person is listening rather than talking) by only assigning a circuit to the call when speech activity is present. Gains of over two to one are possible using this technique, but this figure can be dramatically reduced depending on the number

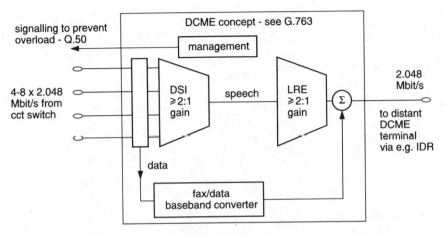

Figure 16.14   Digital circuit multiplication hierarchy (DCME) concept

of calls/circuits being scanned (because of the statistical nature of the technique) and the proportion of data calls (because these calls do not have the silent periods present in speech signals and therefore tend to hold a circuit permanently for the duration of the call).

ADPCM is a more efficient coding scheme than PCM, resulting in a nominal bit rate per call of 32 kbit/s rather than 64 kbit/s. LD-CELP, a recent development, can code at 16 kbit/s, thus enabling up to 8:1 gain with DSI. This technique is under trial for inclusion in the network. These schemes do, however, introduce extra distortion into the signal and this is particularly important for voiceband data calls. Special facilities have been built into the equipment to cope with these calls.

16.6.2 Network synchronisation

The majority of national digital networks operate synchronously at the primary rate (2 Mbit/s in Europe), i.e. the 2 Mbit/s links and switches within a country are locked to a very accurate national reference clock (typically running to an accuracy of 1 in 1011). Interworking across national boundaries is achieved plesiochronously by the use of buffers which compensate for the timing difference between the national reference clocks, see Figure 16.15.

Special care needs to be taken when introducing satellite links into this configuration. This is because satellites introduce Doppler shift into digital signals passing through them owing to the orbital variations which occur even on satellites which are nominally geostationary. The means by which this Doppler shift is taken out of the signal varies and the techniques used are outside the scope of this Chapter (an example of how this can be done is shown in Figure 16.16), but the important thing to note is that timing between digital networks is potentially a major source of impairments and it is particularly important to ensure that the

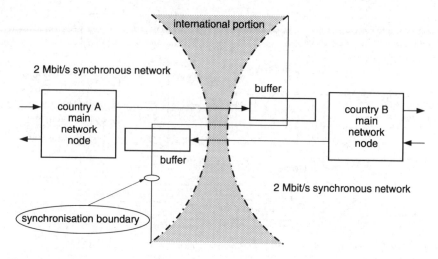

Figure 16.15   Network synchronisation

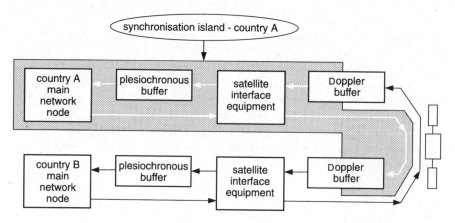

Figure 16.16   Satellite synchronisation

extra complication of Doppler shift which satellites introduce is taken into account early in the network-design process.

In the SDH environment, synchronisation takes place at the SDH primary rate, which is the STM-1 155 Mbit/s rate. The approach is similar to that shown for the PDH. The synchronisation strategy for SDH over satellite has not been determined, since SDH over satellite is not yet beyond the prototyping/development stage. However, the requirement to buffer the SDH signal and the Doppler effects will be similar to those in the PDH environment.

16.6.3 Signalling (routing control)

The signalling systems used between switched network nodes are becoming increasingly sophisticated and are opening up the possibility of steering calls through the network according to the specific requirements of the service being carried on that particular call (this is a particular feature of the ISDN), but the main function of the current signalling network is to connect one customer to another according to the number which the originator of the call inputs to the network. From a routing point of view, the only significant feature which the presence of a satellite link in the network introduces is the delay and associated echo control (see Sections 16.7.1 and 16.7.2). The signalling/routing is therefore arranged to minimise the likelihood that two satellite hops are included in the call routing and to disable intermediate echo control, if two sets of controls are in tandem.

The delay which satellites introduce has another impact on the signalling network, this time related to the function of the signalling systems themselves. Initially, signalling was based on a separate signalling channel associated with each individual circuit. Signalling systems between digital switches are increasingly based on common-channel signalling principles (the ITU-T standard being ITU-T no. 7) where the signalling information for a large number of calls is concentrated into a single data channel. However, whatever system is used, the basic function is the same, i.e. to transfer information from one switch to another. It is important to ensure that the propagation delay between the switches which the transmission link introduces does not adversely affect the signalling-system protocols: common-channel signalling being particular sensitive in this respect because of the large amount of information concentrated onto one signalling channel controlling a high number of calls. All the major international network signalling systems (ITU-T nos. 5, 6, 7 and R2) have been designed/adapted to be compatible with operation via satellite.

16.7 Satellite system performance in relation to service requirements

Network performance can be characterised in a number of ways, but for main network services on the digital transmission network, it can be considered in terms of just three parameters, namely echo, delay and digital errors. Echo is outside the control of the satellite system designer, delay is in theory controllable but in practice not and errors are controllable within certain bounds. Nevertheless, all three must be considered by a network designer intending to use satellites as part of that network.

16.7.1 Echo

Echo occurs on the public switched telephony network (PSTN) from two sources: impedance mismatches at the 2/4 wire-conversion point and acoustic echo from the earpiece to the mouthpiece of the telephone handset. Therefore, all telephony calls suffer from an echo of the original speech but the signal level of that echo will vary. It only becomes a problem when the associated propagation delay of a connection is so long that a talker begins to distinguish the echo as a separate signal from his/her original speech—it then begins to have a very disruptive effect on a conversation.

The precise amount of delay that can be tolerated depends on a number of factors but it is not only a problem for satellite-routed calls, as one-way transmission delays of less than 50 ms have been shown to cause customers difficulty because of echo-related effects. Nevertheless, some form of echo control is always advisable on satellite-based networks carrying voice traffic, irrespective of the associated terrestrial delays (the one-way propagation delay between satellite earth stations via a geostationary satellite is approximately 260 ms—see ITU-T recommendation G.114). This control is either by means of echo suppressors, which introduce extra attenuation into the return direction of transmission to reduce the level of the echo or, more recently, echo cancellers, which model the expected echo and subtract this from the return path to cancel out the echo.

Both these methods of control usually need to be disabled during data calls, which require an uncorrupted path in each direction of transmission.

16.7.2 Delay

Even on networks with perfect echo control, delay can be a problem in its own right. Delays above 400 ms can cause voice customers problems and the level of difficulty which they experience continues to increase as the delay rises. Data protocols are sensitive to delay and, although it is standard practice to design protocols for use on the main network to take account of satellite plus associated terrestrial delays, these delays must be kept within bounds. Providing this is done, satellites can and do play a full part in the main network, including the evolving ISDN. (ITU-T recommendation G.114 (Blue Book) covers this subject in more detail and also shows the delays which can be expected through other elements of the network.)

The net result of this is that double satellite connections should be avoided. It is possible to get a double hop on the current main network but this is very unlikely owing to the call-routing algorithms employed and careful network planning. Typical delay values for UK/USA are 45–75 ms by submarine cable and 260–290 ms by satellite.

16.7.3 Digital transmission errors

The major parameter (apart from delay) which characterises the ability of a digital network to transfer information is its error performance. The existing

Figure 16.17   Typical international 2 Mbit/s voice network with cancellers

main network is nominally designed to ITU-T recommendation G.821, although this recommendation has certain drawbacks which make it difficult to relate it directly to the actual network and a new recommendation, G.826, is currently being drawn up by the ITU-T which is expected to address these drawbacks.

Maintenance of satellite links is also described in ITU-T recommendations. The key recommendations for the main network are M.2100 and M.2101.

16.8 Standards

This Section outlines the major international telecommunications network standards within the ITU-T and ITU-R which relate to satellites in the main network, in addition to those mentioned above. Regional standards also exist from bodies such as ETSI in Europe and ANSI in North America but, wherever possible, these are aligned with the worldwide standards of the ITU-T and ITU-R.

16.8.1 ITU-T recommendations

The current ITU-T recommendations are contained in the Blue Books which were agreed at the end of the 1984–1988 study period at the IX plenary assembly in Melbourne. Owing to reorganisation within the ITU in order to make it more responsive to technology developments, some recommendations are being given accelerated approval, and all new or modified recommendations are now published as white booklets. Where no modifications have been made, the Blue-Book version will remain as the current standard.

It is worth noting that these recommendations are exactly that, recommendations, and are not mandatory, but network operators go to great lengths to conform to them wherever possible as it considerably simplifies network interworking. They are arranged in a series of groups, the ones of most interest to main-network satellite operation are as follows:

E series — telephone network and ISDN;
G series — transmission;
I series — ISDN;
M series — maintenance;
Q series — switching and signalling;
V series — data communication over the telephone network;
X series — data communication networks.

The ITU can be accessed at URL http://ties.itu.ch, where a full database of recommendations is held. Searching is possible against descriptions, keywords and titles. Ongoing work items can be downloaded (the ITU standard is Word for Windows 7), whereas completed recommendations are available either from the online bookshop or by subscription to 'Recommendations online'.

16.8.1.1 Recommendation G.821

This begins by characterising errors by means of two parameters:

(i)   Errored second (ES) — a second containing one or more errors.
(ii)   Severely-errored second (SES) — a second with a BER worse than 1 in 10^{-3}.

These parameters relate to an end-to-end 64 kbit/s ISDN call but, in practice, have been used in general to engineer the whole main network. Percentage limits are specified for each parameter and apportionment rules outlined for splitting these limits across the network. The new recommendation G.826 is being developed to cover performance objectives at higher bit rates and it is expected that this will be more closely aligned to the actual structure of the transmission network.

16.8.1.2 Recommendation M.2100

Maintenance performance limits for both SDH and PDH at all bit-rates are detailed in M.2100. These limits are used when first commissioning transmission systems, when bringing specific circuits (between operators) into service and for day-to-day maintenance activity. Day-to-day maintenance activity is usually performed when systems are carrying traffic, so it is necessary to make measurements of error performance in service.

M.2100 defines how to derive errored and severely-errored seconds from PDH transmission systems in service. The details of these derivations are beyond the scope of this Chapter, however, it should be noted that the maintenance limits for satellite links are given in this recommendation.

16.8.1.3 Recommendation G.826

Errored and severely-errored seconds at 2048 kbit/s and above are defined and a further parameter is given, as follows:

(i)   Errored second (ES) — a second containing one or more errored blocks.
(ii)   Severely-errored second (SES) — a second with thirty per cent or more errored blocks.
(iii) Background block error ratio (BBER) — this is the ratio of errored blocks to the total number of blocks, and is designed for in-service measurement in both the plesiochronous and synchronous hierarchies.

Note that for the PDH, recommendation G.826 refers to recommendation M.2100 for the derivations of ES and SES.

16.8.1.4 Recommendation M.2101

This recommendation provides the same function for the SDH as M.2100 does for the PDH.

These parameters (ES, SES and BBER) are not readily usable by satellite-link designers and so ITU-R Study Group 4 has produced ITU-R recommendation 614 which gives a method of translating the ITU-T objectives specified in terms of ES and SES to a BER against percentage of time mask which can be more readily employed in link design. Error performance, as currently specified, is not an issue for PCM-encoded voice calls as they are very robust to transmission errors. This is not the case for data services (both 64 kbit/s data directly accessing the digital network and, to a lesser extent, voiceband data) and customer expectation of the level of performance of these services is likely to increase as their applications become more sophisticated and widespread.

16.8.2 ITU-R

The current ITU-R standards are documented in a series of recommendations, as in the ITU-T, but much of the background material is in a set of supporting reports. The ITU-R is currently reorganising its documentation to more closely align with the ITU-T practice of putting greater detail in its recommendations but, for the moment, the following recommendations and report are of most relevance to main-network operation (there are obviously a wide range of ITU-R recommendations relating to more detailed aspects of satellite design and operation but these are outside the scope of this Chapter.

16.8.2.1 Recommendation 614

Allowable error performance for a hypothetical reference digital path in the fixed satellite service operating below 15 GHz when forming part of an international connection in an integrated services digital network. This recommendation provides bit-error probability/link-budget masks to achieve the requirement for error performance of the link. It is intended to ensure that a 64 kbit/s channel, when carried by satellite, will meet the requirements of ITU-T recommendation G.821.

16.8.2.2 Report 997

Characteristics of a fixed satellite service hypothetical reference digital path forming part of an integrated services digital network.

This report includes an algorithm for determining the error performance of a satellite link and relating it directly to the periods of BER against time specified in the BER mask of recommendation 614.

16.8.2.3 Recommendation 1062

The function of this recommendation is similar to that of recommendation 614; it is to ensure that paths at bit rates of more than 64 kbit/s (e.g. up to 140 Mbit/s PDH and 155 Mbit/s SDH) will meet the requirements of ITU-T recommendation G.826.

16.9 Future network developments

16.9.1 Optical-fibre-cable developments

Gbit/s optical-fibre systems are now becoming a reality and this means that the per-channel cost of such systems will make it increasingly difficult for satellites to compete on major arteries of the network where these large-capacity systems can be fully exploited. The result of this is that satellite capacity is being released for better use within the network, rather than standing by in case of failure.

Direct ATM over satellite trials were held in Europe recently, in which BT participated. The results and implications of that trial are presently under consideration.

16.9.2 Mobile voice and telemetry

BT's GlobeTrack service provides the capability to monitor vehicles using telemetry almost anywhere within Europe. This has several major customers, and can provide capability such as vehicle status (fuel, temperature etc.), payload status (e.g. refrigeration systems for food) and location/speed etc. Mobile telephony can also be added.

16.9.3 Internet protocol

IP is likely to be transported within the main network using ATM. Satellite links are eminently suitable for this, since both ATM and IP are not significantly delay sensitive. A North American company is considering launching a large number of nongeostationary (i.e. low-orbit) satellites, with intersatellite communications, to provide high-bandwidth capability for IP users.

The growth in IP traffic transported by the main network has been considerable over the last few months, and is expected to increase. This is resulting in great demand for high bit rate connections from Internet service providers. Again, satellites provide a good opportunity for handling such traffic.

16.9.4 Network management

One of the largest concerns for most major operators is the management of the network. The telecommunications markets in many countries are being opened up to competition, which requires operators to reduce their cost base. Improvements in network management will be sought continuously to reduce costs.

The ITU-T has been very active in developing this area (known as TMN), and a visit to the ITU web page will show the vast amount of work going on in this area.

From a satellite viewpoint, it will be necessary to consider integration of satellite-link management with general network management.

16.10 Conclusions

16.10.1 Unique strengths

Satellite communication systems have unique strengths which make them ideally suited to a range of specialised applications of mobile, television and data services. They are also compatible with applications in the main network, provided that certain basic constraints are taken into account during the network-design stage. They are especially cost effective for thin routes where compression can be used, and locations which are generally isolated.

16.10.2 Differing constraints

The constraints differ depending on whether the application is for leased services, where tight control can be exercised over the individual circuit routing, or for switched services, where the routing is less predictable from call to call.

16.10.3 Future developments

Future satellite usage in the main network will have to take into account network developments, both in terms of the new services which will evolve, such as broadband ISDN, and network developments, such as SDH, ATM and the increasing capacity of optical-fibre systems.

Mobile telephony via satellite (using hand-portable earth stations) is presently under development; air and land mobile systems using the INMARSAT network are gradually being introduced, giving European or worldwide coverage using one mobile terminal. Mobile uses for satellite are growing rapidly, and are likely to be the main area of expansion for satellites in the foreseeable future.

Chapter 17

Digital audio broadcasting by satellite

P. Shelswell

17.1 General introduction

The first reaction to this subject is to ask why should anyone want to broadcast radio programmes from a satellite, and why should the broadcasts be digital?

Digital broadcasting is an evolving technology which will provide an enhanced variety and choice of programme, and a rugged delivery mechanism. These are the main driving forces in broadcasting at the moment. Although the advantages of digital broadcasting were identified many years ago, there are at the time of writing very few all-digital transmissions. The reason for the limited availability is that the combination of technical theory and production technology is only just starting to produce products which can compete in the marketplace.

Satellite broadcasting has gained a place in the market already. Satellite television is available in many places, with a wide choice of programmes. On the other hand, there is very little satellite sound broadcasting. Again, it appears to be a matter that is dominated by the needs of the market. Satellite television provides a useful product, whereas satellite radio is not yet offering something which people feel that they must have. To understand why, we need to understand what it is that people expect from radio.

In the early days of radio broadcasting in the 1920s, transmissions used the lower frequency bands and AM (amplitude modulation), opening the door on a wealth of new entertainment and information opportunities. Listening to the radio rapidly developed into a popular activity. Those of us who listen to the early radios in museums often wonder about the low quality of the reproduction in the formative years, but the novelty and excitement of the programmes took priority.

At the beginning, the quality of service offered by the broadcasters matched the available receiving equipment. There has since been a dramatic improvement in the quality of sound reproduction equipment which has shown up the faults in the broadcasters' chain. Partly in response to this, in the 1950s, FM radio broadcasting began in the UK, and only now in the late 1990s are we starting to see the extension to digital broadcasting. In addition to improving the sound quality, the move to digital technology has provided a more rugged transmission mechanism which prevents much of the noise, distortion and interference found in more traditional broadcast systems.

Radio broadcasting has traditionally been a terrestrially-based service. Satellites were not invented until well into the lifecycle of radio, and there is a considerable momentum in the broadcasting industry which makes it difficult to alter an existing service. Even now that satellites are becoming a common part of the television industry, we are still not seeing much of a trend towards satellites for radio broadcasting. The reason lies in the nature of the business.

There are several types of radio broadcaster. They range from small community stations, through local and regional stations, to national and international broadcasters. The smaller radio stations are an important part of the system, with many dedicated listeners.

Terrestrial transmitters and networks will satisfy many of these operators. The community and local services can make use of any option. Whether the service is AM or FM, it is easy to engineer an economic system to provide coverage of the desired community. Satellites will play little part in their business.

However, when we need to broadcast to larger areas, the terrestrial network is not going to provide ideal reception conditions, and the prospect of satellite delivery becomes much more attractive. AM services operate in frequency bands which are affected by changes in propagation conditions during the day (and over longer time periods). This creates problems, well known to those who listen to medium wave at night or short wave at any time of day. During times of good propagation, signals arrive from a long way away and cause interference. During times of bad propagation, even the wanted signal may not be receivable with any certainty. There are some times when the system works as intended, but these times are certainly not as frequent as most listeners would like. As a consequence, many broadcasters have moved from AM to FM services. With FM, any programme area which is larger than about 70 km radius (the typical distance to the radio horizon from the top of a transmitter mast) needs several transmitters to provide satisfactory coverage. The cost of the network and the cost of providing circuits from the studios to the transmitters can escalate rapidly.

For international broadcasters, short-wave transmissions provide the major method of communicating with the public in distant lands. This method of broadcasting is, in many cases, compatible with normal domestic receivers because short-wave transmissions are also used by the national broadcasters in the target countries.

Now this is changing; national broadcasters are changing to FM and the audience is following. As a consequence, the national audience is listening less to both the national and international transmissions on short wave. We are also starting to see the introduction of digital broadcasting in many countries around the world. This could lead to a significant long-term change in the habits of the listening public, which could possibly accelerate the demise of short wave for international transmissions.

For all broadcasters there is therefore a need to find a replacement for short wave. Whether this is terrestrial or satellite, analogue or digital is a choice for the broadcaster and depends on the particular business.

Satellite transmission offers a number of exciting possibilities. It can provide many national or multinational services (a panEuropean service, for example, has some attractions to the European Union and its member states), as well as a direct replacement for international broadcasting on short wave. The choice and the variety could generate a new entertainment force. If used creatively with additional digital services (such as may be found on the Internet, for example) there could be a major untapped audience to be unfolded.

Of course, the listeners are interested in the programme and not in the method of delivery. Thus, provided that there is no unacceptable impairment in the programme caused by the delivery method, the listener will not care how the signal is received. In television, the debate between satellite and terrestrial television does not hinge on the quality of the signal, but more on commercial factors such as the choice of programme offered (key for the viewer) and the potential return (important for the broadcaster). The same logic dominates the discussions about digital audio broadcasting. There is no simple reason why one or other of satellite or terrestrial transmissions should be adopted. There is a complex interplay of financial, programme, technical and time problems.

Any system that is implemented will be subject to very careful scrutiny. It is not the technical proposal that needs to be right, but the complete business plan. It is important that the cost of the satellite, the quality of service and the audience size are all acceptable.

Inevitably, there will be several possible scenarios. One of the most interesting proposals assumes that both satellite and terrestrially-based services will coexist. If this does happen, then it is usually assumed that compatibility between the services is desirable. This inevitably leads to the design of a system which is optimised for one or other of the delivery mechanisms. It is difficult to ensure optimum performance on both. If a combination of satellite and terrestrial services does not occur, then the assumption could lead to a design of system which is suboptimum, and other options may be more suitable.

In this Chapter, both options will be explored. First, the constraints of compatibility with a terrestrial system will be considered, using the Eureka 147 DAB system as an example. Then satellite options for a compatible and noncompatible approach will be discussed.

In the discussion it will become clear that satellite audio broadcasting is not going to provide all the answers. The listener has grown used to reception on portable receivers, without the need for careful antenna alignment. In many of the usual places where radio is used, satellite transmission will not permit this option without some specific reinforcement of the signal.

17.2 The Eureka 147 DAB system

The Eureka 147 digital audio broadcasting (DAB) system was developed by a European collaborative group to provide a radio system which could be used on satellites, cable systems and terrestrial circuits. The assumption was that no listener would be willing to pay three times for similar types of domestic equipment, even if they could receive more programmes as a result. The transmission standard allows the same signal to be broadcast over all types of delivery mechanism.

There is a penalty to be paid for this compromise. The system can only be optimised for one type of bearer. In practice, the terrestrial system is the most difficult to make work properly, and in optimising the system for terrestrial reception, the satellite link is suboptimal.

The Eureka 147 DAB standard is now being implemented in many countries around the world as a terrestrial system. Although there have been several satellite experiments, no satellite system has yet been formally proposed using this standard. The standard encompasses the sound coding, the modulation system and the data multiplex, and the whole standard has now been adopted by the European Telecommunications Standards Institution[1].

17.2.1 The sound coding

To the listener, a key item is the quality of the sound. Although compact discs offer a very high level of quality which has been accepted by nearly all, it requires a bit rate of about 1.5 MBit/s to transmit a stereo sound signal. This bit rate is a consequence of sampling the signal at 44.1 kHz, and coding each of the stereo samples with 16-bit precision. With the current pressure on spectrum, we would not now be allowed the luxury of transmitting such a high bit rate.

Bit-rate reduction systems for broadcasting should ideally introduce no audible noise or distortion. Originally, the coding accuracy of CDs relied on the quantising levels of the system being small enough to provide negligible quantising noise in the absence of signal. Now it is recognised that the quantising noise and distortion can be masked by the signal when it is present. This is the basis of many bit-rate reduction techniques.

The threshold of perceptibility of sound is a complex function of frequency. It varies from person to person and has a strong dependence on the environment. If the environment is quiet, many low-level sounds can be heard (like a pin dropping), but if there is a lot of high-level sound (like the clash of dustbin lids), then many of the low-level sounds are inaudible: i.e. they are masked. This forms the basis of many bit-rate reduction techniques. If we cannot hear the pin drop, there is no point in transmitting the information which reproduces that particular sound. Or if the distortion caused by bit-rate reduction sounds like a pin dropping, we can tolerate it in our transmission standard. Things are not always so simple: the sound of a baby crying is very difficult to miss, even in a noisy environment. The understanding of psychoacoustic models has increased dramatically as a result of the need to improve digital bit-rate reduction standards.

In linear digital systems the quantisation noise is essentially uniform over the spectrum. The masking of noise is frequency sensitive and the system is defined by the most sensitive point on the masking curve.

In systems such as DAB, the signal is processed into a number of frequency bands. The masking thresholds for these bands are calculated and used to define the maximum amount of quantising noise or distortion which can be tolerated in each band. Those bands which contain sounds at levels below the masking threshold contain no useful information and there is no need to transmit anything. It would not be heard anyway. When there is significant energy in the band, it only needs to be coded to an accuracy which ensures that the quantising noise is below the masking threshold. As some bands contain a high level of signal, their masking threshold is high, allowing an increased level of quantising noise and distortion or, in other words, the quantising accuracy is lower and the bit rate can be reduced.

The Eureka 147 group has refined this system, producing a system standardised as ISO 11172-3 (otherwise known as MPEG 1 layer 2) which provides good sound quality at bit rates between 192 and 256 kbit/s for a stereo signal. This is a factor of six improvement over the linear CD coding techniques. With acceptable distortion even lower bit rates are possible. Bit rates as low as 16 kbit/s are being offered with noticeable reduction in quality, but with clear intelligibility. As knowledge of the masking properties of the ear improve, and implementation techniques are refined, the performance has improved.

All modern bit-rate reduction systems use similar principles[2]. It is the nature of the analysis into subbands and the detail of the psychoacoustic model which tends to distinguish them. Even in the short time since the Eureka 147 system was finalised, there have been noticeable improvements in sound quality at low bit rates (64 kbit/s and below, for example). The improvement has been such that the early draft of this Chapter which quoted relationships between bit rate and audio quality for many of the moderate bit-rate options was out of date, and accurate figures to replace the original text are not likely to bear the test of time.

17.2.2 COFDM

Having identified an efficient method of coding the audio, how is it transmitted?

For a satellite system, there would be no problem with a simple arrangement such as QPSK with some form of error-correction coding to improve its power efficiency.

However, one of the accepted requirements is to have a receiver which is capable of receiving digital broadcasts via all delivery systems, whether terrestrial, satellite or transmitted by cable. The choice of system will then be influenced by the problems of all other broadcast paths. The difficulties of the terrestrial broadcasting path influence the decision on which system to use on a satellite.

If the path between a terrestrial transmitter and the receiver is unobstructed, then again a simple modulation system will suffice. Indeed, the NICAM 728 system[3] used for terrestrial television is an example of a QPSK transmission giving stereo sound to receivers in the home. This system does not work well, however, when the receiver is moving, or when there is a lot of multipath propagation affecting reception.

Multipath reception causes frequency-selective fading. Short-wave reception is a good example of the problems which this causes, but similar effects can also be heard on FM sound broadcasts; such fading is one of the reasons that digital transmissions are anticipated so eagerly.

The Eureka 147 group has developed a system called coded orthogonal frequency-division multiplexing (COFDM) which seeks to avoid many of the problems of a channel with selective fading[4]. In COFDM, data is transmitted using a large number of digital carriers which are spaced in a frequency multiplex over a bandwidth which is larger than the coherence bandwidth of the fading channel. Before transmission, a powerful error-correction code is applied to the data. The resulting signal is then modulated on the carriers, interleaving data which were originally sequential over well dispersed carriers, and over a time frame of many milliseconds. Thus, when there is a time-variant frequency-selective fade, although a lot of data may be lost, there is a high possibility that enough data is recovered for the error-correction code to recover the original data. With an appropriate choice of parameters, this provides a rugged transmission system. In the Eureka 147 system, each carrier is modulated using differential QPSK, with convolutional coding applied to the system. For the satellite version of the system there are 192 carriers in a bandwidth of 1.536 MHz. The modulation and demodulation processes can be carried out using signal-processing software, and so should become significantly cheaper if the current trend in integrated-circuit technology continues. In the modulator, the incoming data is used to define each carrier in the frequency domain. The frequency-domain representation is transformed into the time domain using FFT techniques to generate the signal, and the reverse process is applied in the receiver.

The multicarrier system is highly rugged and immune to echoes, Doppler shift and noise[4]. Given the fact that it is difficult to tell the difference between an echo and another transmitter which transmits exactly the same signal, it is possible to introduce cochannel repeaters into a transmitter network with little difficulty. This is a major feature, which allows the introduction of the so-called single-frequency network (SFN) on a terrestrial network and hybrid operation on a satellite system (small local retransmitters which pick up and rebroadcast a satellite signal in places of difficult reception).

In terms of performance, COFDM gives good results in a fading channel, where multipath distortion is the dominant degradation. It gives fairly good performance on a satellite channel, but is suboptimal in this respect, especially when compared with standard QPSK signals.

17.2.3 *The data multiplex*

In order to ensure that the bandwidth exceeds that needed to give some immunity to frequency-selective fading, the Eureka 147 specification provides a signal which is 1.536 MHz wide. With 192 carriers, each modulated with QPSK, in an 8 kHz raster, and with a symbol period of about 156 μs, this provides a data capacity of the order of 2.4 Mbit/s before error correction is applied, or 1.2 Mbit/s after error correction.

A simple calculation shows that such a system can carry several sound programmes: from five very high-quality stereo channels to as many as 64 very low-quality mono channels. There is considerable flexibility possible in the system. The content of the multiplex can be varied, allowing a different range of services to be offered during the course of a day. Different arrangements of sound and data can be signalled, resulting in a very powerful transmission system. It is the flexibility of carrying sound and data within a single multiplex that makes the opportunity so exciting. In the hands of a creative scheduler, the public can look forward to a varied choice of radio programme.

17.2.4 *Receivers*

The Eureka 147 system has now been standardised[1] and we are starting to see the first receivers coming onto the market. These operate over both the 1452 to 1492 MHz satellite band, and over the alternative VHF ranges which have been adopted by CEPT in Europe.

The receivers are, therefore, capable of both satellite and terrestrial reception. Good quality reception of sound has been demonstrated in many countries, and so there is confidence in the system. Many researchers are investigating new ways of using the data applications to provide increased value to the listener.

17.3 Satellite-specific options

Of course, the Eureka 147 system is not the only possibility. The Eureka system assumes that there are likely to be both digital terrestrial and satellite services operating in the same place together. On that assumption, the logical conclusion is that the same system should be used for satellite and terrestrial broadcasting.

If there is no terrestrial digital service, this assumption is not valid, and a dedicated system can offer technical advantages. The difficulty is that decisions cannot be made on a country by country basis. The nature of satellite broad-casting is that the same system is likely to provide a service to a whole continent.

One of the reasons that sound broadcasting from satellites has been slow to develop is that the link budget is marginal. The amount of power available to the receiver from a satellite transmission is going to be at the low end of the acceptable range. To overcome this, most workers suggest a single-carrier QPSK modulation system with both convolutional and block coding to reduce errors from noise. The use of a higher-level modulation, such as 8PSK or 16QAM, may be more spectrum efficient, but cannot provide the ruggedness that is needed.

In the USA, both the Voice of America and a company called WorldSpace have proposed satellite systems, based on QPSK, with a rate-convolutional code, reinforced with a Reed–Solomon code to improve the error perfor-mance.

Compared with the Eureka 147 system, there are some significant differences. The Eureka system uses differential coding of the data. This makes sense in a mobile channel, but leads to about 3 dB worse performance in the simple gaussian channel of a satellite system. The additional use of a Reed–Solomon code gives a small improvement, at the expense of greater bandwidth require-ments. Also, the multicarrier nature of the COFDM system has a nonconstant amplitude. The signal can be distorted by amplifiers unless there is a small reduction in power level through the nonlinear stages of the satellite system (known as output back off). This, of course, results in a lower permissible average transmit power. QPSK is not so sensitive as COFDM, and so the US systems should be able to provide about 6 dB better performance over the satellite than does the Eureka system.

In other respects, the audio coding is similar to that of Eureka, but more complex versions are used to work better at the lower bit rates. The multiplexes are also different, enabling a different, but still broad, range of broadcasters to share the same transmission. In the case of the WorldSpace satellite, onboard processing will be used to enable uplinks on different frequencies to be combined into a different time multiplex, broadcast as a multiple-channels-per-carrier (MCPC) signal. This provides a useful operational feature for broadcasters who may share access to a common satellite.

17.4 System engineering

So far as the listener is concerned, there should be no distinction between a ter-restrial and satellite system. The satellite just provides a transmitting tower in the sky. To enable a satellite system to work satisfactorily, the complete system has to be engineered to ensure that reception is reliable. As will be seen, this is not easy, and as a consequence each part of the system needs to be carefully designed. There are several technical areas of interest when defining a satellite system:

- what receivers are assumed?
- what satellite orbits should be used?
- how many programmes should be broadcast and to where?
- what frequencies are appropriate?
- what power is needed?
- what is the cost?
- what is the time scale?

There are many nontechnical factors in the decision.

17.4.1 Receivers

The main use of radio is in the home and, in the developed world, in the car, and the vast majority of listening is on portable receivers. We therefore need to consider a receiving installation which requires little setting up, which is insensitive to movement and which may easily move into areas where reception is subject to blockage from trees, bridges and, of course, buildings.

As a result, it is desirable to use a receiver which does not require a steerable antenna. If the antenna is directional, it could easily be moved, by accident, to point in the wrong direction. Some gain may be acceptable, but there is a compromise to be reached. The usual option to enhance reception if the antenna gain is low is to improve the noise performance of the front end. However, if a low antenna is used, the antenna will see a lot of the surrounding ground and buildings, and these will contribute noise at about 300 K. This restricts the G/T that the system can achieve. Compared with television systems, the lower antenna directivity of the sound system limits the ultimate power which can be received.

Portable and mobile receivers are not necessarily placed in the best position for reception. There are often obstructions in the path to the satellite, and this path blocking requires a relatively large margin to overcome the signal attenuation. At the frequencies used for DBS television, any large building or vegetation gives considerable loss of signal. As the frequency drops, so does the loss. One of the interesting results coming from studies at the University of Bradford[5] is the way in which the loss becomes more acceptable as the frequency drops and as the elevation of the satellite increases. Using these results it is possible to design a system which would give acceptable reception out of doors with mobile receivers.

Even so, there is still some difficulty in providing reliable reception indoors, except to radios near to windows on the sunny side of the house.

17.4.2 Propagation and satellite orbits

The message that comes from studies at the University of Bradford is that the geostationary orbit is not really suitable for transmitting to countries such as the UK because the satellite elevation is so low (less than 30°). As a consequence,

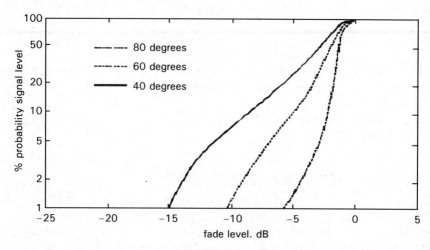

Figure 17.1   Statistics of fading. When a vehicle is moving along a road there is a chance that reception from a satellite will be restricted by tree and building shadow. This set of curves shows how the statistics of the depth of fade vary with the elevation angle of the satellite. (The curves are taken with permission from Reference 5, which provides figures for a wide range of possible types of environment)

there is considerable blockage of signals by modestly-sized buildings. The increase in link margin needed to overcome the blockage results in satellites which are too expensive.

Measurements of the typical variation of signal strength show that there is a greater variation as the elevation angle drops. For satellites which are low in the sky, a large margin is needed to compensate for the attenuation caused by trees, bridges and buildings. This makes the geostationary orbit unsuitable for the higher latitudes.

On the other hand, highly-elliptical orbits (HEO) are much more suited to this type of application. The Molniya or Tundra type of system provides a satellite which appears to be nearly stationary and overhead for several hours. By using a constellation of such satellites, it is possible to provide continuous coverage. The benefit of this is that for reception in cars etc. there is little blockage from buildings. The margin required to cope with trees is smaller and, as a result, there is more flexibility in the system. Indoor reception becomes more of a problem, however, as the signal now has to penetrate the roof and possibly one or more storeys of a building. Studies by the European Space Agency do indicate that the HEO option is better for many of the countries in the northern part of Europe. Equatorial countries, of course, still benefit from GEO satellites, and countries such as Spain would have no obvious preference for one system or the other.

There are also options to use low-earth orbits (LEOs). These systems are closer to the earth, generally use a directional transmit antenna and so provide a

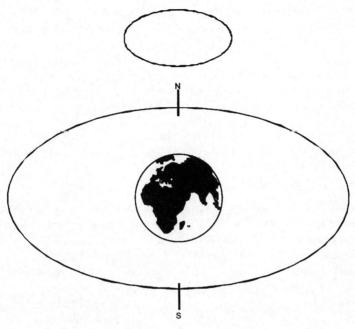

Figure 17.2a   The ideal satellite orbit is directly overhead. The standard geostationary orbit is suitable only for equatorial countries. For countries which are non-equatorial, a rather impractical orbit is needed

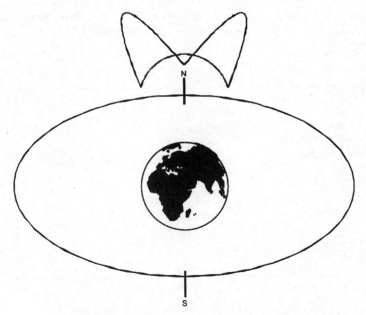

Figure 17.2b   The impractical orbit of Figure 17.2a can be approximated by using parts of the arcs of three (or more) separate highly-inclined orbits

higher signal level on the ground. More than one satellite may be visible at any one time, offering diversity transmission and so reducing the effects of blockage. To balance this, there is a need for many more satellites: typical numbers required being around 70. To make this commercially attractive requires the satellites to be used by the vast majority of the countries over which they pass. This may be achievable in the telecommunications market, which may be the first to use the LEO systems in significant public ventures. It is too early to say how the broadcast market will respond.

17.4.3 Coverage requirements

Originally, studies of sound satellite broadcasting assumed that broadcast transmissions would be national, with little overspill. However, experience with television transmission shows that there is a strong interest in crossborder transmissions, and so wider beams are desirable. The requirement appears to be such that the beam at least covers an area in which one language is dominant. This does not necessarily lead to a simple statement of optimum beam size.

The transmitting antenna on the satellite is restricted in size. Any antenna larger than the diameter of the launch vehicle needs to be built in space. This is difficult, not necessarily perfectly reliable and, as a result, expensive. The size of an antenna that provides country-wide coverage (in Europe) is just around the upper limit of technology. It certainly makes wider coverage areas a more attractive technical proposition. On the other hand, if the antenna is too small, then the beam may be too wide, and then the required transmitter power becomes excessive. This prevents the widespread adoption of global beams.

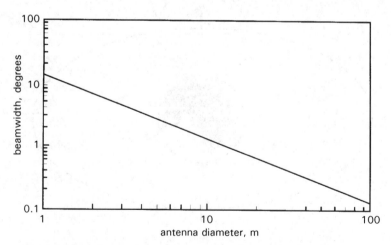

Figure 17.3   *The variation in beamwidth of the signal as a function of the antenna diameter (for 1.5 GHz)*

As a consequence, there is a convergence towards beams of the order of 1.5 degrees or so.

There are also some political and technical constraints. Not all countries are prepared to accept international transmissions over their territory, and put pressure on broadcasters to do all they can to prevent overspill. Also, not all countries have adopted the same frequency band for sound broadcasting, and so there is a potential need to restrict overspill to avoid interference.

17.4.4 *Number of programmes*

The number of programmes that the market will require is difficult to forecast. However, most studies predict that there will be a significant demand for capacity if the financial climate is right. Figures often heard are of the order of 12 programmes per country as a minimum at the start of service, expanding as time progresses.

17.4.5 *Frequency*

Before 1992 there was considerable debate about which frequency should be used for digital audio broadcasting by satellite. Now that the ITU has discussed the matter, the range 1452 to 1492 MHz has been allocated, but only with effect from the year 2007 in many countries. Before then, considerable diplomacy would be needed to start a service because of the large number of existing users of that range of frequencies. This simple picture is somewhat complicated, because the USA has not adopted this frequency range, having, alone, declined its use for satellite sound broadcasting. In its place the USA has adopted 2.3 GHz, and some countries have also accepted this as an option, in addition to the use of 1.5 GHz.

The use of 1.5 GHz is about optimum. Lowering the frequency is not going to give much benefit because of the excessively large transmit antennas which would be needed. On the other hand, significant increases in frequency are ruled out because of the high powers that would be necessary. Even moving to 2.3 GHz appears to require a significant increase in fade margin (and hence desirable transmitter power).

Use of the 12 GHz satellite band is really not suitable for systems with mobile reception.

17.4.6 *Power*

As with all satellite systems, the link budget is one of the most useful tools for analysis. In this Section, two link budgets will be analysed, one for a COFDM-type system and one for a QPSK-based system[6]. This latter example is not necessarily that proposed by either of the USA groups, but is provided in a format which makes comparative judgement a little more easy.

From this budget we can see why it has taken so long to start satellite sound broadcasting. The parameters are all biased towards their acceptable limits.

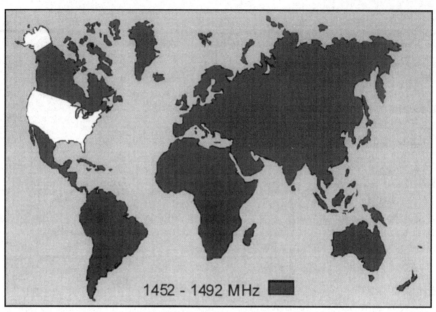

Figure 17.4a   Parts of the world permitted by the Radio Regulations to use L-Band for satellite sound broadcasting

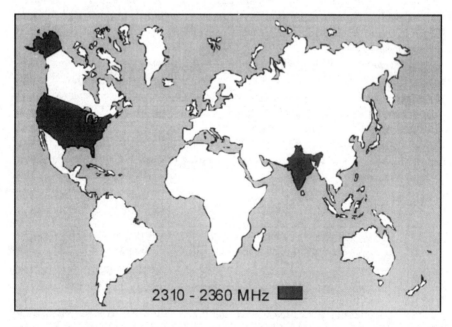

Figure 17.4b   Parts of the world permitted to use the 2.3 GHz band

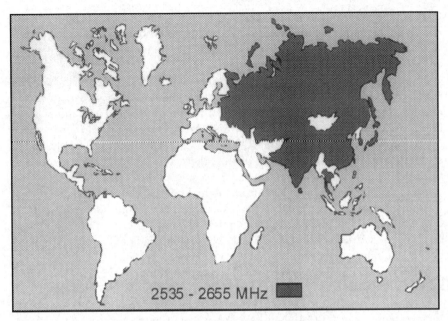

Figure 17.4c   Parts of the world permitted to use the 2.5 GHz band

Table 17.1   Example link budget for digital audio broadcasting by satellite

	COFDM	QPSK	
Required E_b/N_0	6.0	2.3	dB
Bit rate per stereo channel	224	224	kbit/s
Number of channels	6	6	
Required C/N_0	67.3	63.5	dBHz
HPA power	100	100	dBW
HPA back-off	3	0.5	dB
Useful HPA power	17	19.5	dBW
Output losses	−1.5	−1.5	dB
Antenna gain	36	36	dBi
Satellite EIRP	36	36	dBW
Spreading loss	−163	−163	dB
Polarisation loss	0.5	0.5	dB
Receiver G/T	−19	−19	dB/K
Effective area of isotropic antenna	−25	−25	dB (m²)
Boltzmann's constant	−228.6	−228.6	dBW/K/Hz
Downlink C/N_0	−72.6	75.1	dBHz
Margin	5.3	11.6	dB

Even so, the margin that is provided is not enough for indoor reception. It will not be adequate for reliable outdoor reception when the satellite is low in the sky. Only with high elevation angles is the margin high enough to cope with the obstructions from buildings.

The transmit power suggested is of the order of 100 W. In the case of satellite television, the use of higher powers was found to be more expensive and not necessarily reliable. This is one reason for the slow implementation of the broadcast plan adopted by the ITU, and the reason why the lower-powered options provided by ASTRA and EUTELSAT in the communications band, for example, are so popular. Because of this, there is some reluctance to propose systems for sound broadcasting which have higher transmit powers.

Some improvement in link margin might be feasible if a more directional transmit antenna were used. This example uses a 5 m antenna, with a beamwidth of 1.5°, which is ideal for a relatively large country. Larger antennas could be used, at a cost, but would only be suitable for small countries. Several beams would have to be transmitted to cover a larger country.

Some increase in the receiver's G/T is possible if more directional antennas are used. Unfortunately, the assumed G/T is based on an antenna with 6 dB gain, which is reasonable for a satellite which can be guaranteed to be overhead. If the satellite is in GEO then a lower gain would be needed to cope with the variable direction of arrival of the signal. This leads to a lower G/T. There is still considerable discussion about the achievable level of noise temperature which is appropriate to assume. The low-gain antennas will, of course, see a significant proportion of the environment at around 290 K. The assumed noise figure in this example is 300 K.

17.4.7 The cost

The results of studies into digital audio broadcasting from satellites have all indicated similar link budgets. It is possible to consider minor changes in parameters, with some change to the operational performance.

The key problem is the cost. In order to set the system in motion we need to know the cost and the likely chance of recovering the investment after a given period of time. When making investment decisions, the cost of financing has to provide a return which matches other opportunities. Figures up to about 40 per cent have been suggested as suitable targets!

Studies for the European Space Agency by British Aerospace, Touche Ross and Nokia have all led to much the same conclusion. With a large enough satellite it should be possible to provide a service to much of Europe. If there are sufficient programmes, and people buy receivers in sufficient numbers, then digital audio broadcasting from satellites could be financially viable. There is a small risk in the development of the technology, but this is outweighed by the risk of uncertainty in the development of the market. The optimistic analysis

Table 17.2   The financial budget

The cost of financing is		
		interest to bank
	plus	dividends to shareholders
	plus	loss on alternative investments
	plus	required return within business
	equals	cost of financing

shows that there is certainly a place for digital audio broadcasting from satellites.

It would appear that the cost of a satellite system is comparable to that of a terrestrial system. Much depends on the assumptions made in the analysis, and there is no clear-cut decision which can be made on purely financial grounds. Many of the considerations are subject to uncertainty and variation with time.

17.5 Experimental evidence

One of the main problems with doing work in the area of satellite sound broadcasting is that there are few opportunities to test the theory. Until the theory is developed, no-one can afford to launch a suitable satellite, and until someone has launched the satellite no-one can prove the theory. It is a chicken and egg problem.

However, some work has been done with satellites which have some suitable characteristics. Experiments with OPTUS B, an Australian satellite for mobile communications, first demonstrated the Eureka 147 system working to mobile receivers, even though the transmit power was relatively low (of the order of 40 dBW). These experiments were confirmed by the BBC with help from the Mexican Telecommunicaciones de Mexico and IMC, using the Mexican Solidaridad II satellite[7]. In America, the Voice of America has carried out tests of the QPSK-based system using the satellite TDRS (tracking data relay satellite).

All these tests have shown that it is indeed possible to broadcast to receivers with low-gain antennas. There is, therefore, a good chance that portable reception is a realistic goal. All the tests have shown that buildings cause severe attenuation. Without a direct path between the satellite and the receiver, reception will be poor.

This indicates that satellite broadcasts need to be considered as part of a broader delivery strategy which incorporates some form of signal enhancement. This may be large transmitters feeding small towns, or perhaps domestic repeaters feeding indoor receivers.

17.6 The way forward

The real issue with satellite sound broadcasting is finance. To justify the investment in a satellite system, the financial budget (which is, of course, somewhat different from the engineer's link budget) needs to be balanced. Financiers need to be convinced that they can recover the outlay required to set up a satellite service.

The cost of the satellite system itself is a small part of the total outlay. As an example, typical television transponders are priced commercially at about £1 to £5 million per annum, depending on what is being offered in the package. Let us assume that radio prices will be a little cheaper as they use less capacity.

For a country such as the UK, where there are over 50 million people, each person may own at least one receiver. The investment in receivers works out at considerably more than the cost of the satellite system.

Similarly, if we look at the cost of programme making, this can be an order of magnitude greater than the cost of the satellite capacity. It depends on whether the programmes are simple music programmes, based on the popular charts, or more serious talk shows with considerable detailed research and scripting, or even drama. There is a wide range of programme costs, but they all make the satellite costs seem small.

Therefore, there are three groups of people who have to take ownership of the system to make it succeed. The broadcasters need to invest in programmes, and see a way of recovering their investment through advertisement income, licences etc. The satellite operator must see a profit in the provision of capacity and the listener must be prepared to buy a new receiver. Only if there is an exciting variety of programmes available, which are not available elsewhere, is he or she likely to spend any significant amount for such entertainment.

It is this last emphasis on the listener that is the basis for the Eureka 147 group's plan to base satellite sound broadcasting on a terrestrial system. Terrestrial broadcasting is an easier and more certain development. Satellite broadcasting can ride on the back of an existing receiver base.

On the other hand, there is at least one organisation which believes that satellite sound broadcasting can be made to work without reliance on terrestrial options. The future should be exciting.

17.7 Conclusions

Satellite sound broadcasting is technically feasible, and appears to be financially viable when considered separately from other methods of delivery.

The big question is whether the business plan is sufficiently attractive to launch a service. With terrestrial broadcasting acting as a competitor, there is no firm conclusion.

17.8 References

1 'Radio broadcasting systems; digital audio broadcasting (DAB) to mobile, portable and fixed receivers'. European Technical Standard ETS 300 401, ETSI, Sophia Antipolis, France, February 1995

2 GILCHRIST, N.H.C., and GREWIN, C.: 'Collected papers on digital audio bit rate reduction'. Audio Engineering Society, September 1996

3 BOWER, A.J.: 'NICAM 728—digital two-channel sound for terrestrial television'. BBC Research Department Report, RD 1990/6

4 SHELSWELL, P.: 'The COFDM modulation system: the heart of digital audio broadcasting', *Electron. Commun. J.*, 1995, **7**, (3), pp. 127–136

5 SMITH, H., GARDINER, J.G., and BARTON, S.K.: 'Narrowband measurements on the high elevation satellite-mobile channel at L and S bands'. University of Bradford report on RD EC/0693/87

6 'Terrestrial and satellite digital sound broadcasting to vehicular, portable and fixed receivers in the VHF/UHF bands'. International Telecommunications Union special publications, Geneva 1995

7 ZUBRZYCKI, J.T., EVANS, R.H., SHELSWELL, P., GAMA, M.A., and GUTIERREZ, M.: 'Experimental satellite broadcast of Eureka 147 digital audio broadcasting from Solidaridad 2'. BBC Research and Development Department Report 1996/5

Chapter 18

Digital video broadcasting by satellite
G.M. Drury

18.1 Introduction

The space age began when the Russians launched the first man-made earth-orbiting device, SPUTNIK, on October 4 1957; this feat was followed, on October 26 1959, by the transmission to earth of television pictures of the far side of the moon by the Russian vehicle Lunik III.

When these space technology achievements took place, television was a mere 21 years old in practical terms[1]. At the end of the 1920s, when economic conditions were poor, the motion-picture industry was surviving through the appeal of the talkies which had been recently introduced. Several experiments and demonstrations of television had been conducted in Europe and the USA during the 1920s and early 1930s, using Baird's 30-line electromechanical scheme (UK) and various electronic systems using several hundred lines (various countries including the UK), these latter exploiting the recent development of electronic imaging devices such as the iconsoscope and the work of Zworykin, Farnsworth, Schoenberg and Blumlein among others[2]. Regular terrestrial transmissions using a standard high-quality electronic system, developed by the newly formed EMI company and Marconi, were began by the BBC in London in 1936 but were suspended when the Second World War began in 1939. This service was resumed on June 7 1946 by which time the pioneering article by Arthur C. Clarke, published in 1945[3] in the 'Wireless World' magazine, had envisaged broadcasting as one application of satellites operating in the geostationary orbit. Clarke's ideas took almost 20 years[4] to implement and the technical advances involved owed much of their success to rocketry development and funding from the military and allied sources as well as developments in communications technology. One major event on the way to realising Clarke's ideas occurred on July 11 1962, when the AT&T satellite TELSTAR was used to transmit live television pictures between the USA and Europe (France and the UK). This event brought satellite and television technologies significantly and directly to public notice[5]; one example of how the popular imagination was inspired by the event was the very successful popular music recording, named after TELSTAR, which was top of the best-selling popular music charts for five weeks and remained in the top twenty for almost five months during the period September–December 1962. This presaged the more recent success of satellites in delivering popular music directly to the home.

Early satellites such as SPUTNIK and TELSTAR were of low power and orbited subsychronously and so had to be used with very large and expensive earth-station equipment, which also had to be sited very carefully and be able to track the satellite as it moved through the sky. Both these factors clearly limited the use of TELSTAR for regular television transmissions except for very special events and for news exchanges—such was its trajectory, it was only simultaneously visible on both sides of the Atlantic for 18 minutes out of its 157 minute orbit period. Following on from the success of TELSTAR as a demonstration of the feasibility of building, launching and operating spacecraft for communications, several other experiments were conducted over the next two years, some attempting to exploit the more valuable geostationary orbit. Eventually another satellite, called SYNCOM, was launched in August 1964 and this used an Atlantic region geostationary orbit position. This craft was used successfully for television and telephony, as was a larger and more powerful successor called Early Bird launched on April 6 1965; this latter satellite was used for a number of television events during May 1965 but it was then acquired by the newly formed INTELSAT organisation (see Chapter 2) and was renamed INTELSAT I. Further generations of INTELSAT satellites, each more powerful and of greater capacity and reliability than its predecessor, ensued during the 1960s and 1970s, and each stage supported the rapidly growing international telephony network but also expanded the global reach of the television medium in all three oceanic regions of INTELSAT. Major political and sporting events were televised live on a global basis and soon such exchanges became commonplace, despite the intercontinental time differences involved. Exchange of programming between the USA and Europe in these early days was also considerably hampered by the different television scanning and colour standards used in the two regions (see also below). At first, the conversion process used analogue technology and equipment was expensive and unreliable but, when digital techniques were applied to the problem in the late 1960s and early 1970s, the cost, picture quality and operational reliability improved dramatically.

During these practical developments of satellite technology during the 1960s the close interest of the general public was maintained through the use of television coverage, often live, of launches of various space vehicles and other major events. For example, in preparation for a manned moon landing, targeted for achievement by the US administration by the end of the decade, a series of probes, moon encirclements and hard landings culminated with the soft landing on the moon of the exploration vehicle Surveyor I. BBC 1 viewers willing to rise and watch at 6.30 on the morning of June 2 1966 were rewarded by live pictures from this landing. This successful landing, and the experience gained in making it possible, paved the way for the event that will probably be remembered as being among the greatest of live television broadcasts; it took place on July 21 1969 when man first stepped, and was seen to step, onto the surface of the moon, an event impossible without a global network of communications satellites and the joint development of space and television technologies.

The activities of INTELSAT, and its various regional successors such as EUTELSAT, included from the start special facilities to deal with television transmission where each major earth station had a separate transmit and receive chain together with operational procedures and performance standards agreed throughout INTELSAT and the broadcasting community. Over the years the power and bandwidth available from spacecraft have increased considerably, as has the performance of low-noise amplifiers, and this has allowed the use of simpler, more compact and cheaper earth stations. This has been particularly useful for television newsgathering and the deployment of television receive only (TVRO) antennas at broadcasters' premises as well as mobile news teams with compact uplinks. The trend has continued towards very small aperture terminals (VSAT) and wider access to satellites for smaller organisations, and this has expanded the range of purposes to which satellites can be put in relation to television and to the organisations and companies able to take benefit from them.

In the late 1960s the satellite community became interested in digital transmission techniques which had been used for terrestrial telecommunications for some years. By the early 1970s the idea of time division multiple access (TDMA) on satellites and television in digital form had made great progress. The INTELSAT system had sponsored work at COMSAT (the organisation which managed the INTELSAT system) on the US TV standard NTSC (see below for more details) at a bit rate of about 45 Mbit/s. In 1980 experiments were performed in Europe with the orbital test satellite (OTS) to investigate the feasibility of transmitting PAL television signals in digital form at a bit rate of about 60 Mbit/s and using a transponder bandwidth of about 36 MHz. The bandwidth was defined by the standardised frequency plans and used spacings of 40 MHz between transponders, and this in turn defined the bit rate since the number of bits per Hertz of bandwidth is restricted to a value between 1 and 2 for the simple modulation systems, e.g. BPSK and QPSK, used at that time. The television signal processing involved reducing the bit rate from more than 100 Mbit/s (using PAL as a source at that time) by simple coding techniques such as differential PCM. These techniques have improved and diversified over the years since and now reduction ratios approaching 100:1 are feasible (see below).

The purposes for which these technologies were being developed included professional applications in transmitting television signals between broadcasting studios. The state of the art was such that the cost and complexity of this equipment made it inappropriate for other uses. More recent developments have changed this situation and digital satellite television directly to the home is now an established fact, as will be described below. Directness in this context means that there are no intermediaries between the programme source and the viewer or listener. Generically, this kind of approach is known in some circles as direct to home (DTH); clearly, in the satellite context, DTH can be synonymous with direct broadcasting by satellite (DBS). However, different operators have become established in different frequency bands where the frequencies allocated by international agreement (see Section 18.2 below) for the fixed satellite service (FSS) and the adjacent broadcast satellite service (BSS) are both used to deliver television directly to the home. Any services using the FSS band are planned to be shared with others, especially terrestrial telecommunications, and so are subject to some interference; they are also limited in power level to minimise interference from them to other services. The FSS bands were not planned for television services but advances in technology have made it possible to reach domestic receive-only installations by this means. The BSS bands are dedicated to broadcasting and so are not subject to interference from, and will not interfere into, other services and so may have higher transmit power levels than for the FSS bands, typically by 10 dB. Some television services delivered by satellites using the FSS bands are described as DTH, and those delivered in the BSS bands are known as DBS.

Direct broadcasting by satellite has been considered feasible for almost 20 years using analogue television standards and frequency modulation (FM) of the satellite carrier frequencies. Indeed, plans had been made in 1977 for European regional implementation of DBS but these were slow to materialise mainly for commercial reasons. From a perspective centred in 1997, much of the promise of television by satellite has already been realised since commercial services exist using both analogue and digital techniques; digital technology offers much more scope for the future and is therefore developing well towards more sophisticated and technically advanced systems all over the world. The lesson from the early experiences is that technology alone will not guarantee commercial success and, regardless of the inherent worth of any new digital technology, it will only succeed if it meets commercial needs.

Given this essential fact, this Chapter begins with a broad perspective of broadcasting, including the commercial and regulatory aspects, as well as a brief review of current and developing television standards, in order to place in context a description of current digital systems, both commercially available and in development. Although this Chapter is primarily concerned with television, it should be noted that the audio broadcasters also have some interest in the development of new technologies. The use of satellites for broadcasting sound has two manifestations: sound broadcasting in its own right and sound as part of television. Interest in Germany has produced the digital satellite radio (DSR) scheme (now in decline after being superseded) and a similar system in Japan, both ratified by CCIR[6], and there is a pan-European system proposal developed under the EUREKA 147 project for digital audio broadcasting (DAB), now an

ETSI standard[7], which has also attracted considerable interest worldwide. This latter scheme has been designed with satellites particularly in mind, but allocated spectrum is available to deploy the technology terrestrially with services already begun in the UK. A more comprehensive assessment of digital audio broadcasting is given in Chapter 17.

18.2 Standards and regulation

18.2.1 The infrastructure of broadcasting

A complete broadcasting chain comprises many parts; the significant ones are:

1 Programmes and their production. These can be preprepared and purchased externally, e.g. films, or specially commissioned and produced for television. Programmes are often produced in segments in different places over a period of time and transmitted to a studio using high-quality links, or they can be live. This is the contribution process. In North America this is called back haul.

2 The compilation of the completed programmes into an advertised schedule and their presentation in an orderly sequence. This is often called playout. The location of the playout facilities need not coincide with the production location.

3 Network transmission from the playout centre to the terrestrial transmitter, satellite uplink site or cable head from where the signal is transmitted. This is the first stage of the distribution process and is known as primary distribution.

4 The emission or radiation of the signal from terrestrial transmitter or satellite. This second stage of distribution is known as secondary distribution. Technically, the satellite is only a repeater but actually does the broadcasting; transmitting to the satellite is part of the distribution process, but normally is deemed to be part of broadcasting. If, in the future, satellites with onboard processing capability, e.g. switching or multiplexing, become available then broadcasters will surely be among the early users of such techniques to enhance their services.

5 An industrial infrastructure which makes available to the viewers a range of realistically-priced receivers.

18.2.2 Regulation

The need for regulation stems from the need to control access to scarce spectrum. On a world perspective, however, different regions have adopted contrasting approaches; a commercially-driven regime, such as that in the USA, or a public-service regime common, until recently, in Europe generally. A third approach is used in some regions and countries where there is complete state control and the broadcasting function has a more overt political purpose.

Other media now offer a range of services and types of programming such that there is not the same need for regulation to protect and share the limited medium capacities among the players. The following lists the alternative media which could be used to provide video services:

- cable TV;
- optical fibre and copper pairs;
- disc/tape;
- terrestrial UHF/VHF;
- computer systems — multimedia

The trend towards deregulation is gathering momentum and will encourage the commercial exploitation of opportunities to develop these media in an environment which will be very competitive. The satellite medium will, in the next few years, be an important means of delivery and the investment in system and equipment design is already taking place, all over the world. The close traditional connection between satellite and cable will be required to continue if the success of its past combination is to be exploited. This will mean, among other things, the harmonisation of standards so that digital services, designed for satellite, can be passed through cable systems with minimal modification.

The technical aspects of spectrum allocation are, of course, the same everywhere but the administrative processes, criteria and controls can be and are different. A complete discussion is not relevant here, but it is essential to realise that regulatory matters still have a significant influence on broadcasting; technological change can bring about regulatory change and this has been a dominant feature of recent years in Europe, largely connected with the developments in satellite-delivered television. The existing regulatory regimes in Europe vary from country to country and there is an attempt by the EEC to harmonise the rules and procedures throughout the community. This will take some time to have any effect on traditional terrestrial broadcasting, but is somewhat easier to manage for satellite systems thus making it possible for these systems, for which uniform regulation is vital for obvious reasons, to develop more quickly.

For the UK, the relevant departments of government are the Department of Trade and Industry (DTI) and the Department of Culture, Media and Sport (DCMS). The DTI, through the Radio Communications Agency (RCA) is responsible for all national and international technical aspects of radiocommunications such as the frequency spectrum and its allocation; all internal issues relating to the administration of broadcasting such as which organisations are permitted to broadcast, their financial structures and ownership and the control of their activities is the responsibility of the DCMS. Thus it is the DCMS which sets up, usually by means of Acts of Parliament, UK broadcasting regulators such as the Independent Television Commission (ITC) and the Radio Authority (RA). The governors of the BBC are also responsible to the DCMS.

In the USA the Federal Communications Commission (FCC) has responsibility to regulate aspects of television broadcasting, among other things, and has recently been involved in an exercise to define a new digital terrestrial broadcasting standard for the USA. Already, satellite systems have been in successful commercial operation with conventional digital television services in the 12 GHz band for some years.

18.2.3 Administration of technical standards

For over a century the majority of the world's nations have subscribed to international agencies which are responsible for providing facilities for the discussion of matters of global importance. When the United Nations was established, a division entitled the International Telecommunications Union (ITU) was appointed to absorb the existing International Consultative Committee for Radio (CCIR) and International Consultative Committee for Telephone and Telegraph (CCITT). Signatories to the UN and, hence, the ITU are bound as in treaty to the terms of reference and are obliged to respect the decisions and procedures of its committees.

In a recent reorganisation of the ITU the CCIR is now known officially as the ITU Radiocommunications Bureau (ITU-R) and the CCITT as the ITU Telecommunications Bureau (ITU-T). Broadcasting, as a radio-spectrum user, is administered under the CCIR/ITU-R via its study groups 10 and 11. A further group of relevance—formerly the Mixed Committee for Television and Telephony (CMTT)—deals with the interaction of telecommunications standards with those for television as they come together when television signals are transmitted via telecommunications systems designed primarily for telephony. This group is now administered via the CCITT/ITU-T as its study group 9. Within the ITU there are other groups of relevance to broadcasters using satellites, and these include the International Frequency Registration Board (IFRB) and the World Administrative Radio Conference (WARC). The results of these conferences find their way into the radio regulations which provide separately for three regions of the world: region 1—Europe, Africa and the Middle East; region 2—the Americas; region 3—the Far East, China and Australasia.

In Europe, the European Telecommunications Standards Institute (ETSI) has the task of producing European standards for telecommunications systems and this includes broadcasting by means of a formal liaison with the European Broadcasting Union (EBU). There has, for the last several years, been considerable activity focussed on progressing new digital broadcasting standards in ETSI. A panindustry group called the digital video broadcasting (DVB) project was formed in Europe with strong representation from both technical and commercial interests and expert groups have discussed the several aspects of a set of new standards for digital broadcasting. A standard for satellite broadcasting[8] has been agreed along with one for cable television[9] and others to address multichannel microwave distribution systems (MMDS)[10] and terrestrial television[11] have been published.

In the USA, some of the official standards groups are the American National Standards Institute (ANSI), the Institution of Electrical and Electronics Engineers (IEEE) and the Society of Motion Picture and Television Engineers (SMPTE), and these have been in existence for many years. Similar technology to that identified by DVB has led to DirecTv, a proprietary satellite broadcasting system currently operational in the USA, but, more recently, the use of DVB standards by newer broadcasters has occurred. There are now five digital satellite broadcasters operational in the USA with more planning services.

Similar services, operational in using DVB, are parts of the Far and Middle East, Australasia and South and Central America.

Organisations like DVB and others have come into being as a result of the lack of response of traditional standards groups to the rapid development of new digital coding and transmission technologies. Indeed, some existing standards groups have proved more capable of responding to this challenge than others— for example, the ISO/IEC joint work in the moving picture experts group (MPEG, see below). The ATM forum (ATMF) has appeared to develop and define asynchronous transfer mode (ATM), a fast packet-switching and transmission technology, and the digital audio-visual council (DAVIC) has been convened to exploit all these new technologies and define operational protocols and procedures to permit interactive services such as video on demand to be supplied in a unified way over a number of media. Similarly, the Internet has caused the convening of the Internet engineering task force (IETF) with objectives which include the harmonisation and further development of the many aspects of the network, such as interoperability, in order to make it easier to use and to address the demands made upon it by a wide range of services. DAVIC and IETF are not standards bodies, neither is the ATMF, and these informal organisations derive their authority from their support by all the important elements of the communication industry; their objective is to prepare specifications which then become accepted practice and are adopted and administered by legitimate standards bodies, perhaps on a world or, at least, a regional basis. This, despite the speed with which these groups work, seems to be succeeding and could be a future pattern, where standards groups are formed as required when a particular technology reaches sufficient maturity that it needs development. Figure 18.1 illustrates some of the many standards groups involved in the development of digital-television specifications for a wide range of applications.

18.2.4 *Implementing new television services*

In the past, new broadcasting services were introduced as a result of new spectrum allocations. Even now, if a new delivery technology has been developed to an advanced stage its use will be limited by the degree to which new spectrum is available or existing spectrum can be further exploited. The existing fixed satellite service (FSS) band and the contiguous broadcasting satellite service (BSS) band between 10.7 and 12.75 GHz have both been used in Europe for satellite broadcasting applications and are now designated for use in other ITU regions. When the possibilities for satellite broadcasting in Europe were studied during the late 1970s it was clear that there would be new spectrum, the BSS band from 11.7 to 12.5 GHz, and a service to exploit the features of the new medium would be significantly different from the current terrestrial services. To emphasise the importance of this rare opportunity, a new video and audio coding scheme was developed for use in this new band—the MAC system (described below)—which was intended to optimise television system performance on satellite channels.

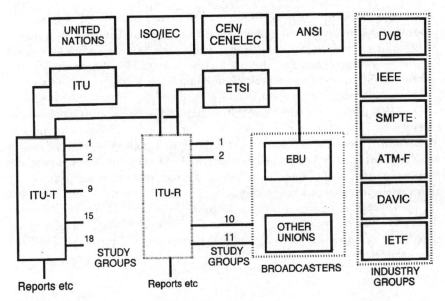

Figure 18.1   Standards bodies affecting broadcasting

Any proposal for a new channel is also subject to the judgement of its commercial potential and its effect on existing services. Recently in the UK a fifth terrestrial channel has been launched, but the initial regulatory position taken was that there was no commercial justification. One of the interesting features of Channel 5 is that the service cannot be a national one with full coverage available everywhere owing to frequency planning constraints.

By far the greatest problems relate to the need to make available large quantites of low-cost receivers for the start of any new service and to provide motives for a rapid uptake of the new receivers by the public. One way of ensuring that this happens is to design receivers which can take the form of set-top boxes, thus using the existing television receivers' display and audio facilities and enabling connection to existing videocassette recorders. Correct introduction scenarios are crucial to commercial success.

18.2.5 Funding matters — conditional access (CA)

Traditional methods of funding are by means of licence-fee revenues and by advertising, both well used in the past. More recently in Europe, and for many years in the USA, subscription schemes, supported by methods of access control, have been used to provide the exploitation of viewers' specialist needs. This will be the trend for the future where increasing competition and deregulation, as well as the availability of many more channels, will need more focussing of programme streams to small groups of specialist viewers rather than a range of programme types in one channel. The ability to pay as you view may be the ultimate facility where the receiver stores credit units, probably in a smart card, which the viewer can use at will to maintain an uninterrupted flow of wanted programmes with occasional topping up as credit is used. However, in trying to avoid total confusion of the consumer with a plethora of different systems operating on different media, there is a clear need to identify means whereby common standards for some aspects of the conditional access system so that they can have application across all media.

The DVB project has addressed this matter and some useful standards have emerged which specify a scrambling process and a common receiver interface to allow future extensions of the system[12,13]; the technical elements of a CA system are described further below (see Section 18.11). Because of the commercially sensitive nature of this subject, including the fact that there are proprietary rights issues involved, and the national and regional security aspects such that the information is subject to export controls, the standard and the algorithm details are subject to close control over distribution. A licensing procedure, together with terms, has been agreed and is administered through ETSI from where the relevant details can be obtained[14].

18.3 Current television standards

18.3.1 Introduction

Before proceeding towards a discussion of standards for digital television by satellite, it may be useful to review some technical fundamentals of television as an aid to better appreciating some of what follows.

A commercially-developed all-electronic system was chosen for the BBC to begin the world's first regular television service in 1936; this system was black and white only and used 405-line interlaced raster scanning with a 25 Hz picture rate and was transmitted in Band I spectrum at about 50 MHz. So durable was this standard that it was only discontinued in the UK at the end of 1984. In the USA, before the war, experiments were carried out and afterwards a system using a scan of 525 lines with a 30 Hz picture rate was introduced. In the 1950s colour was introduced (see below) and the system is still used in the USA, Canada, Central and South America, Japan, Korea, the Phillipines and other parts of the Far East.

In 1964 the CCIR agreed a new scanning standard of 625-line rasters with 25 Hz picture rate and this was introduced promptly in the UK for BBC2 and this system, together with colour, introduced in the late 1960s using the PAL scheme (see below), is also still in use today in Europe, including Scandinavia and the former USSR, China, Australasia, the Indian subcontinent and most of the Middle East. The differences in these standards cause the need for standards-conversion equipment to be used for programme interchange between the regions of the world, particularly when this is live via satellite.

18.3.2 Colour — NTSC, PAL and SECAM

Colour television is an illusion. True natural colours are defined by the wavelengths of the electromagnetic radiation in the visible spectrum between about 400 and 800 nanometres, for example yellow is in the region of about 450 nanometres. However, the eye will accept an appropriate mix of red and green light as a direct substitute for yellow; similarly a mix of red and blue causes the sensation equivalent to magenta. In practice, most of the colours found in nature can be approximated by a colour reproduction system developed and codified during the 1920s and 1930s by the Commite International d'Eclairage (CIE). The CIE basis was studied in the 1950s by the National Television Standard Committee (NTSC) during the development of colour television in the USA, and NTSC was adopted by the FCC in December 1953. Over a decade later, in the mid-1960s, Europe adopted the same basic scheme for colour television in which the camera resolves incoming light into its component red, green and blue parts[15].

In order to make a coded colour signal compatible with existing monochrome receivers, a brightness signal is generated and transmitted as if it had come from a black and white camera. Fortunately, another illusion can be invoked since an appropriate and constant mix of red (30 per cent), green (60 per cent) and blue (10 per cent) light appears colourless. Thus, at the colour-coding stage at the studio, a brightness or luminance signal is made together with two other colouring signals — chrominance — by taking appropriate different proportions of the red, green and blue signals. These three intermediate luminance and chrominance signals are called components and, as illustrated by Figure 18.2, are a stage on the way to generating the colour-televison signals most commonly encountered.

In all of western Europe but France the colour coding standard is phase alternate line (PAL); in France the sequential couleur a memoire (SECAM) system is used. Although most of the fundamentals of the two systems are the same, one cannot be received completely on a receiver made for the other. The CCIR Report 624 describes all the essential detailed features of the standards used worldwide; the UK version of the PAL video signal is known as PAL-I[16,17] and differs only in detail from other forms of PAL video. Differences between versions of PAL (and for other formats) emerge when the complete broadcast signal itself is examined; an example is the sound carrier-frequency offset from the vision carrier. The nomenclature should be noted here: the term video is used for the baseband video but vision is used for the component of the modulated RF signal which is caused by the video. The carrier-frequency offset is fixed by the bandwidth chosen for the vision; in the UK this offset is 6 MHz whereas in continental Europe it is 5.5 MHz and in China it is 6.5 MHz. In the PAL standard the colour information is conveyed by amplitude and phase modulation of an in-band subcarrier (the frequency of which is the same for all PAL versions) added to the luminance thereby producing a frequency division multiplex of the separate signals as illustrated by Figure 18.3. This scheme needs

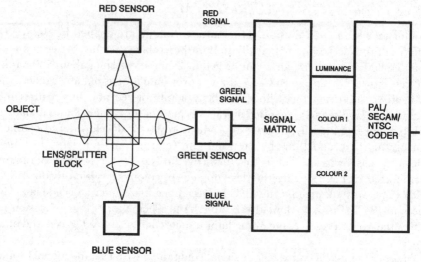

Figure 18.2   Measurement-system model for digital TV systems

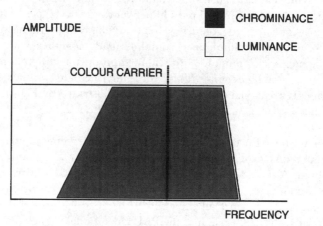

Figure 18.3   PAL FDM spectrum

a phase reference which is sent during the line-blanking periods as a burst of ten cycles of unmodulated subcarrier. SECAM operates in a similar way but uses frequency modulation instead of phase modulation. For PAL the limiting of the vision bandwidth reduces brightness resolution directly through reduced bandwidth; however, there is also a restriction of the upper sideband of the modulated colour subcarrier which has some effect on the quality of transmission of the colour signals[17].

There is an agreed international standard for digital coding of the component signals and this can be found in CCIR recommendation 601 which defines the coding of component television signals which are the brightness, or luminance, and two colouring, or chrominance, derivatives of the original independent red, green and blue signals resolved in the camera. There is an additional specification—recommendation 656—which provides for digital interfaces between equipment supporting recommendation 601 coding parameters. There are no equivalent publicly defined standards for the digital coding of the composite PAL, NTSC or SECAM signals defined in their analogue forms by CCIR/ITU-R recommendation 624. CCIR 601 generates 216 Mbit/s, or 27 Mbytes/s. There are already systems using the provisions of recommendations 601 and 656 in widespread operation on a professional basis in studios and, for contribution and distribution applications, associated video-compression technology has been developed[18,19].

18.3.3 *The MAC family*

The MAC system, first proposed in 1981[20], was introduced as a hybrid analogue video/digital audio format which would be component-based and avoid some of the problems associated with the processing of PAL and SECAM signals. It was adopted in Europe as a standard for satellite television in 1983 and has found uses in other parts of the broadcasting chain[21,22]. The MAC system was stimulated directly by the coming of satellite television and was a development which had three main driving forces:

(i) The establishment of the WARC 1977 DBS plan for Europe, giving a very rare opportunity to use new technical standards in the newly created medium.

(ii) The trend towards digital signal processing throughout much of production and distribution and leading in particular towards the processing of the video signal in component rather than composite PAL form.

(iii) The need to consider the future evolution of standards towards better definition television (see below).

It was recognised very early in the evolution of MAC that its new start made it an ideal vehicle to carry enhanced and higher-definition television pictures in a processed form, making full and efficient use of a satellite FM channel, with receiver processing to restore a high-quality picture for display on a suitable device. Existing composite signal formats such as PAL, having been developed in the early 1960s, were not capable of those features which were required to take television into a new era.

Unlike PAL, which uses a frequency division multiplex format, MAC takes a time-division multiplex form which has advantages for basic picture quality and resilience to satellite channel noise. A comparison of the time waveforms of PAL and MAC for a single television scanning line is illustrated by Figure 18.4; for PAL (and similarly for NTSC and SECAM) the picture information is conveyed in about 81 per cent of the line time and the remainder is used to carry synchronisation information for the line scan (the sync pulse) and for the colour decoders (the burst of colour subcarrier—not SECAM). In MAC there is no subcarrier and the line synchronisation is done digitally using the data burst shown in Figure 18.4.

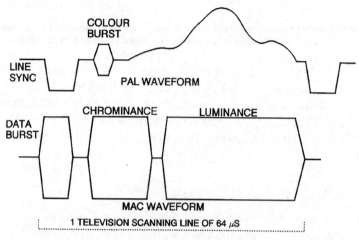

Figure 18.4   PAL FDM and MAC TDM formats compared

It should be noted that, despite the commercial difficulties which caused the demise of the official UK satellite broadcaster, British Satellite Broadcasting (BSB), and the cessation of MAC transmissions in 1980, the MAC system is still in use and the satellites constructed for BSB have been redeployed in Scandinavia.

18.4 New television standards

18.4.1 Introduction

The following is a brief description of all the new television formats developed over the last decade or so; its purpose is to illustrate the choice facing a broadcaster during this period when contemplating new services. In the context of satellite television not all of these standards have proved to be useful or practical but they certainly have influenced the decisions of the programme providers and the receiver manufacturers who supply the essential parts of a practical system. In the broadcasting chain the fundamental commercial transaction is between the programme provider and the viewer; all other parties are middle men in the process. The choice of medium and technology standard will be influenced strongly by the extent to which any system helps or hinders this basic transaction.

18.4.2 High-definition television

Unfortunately, HDTV is a term which can, and does, mean all things to all men. From a technical viewpoint, HDTV is about image resolution, where the number of lines is a primary parameter from which others are derived, and resultant picture quality must be in some sense better than what is already available[23].

The CCIR, when it began studying HDTV in the mid-1980s, defined HDTV systems as those with more than 1000 scanning lines. At that time the world broadcasting community was seeking a single world standard, a very laudable and worthy objective, but one fraught with political and practical difficulties. There were several proposals for HDTV standards and these included:

- 1250 lines (Europe);
- 1125 lines (Japan);
- 1050 lines (USA).

These proposals were hotly debated until the end of the 1980s when it was realised that early progress was not going to be made. The rapid emergence of conventional-definition digital television systems changed the direction of standards and service planning work such that, for the time being, the issue of HDTV standards has abated. However, in the USA, where there has been a prolonged process to define a terrestrial HDTV system since its start in 1990, some progress has been made with HDTV. Perhaps, when the conventional digital systems now being used have become well established, broadcasters will return to the issue of HDTV and reconsider the means to acquire programmes and to support their transmission.

The most important features required of HDTV are:

Better resolution

- spatial;
- temporal, i.e. no visible motion artifacts;
- removal of scan and colour-coding artifacts.

Wider aspect ratio

- 16:9 rather than 4:3.

Improvements in transmission performance

- signal-to-noise ratio;
- linearity etc. for analogue schemes;
- low bit-error rate and quantisation or coding defects in the case of digital systems;

so that the additional signal resolution is not masked by transmission defects.

Compatibility with existing context

- ready conversion to/from existing scanning standards;

- awareness of convergence of technologies, e.g. the computer industry;
- realistic commercial introduction scenarios;
- spectrum efficiency;
- spectrum-sharing options.

18.4.3 W-MAC

Wide screen or W-MAC is simply a slightly modified form of MAC which deals with the wide-screen transmissions using an aspect ratio of 16:9[24].

18.4.4 HD-MAC

The HD-MAC system is a method of compatible HDTV delivery such that normal MAC receivers can operate satisfactorily with an HD-MAC input, and specially equipped receivers with the decoder circuits installed and a 16:9 screen can obtain the full benefit of HDTV at 1250 lines resolution. The MAC[25-27] and HD-MAC[28] standards were developed for use in Europe and, although expressed as an analogue signal format, the majority of the signal processes were realised digitally[29]. The original plan was to make the system available by 1995 but this was curtailed owing to practical and economic considerations.

18.4.5 MUSE

The Japanese HDTV proposals chose 1125 lines with a 60 Hz field rate. By using the same signal-processing techniques as were proposed for HD-MAC the system known as MUSE was proposed[30] for the 1125 environment and has actually been in service (a few hours a day) in Japan for a number of years, although the number of viewers has been small owing to the large size and cost of the receivers.

18.4.6 Enhanced PAL — PALPlus

As a result of the proposals for higher-quality resolution standards for satellite television, terrestrial broadcasters gave serious thought to the new competitive environment in their industry in the wake of reregulation where market forces were being allowed to have their natural effects. Although not attempting to produce HDTV standards of quality, which some believed was neither practical nor necessary, some terrestrial operators have supported the development of, and now adopted, enhanced PAL to place themselves better to compete with any threat from satellites or cable.

The main features of the PAL enhancement processes are:

(i)   Wider aspect ratio, but with acceptable effects on the normal 4:3 screen.
(ii) Reduced levels of coding artifacts such as crosscolour etc.
(iii) Better sound system.
(iv) Mitigation of propagation and multipath effects e.g. echo cancellation.
(v)   Improved resolution.
(vi) Compatibility with existing receivers.

The PALPlus system has been standardised[31] and services are being transmitted by a number of European broadcasters. The marketing of the receivers has been tied with that of wide-screen television sets and, in some countries in Europe, this has been successful.

18.5 The WARC 77 DBS plan for Europe

No review of satellite-television applications would be complete without some reference to the World Administrative Radio Conference (WARC) satellite-broadcasting plan of 1977. What follows here is a brief account of the plan and its current status.

The introduction of DBS in Europe had been planned on a technical level for several years during the 1970s. The CCIR had, in 1977, agreed a plan which allowed each European country to have spectral and orbit allowances. The main features of the plan[32,33] are illustrated in Figures 18.5 and 18.6. It was not until 1987 that the five UK channels were considered to be commercially viable and, hence, given regulatory approval. Although, as a result of the commercial failure of British Satellite Broadcasting (BSB) in the UK, the only truly commercial satellite broadcaster in Europe at the time using DBS frequencies, there is now no significant activity based on the plan, the spectrum portion, 11.7–12.5 GHz, designated for the use of MAC, has been replanned for digital applications and has thereby increased the number of channels available, but has retained much of its original basic form.

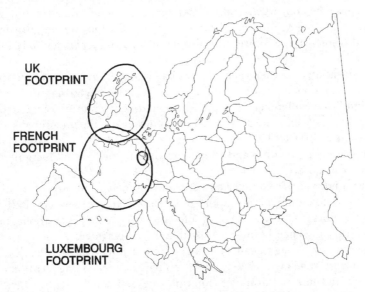

Figure 18.5   WARC 1977 footprints for UK, France and Luxembourg

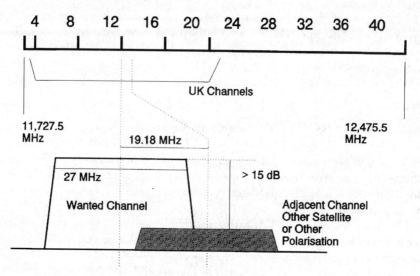

Figure 18.6   The main features of the WARC plan of 1977

18.6 Digital coding

18.6.1 Introduction

During the early to mid 1960s the coding of various signals, including television, had been studied with a view to their transmission via digital telecommunications transmission networks. The stimulus for these networks derived from the coming together of semiconductor technology and pulse code modulation (PCM). Now, the convergence of technologies capable of processing images is such that computing, telecommunications and broadcasting are merging into an integrated whole. In the future, this distinction will disappear, not only for the image producers but also for the viewers.

The digitisation of broadcasters' signals had a major advantage for contribution and distribution applications in that, once in the digital format, there would be no significant difference between them and, say, multiplexed telephony signals and so these early ideas were a step on the way to the integrated services digital network (ISDN). They would mean the removal of separate link systems and procedures for television and telephony with an attendant reduction in cost and network flexibility[34].

The remainder of this Chapter concentrates on the various aspects of digital video compression and its applications to satellite-delivered services. Both professional and domestic services are considered but the DTH/DBS application is given some prominence because of its greater public interest and because it represents the most significant technical and commercial challenges. The aspects considered below include: source coding, channel coding, error control, modulation and link performance, multiplexing and satellite-access methods.

Finally, there is a summary of implementational aspects to give a practical conclusion.

Digital transmission systems for broadcasting draw heavily on the experience gained in other applications and use the same classical transmission-coding systems and techniques; the special features of broadcasting can be accommodated readily because of the flexibility of the digital format.

Figure 18.7 illustrates the generic structure of the processing chain; the source coding takes the source information, analogue video and audio signals, and transforms them into convenient standard digital formats for interfacing with subsequent processes. There is good reason to try to make this interface an industry standard to allow for harmonisation across media so that the medium-specific elements are separated from the coding aspects.

The channel-coding process is peculiar to each different channel medium and its characteristics have to be accommodated by means of specific coding. In satellite channels employing modulation, which should be seen as a form of coding, the main additive source of transmission impairment is receiver noise, which is virtually random, and interference with characteristics which are not definable accurately. The design of the channel-shaping filter bandwidths in relation to the symbol rate is assumed to be optimised. There will be the usual intermodulation factors to consider owing to the nonlinearities of the travelling-wave tubes at the uplink and the satellite when multicarrier operation is required.

In addition to the transmission system design and its ability to deliver digitally-coded programme data, there is a need to make the system user friendly and flexible in its use by both the consumer and the broadcaster. A complete system, representative of that considered for the European DVB standard for digital satellite broadcasting[8], is illustrated by Figure 18.8, from which it can be seen that multiplexing arrangements are also required to combine the component parts of the transmission. Each bit stream formed in

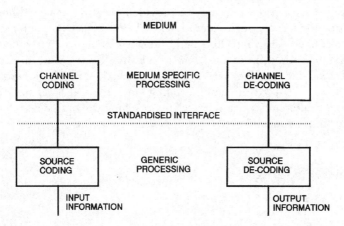

Figure 18.7   Generic coding processes in digital transmission for television

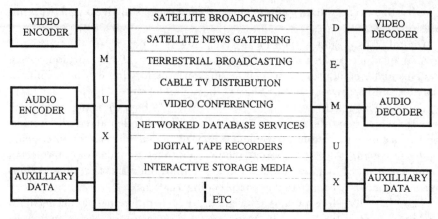

Figure 18.8   DVB satellite-channel adaptation

this way will involve many more than one television programme per transponder and will enable the flexible reconfiguring of the multiplex structure to accommodate different modes of programme presentation. For professional applications, the same degree of user control is not needed and the multiplexing need not be so flexible or provide as many channels. Such schemes have already been implemented in systems which are commercially operational in many parts of the world.

The ability to reconfigure dynamically the programme multiplex from time to time and to inform the viewer of the contents of the multiplex will be achieved by adding a specific data stream to the multiplex which contains the appropriate information. This service information (SI) will be an essential part of a practical system for broadcasters where the public are faced with complex technology. More information about this aspect of the system is given below in the section describing implementational matters. Conditional-access data will also have to be transmitted in the multiplex.

A degree of error protection may be provided for at this stage, although this is more properly a channel consideration (see below), since the source decoding process must have some protection against uncorrectable errors or other failures in the channel-coding process e.g. synchronisation losses owing to cycle skips during demodulation. The nature of the protection required here is very different from that used in channel coding since the statistical nature of the errors is very different; typically the errors reaching the source decoder can be extremely nonrandom and need burst-correction capability.

18.7 Source coding

18.7.1 General

Broadcasters have been using digital transmission techniques for the distribution of audio and video for some considerable time already, but only in the last four or five years has it been possible seriously to consider digital transmission of high-quality television signals directly to the home at low bit rates, i.e. about 10 Mbit/s or less. This is because of developments in video and audio compression techniques and, already, widespread consumer use of these techniques is established, largely because the complex integrated-circuit developments needed to make the receiver equipment cheap enough for consumer applications have been completed. When the WARC plan of 1977 (see above) was being studied, digital coding was not seriously considered because coding technology was not advanced enough.

For many years broadcasters have been developing (in CMTT and EBU) digital bit-rate reduction techniques for use in transmitting television between studios on telecommunications networks for both conventional and HDTV[18,19].

A large number of techniques exist for reducing the bit rates needed to remove inessential information from television pictures in order to reduce the amount of data needed to be transmitted or stored. Differential PCM (or predictive coding) was the first to be used but this was followed by transform coding methods in the mid-1980s, in particular, the discrete cosine transform (DCT). Motion-compensation techniques are now used in hybrid schemes combining predictive and transform techniques, and statistical (Huffman) coding is used to extract any redundancy remaining in the bit stream transmitted to the channel, see Figure 18.9. Packet techniques are also considered for certain types of network where the variable rate of information in a real signal can be matched by a dynamically varying transmission capacity allocated to it — asynchronous transfer mode (ATM).

There are already digital audio/visual services which give nonbroadcast standards of quality; these are videoconferencing and videotelephony (see CCITT recommendations H.261 and H.32x). The coding equipment is still expensive but the integration on silicon of the essential processes is causing prices to fall. For use in consumer applications such as broadcasting, the receivers must be cheap, typically less than £250 in the shops, complete with antenna and low-noise block converter (LNBC). These receivers will rely very

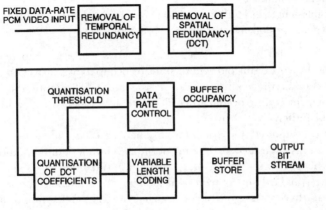

Figure 18.9   Video compression coder block diagram

heavily on the production of cheap but powerful integrated circuits to make them possible at all. The promise of the enormous market traditionally associated with consumer electronics is a great attraction to semiconductor manufacturers, which have been very active in the standardisation process.

18.7.2 ISO/IEC MPEG

For a system specifically for broadcasting, the source coding and multiplexing portions are defined via the standards developed jointly by the International Standards Organisation (ISO) and International Electrotechnical Committee (IEC). Subcommittee 29 of the 1st Joint Technical Committee of the ISO/IEC — ISO/IEC/JTC1/SC29 — is responsible for the coding of moving pictures and associated audio. The 11th working group of this committee is known as the moving picture expert group — or MPEG. A sister group — the joint photographic expert group (JPEG) — has developed a standard for coding still images (see ISO standard 10918-1). It is vital to understand that MPEG does not specify any transmission channel parameters; this has been left deliberately for other groups, such as the DVB, to deal with.

The MPEG has developed appropriate generic source coding standards for video and audio coding, and these have now been accepted and adopted worldwide by broadcasting experts. Beginning at low-quality transmission at bit rates up to about 1.5 Mbit/s (MPEG I — see ISO standard 11172), the group has also evaluated compatible improvements at rates up to 10 Mbit/s (MPEG II) which will deal with broadcast quality, at least at the consumer level. A European Eureka project (VADIS or EU625) was convened to consider the application of MPEG II-type processes in Europe and had considerable influence on the design of the video and audio coding algorithms chosen in MPEG. The MPEG algorithm is a hybrid of three processes (see Figure 18.9):

(i)   Predictive coding, which also uses motion estimation and compensation, to exploit temporal redundancy in the moving images.
(ii) Transform coding which uses the discrete cosine transform (DCT) to exploit spatial redundancy.
(iii) Variable-length or Huffman coding to remove remanent redundancy from the bit stream produced by the first two processes.

The algorithm also relies upon a buffer store which is used to regulate and smooth the flow of data and whose state of occupancy controls, by means of a feedback path, the coding accuracy; the more full/empty the buffer the coarser/finer the coding becomes in order to reduce/increase the amount of data entering the buffer.

The bit rates quoted above should only be treated as guides since, although the applications envisaged at the time of the development of the standards were indeed at these bit rates (see CD-I, for example), time has passed and the provisions of the standards have been used at higher rates. For example, products existed about five years ago which employed the MPEG-1 standard at bit rates from 1 Mbit/s up to 15 Mbit/s; MPEG-2 can be used from under 1 Mbit/s to well beyond the 15 Mbit/s nominal limit. The labels are now somewhat historical but, nevertheless, in order to promote interoperabilty, some outline at least of the MPEG specification must be clear for a number of applications. The specification allows for this through its so-called profiles and levels; the latter are simply resolution options and the former allow for conventional television parameters e.g. main profile; other profiles allow for scalable versions of this main profile where higher definition can be included in a transmission which can be used by both normal and specially-equipped high-definition receivers using hierarchical coding methods. The high profile allows for full HDTV. Contrary to what might be expected, this flexibility does not render these latest developments nonstandard since the standards are very generic and publish only the rules by which coded bit streams are derived and constructed, and are therefore largely independent of practical factors such as bit rates or coder implementations.

There is also within the MPEG scheme a system layer which provides for the means to transmit an organised stream of data representing a programme or a set of programmes in multiplex. The data is organised into packets which are 188 bytes long. There are transport streams, which are responsible for carrying complete programme assemblies which may comprise more than one television programme; these streams also comprise means of synchronising receivers, including clock references and presentation time stamps (PTS), as well as the data. There are also programme streams that carry data from video, audio and data sources which belong together as a service and so need to be maintained as a common entity. Data from individual coding processes, say a video channel or a data channel, are called an elementary stream.

The practical use of the wide degree of choice available to a user within the rules of the MPEG standards is beyond the scope of this Chapter; ISO 13818-1, 13818-2 and 13818-3 are the appropriate specifications for the systems, video and audio aspects, respectively. A fourth document, ISO 13818-4, gives the rules governing compliance. The video part is jointly published with the CCITT/ITU-T, reference recommendation H.262. Other parts of the MPEG 2 specification, beyond part 4, deal with practical matters such as digital storage media command and control—DSM/CC—interfaces and protocols, part 6, and a real-time interface, part 9, for defining the tolerable variation in the timing and synchronisation parameters of transport streams as they pass through systems. Part 7 deals with nonbackwards-compatible audio coding.

It should be noted that although MPEG has defined a complete solution for audio coding, including multichannel systems for surround-sound and home-theatre applications, other proprietary systems have been adopted, the most notable being Dolby digital AC3 which has been chosen for the US digital terrestrial HDTV system and will be used by some US satellite operators, even those using DVB specifications. In practical terms, AC3 and MPEG audio perform similarly well, as would be expected since some of the processing techniques are very similar, but the choice has been made on a wider basis than performance.

18.8 Channel coding

18.8.1 Transmission performance of media

One of the consequences of removing redundancy from television images is that, when transmitted by digital means, the images become more susceptible to channel errors. Some media are much cleaner than others, in that they generate fewer errors and the rates are consistently low, although some have high rates and are not consistent. One example of a clean medium is optical fibre as used for telecommunications links where the guided nature of the medium virtually guarantees consistent error rates as low as perhaps one error in 10^{10} per link; the statistical distribution of these errors is usually far from ideal and gaussian, as assumed by most textbook approaches.

An example of a medium that has an error rate which is inconsistent over time is that which would be used for terrestrial digital television in the UHF spectrum. The presence of noise, multipath and other disturbances made worse by the poor antenna performance possible in the environment, where portability is important to the user, make modulation design for this medium very difficult. The failure characteristic of complex digital modulation will be very different from the graceful failure of existing analogue systems, hence the search for a scheme which is scalable to provide a hierarchy to soften the failure mode if required.

18.8.2 Harmonisation of standards

Although it is possible to design in isolation for digital television transmission over different media, there is a need to consider the harmonisation of standards so that receiver components can be common, which helps to reduce costs, and also could allow a common receiver core for several media. It has become clear over the last few years that there is a special relationship between satellite and cable television distribution systems which, in order to continue, would require some special commonality. There will almost certainly need to be different modulation schemes, and even different error-correction schemes may be used, but the net bit rate could be similar or even identical in each case. At a cable head the satellite modulation could be replaced by something more appropriate (see below). The cases of remodulation at cable heads, including satellite master antenna TV (SMATV) systems, and remultiplexing are illustrated by Figures 18.10, 18.11 and 18.12.

However, although the technical interests of the satellite and cable systems have much in common, it must be remembered that they are operated by commercial companies competing for viewers. Direct remodulation at the cable head end, although technically possible, is probably commercially not likely in general. The cable operator may wish to select some of the channels received from the satellite, add some new ones to make a new multiplex, and add its own conditional-access signals to the final bit stream, before transmitting the new signal along the cable.

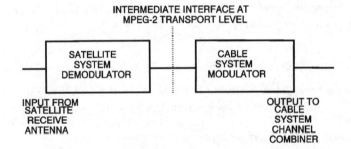

Figure 18.10   Remodulation from satellite to cable channels

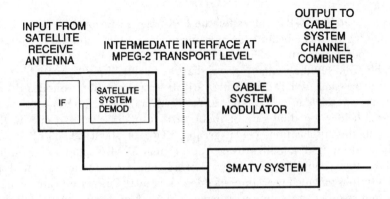

Figure 18.11   DVB SMATV application

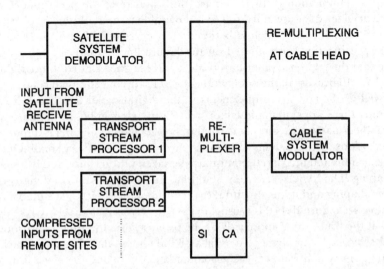

Figure 18.12   Remultiplexing of DVB services

The system must therefore have flexibility to allow this kind of repackaging to be done at interfaces between media.

18.8.3 Satellite channels — link budgets

A figure of merit used for modulation system evaluation is E_b/N_0 — the ratio of energy per transmitted bit of information (E_b) to the spectral density of the channel noise (N_0). The N_0 is simply related to the channel noise power and E_b is simply related to the signal power so that E_b/N_0 is related to carrier-to-noise ratios in the transmission channel. For QPSK, E_b/N_0 has a value of 8.4 dB at an error ratio of 10^4; this is a reference point which can be used for performance comparisons between different systems. Normally, the performance of satellite transmission systems is limited by carrier-to-noise considerations rather than any other.

Typical satellite EIRP values for the FSS band in Europe are restricted to less than 50 dB at the beam edge, and the power budget will tend to make the system operate in a power-limited regime. This, together with G/T figures available from small antennas, means that the threshold E_b/N_0 available to the digital processing in a DTH receiver will be low, typically around 5 dB and therefore extensive use will have to be made of the error-control schemes discussed below; a coding gain of about 4 dB at an error ratio of one in 10^4 is sought in this application. Given greater EIRP of about 60–63 dBW, as is available from BSS satellites using the original WARC plan bandwidth of 27 MHz, there could be more degrees of freedom available to operate differently. Options include lower antenna sizes, more available net bit rate by means of a higher-order modulation scheme which then allows more television channels, less FEC which reduces receiver cost and complexity, poorer quality front-end noise performance, including LNB which will also reduce receiver cost, or a combination of these. Figure 18.13 illustrates the performance of a modem developed for use at 8 Mbit/s; it shows that a threshold E_b/N_0 of around 4 dB can be achieved, which is better than a current INTELSAT IESS 308 modem specification, also plotted on the Figure for comparison. Note in this Figure that the reference point for QPSK at 10^4 has been shifted to the right by about 1.25 dB because of the application of a 3/4 rate convolutional code.

As well as the additive impairments to the digital signal which will lead to bit errors, there are systematic factors caused by intersymbol interference and jitter in the usual way[35]. The choice of the correct channel shaping is important, especially for satellite systems where the power margins are very small. Out-of-band emissions and interference must also be taken into account in the design of the shaping. The Nyquist criterion establishes the basic rules for the design of the channel shaping and, typically, it takes the form of the choice of a suitable roll-off factor (see Figure 18.14). It is usual to have the Nyquist channel filtering done partly at the transmitter and partly at the receiver, usually equally distributed. The filtering can be implemented using a digital filter so that automatic tracking of changes in symbol rate can be accommodated by taking the clock from the

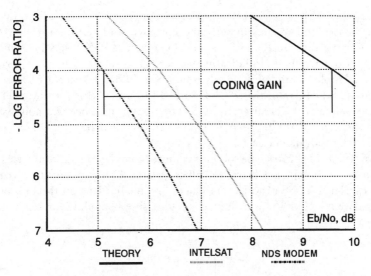

Figure 18.13   Error ratios as a function of E_b/N_0

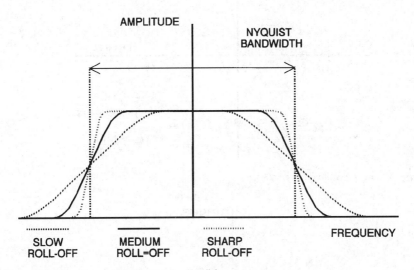

Figure 18.14   Channel shaping and roll-off factor

extraction process in the demodulator. This allows the production of a receiver which is insensitive to bit-rate changes over a wide range.

18.8.4 Digital modulation systems — PSK family

One natural choice for modulating radio-frequency carriers with digital baseband signals is the phase-shift keying (PSK) family and this has now been in practical use for at least two decades in telecommunications and other fields.

The family of PSK begins with its simplest member, which uses only two separate phases and is consequently most rugged and returns the best C/N against bit error rate (BER) performance. Quadrature PSK (QPSK) is efficient in using two carriers in quadrature which gives more bits per Hertz of bandwidth employed and a useful advantage over 2PSK, but at the expense of a C/N penalty which is illustrated by Figures 18.15 and 18.16. The latter Figure shows the source of the error event in the corrupting influence of a noise vector causing the symbol 0,1 to become either 1,1 or 0,0 depending upon which quadrant the resultant enters.

Figure 18.17 illustrates one of many more complex modulation schemes which increase the bits per Hertz to values greater than two, but these require greater carrier-to-noise ratios which places greater burdens on the link power budgets available. For a power-limited satellite, simple modulation schemes are

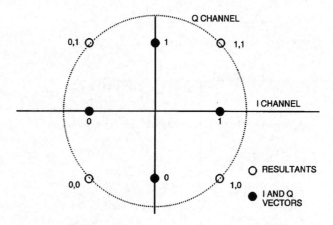

Figure 18.15   PSK modulation scheme

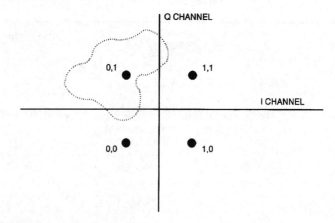

Figure 18.16   Noise disturbances to PSK vectors

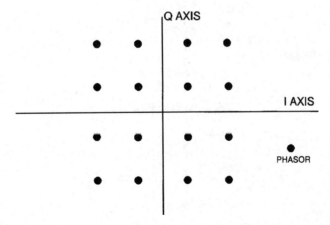

Figure 18.17   16QAM phase/amplitude state diagram

the only feasible solution; if power were more plentiful, as for so-called high-power DBS where EIRPs are more than 10 dB higher than for the FSS band (by limit of CCIR), more complex schemes would allow greater bit rates to be sustained at more than 3 bit/Hz. Schemes such as this should ideally be of constant-envelope type, to avoid the effects of TWT nonlinearities, but multi-amplitude schemes combining phase and amplitude modulation have been used successfully to improve the bit/Hz ratio. Examples are the variants of PSK, e.g. 8PSK, 16PSK, MSK etc. and 8QAM, 16QAM, etc. to NQAM where N is usually a multiple, if not a power, of 2; these are numerous and effectively offer a trade of carrier-to-noise ratio for more gross bit rate at a given error rate. It should also be noted that the descriptor 32QAM, say, is not complete since there are alternative ways of placing the channel symbols in amplitude/phase space and the error susceptibility in each case is different (see Figures 18.18 and 18.19 for examples). The modems needed for these complex schemes are also complex

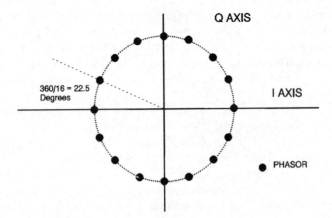

Figure 18.18   16PSK phase state diagram

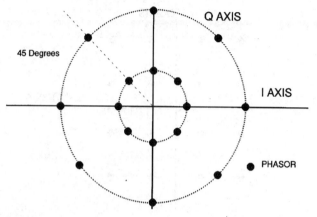

Figure 18.19   Alternative 16QAM state diagram

and expensive compared with the simple PSK schemes, such as QPSK, normally used. Some of these complex schemes are used widely in telephone-line data-transmission modems where, provided that a good analogue line is available, data rates of up to 56.6 kbit/s can be sustained.

18.8.5 Modulation schemes — spread spectrum

One of the most effective modulation techniques for dealing with hostile propagation environments is that which spreads the signal energy over a wide spectrum so that the effect of any defect in transmission in any part of it is avoided. The clear application to military systems where jamming can be defeated is an obvious one. For entertainment or business applications the value of these techniques is to allow more robust transmission in order to enable the possibilities for simple and cheap receivers using simple antennas, reduced power levels in transmitters, better spectrum efficiency, as well as maximising the channel capacity in terms of bit rate or bits per Hertz, to be exploited. The nonintrusive nature of these techniques also allows them to be deployed where other, more conventional, systems simply could not work, perhaps because of interference effects. Indeed, it is possible to bury the new digital signals among the existing conventional transmissions, thereby using the same spectrum for new services.

One modulation technique with suitable features is that known as orthogonal frequency-division multiplex (OFDM)[36,37]; it relies upon the use of a large number (several thousand) of harmonically-related carriers spaced only a few kHz apart (see Figure 18.20) with their modulation sidebands overlapping by design. Just as for digital transmissions in the time domain where successive impulse responses are ideally mutually orthogonal, thus resulting in zero inter-symbol interference, each carrier in OFDM is orthogonal in the frequency domain. This is achieved by ensuring that the spectral nulls, associated with

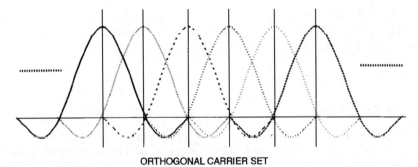

ORTHOGONAL CARRIER SET

Figure 18.20   Orthogonal carriers in OFDM

harmonics of the symbol modulation rate and a feature of the $(\sin X/X)$ spectral shape, coincide with the centre frequencies of the series of adjacent carriers. Such a structure can contain as many carriers and as dense a concentration of carriers per Hertz as the application demands. The ruggedness of OFDM arises from its use of low symbol rates on each carrier of the array; as well as the natural properties of the scheme, performance can be enhanced by the application of error-control coding when the tehnique is known as coded OFDM (COFDM).

The following is an illustration of the way in which OFDM operates. Each carrier is modulated by a low bit rate which is precisely related to the frequency separation and which is derived by demultiplexing a serial input stream into very many separate streams. The individual spectra of the many carriers combine without mutual interference in the same way as for intersymbol interference in the time domain for normal digital transmission channels. For example, with 1000 carriers an input bit rate of 1000 kbit/s reduces the modulation on each carrier, using 2PSK, to 1 kbaud. The consequence of this modulation is that the symbol period for each carrier is very long (one millisecond) and so any multipath or echo effect is very small compared with the symbol period. This means that the effect of the interference is largely mitigated and confined to only some of these many carriers during the early part of the symbol period, leaving the latter part of this period unaffected and suitable for decoding.

A more practical case would be a 20 Mbit/s input being separated into, say, 2000 streams of about 5 kbauds if each carrier is also modulated using four-phase PSK. Higher orders of modulation are possible to increase the bit rate or reduce the symbol rate, as required. The process of performing the modulation does not involve several hundred modulators; rather, it uses commercially available digital signal-processing (DSP) chips to transform the input digit stream using mathematically-defined algorithms based upon the fast Fourier transform (FFT). The technique is well understood and has been chosen for use in the digital audio broadcasting (DAB) system as well as for digital-television applications[11]. The technique can be used for satellite transmission systems but care must be taken with its power level as it has many of the spectral properties of noise and thus can have a peak-to-mean power ratio of approaching 12 dB which cannot be ignored in link-budget planning.

18.8.6 Error control techniques

18.8.6.1 Introduction

Owing to the fundamental nature of digital transmission processes, the failure characteristic is very rapid and the application of error-correcting codes[38] causes the failure mode to be faster still. The management of the errors in bit-rate-reduced television applications is a very important aspect of system design. Details of coding are given in Chapter 10 and herein we concentrate on the applications.

Many picture and sound-coding algorithms respond very poorly to errors and, despite performing well in back-to-back tests, fail in real systems. One remedy for bringing the error problem under control is to use appropriate error-correction schemes. It is common to use two layers of scheme, one to deal with the channel errors directly, often called the inner code, and one, the outer code, to deal with the residual errors with statistical properties which are usually highly nonrandom and can cause difficulties with error-correcting codes designed for randomly-occurring single errors. There is often a stage of data shuffling or interleaving in between the inner and outer coding processes to assist in dealing with the statistical nature of the residual errors. Figure 18.21 illustrates the processes and their relationships.

18.8.6.2 Outer code

As has been stated already, the outer code is required to deal with the errors which remain in the output of the inner decoder and the interleaver. The statistics of these errors is nonrandom, and bursts of errors caused by the failures in

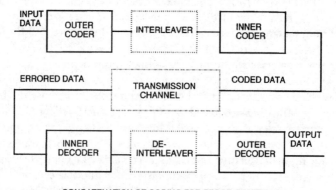

CONCATENATION OF CODING FOR ERROR CONTROL

Figure 18.21   Concatenation of coding methods

the inner code are included. An appropriate system for dealing with such errors is provided by the Reed–Solomon (RS) family of codes which is itself an extension of the wider Bose–Chaudhuri–Hocquenghem (BCH) code[39,40]. The BCH family was discovered simultaneously by Bose and Chaudhuri and by Hocquenghem, but only in its binary form; that is, its individual code elements are represented as binary numbers and manipulated in binary arithmetic; in RS codes the mathematics is extended to elements which are represented and manipulated as powers of 2, i.e. modulo 2^n where n is integer. Typically, individual RS elements are calculated in modulo 2^0 arithmetic as this is convenient for data byte structures used in signal processing; silicon devices are available to implement RS codes in 2^8 arithmetic.

The systems layer of the MPEG-2 packet structure (see Figure 18.22) allows for packets of length 188 bytes; by adapting the frequently-used 255, 239 RS code the protected data block size can be set to 188 bytes which is the size of the MPEG-2 packet, including its four header bytes. The additional bytes required to protect the data bytes depends on the number of bytes allowed simultaneously to be in error. If eight bytes are to be allowed in error then a 204 byte total block size is needed, giving 16 bytes of protection. The rate of the code is 204/188 or about 8.5% overhead. The addition of the 16 control bytes is illustrated by the second layer of Figure 18.22.

18.8.6.3 *Inner code*

The inner code deals directly with the channel errors. The satellite channel is dominated by gaussian noise generated in the receiver front end; this is a classical channel to model and several error-protection systems can be used. An appropriate inner code for this satellite case is a convolutional code using the commonly-applied decoding methods of Viterbi[41]. Other channels, for example

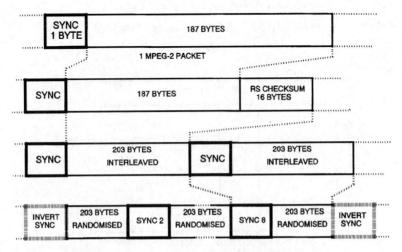

Figure 18.22   MPEG-2 packet and transmission frame structure

cable or terrestrial VHF/UHF, have different dominant impairments and may therefore require more or less FEC. A useful form of convolutional code is based upon a natural code of rate 1:2 which means that for every input data bit two bits are transmitted to the channel, which clearly generates 100 per cent overhead and so doubles the bit rate. In general, the rate is N:M where M output bits are generated from N input bits taken as a group. The output bits are generated according to the mathematics defining the code. Figure 18.23 illustrates the process of producing the two bit streams from the input by means of two modulo 2 additions; to produce a code of a different rate from this structure it is possible to delete bits from the output streams in a controlled fashion using a puncturing process, as described below, to produce streams A and B which may thus have rate N:M where M output coded bits are derived from N input data bits.

An important parameter which determines the number of input bits required to be processed simultaneously by the mathematical basis of the code is the constraint length; for the DVB satellite channel the constraint length is seven. Figure 18.24 illustrates the code in rate 1:2 form; note that there are seven

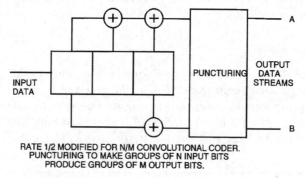

Figure 18.23   Convolutional coder: rate 1:2 punctured for rate N:M

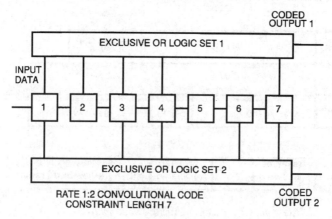

Figure 18.24   Convolutional coder for DVB satellite channel

data bits available from the input data-stream shift register and that the two output data streams required for the rate 1:2 case are generated by the two mathematically-defined exclusive OR logic processes. Note also that the two logic processes are different in that they use different sets of input data bits to produce their outputs. The convolutional code of rate 1:2 can be adapted to other rates by a defined process of selective deletion of bits from the output of the rate 1:2 coder. This process systematically reduces the power of the resultant code but also reduces its overhead, thus releasing more bits for payload. The inner code rates chosen by DVB for the satellite case are: 1:2, 2:3, 3:4, 5:6 and 7:8; these rates progressively degrade the code in its power to correct errors. The puncturing is also defined mathematically[38].

The convolutional decoding process at the receiver may be implemented by means of any algorithm based on trellis searching. The algorithm due to Viterbi[41] is very commonly used and is available in integrated circuits. The decoding process requires the use of memory to retain decoded data from previous parts of the bit stream and, depending upon the relative probabilities of the available candidates for decodes, choices can be made of the output bits most likely to be correct at any particular instant. This is the so-called maximum-likelihood algorithm. The algorithm performance can be enhanced by using soft-decision techniques to improve the information available to the decoder concerning the quality of the decisions made by the hardware data slicers from which the errored bit stream comes.

Once the decoded data has been produced it can be depunctured, deinterleaved and passed on to the RS decoder. When the convolutional decoder fails to obtain the correct result there is a period of time before it can regain the correct route through the decoding trellis; there is a finite but small probability that the decoder will never regain this path, and will have to wait for a periodic reset when a known output bit pattern, such as a sync block or unique word, occurs in the data. The periods of decode failure can therefore be protracted and will lead to error bursts at the decoder output. This is one reason why a burst-resilient outer code preceded by interleaving is needed.

18.8.6.4 *Interleaving*

Although the processes described above for the control of errors in the satellite channel can be satisfactory when concatenated, the level of performance required means that other protective methods are also needed. Interleaving is a very useful technique for assisting normal protective codes to maintain their performance by breaking down burst patterns into shorter patterns which have a more random nature and are thus less likely to exceed the correction capacities of any given code. The choice of the outer coder can be optimised by applying interleaving which simply reorders the data bits at the transmitter according to a defined pattern. At the receiver this reordering is reversed to restore the original data order. The consequence for any error bursts in the channel is that the errored bits experience only the receiver reordering which, by design, causes contiguous errors to be dispersed thus reducing the burst lengths and, hence, the power of outer code needed to deal with them. Note that the interleaving process does not increase the bit rate and so requires no transmission capacity overhead. The amount of interleave depends on the behaviour of the burst patterns but, given that the coding gain of interleave is a nonlinear function of its depth, there is a practical optimum. For the case employed in the satellite systems under consideration an interleave depth of 12 is sufficient to satisfy performance and practical criteria. The interleaving can be performed bit wise or byte wise.

Interleaving can be implemented in a number of ways[42,43]; two are: convolutional interleave or block interleave. The difference between them in performance terms is not great and the choice is made as much on practical criteria as any other, for example, memory required to implement. The convolutional case is more resilient to periodic error bursts because if, fortuitously, the error burst length and frequency coincide with the block length of a block interleave, for example, then there could be undesirable effects which could not be avoided. Convolutional interleaving can itself be implemented in more than one way.

The DVB specification requires a 12-byte interleave and is realised using short memories, one each at transmitter and receiver, the read address sequences of which are modified to change the output order of the contents. The cycle of interleave is chosen to synchronise with the MPEG packets such that the MPEG synchronisation byte is not affected; the 204 bytes of the modified MPEG-2 transport packet (see third layer of Figure 18.22) is convenient since 204 factorises as 17×12. This allows the interleave to have a depth of 12 bytes with 17 bytes as the delay unit size. Figure 18.25 illustrates the DVB interleaver where the 12 paths through the array are shown and the clear undelayed path

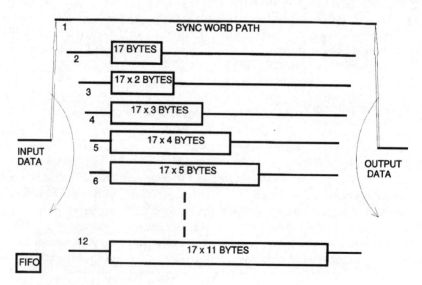

Figure 18.25   Interleaving process

for the synchronisation byte can be seen, requiring the output to take the synchronisation byte as it is entered; this synchronises the reading and writing of the array. The process is symmetrical so that complementary arrays can be used to interleave and deinterleave the data. The data enters the array and every consecutive byte is routed through a different delay element in a cycle of 12 bytes. The 11 delays, each an integer number of 17 bytes, ensure that each byte of a sequence of 12 experiences a different delay through the array and so comes out in a different place in the byte sequence from where it entered and from its original neighbours.

18.9 Transmission aspects

18.9.1 Satellite access methods

For the satellite medium there are two main methods of accessing and sharing the capacity of the transponders—frequency-division multiple access (FDMA) and time-division multiple access (TDMA)—both of which are well known. In addition, the combination of several signals can be done on a frequency-division multiplex (FDM) basis, where different carriers are used with defined spectrum capacity for each signal (often called single-channel-per-carrier or SCPC) or a time-division-multiplex (TDM) approach can be adopted where a single carrier is modulated by a single signal comprising the several signals in a digital multiplex (see Figure 18.26). TDMA is an emerging contender for applications in broadcasting in conjuction with on-board processing.

The FDM/FDMA approach has advantages in broadcasting applications where point-to-point transmission is required; for contribution and distribution

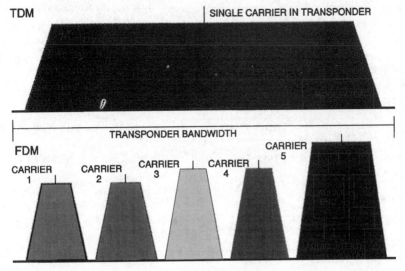

Figure 18.26   FDM and TDM methods of transponder sharing

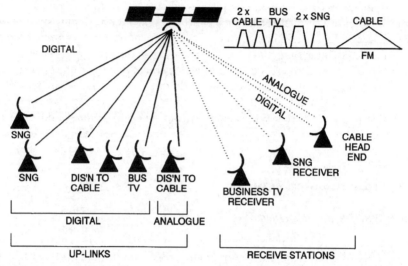

Figure 18.27   Transponder access for digital TV services

where several simultaneous transmissions are needed, perhaps only for short periods, with only a few channels at a time, FDMA techniques are sufficient and appropriate, working with partial transponder bandwidths as required. Clearly in this case, which is a point-to-point application, several carriers at various power levels may exist in any transponder at one time; this is illustrated by Figure 18.27 and discussed further below.

In this application, particular attention has to be paid to the power budget and the intermodulation noise, since the satellite power is shared among several carriers, and to the practical factors such as the receiver mixer phase-noise performance when the channel bandwidth is small. If a low-noise block converter (LNB) is used, its output intermediate frequency (IF) tolerance is important because it may be comparable to or even exceed the bandwidth of the channel and so an appropriate acquisition and synchronisation algorithm will be needed.

A number of commercial products can support a wide range of FDM/FDMA and TDM/FDMA service options[44]. These include:

- four carriers, each of two TV channels, with each channel at, say, 8 Mbit/s in a 72 MHz transponder;
- four carriers, each of one TV channel, at 8 Mbit/s in a 36 MHz transponder;
- two carriers, each of two TV channels, each channel at 8 Mbit/s in a 36 MHz transponder;
- one carrier of two TV channels, each channel at 12 Mbit/s in a 24 MHz transponder.

Carrier bit rates may be mixed within a transponder. Practical experience with these configurations is now widespread.

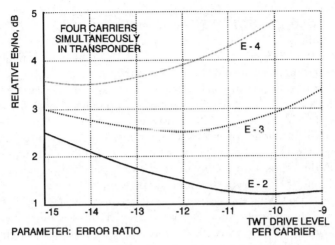

Figure 18.28 *Optimum back off for four simultaneous digital transmissions per transponder*

Normally, the transponder will be loaded on the basis of a defined value of energy/bit $[E_b]$ per carrier. The optimum transponder travelling-wave-tube (TWT) drive power per carrier is defined by the trade off between received carrier power and intermodulation in the transponder TWT; for multicarrier operation, the optimum total transponder input drive power back off is about 10 dB. It can be seen that the optimum operating point is a relatively insensitive function of input back off and error ratio. Figure 18.28 shows the C/N penalty (vertical axis) for various levels of input back off (horizontal axis) with the BER as a parameter; the broad minimum for each fixed BER indicates the desirable minimum C/N penalty. The value of the penalty is seen to be a weak function of BER. The use of the transponder in this way means that each digital carrier operates in a substantially linear channel, with intermodulation noise generated by the full transponder carrier load. Ideally, the system operates at or close to simultaneous power and bandwidth limit (for the intended receive applications). For operation with digital carriers in FDM/FDMA and fixed total transponder bit rate, the uplink power per carrier is proportional to the carrier bit rate. In multicarrier FDM operation with fixed total transponder bit rate, the required receive G/T ratio for a given link availability is independent of the bit rate of the received carrier and depends on the total transponder bit rate. A transponder bandwidth consistent with the total bit-rate requirements is therefore used.

18.9.2 Point-to-point applications

This case is exemplified by conventional contribution between studio centres where fixed network infrastructure may exist. The reduction in bit rate reduces transponder bandwidth occupancy and hence cost. The practical limitations are caused by factors described in Section 18.9.1 above, and relate to transponder loading. For best quality a high bit rate may need to be used and the EBU/ETSI and the CCIR/ITU-R have defined standards such as ETS 300 174 (CCIR recommendation 723) for 34 Mbit/s transmission and recommendation 721 at 140 Mbit/s. Such facilities can be provided by telecommunications networks rather than operated by broadcasters themselves.

This case is also exemplified by distribution to cable heads, satellite uplinks or terrestrial transmitters. Currently, a full transponder is used for analogue technology thus incurring the full transponder cost for a single programme although half-transponder operation is sometimes feasible with lower quality. Considering that the cost can be as much as £4 million per annum, any reduction in this figure by allowing more channels per transponder is immediately and significantly cost effective and compression equipment is instrumental in achieving this saving by as much as four times. It has the additional advantage that more programme choice can be offered to the subscribers. Often cable-head installations already have appropriate satellite receiver equipment including a large dish antenna which can provide more than enough $C/\mathcal{N}$ to permit good digital operation. This is a very successful application commercially all over the world.

18.9.3 Point-to-multipoint applications

18.9.3.1 Satellite newsgathering (SNG)

Strictly, the satellite newsgathering (SNG) case is not a pure point-to-multipoint application since several transmitters could be mobile within the footprint and the receivers are fixed at the news studios; nevertheless, it can be treated as a point-to-multipoint case. Digital SNG channels can access the transponder using a lower value of uplink EIRP, in some cases by as much as 16 dB, than for fixed analogue uplinks, with reception on a proportionally higher G/T ratio receiver. Current uplink EIRP for analogue FM systems is about 76 dBW whereas a digital scheme may only require 60 dBW. This might translate into a combination of an antenna size reduction from, say, about 1.5 M to 1 M, and an RF power reduction from, say, 400 W to about 25 W. SNG systems may also have less demanding nonavailability figures than other broadcast applications.

The advantages envisaged for digital SNG[45] are not only in picture-quality performance or reduced costs of space segment but also in the very practical area of the costs of shipping the equipment. Current SNG teams have to transport bulky and heavy equipment, mainly the TWT and its power supply, with consequentially high shipping costs. A typical weight of an analogue flyaway SNG unit, designed for one-person operation, is about 180 kg with a power requirement of about 5 kVA. Reduced uplink EIRP means that the TWT and its PSU can be replaced by lower-power solid-state amplifiers together with smaller antennas; this, in combination with ruggedised compression and modem equipment of broadcast performance (i.e. not using videoconferencing equipment), allows some significant operational benefits to be achieved. A typical weight of a digital SNG unit might be around 80 kg, a reduction of 100 kg made mostly in reduced antenna mount and generator sizes, with a total power requirement of 500 W, a reduction of ten times. Although for normal point-to-multipoint operation the cost of the relatively expensive encoder can be spread over many receivers, for SNG the encoder must be small and as inexpensive as possible which is against the current trend. However, integrated circuits for the MPEG encoding process are making smaller and cheaper encoders available to the SNG market.

18.9.3.2 Direct broadcasting

Where DBS applications are concerned, many channels in a point-to-multipoint broadcast mode are needed and a combination of time-division multiplex (TDM) of the several programmes with an FDMA access mode can be made, where complete transponders can be occupied by the multiplexed and modulated signal. Only a single carrier exists in the transponder in this case with the attendant benefits which this brings by maximising downlink EIRP. The cost of receiving installations for DBS has to be minimised, in particular, the receiver antenna must be small and unobtrusive. This arrangement may mean that all the programmes have to be assembled and multiplexed at the same geographic location. This may not be completely convenient for all broadcasters and there may be a need to allow access to the satellite from different places but with the result that receivers are largely unaware of this. This is a typical case where TDMA could be used; currently, this is being considered by some satellite operators but, because of its complexity, has limited future uses. It is possible, however, to use telecommunications links to bring programmes together at an uplink site for final multiplexing; this case is operationally complex, but is under consideration, and it would require the satellite bit rates to be chosen with due regard for bit rates supported by telecommunications links such as those set out in CCITT recommendations G.702 for the plesiochronous digital hierarchy (PDH), 707, 708 and 709 for the synchronous digital hierarchy (SDH) or those proposed for asynchronous tranfer mode (ATM) technology.

In the DBS case, the parameter set and system configuration need to be chosen appropriately and best results are achieved when the satellite is operated such that the power and bandwidth limits give identical operating points. In practice a low G/T (i.e. small dish size, say less than 60 cm diameter) is the main limiting factor which then drives the remaining system parameters. Given the remarks made above about satellite loading, studies show that a single bit stream of around 55 Mbit/s gross bit rate could be sustained in a 36 MHz transponder having an EIRP of 50 dBW. The downlink frequency would be about 11.5 GHz and receive noise figure would need to be better than 1.5 dB; rainfall figures for southern UK have been assumed for this example giving an unavailability of 0.1% of the time [about nine hours per annum]. By using more EIRP, bandwidth, better G/T or greater unavailability the bit rate could be increased.

The 55 Mbit/s quoted in the above example could be used in a number of different ways and could, following the techniques embodied in the MAC system, use an embedded signalling system to enable the broadcaster to change at will the configuration of the receiver to suit the transmission. As an example, 55 Mbit/s gross bit rate, assuming a simple rate 3/4 convolutional code (see above for error-control discussion), gives a net bit rate of about 42 Mbit/s. This could support ten channels of video at 4 Mbit/s with associated digital stereo for each service at 256 kbit/s each, or 18 channels of video at 2 Mbit/s with an associated digital stereo channel for each service. At the opposite extreme, 55 Mbit/s could support, say, two high-definition television channels at a bit rate of 20 Mbit/s per channel. Different broadcasters could use different configurations to suit their business needs. The technology now exists to permit DTH/DBS satellite services to be launched; what remains is to finalise system parameters so that low-cost receivers can be made available and that these receivers can be made using industry standard VLSI components which will allow other media to follow satellite in realising the potential of video compression.

18.9.3.3 Video on demand (VOD)

This service has been identified as one which will be significant in the future for a number of medium operators from satellite broadcasters to cable TV and telephone companies. British Telecom in the UK and other similar organisations, especially in the USA, have advertised trials of VOD services down the telephone line using a technology called asymmetric digital subscriber's loop (ADSL)[46,47] which allows interactivity with the subscriber in order to set up the service. VOD is the transmission of primarily precoded material to the viewer on request and is the electronic equivalent of the video-tape rental business. Figure 18.29 illustrates how such a service would be supported.

Broadcasters using satellites can provide the video but clearly have difficulties competing with the interactivity element. Nevertheless, there are means of pre-

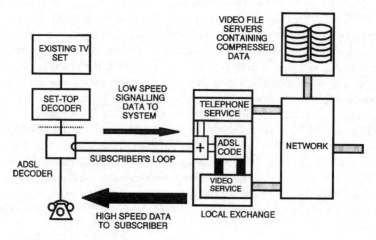

Figure 18.29   Video on demand using ADSL

senting the service which can avoid the need for complete interactivity such as near video on demand, where the multiple channels (say 12 at 3 Mbit/s each) available in the capacity of the digital broadcasting systems described above can be deployed to transmit the same programme, a film say, in 12 time slots separated by a few minutes. This means that a viewer would never be more than these few minutes from the start of the programme during an evening. This kind of service is not possible with current analogue technology and will be one of the differentiators which will help to sell new digital systems to viewers.

18.10 Applications

18.10.1 Professional systems

Professional systems are characterised by high performance and equipment ruggedness. They tend to be the point-to-point applications identified in Section 18.9.2 above. The implementation for professional applications has been governed by particular standards developed by CCITT/ITU-T[18,19]. However, in the case of the video and audio compression processes, advantage has also been taken of standards developments such as MPEG which have enabled the production of VLSI devices for the very large markets anticipated for a wide range of video services. Whereas some suppliers of professional compression systems have taken a nonstandard approach, there is now a clear preference in the market place for MPEG-based schemes since they provide a more secure means of ensuring that, as technology develops further, there will be no short-term obsolescence and a clear upgrade path.

Early commercial product from some suppliers adhered closely to the MPEG specifications, first MPEG-1 and later implemented MPEG-2 video processing, including full resolution, B frames and field/frame processing. Full MPEG-2, complete with the systems layer multiplex, has been available in product since 1994. The devices to perform the video decompression are available from several sources, some public and some tied to particular projects such as DirecTv (see above). There are also VLSI devices which can realise the Viterbi decoding of the convolutional forward-error-correction scheme and which allow digital demodulation of QPSK. The realisation of the MPEG-2 systems layer multiplexing functions took some time to appear in VLSI since the specifications were only finalised at a meeting of MPEG in November 1993. There are several developments in progress to make ASIC devices perform the processing required.

For applications involving many channels (see DBS, Section 18.10.2) there will be great emphasis on the multiplexing process but, for few channels, and for specific cases, proprietary multiplexing schemes are acceptable since they are simple and no great flexibility is called for by the user. One simple and obvious scheme is to put two separate coded bit streams on the I and Q phase pairs of a QPSK system.

Early encoding was done with discrete components in relatively bulky equipment since no MPEG-2 compliant devices yet existed to do the encoding. There were, however, developments proceeding and VLSI encoding devices have been available since the end of 1994 and others are appearing. For MPEG-1, devices do exist at low bit rates but the market is not yet so sensitive to the encoding problem since, for applications such as films on CDs, the coding need only be done once and can be done off line in nonreal-time. This also has the advantage that software coding can be done and individual pictures which cause coding difficulty can be adjusted to avoid visible defects. For real-time broadcasting, the software off line solution is clearly not viable. It is, however, possible to code a film and copy the resultant file onto a disk store from where it may retrieved, ready compressed, at the time of transmission. This applies equally to the professional and DTH/DBS applications.

Modems suitable for use with the FDMA access method are very similar to those used for data transmission by satellite and share many of the design criteria. One feature in particular is the case where the bit rate is low, say less than 2 Mbit/s, and the occupied bandwidth is also low. The bandwidth in this case may be such that LNB frequency errors are greater than the bandwidth, thus making initial carrier acquisition difficult; this will need special measures in the acquisition loop. The phase noise of the PLL deriving the local oscillators will also have to be very good in order not to impair performance; although the normal thermal noise calculations can be made, when operating close to threshold with a low bit rate there is a danger that phase noise actually dominates and causes the performance to diverge from theoretical predictions. The practical implementation will have to take special measures to avoid EMC problems or in-circuit noise, generated by the digital circuitry, affecting the phase noise. This is not easily predictable from theory and has to be contained by empirical pragmatic methods. The effects of cycle skipping in the demodulator can trigger error-extension processes which may tax the error-control system.

18.10.2 DBS systems

18.10.2.1 Overview

The configuration shown in Figure 18.8 is a simple form of the reference system developed by the DVB project (see Section 18.2.3) in its formative period when the satellite medium was its preoccupation[48,49]. In the period since DVB has addressed many issues in producing a set of standards not only for the channel-coding aspects[8-11] but also for other important matters such as service information[50] (see below), carriage of teletext data[51] and interfacing with existing systems such as satellite master antenna TV (SMATV)[52]. DVB continues to develop these enabling systems in support of the complete broadcasting capabilities offered by digital technology.

A complete system for DTH/DBS applications is illustrated by Figure 18.30. The interfaces marked A are amenable to being harmonised with those used for telecommunications networks, for example CCITT recommendation G.702. The value of this is to allow a geographic separation between the uplink facility and the multiplexing point. The bit stream leaving the multiplexer may, in this case, need to have a degree of error-control overhead added to cover the effects of the telecommunications network. The broadband digital signal conveys a number of video and audio services which are coded and multiplexed according to MPEG and DVB rules. The capacity of the complete channel varies depending upon the specific medium, but for satellites using 36 MHz transponders the typical payload is about 50 Mbit/s (see Section 18.9.3.2) and this can be used in a wide variety of ways. Figure 18.31 illustrates the scale of this variation and also illustrates the trade off available to the broadcaster in choosing the bit rate per video service and the number of services in each multiplex.

18.10.2.2 System management—service information (SI)

A system as complex and flexible as a digital DTH/DBS system, whether it is proprietary (say, DirecTv) or an open standard (DVB), cannot be managed in a simple fashion. The wide range of facilities offered to the broadcaster, both in manipulating the coding processes—MPEG parameters—and the transmission parameters—DVB—requires means of signalling to the receiver/decoder, at least, details of the current contents of the transmission so that the viewer can operate the receiver correctly and acquire the programmes of choice. Data of this kind has to be generated by the broadcaster to a large extent already because of the need to configure, almost certainly automatically, the transmission system to deliver the chosen programmes at any particular time. There is,

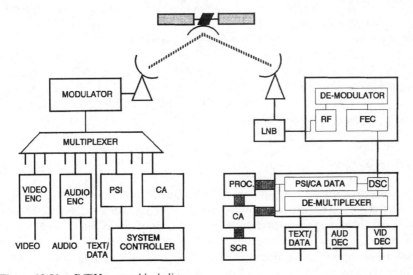

Figure 18.30   DTH system block diagram

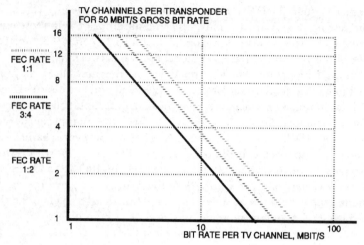

Figure 18.31   Trade of bit rate against number of TV channels

therefore, a natural connection between the SI control data, the programme schedules and the automation system files which configure equipment.

MPEG foresaw some of this requirement for the picture and sound-coding elements and defined its programme-specific information (PSI) as part of the MPEG system layer (see ISO 13818-1), but left for others, such as DVB, the extension of the idea to the broadcasting process itself. DVB has defined[50] wide and flexible extensions which allow not only the desired effect to be achieved on a given delivery medium such as satellite, but also harmonisation across a range of media such as cable, terrestrial, MMDS etc. so that receivers with strong common elements can be designed and produced cost effectively and used by viewers in a coherent way.

Because both MPEG and DVB have wide-ranging facilities there is no expectation that all broadcasters will use them all, certainly not all at once. It is therefore usual to extract subsets for particular applications; DVB has done this and published guidelines so that broadcasters can see how DVB envisaged the use of its own SI specification as well as the use of MPEG PSI[53] and how it envisaged the use of MPEG itself[54].

18.10.2.3 Receivers

By far the most significant aspect of DTH/DBS system implementation is that of the receiver. To be a success any new digital broadcasting system has to gain significant market penetration from the early days, and the parameter of greatest importance is the cost of the receiver. To achieve the price targets, which must be in the traditional consumer electronics region of about £200– £300 in the shops, the receiver must be as simple as possible to operate, allow access to a range of services, give reasonable picture and sound quality, contain a conditional-access subsystem and interface easily to the consumer's existing television equipment.

To pack that amount of technology into a receiver, and more, there must be a high degree of circuit integration in the form of a chip set. Video and audio decoding VLSI is already available and development of the demultiplexing and FEC VLSI is already substantially complete. Appropriate LNBs and tuner assemblies of suitable performance are also becoming available.

Figure 18.32 illustrates a block diagram of a receiver where it can be seen that, in addition to the usual transmission-system elements, there is a processor to manage the interfaces to the viewer's remote-control handset and transfer the commands to the hardware and software and to the CA subsystem with its associated smart card. The presence of the telephone modem should be noted as a key element in allowing interaction with the service provider. The video VLSI implementing the MPEG decoding will need a substantial amount of memory which is also shown in the diagram. Depending on the options available within the MPEG system, the amount of memory needed can vary; for full resolution as much as 20 Mbits may be required just to support the video decoder. The MPEG audio will be decoded using DSP techniques and a dedicated processor. There must be means to enable messages to be directed to the screen and so the processor has direct access to the video memory. There will be ample opportunity for receiver manufacturers to use the memory and other hardware and software to provide attractive features which will enhance the service available to the viewer.

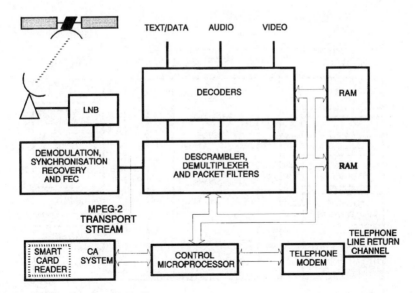

Figure 18.32   Receiver block diagram

18.10.2.4 Software

It should be clear from Figure 18.32 that software will be a major enabler for the successful implementation of MPEG/DVB broadcasting systems. The organisation of this software in a receiver requires significant attention since much of it will need to be hidden from the user and that which is visible will need to be well presented to make use of the receiver easy, especially for the untrained. The notion of an electronic programme guide (EPG) is widely accepted as the means to deal with this latter issue.

The software is expected to be in three main layers, as illustrated by Figure 18.33:

(i)   A lower layer which comprises device drivers for all the silicon devices and other functional hardware items in the receiver.

(ii) A middle layer of core systems routines for dealing with specific processes such as managing SI transactions.

(iii) A higher layer of applications that will require linking into the lower levels by means of some application programme interface (API) which may need to be standardised. Some of these applications may have downloaded elements.

The facilities for external communications provided by the telephone modem allow a small degree of interactivity to be supported by the receivers. Even a small capacity in the return path—perhaps 9600 baud—allows a surprising amount of support for CA systems, or even for Internet access, which then expands the possibilities for new services. This kind of consideration illustrates the fact that convergence between broadcasting, telecommunications and computing is real and actively being embodied in new generations of television receivers.

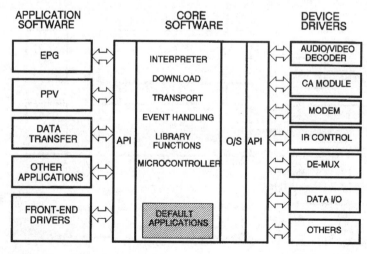

Figure 18.33   A receiver software model

18.11 Conditional access

Figure 18.34 illustrates a CA system; the transaction supported is the exchange of programmes for funds, i.e. payTV. The system to support this requires a database to manage the subscribers, their addresses and programme requirements which, provided the subscriber is still in credit, will be sent to the receiver using an appropriately structured message—the entitlement management message (EMM). Such a database is known as the subscriber management system (SMS). The CA system timing and synchronisation, together with the current encryption key, is sent in a separate message—the entitlement control message (ECM). These terms apply to the system adopted by the DVB which has also defined a scrambling algorithm and receiver interface to allow an open-system approach to the normally proprietary CA system architecture. The conclusion of such agreement among highly competitive CA system vendors is a significant achievement and one which is essential to the future commercial success of digital broadcasting systems.

CA systems comprise two stages: scrambling and encryption. The former is required to render the transmitted signal unintelligible to the receiver unequipped with the means to descramble the transmission. The means to enable the receiver to perform the descrambling is conveyed to the receiver in the form of a key or secret digital number and an encryption process is needed to make this key secret. Key generation and management are the essential features of the encryption process as are the service entitlements of the viewer. Thus, the encryption process carries not only keys in secret but also the viewing entitlements on an individual basis.

Figure 18.35 illustrates the common interface adopted by the DVB project for the receivers so that several different CA systems' security modules, with or without smart cards, could be attached to them in various ways. The integration of the different CA systems with MPEG/DVB transmission equipment is a

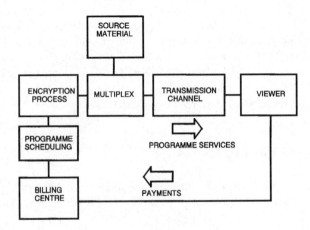

Figure 18.34   Conceptual conditional access system

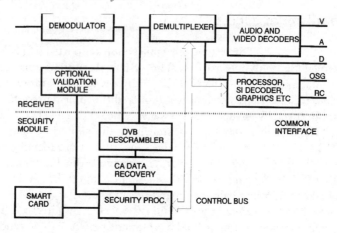

Figure 18.35   DVB common CA interface

challenge and two methods of doing so have emerged: simulcrypt, which allows more than one CA system to use the same scrambler mechanism in the receiver, and multicrypt, which would need the existence of more than one completely separate system in the receiver for which the common interface is essential to allow each CA system access to the receiver circuitry. Each method allows several different CA systems to transmit their individual messages in separate bit streams in a multiplex with the programme data; at the receiver the appropriate streams are found and decoded to release the data needed by the descrambler and to provide the viewer with information about entitlement to programmes.

For a business to succeed a policy on CA inclusion in receivers is essential and this, rather than the receiver MPEG and DVB decoding elements, is the factor which determines the economics of broadcast services. To this end, governments and the Commission of the European Union will be defining directives and local rulings, respectively, to ensure open competition in the provision and operation of CA systems for broadcasting. The technology of CA systems does not greatly affect the conduct of business; insofar that technical solutions need to be agreed among the various entities in a broadcast chain, the DVB has fostered such agreement and ETSI and CEN/CENELEC have published, or are in the process of publishing, specifications. The greatest issues in CA are commercial and as such are broadly beyond the scope of this Chapter[12–14].

18.12 Picture quality issues

For analogue systems impairments such as noise and distortion are always present more or less to the same degree all the time and all over the picture area, so that the eye may get used to them to some extent. In contrast to this, the performance of compression systems is picture dependent and digital impairments vary with time (i.e. picture-coding difficulty) and within the picture area. The viewing experience with compression is therefore quite different from that with analogue and the limits of acceptability are largely defined by those rare, critical picture sequences which are mostly unpredictable, certainly for live programmes. This probabilistic element in coding design can be avoided in the case of compressing existing recorded material since there can be an element of adjustment during multiple passes augmented perhaps by human or computer intervention during the transfer process. Indeed, this is the method used to prepare films and other material for transfer to optical or magnetic disc stores and for CDI and CDV.

The assessment of video transmission systems[55,56] is usually done using a range of agreed test waveforms and methods which have evolved over the last 40 years and are based on proven correlations between the observed subjective picture quality and objective measurement of impairments; see, for example, CCIR recommendations 500-4 and 567-3. A similar approach has been used for audio assessment. The success of these methods has led to the routine use of insertion test signals (ITS) in the vertical blanking interval (VBI) of an analogue video waveform which allows in-service quality monitoring, provided the ITS/VBI undergoes exactly the same transmission experience as the picture. Some networks reinsert the ITS/VBI occasionally and so destroy the integrity of the overall measurement and, hence, its value.

If digital techniques are employed in the transmission of video signals then the impairments caused cannot necessarily be measured using methods designed for analogue waveforms and impairment types. For example, the effect of digit errors during transmission cannot be measured meaningfully using the ITS signal. The correlations of the subjective effects of digitally-caused impairments with objective measures of the quality using the ITS are neither clear nor proven. Nevertheless, some compression-system evaluation can be done with normal test procedures using full-field test signals, i.e. the same test line repeated frame by frame, but the results need to be treated with some caution.

The full-field test mode (i.e. still picture) means that any compression system is only ever evaluated objectively in its still-picture mode; such a test is also done out of service. Any other modes, for example used for moving pictures with rapid motion (cuts etc.), are left unused during these tests. The only occasion when this is not the case for test signals is if enough noise is added to the test signal to cause the system to select the moving mode. Were this to be done, the $2T$ pulse, for example, would be poorer but the attendant subjective picture quality would be judged acceptable. Practical evaluation tests of this kind confirm this to be so. This illustrates the problem of evaluating compression systems with any degree of certainty and should be a warning to the unwary; compression technology is breaking new ground and there is bound to be controversy surrounding its evaluation at this stage. Subjective assessments are currently an essential part of quality measurement, and this will not change significantly until some objective method of assessment has been devised which reliably reflects subjective methods. Figure 18.36 illustrates the test environment which digital video

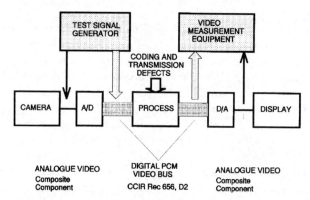

Figure 18.36   Measurement system model for digital TV systems

systems present where the processes which impair the video signal quality cannot yet be fully characterised and so there is no defined solution for their measurement.

This problem is made worse when several compression systems, either similar or not, may be cascaded in a transmission chain so that impairment builds up; this is not fully understood as yet and the rules by which predictions of performance can be made with some confidence are not clear. There is reason to suppose that, once the first compression process in such a chain of similar systems has processed a given video signal, no further degradation can occur, especially if the signal interface format between the compression systems does not return to analogue and provided that each compression process is identical in type. There is some evidence to support this assertion for some compression schemes, but it cannot be assumed to be universally true. If the compression systems are different in type, or analogue interfaces exist between the systems, then prediction of overall quality is difficult and subjective assessments are at present the only reliable method of dealing with this matter.

Applications such as contribution, where further signal processing may take place at a destination studio, require a higher degree of numerical fidelity from a compression system than do distribution applications. Signals destined for viewing only can tolerate a degree of impairment provided that this is matched to the incapabilities of the human eye. These can be expressed through psychovisual models which can perform calculations on the impairments and assess their visibility. This work is part of progress towards establishing objective assessment processses for compressed video[57-65].

While such studies are being performed and results assessed, there are practical developments occurring in areas such as multichannel DVB direct broadcasting systems whereby channel capacity sharing can be arranged between the several individual video services on a short-term basis. This process is called statistical multiplexing and involves a control system monitoring the allocation of channel capacity to the individual services which aims to ensure that picture quality is maintained at the highest level for all services present in the multiplex. The variable bit rate per service so provided has influence on picture quality and so needs careful management[66].

18.13 Digital audio broadcasting (DAB)

18.13.1 Some history

Broadcasters have been interested in the transmission of both audio and video since the feasibility of the technology was established in the late 1960s. The BBC had done some early work on a system known as 'sound in syncs' which digitally coded the mono television sound of the time and placed it in the 4.7 μs synchronisation pulse of the television waveform. This scheme was introduced during the 1970s and 1980s mainly for use on the programme distribution links which convey the signals to the terrestrial transmitters. It has been possible to use very similar techniques to allow digital transmission of outside broadcasts (OBs) from all over the world via satellite. The use of digital coding of the audio for other reasons has fitted well with these developments in transmission which now allow a complete digital production and transmission chain to be provided. The last remaining analogue portion of the chain between production and listener is the final stage to the listeners and their receivers.

The system used for digital distribution of sound has been redesigned for stereo sound in the last few years in order to support the introduction of stereo sound to television. The method of coding is called near instantaneous companding or NICAM (also used for the MAC system) and is now available directly to viewers who choose to purchase an appropriate receiver, so there is now the means for a completely digital channel between production and listener.

For sound radio programmes (as opposed to the sound accompanying television) which the BBC operates there is a need to convey the several different programmes (national radios 1, 2, 3 and 4 plus any local channels) to the terrestrial transmitters. By coding the stereo sound and multiplexing the streams so produced, a telecommunications link can be used to carry the complete assembly in an effective manner. This has been done for many years using BBC proprietary designs and operating in a 6 Mbit/s channel, although it has taken some time to take this service to all parts of the country. The nature of local independent (commercial) radio stations is such that this arrangement is not appropriate and analogue circuits have been used until very recently.

For satellite broadcasting it has already been indicated above in the description of the MAC system that digital audio of a high quality, comparable with compact disc sound at its best, is available to the viewer and listener. Other schemes have been used for satellite systems such as ASTRA which rely upon the addition of subcarriers above the video baseband (say about 7 MHz) to carry stereo sound or additional language services. These can also be used for audio not connected with the video and so support sound radio broadcasting by satellite. The Wegener system, which uses a companding technique, is one such scheme[67]. SES/ASTRA has recently announced an all-digital audio broadcasting system.

There are already available digital audio distribution schemes, especially in North America, which deploy satellites to distribute audio services such as background music to businesses and stores.

18.13.2 Digital satellite radio systems

The German digital satellite radio system (DSR), now believed to be defunct, was operationally deployed on both the German Kopernikus (FSS band) and TVSat (BSS band with 10 dB more EIRP) domestic satellites. This scheme uses a very modest NICAM-like companding technique for source coding and can take digital compact-disc signals directly from the player to the channel with no intermediate analogue stages; the received audio quality is superb. The sampling frequency is 32 kHz and there are 14 bits per sample; a 63:44 BCH error-correcting code is applied to the 11 most significant bits of the audio data. The system combines 16 stereo audio signals in time-division multiplex to make a single bit stream which, with added synchronisation, error-control bits and service-description information, modulates a single carrier using 4PSK at 20.48 Mbit/s. This carrier occupies part of a transponder; several such carriers would occupy a complete transponder thus giving access to a large number of high-quality audio services[6]. Although planned for reception in Germany, where a 19 cm squarial-type antenna is sufficient, these transmissions can be received satisfactorily in the south-eastern part of the UK with a 40 cm antenna, and receivers costing about £500 (complete with dish) have been available for several years.

There is a very similar Japanese system[6] with an additional mode to allow 48 kHz sampling and 16-bit-per-sample linear coding. The Japanese scheme mode A uses ten bits per sample, as for current terrestrial digital sound in the UK, and a bit rate of 640 kbit/s per stereo pair. The parameter values are slightly different from the DSR scheme so that the final bit rate is 24.576 Mbit/s and the BCH code used is either 63:50 or 63:56 depending on the mode; the modulation is MSK. Either system needs a C/N of about 8 dB to give a bit error ratio of 1 in 10^3.

18.13.3 Audio coding

One of the main difficulties with the schemes mentioned above, despite the fact that they work quite well, is that they take up quite large amounts of capacity; DSR uses 896 kbit/s per stereo pair. The compact disc itself, having a capacity of some 80 minutes, uses a source bit rate of about 1.4 Mbit/s per stereo pair; the sample rate is 44.1 kHz with 16 bits per sample quantisation. After error-correction coding and other signal processing, the bit rate on the disc is near to 4 Mbit/s. In sound production and recording studios the sample rates are 48 kHz, sometimes more in esoteric cases, with up to 24 bits per sample allocated in the coding formats in order to allow for signal processing e.g. arithmetic round-off errors etc.

If ways could be found to compress the audio to a lower bit rate the capacity of the recording or transmission could be greatly increased; several systems exist to do this and are commercially available but are plagued by standards arguments. For the same basic reasons, broadcasters are keen to enable the expansion of the capacity of their allocated spectrum by compressing the audio data so that they can be carried in less bandwidth. The consumer could then be offered more choice of higher quality digital audio with no more spectrum than is used today.

There has been considerable activity among European broadcasters to devise a scheme which will be useful all over Europe; a compression system called MUSICAM (it was originally called MASCAM but has been renamed after modifications) has been selected which reduces a stereo pair of channels from about 2.3 Mbit/s (the studio coding standard for audio is more demanding than that for compact discs) to about 300 kbit/s. The reduction ratio is carefully chosen since the ear is very sensitive and the impairments introduced by the processing must be handled very carefully indeed.

18.13.4 Transmission media

The medium chosen to convey the audio in the reduced bit-rate form influences the modulation scheme significantly; in the case of satellites, where the transmission medium is fairly well behaved and stable, a conventional scheme based on, say, PSK could be used. However, the terrestrial transmissions in the current allocations at Band II (88 to 108 MHz in Europe) suffer from multipath and echoes, particularly if the receiver is portable or mobile. In order to make better use of the spectrum and to serve better the community of mobile listeners an appropriate modulation system is needed. Such a system is orthogonal frequency-division multiplex (OFDM)[36,37] which is also planned for use with digital television as described in Section 18.8. This scheme is a form of spread spectrum and is, in particular, very resistant to multipath. On satellites the system can bring advantages in that, because of the signal processing involved, the spectrum can be controlled and this can help in controlling interference.

18.13.5 The European digital audio broadcasting (DAB) project

The work in Europe mentioned above has been organised under the Eureka 147 collaborative development project and has many European partners including broadcasters, equipment suppliers, transmitter manufacturers, semiconductor manufacturers and aerospace companies. The project has been running for several years[68-72] and the system specification has been published[7]. Practical systems are now entering service.

18.14 The future

This Chapter has reviewed low-bit-rate digital transmission of television signals on satellites and placed it in a context of existing and developing television standards and a changing regulatory environment. Some of the systems aspects have been discussed and particular details of practical satellite systems already available commercially have been given. For completeness some brief discussion of digital audio broadcasting has been included (see Chapter 17).

There are areas yet to be explored more fully which relate to the completion of the development of standards for video compression, audio compression, medium-specific aspects such as multiplexing and a policy on conditional-access systems.

There has never been a more exciting time in the broadcasting industry. Every aspect of the industry has been questioned and challenged seriously in the last few years, to the extent that it is almost unrecognisable compared to only a few years ago. A market-led regime exists and this means risks have to be taken by all players, and by technology suppliers in particular.

These developments, together with attention to practical design details and particularly receiver costs, will eventually result in a variety of broadcast systems having lower overall cost to the end user with significantly increased facilities and choice over analogue systems. These costs depend on many factors, not the least of which is the functionality required in the processing; aspects of the design of the systems and particularly the receivers have been discussed above where it can be seen that the designer's choice is wide and needs to be matched to market needs. It will be the reduced costs and increased functionality of digital systems that will lead to their success.

18.15 References

1 BURNS, R.W.: 'A history of television broadcasting'. (IEE History of Technology Series, Peter Peregrinus, London, 1986)
2 ABRAMSON, A.: 'Pioneers of television—Philo Taylor Farnsworth', *SMPTE J.*, November 1992, pp. 770–784
3 CLARKE, A.C.: 'Extra-terrestial relays', *Wirel. World*, October 1945, pp. 305–308
4 WILLIAMSON, M.: 'How satcoms took off', *IEE Rev.*, 1991, **37**, (6), pp. 379–383
5 MARSHALL, P.: 'Satellites and television broadcasting' *in* PELTON, J.N., and HOWKINS, J. (Eds.): 'Satellites international' (Stockton Press, 1987) pp. 95–98
6 CCIR report 1228, CCIR documents, XVIIth Plenary, Dusseldorf, 1990, Annex to volumes X and XI—Part 3, pp. 310–337. Annexe 1 gives DSR specification and Table 1 lists the details of the Japanese MDSD system
7 'Radio broadcast system: digital audio broadcasting (DAB) to mobile, portable and fixed receivers'. ETSI specification ETS 300 401, February 1995
8 'Digital broadcast systems for television sound and data services—framing structure, channel coding and modulation for 11/12 GHz satellite services'. ETSI ETS 300 421, November 1994
9 'Digital broadcast systems for television sound and data services—CATV systems'. ETSI, ETS 300 429, November 1994

10 'Digital broadcast systems for television sound and data services—MMDS systems using frequencies below 10 GHz'. ETSI, ETS 300 749, January 1996; 'Digital broadcast systems for television sound and data services—MMDS systems using frequencies above 10 GHz'. ETSI, ETS 300 748, January 1996

11 'Digital broadcast systems for television sound and data services—framing structure, channel coding and modulation for digital terrestrial television'. ETSI specification ETS 300 744, February 1996

12 'Digital video broadcasting (DVB) common scrambling (CS) system description'. ETSI report ETR 289, draft July 1996

13 'Standard for common receiver interface'. CEN/CENELEC, EN 50221, draft D, April 1996

14 'Common scrambling/de-scrambling algorithm, distribution agreement'. DVB document A011, October 1995

15 DARBY, P.J.: 'Colorimetry in television', *IBA Tech. Rev.*, 'Light and colour principles', (22) pp. 28–39

16 'The white book'. UK PAL specification, revised edition, 1984, published jointly by DTI/IBA/BBC/BREMA

17 TOWNSEND, B., 'PAL colour television' (Cambridge University Press, 1985, 2nd edn.)

18 'Network aspects; digital coding of component television signals for contribution quality applications in the range 34–45 Mbit/s'. ETSI, ETS 300 174, November 1992. Also published as: 'Transmission of component coded digital television signals for contribution quality applications at the third hierarchical level of CCITT Recommendation G.702'. CCIR recommendation 723, CCIR documents, XVIIth plenary, Dusseldorf, 1990, **XII**, pp. 54–67. Now adopted by ITU-T as recommendation J.81

19 'Transmission of component coded digital television signals for contribution quality applications at bit rates near 140 Mbit/s'. CCIR recommendation 721, CCIR documents, XVIIth plenary, Dusseldorf, 1990, **XII**, pp. 68–79. Now adopted by ITU-T as recommendation J.80

20 LUCAS, K., and WINDRAM, M.D.: 'Standards for broadcasting satellite services', *IBA Techn. Rev.*, 'Standards for satellite broadcasting', (18), March 1982, pp. 12–27

21 MALCHER, A.T.: 'A TDM system for ENG links'. *IEE Conf. Publ.* 286, International Broadcasting Convention (IBC), 1984, pp. 271–274

22 DALTON, C.J., and MALCHER, A.T.: 'Communications between analogue component production centres', *SMPTE J.*, **97**, pp. 606–612

23 DRURY, G.M.: 'HDTV—how many lines?', *Image Technol.*, January 1990, pp. 16–20

24 GRIFFITHS, E., and WINDRAM, M.D.: 'Widescreen TV—a commercial reality'. *IEE Conf. Publ. 327*, International Broadcasting Convention (IBC), 1990, pp. 137–144

25 EBU specification Tech 3258, October 1986

26 ETSI specification for the D2-MAC system: ETS 300 250

27 ETSI specification for the D-MAC system: ETS 300 355

28 ETSI specification for the HD-MAC system: ETS 300 350

29 ANNEGARN, M. *et al.*: *Philips Tech. Rev.*, August 1987, **43**, (8), pp. 197–212

30 NINOMIYA, Y. *et al.*: *IEEE Trans.* December 1987, **BC-33**, (4), pp. 130–160

31 ETSI specification for the PALPlus system: ETS 300 731

32 WITHAM, A., and HUNT, K.: 'The ITU plan for space broadcasting', *IBA Tech. Rev.*, 'Satellites for broadcasting', March 1979, (11), revised version, pp. 14–17

33 LONG, T.J.: 'Satellite developments and opportunities', *IBA Tech. Rev.*, 'Standards for satellite broadcasting', March 1982, (18), pp. 5–11

34 DRURY, G.M.: 'Digital transmission of television signals', *EBU Rev. Tech.*, October 1987, (225) Geneva, pp. 3–15

35 BYLANSKI, P., and INGRAM, D.G.W.: 'Digital transmission systems' (Peter Peregrinus, 1980)

36 ALARD, M., and LASSALLE, R.: 'Principles of modulation and channel coding for digital broadcasting for mobile receivers', *EBU Rev. Tech.* August 1987, (224), pp. 168–190

37 MASON, A.G., DRURY, G.M., and LODGE, N.K.: 'Digital television to the home—when will it come?', *IEE Conf. Publ. 327*, IBC 1990, pp. 51–57

38 SWEENEY, P.: 'Error control coding' (Prentice-Hall, 1992)

39 BOSE, R.C., and RAY-CHAUDHURI, D.K.: 'On a class of error correcting binary group codes', *Inf. Control*, March 1960, **3**, pp. 68–79

40 FORNEY, G.D.: 'On decoding BCH codes', *IEEE Trans. Inf. Theory*, 1965, **11**, pp. 549–557

41 VITERBI, A.J.: 'Convolutional codes and their performance in communication systems', *IEEE Trans.* October 1971, **COM-19**, (5), pp. 751–772

42 RAMSEY, J.L.: 'Realisation of optimum interleavers', *IEEE Trans. Inf. Theory*, May 1970, **16**, (3), pp. 338–345

43 FORNEY, G.D.: 'Burst correcting codes for the classic bursty channel', *IEEE Trans.*, October 1971, **COM-19**, (5), pp. 772–781

44 WINDRAM, M.D., and DRURY, G.M.: 'Satellite and terrestrial broadcasting—the digital solution', *IEE Conf. Publ. 397*, International Broadcasting Convention (IBC), Amsterdam, September 16–20 1994, pp. 366–371

45 DRURY, G.M.: 'Digital satellite news gathering', *IEE Conf. Publ. 413*, International Broadcasting Convention (IBC), Amsterdam, September 15–19 1995, pp. 570–575

46 *IEEE J. Sel. Areas Commun.*, special issue on *High speed digital subscriber lines*, August 1991, **9**, (6)

47 YOUNG, G., FOSTER, K.T., and COOK, J.W.: 'Broadband multimedia delivery over copper', *BT Technol. J.*, October 1995, **13**, (4)

48 REIMERS, U.: 'The European project on digital video broadcasting—achievements and current status', *IEE Conf. Publ. 397*, IBC 1994, Amsterdam, September 16–20 1994, pp. 550–556

49 COMMINETTI, M., and MORELLO, A.: 'Direct-to-home digital multi-programme television by satellite', *IEE Conf. Publ. 397*, IBC 1994, Amsterdam, September 16–20 1994, pp. 358–365

50 'Digital broadcast systems for television sound and data services—specification for service information (SI) in digital video broadcasting (DVB) systems'. ETSI specification ETS 300 468, February 1996

51 'Digital broadcast systems for television sound and data services—specification for conveying ITU-R system B teletext in digital video broadcasting (DVB) systems'. ETSI specification ETS 300 472, December 1994

52 'Digital broadcast systems for television sound and data services—framing structure, channel coding and modulation for SMATV', ETSI specification ETS 300 473, December 1994

53 'DVB guidelines on the use of MPEG programme specific information (PSI) and DVB service information (SI)'. ETSI report ETR 211

54 'DVB guidelines on the use of MPEG'. ETSI report ETR 154

55 WEAVER, L.E.: 'Television neasurement techniques'. (IEE Monograph Series, Number 9, London, 1971)

56 DRURY, G.M.: 'Picture quality issues in digital compressed video', *IEE Conf. Publ. 413*, International Broadcasting Convention (IBC), Amsterdam, September 15–19 1995, pp. 13–18

57 RAYNER, A.: 'A quality of service view of digital television network distribution', *IEE Conf. Publ. 358*, IBC 1992, pp. 498–503

58 LODGE, N.K., and WOOD, D.: 'Subjectively optimising low bit rate television', *IEE Conf. Publ. 397*, IBC 1994, Amsterdam, pp. 333–339

59 VORAN, S.D., and WOLF, S.: 'The development and evaluation of an objective quality assessment system that emulates human viewing panels', *IEE Conf. Publ. 358*, IBC 1992, pp. 504–508

60 HISTED, C.: 'How to evaluate on-line compression', *TV Buyer*, **5**, (6), pp. 14–18

61 MITCHELL, J., and FERNE, A.: 'Cascading data compressors in the broadcast chain', *Int. Broadcast.*, 1994, **17**, (5), pp. 26–30

62 WILKINSON, J.H., and STONE, J.J.: 'Cascading different types of video compression systems'. IEE colloquium 1994/055, March 1994
63 WOOD, D.: 'European perspectives in digital television broadcasting—quality objectives and prospects for commonality', *EBU Rev. Tech.*, Summer 1993, (256), pp. 9–15
64 ZOU, W.Y.: 'Performance evaluation—from NTSC to digitally compressed video', *SMPTE J.*, 1994, **103**, (12), pp. 795–800
65 ZOU, W.Y. *et al.*: 'Subjective testing of broadcast quality compressed video', *SMPTE J.*, 1994, **103**, (12), pp. 789–794
66 DRURY, G.M., and BUDGE, M.R.J.: 'Picture quality and multiplex management in digital video broadcasting systems', *IEE Conf. Publ. 428*, International Broadcasting Convention (IBC), Amsterdam, September 12–16 1995, pp. 397–402
67 JACKSON and TOWNSEND (Eds.): 'TV and video engineers reference book (Butterworth, 1991) pp. 48/7–48/8
68 DOSCH, C., RATLIFF, P., and POMMIER, D.: *EBU Rev. Tech.*, December 1988, (232), pp. 275–282
69 GILCHRIST, N.: BBC research report RD 1990/16, 1990
70 SHELSWELL, P. *et al.*: BBC research report RD 1991/2, 1991
71 PRICE, H.: 'CD by radio', *IEE Rev.*, April 1992, **38**, (4), pp. 131–135
72 O'LEARY, T.: *EBU Rev. Tech.*, Spring 1993, (255), pp. 19–26

Chapter 19

Mobile satellite communications

I. E. Casewell

19.1 Introduction

Those concerned with maritime and aeronautical communications have, since the first experiments in satellite communications, realised the potential benefits which satellite communications can bring to their respective communities. Clearly, the prospect of near global, highly reliable communications made possible by satellite technology is very attractive to mariners and aviators alike. The arrival of cellular radio services has awakened interest in land mobile satellite services. The exploitation of this technology to satisfy these applications has largely taken place during the last two decades.

These developments have been enabled by the advances in spacecraft and communications technology which have allowed systems to be developed that permit the use of acceptably small user antennas, while providing sufficient system capacity to remain economically viable. Inevitably, these constraints have meant that, in the early 1990s, user voice communications capacity has been limited to, typically, one voice channel when user antenna gains of between 12 and 22 dBi are used. Similarly, data communication rates of approximately one kilobit per second when user antenna gains of about 1 dBi are employed. To facilitate the use of economic satellite RF powers, of the order of 1 W per voice channel, low-rate vocoder and forward-error-correction coding techniques are now extensively employed. Furthermore, to make the best use of the available satellite power, capacity is always demand assigned in some way. As is implied by the preceding sentences, most systems currently in use or under development are power limited. However, with the meagre frequency allocations that are currently available for use by the mobile satellite service, more emphasis is now being placed upon spectral efficiency and it is likely that during the 1990s systems will become frequency limited. This is particularly true of the new generation of satellite personal communications systems (PCS) that are currently under development.

Until the mid-1980s mobile satellite systems and services were predominantly aimed at maritime users. Hence, the organisations and infrastructure to provide a near global service were created.

This international emphasis also remained well suited to the aeronautical services which were introduced in the early 1990s. However, the introduction of the younger land mobile satellite services has led to the development of regional systems which limit coverage to continents such as Australia or North America. Although these systems are aimed at large perceived land mobile markets, the operators also wish to provide services to mariners in coastal waters and over-flying aircraft. For this reason, these emerging regional operators prefer to no longer differentiate between the maritime, aeronautical and land mobile satellite services but to combine them as a single mobile satellite service (MSS). The main applications of the land mobile systems are initially perceived to be messaging and vehicle location, thus enabling efficient wide-area vehicle management systems to be introduced. With the introduction of spot beams and the new satellite PCS systems, the range of voice and data services familiar to GSM users will soon be provided via satellite.

The foreseen scarcity of spectrum at L-band and the emerging concept of satellite personal communications, which will require significant spectrum, is increasing the interest of the research community in the yet unexploited allocations in the 20/30 GHz bands.

This Chapter aims to provide an overview of the current and emerging mobile satellite communications systems and an appreciation of the constraints imposed upon them.

19.2 Frequency allocations[1]

One of the major functions of the International Telecommunications Union (ITU) is the administration of frequency allocations. As a result of the work of the International Maritime Organisation (IMO) and the International Civil Aviation Organisation (ICAO), the ITU World Administrative Conference (WARC) for space telecommunications, held in 1971, allocated frequencies in the 1.5/1.6 GHz bands to the maritime mobile satellite service (MMSS) and the aeronautical mobile satellite service (AMSS). These allocations form the basis for the bands currently used by the mobile satellite service (MSS). The original 1971 allocations have been modified over the years with the most significant changes taking place at the ITU mobile WARC, held in 1987. This conference was significant because of the reallocation of part of the previous AMSS band to the land mobile satellite service (LMSS), so providing L-band allocations for the LMSS.

Although the major band of interest is at L-band, other allocations exist. These include various VHF and UHF frequencies, in particular 121.5 MHz,

243 MHz and 406 MHz for use by emergency position indicating radio beacons (EPIRB). Historically, the allocation of 806–890 MHz and 942–960 MHz, principally in ITU regions 1 and 2, provided a significant stimulus in the development of the LMSS in North America.

An earlier WARC, held in 1979, allocated the 14.0 to 14.5 GHz band to the LMSS for earth–space transmissions on a secondary basis, the primary allocation being to the fixed satellite service (FSS). No corresponding band was allocated to the space–earth LMSS because it was assumed that signals in the LMSS would appear identical to those operating in the FSS band of 11.7 to 12.2 GHz.

Primary allocations at 20 and 30 GHz and at higher frequencies have been made but the exploitation of even the 20/30 GHz allocation is only now starting to be considered.

For the 1.5/1.6 GHz bands to be effectively used for services of an international basis (e.g. the AMSS) they must be allocated on a global basis. However, this need not be the case for the LMSS which may be operated on a regional basis.

Figure 19.1 shows the current frequency allocations for the 1.5/1.6 GHz bands. The earth-to-space band extends from 1626.5 MHz to 1660.5 MHz, and the space-to-earth band extends from 1525 MHz to 1559 MHz. The majority of each band is subdivided in a similar way to allow the use of duplex channel pairing with an offset frequency of 101.5 MHz.

Allocations may be on a primary basis or a secondary basis. Consider, initially, the subdivision of the bands on a primary basis. The band is subdivided into four subbands, one each for the MMSS, the joint MMSS and AMSS for distress and safety use, the AMSS and the LMSS.

The aeronautical mobile satellite service band was originally allocated for safety services and airline operational correspondence purposes. This is denoted by the designation R. However, with the rise in interest in aeronautical public correspondence (APC) services, WARC-MOB 87 agreed that the AMSS band may also be use for APC, but with safety-related traffic retaining absolute priority.

To enable INMARSAT to provide land mobile services using its first- and second-generation spacecraft, the MMSS sections of the band were allocated in 1987 to the LMSS as follows. The bands 1525 to 1530 MHz, 1533 MHz to

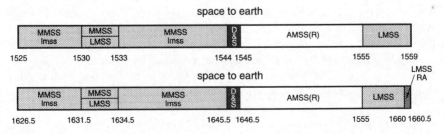

Figure 19.1   L-band MSS frequency allocations following WARC-MOB 92

1544 MHz, 1626.5 MHz to 1631.5 MHz and 1634.5 MHz to 1645.5 MHz were allocated to the land mobile data-only service on a secondary basis. At the same time bands 1530 to 1533 and 1631.5 to 1634.5 MHz were allocated on a coprimary basis to the MMSS and the LMSS with no restrictions on the type of traffic.

The 1987 meeting of WARC was only mandated to address the spectrum already allocated to the mobile services, hence it could only 'share out the cake' in a different way. WARC 92, held in Malaga, was not constrained in the same way and so significant new spectrum was made available to the mobile satellite services. This meeting of WARC had the difficult task of satisfying the demands of the newly-arrived little and big LEO system operators and the expanding requirements of the existing and proposed geostationary MSS.

The major changes agreed at WARC 92 are as follows:

- coprimary allocations made for the little LEO systems in the 137/148 MHz and 401 MHz bands;
- coprimary allocation in region 2 of 1429 to 1525 and 1675 to 1710 MHz for the generic MSS;
- the frequency bands of 1525 to 1530 and 1626.5 to 1631.5 MHz allocated on primary basis to the MSS in regions 2 and 3; in region 1, the allocation is on a coprimary basis for the MMSS and a secondary basis for the LMSS;
- coprimary allocation to the MSS in the bands 1610 to 1626.5 and 2483.5 to 2520 MHz; this is intended for use by the big LEO systems;
- coprimary allocations to the MSS, intended for use by the FPLMTS, in the bands of 1980 to 2010 and 2170 to 2100 MHz;
- coprimary allocation from 2005 onwards to the nongeostationary MSS in the bands 2500 to 2520 and 2670 to 2690 MHz.

Initially, the 1.5/1.6 GHz bands were used on a global basis by INMARSAT and so the problem of intersystem coordination did not arise. This somewhat idyllic situation has not lasted long following the emergence of the regional MSS and PCS operators, concerned with the provision of a voice service for land mobiles which are located in the remoter regions not adequately covered by terrestrial cellular systems. Hence, there will be great demand for the 4 MHz allocated to this use. This situation has led to the adoption of antenna gains of at least 10 dBi for use by voice terminals on the hypothesis that the necessary intersystem isolation may be provided by the earth segment alone. This is based on the assumption that cochannel satellites are separated by at least 30° of longitude. Only now, with the introduction of truly spot-beam satellites, is it becoming possible to rely upon the spacecraft for the necessary isolation to allow significant frequency reuse with low-gain mobile antennas.

19.3 INMARSAT system and standards

19.3.1 Introduction

At its formation in 1979, INMARSAT was mandated to provide systems capable of supporting services for maritime users alone. During the 1980s, INMARSAT extended its field of interest to include aeronautical and land mobile systems. To accommodate these intentions, INMARSAT developed a space segment which is common to all systems. The space-segment configuration and the major system standards are briefly described below. In the early 1990s, INMARSAT formed an affiliate company, known as ICO Global Communications, specifically to develop a satellite PCS on a more commercial footing.

19.3.2 Space segment

The INMARSAT space segment is deployed in a four-ocean-region configuration which was developed from the earlier three-ocean-region arrangement designed to support maritime services. Figure 19.2 shows the 0° and 5° elevation contours for the four-ocean-region configuration. The first generation of satellites were leased from COMSAT, ESA and INTELSAT, but these satellites have now been replaced by a second and third generation of satellites. Unlike the first generation, these satellites are owned and operated by INMARSAT.

Four INMARSAT-2 satellites, which made up the second generation of the space segment, were launched, one into each ocean region, during the period

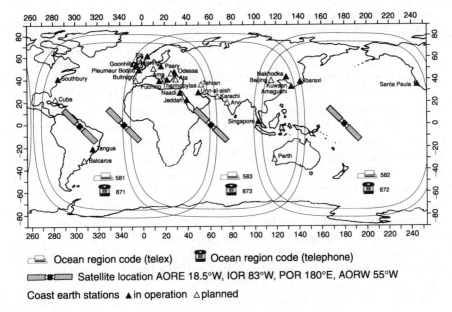

Figure 19.2 INMARSAT coverage showing 0° and 5° contours

March 1991 to April 1992. These satellites are now being used as in-orbit spares following the launch of the third-generation satellites starting in April 1996. The final INMARSAT-3 satellite was successfully launched into the Atlantic (west) orbital slot in June 1997.

The second-generation satellites, like their predecessors, provided global beam coverage, and the third-generation satellites provide both global beam and spot-beam coverage. In addition, the INMARSAT-3 satellites also contain a navigation payload which will form part of the global navigation satellite system (GNSS). Details of INMARSAT-2 and -3 satellite payloads can be found in Sections 19.7.2 and 19.7.3, respectively.

19.3.3 INMARSAT-A

The INMARSAT-A system, previously known as standard-A, is derived from the COMSAT MARISAT system which became operational in 1976. The system enables signatories to provide circuit-switched telephony and telex services between the PSTN and suitably equipped ships. Around 18 000 standard-A terminals are installed on vessels and approximately a further 7600 transportable terminals have been registered. The system consists of a large population of mobile terminals known as ship earth stations (SES), a number of coast earth stations (CES) and a network coordination station (NCS) configured around the space segment in each ocean region. The CESs act as gateway stations between the satellite network and the PSTN. Figure 19.3 shows this system configuration.

The standard-A terminals are relatively large, with the antenna radome measuring approximately 1.8 m in diameter. They are also relatively expensive, costing approximately US$50 000 each. The SES specification stipulates a G/T of at least -4.0 dB/K and an EIRP of at least 36 dBW. Such performance requires the use of an antenna of between 0.7 and 1.0 m in diameter and so stabilisation, to correct for the ship's motion, is clearly required.

The L-band, global-beam satellite EIRP required to support a standard-A telephony circuit is about 17 dBW and so a second-generation satellite could support up to about 200 simultaneous channels. To support a large population of SESs, demand assignment of capacity is clearly required, and this is common to all of the systems described below. So, in addition to the CES which functions as the gateway between the INMARSAT-A system and the PSTN, a NCS is required in each ocean region. In the case of the telephony subsystem, the NCS is responsible for allocating channels on a call-by-call basis. The telex subsystem employs a more distributed form of control with part of the allocation procedure being the responsibility of the appropriate CES. The INMARSAT-A transmission formats are shown in Table 19.1.

The access-control equipment needed to provide the demand-assignment functionality at a CES accounts for a substantial proportion of the total CES

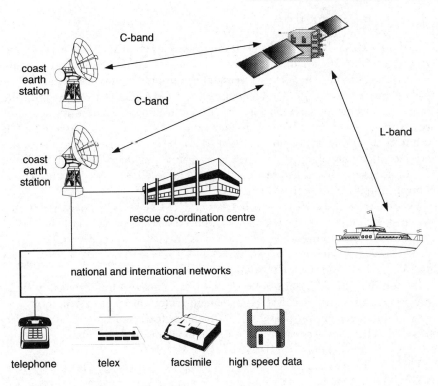

Figure 19.3 *INMARSAT-A system configuration*

Table 19.1 *INMARSAT-A transmission formats*

Direction	Telephony	Telegraphy	Signalling
CES to SES	FM/FDMA/SCPC	BPSK/FDMA/TDM	BPSK/FDMA/TDM
SES to CES	FM/FDMA/SCPC	BPSK/FDMA/TDMA	BPSK/FDMA/TDMA

costs and is always a critical part of the CES procurement. The RF head for a typical CES employs a 15 m diameter antenna and is required to operate both at L- and C-band.

The INMARSAT-A system is now obsolete, following the introduction of digital techniques for voice coding and modulation, and a new system, known as INMARSAT-B, has been introduced. To help precipitate the changeover from A to B, it was decided that no further A SES designs would be type approved after July 1989.

19.3.4 INMARSAT-B

The INMARSAT-B system has been under development since the early 1980s, with the initial objectives of reduced terminal cost and improved satellite resource utilisation. After several years of development, this system has effectively evolved into a digital equivalent of the INMARSAT-A system since the SES G/T specification remains at $-4\,$dB/K. To allow for the introduction of spot beams and to provide for a greater range of services, the INMARSAT-B protocols are considerably more complex than those of the earlier system. In the interests of commonality, it has been agreed to adopt the B protocols as far as possible in the newly-defined M system to enable the CES to use the same access-control equipment for both systems.

The majority of services supported by the INMARSAT-B system are circuit switched. In addition to the telephony and telex services provided by the earlier INMARSAT-A system, low speed (300 bit/s) asynchronous data and medium speed (9.6 kbit/s) data services can be supported. The 9.6 kbit/s data service may also act as a bearer for CCITT group 3 facsimile services.

In addition to the point-to-point services, shore-to-ship broadcast telex services are provided for fleet management, safety-message distribution and network-management purposes. A ship-to-shore distress-alerting facility is also provided. These services satisfy the requirements of the IMO global maritime distress and safety system (GMDSS).

The telephony channel employs adaptive predictive speech coding at a 16 kbit/s rate. A subband data channel operating at a rate of up to 2400 bit/s is included with the telephony channel. To provide the required bit error rate, when at the operating C/N_0 of 47.3 dBHz, rate three-quarters convolutional coding is employed, resulting in a transmission rate of 24 kbit/s. The RF carrier is modulated using filtered OQPSK and the satellite access scheme is SCPC/FDMA. The L-band, satellite EIRP per channel is nominally 15.5 dBW.

The medium-speed data channel is similar to the telephony channel, except that it uses rate one-half convolutional coding which, along with the subband data channel, provides the same channel symbol rate of 24 kbit/s.

To make best use of the limited L-band satellite EIRP, power control will be used with the SCPC/FDMA channels. Additionally, voice activation will be employed with the telephony channels.

The telex and low-rate data channels use TDM/FDMA in the shore-to-ship direction and TDMA/FDMA in the ship-to-shore direction. Initially, when traffic levels are low, a single TDM carrier will accommodate signalling traffic in addition to telex and low-speed data traffic. The transmission rate is 6 kbit/s and BPSK with rate one-half convolutional coding used in the shore-to-ship direction. In the opposite direction, the transmission rate is increased to 24 kbit/s and OQPSK modulation is employed.

In addition to the various communication channels, out-of-band signalling channels are required between the SES, NCS and CES to allow call set up and the dissemination of system status information. These various signalling logical channels make use of the same TDM(A)/FDM channels as the telex and low-rate data services. The main RF characteristics of both the SESs and CESs are similar to their INMARSAT-A counterparts.

The INMARSAT-B system became operational in 1993 and today some 3500 maritime and land transportable terminals have been registered.

A 56/64 kbit/s high-speed data service is now also available with suitably equipped INMARSAT-B terminals. Independent of data rate, the channel rate is 132 kbit/s following $R = 1/2$ convolutional encoding and framing overhead. The modulation scheme is OQPSK and the channel spacing is 100 kHz. The link budget is designed to provide a bit error rate of less than 1 in 10^6 for 99 per cent of the time.

19.3.5 INMARSAT-C

Like the INMARSAT-A and B systems, the INMARSAT-C system was originally designed for the maritime service. However, with the increasing interest in land mobile services, the INMARSAT-C service is now intended to support a low-speed, store-and-forward, two-way message service for land and maritime applications. In addition to the two-way service, a shore-to-ship message broadcast service, known as enhanced group call, may be supported. This system provides for individual ship, fleet and geographical-region addressing and so satisfies the requirements of the maritime safety information broadcasting function of the IMO GMDSS. The system, being primarily intended for short messages, employs packet-switching techniques.

Further protocols have now been developed to support specialist applications such as position reporting and polled data transactions. The system configuration is essentially the same as for the INMARSAT-A and B systems. However, the low transmission rate of 1200 symbols per second is used to enable very low G/T SESs employing unstabilised, hemispherical radiation pattern antennas to be used. This results in a much smaller and simpler SES costing around US\$5000. It is therefore well suited to installation on all types of vessel down to about 5 m in length as well as larger vans and trucks.

Each NCS or CES transmits one or more carriers modulated at 1200 symbols per second using unfiltered BPSK. A TDM format is used to accommodate both fixed and variable-length data packets for all services and signalling. In the ship-to-shore direction SESs may transmit in one of two modes; for signalling and short messages, a TDMA scheme is used to support short messages using a time-slot-reservation protocol or signalling using a slotted-ALOHA protocol. For longer messages, a channel is uniquely assigned to one SES/CES pair.

To ensure a reasonably low packet error probability, under the anticipated fading conditions, bit interleaving and rate one-half convolutional coding is employed. The interleaver depth is fixed in the shore-to-ship direction and variable in the opposite direction, the maximum depth is equal to that used in the forward direction and is equivalent to the frame duration of 8.64 seconds.

With the exception of the broadcast services, which may be used by receive-only terminals, an ARQ protocol based on a 16-bit cyclic redundancy check (CRC) sum is employed to ensure a high probability of error-free message transfer.

The INMARSAT-C terminal consists of a small above-decks unit and a below-deck unit a little larger than a desk-top voice-band modem. The terminal mimics any other data communication equipment, with traffic and control information passing over an RS232C serial link. The terminal must provide a G/T of at least -23 dB/K and an EIRP of 12 dBW. The nominal satellite L-band EIRP is 21 dBW per channel and the CES requirements have been defined so that existing INMARSAT-A or B RF heads may be used.

During 1991, commercial services were introduced in all four ocean regions. At the end of 1996, about 19 000 maritime terminals and about 10 000 land terminals had been registered on the INMARSAT system. All of the CESs are capable of providing the mandatory services of two-way store-and-forward messaging, distress alerting and priority-three (distress) messaging and the SafetyNET enhanced group call broadcasts. The mobile terminals are available from over 20 manufacturers. The terminals fall into the basic classifications of: maritime, land and portable two-way equipment and standalone EGC or INMARSAT-A add-on EGC receivers. In addition, some manufactures produce integrated INMARSAT-C and GPS navigation receivers which are used in position-reporting systems of all types.

Following the introduction of the third generation of INMARSAT satellites, an enhanced version of INMARSAT-C is being developed. This new standard, known as mini-C, will employ a lower power HPA and the system protocols will be improved in light of experience gained with the existing system. The target price for a mini-C terminal is believed to be around about one half of the current terminal price.

19.3.6 INMARSAT-M

The INMARSAT-M system is relatively young, with development work commencing in 1988. Its inception largely results from the observation of developments in Australia and North America and, fortunately, a cooperative approach has been adopted. So far, this cooperation has not yet resulted in a common global standard for M-type systems, however there is now considerable commonality between the various mobile earth-station (MES) specifications.

The system is conventionally configured around the INMARSAT space segment, with an NCS per region and coast or land earth stations which form the gateway to the national PSTNs. The mobile terminals are now known as mobile earth stations (MES), reflecting their application in both land and maritime roles. The land MES is required to have a G/T of at least -12 dB/K and the maritime MES has a G/T of at least -10 dB/K.

The INMARSAT-M system is designed to support circuit-switched, medium quality telephony and full duplex, medium-rate data services. The latter is expected to find application as a bearer for CCITT group 3 facsimile. The system also supports the necessary distress-calling facilities for maritime users. As was noted earlier, the system call set-up protocols are common to the INMARSAT-B service, so easing the introduction of the INMARSAT-M system.

The traffic channels are based on an FDMA/SCPC arrangement using a transmission rate of 8 kbit/s. Filtered OQPSK is used to allow a reduced channel spacing of 10 kHz. In the telephony mode, the vocoder operates at a rate of 6.4 kbit/s with integral forward error correction. The voice-coding algorithm was selected as part of a joint exercise with the Australians; it is known as the improved multiband excitation (IMBE) and was developed by DVSI Inc. The collaboration with the Australians will ensure that the same algorithm will be used by both systems. Subband data and framing information, at a rate of 1200 bit/s, is multiplexed with the vocoder data following rate three-quarter convolutional coding. The system has been designed to allow voice activation and power control on the shore-to-ship link to conserve satellite power.

In the duplex data mode, information at 2400 bit/s is combined with subband data and framing information prior to rate three-quarters convolutional encoding. The coded symbols are arranged into a frame structure such that each symbol is transmitted a second time following a delay of 120 ms. The resulting transmission rate is 8 kbit/s. This arrangement is used in an attempt to provide resistance to the long fade durations encountered in the land mobile environment. A traffic channel requires a satellite EIRP of no more than 17 dBW.

An inband signalling channel mode is also provided for use during call set up and clear down. It also operates at a transmission rate of 8 kbit/s but with twelve-fold repetition of signalling units to provide resistance to fading.

The out-of-band signalling channel uses a transmission rate of 6 kbit/s in the fixed-to-mobile direction, and 3 kbit/s in the opposite direction. OQPSK and BPSK are used, respectively, and information is conveyed using a TDM/TDMA scheme similar to the other systems described above.

Mobile terminals for the INMARSAT-M system also fall into two main types; maritime and land transportable. The antenna for the land MES is a manually-pointed phased array, and the complete terminal is about the size of a small briefcase. The maritime mobile terminal is likely to employ a conventional symmetric beam antenna and a two-axis, actively-stabilised mount. The nominal EIRP of the land mobile terminal is controlled by the system, two nominal levels of 19 and 25 dBW are specified. The maritime mobile terminal is required to produce EIRPs of 2 dB higher than its land mobile counterpart.

Following the approval of NCSs in all four ocean regions during 1992, INMARSAT-M services were introduced in 1993. To date about 2250 maritime terminals and 11 000 land transportable terminals have been registered onto the system.

With the introduction of the spot-beam capability provided by the INMARSAT-3 satellites, a derivative of the INMARSAT-M, known as mini-M, entered service in late 1996. These terminals, the land mobile variant of which is about the size of a lap top computer, differ from the full-M in that the voice coding rate, after FEC, has been reduced from 6.4 to 4.8 kbit/s and the antenna gain has been reduced to about 9 dB. By the end of 1996 some 300 land mini-M terminals had been registered.

19.3.7 *INMARSAT-Aero*

This system, as its name implies, is designed to cater for the needs of aeronautical users. It is arranged to enable both low-gain (0 dBi) and high-gain (12 dBi) antennas to be used on the aircraft. In a low-antenna-gain installation only low-rate data services can be supported, although a high-antenna-gain installation will be capable of supporting multiple-channel telephony, medium-rate data and low-rate data services. The system is configured in the normal way; however, the mobile terminal is known as an aeronautical earth station (AES) and the land-based gateway station is known as a ground earth station (GES). Network coordination facilities are provided by the NCS in the normal way. INMARSAT space segment is used. The main system specifications are contained in the INMARSAT aeronautical system definition manual. This document covers those issues which are essential to the operation of the system; a further specification, ARINC 741, defines the form and fit of the aircraft installation.

A number of channels are provided for communications services and signalling between the AES and GES, they are:

(i)   *P-channel:* this channel operates in a TDM packet mode in the ground-to-air direction at a transmission rate of 600 bit/s. Rate one-half convolutional coding, bit interleaving (384 bits deep) and ABPSK is used. ABPSK is defined as 40 per cent raised-cosine filtered symmetrical DPSK. The P-channel may also operate at the higher transmission rates of 1200, 2400, 4800 and 10 500 bit/s in enhanced systems.

(ii) *R-channel:* the random-access channel is used in the air-to-ground direction for signalling and short traffic messages. The transmission format is similar to that for the P-channel.

(iii) *T-channel:* the TDMA channel operates using a reservation protocol. The frame duration is 500 ms, with 63 time slots per frame, an 8 s super frame is also used. The transmission format is similar to that for the P-channel.

(iv) *C-channel:* the circuit-mode channel operates in a SCPC mode in both directions and may be used for voice or data traffic. In a voice mode, a 9.6 kbit/s RELP vocoder is used. Subband data and framing information is added to produce an aggregate data rate of 10.5 kbit/s. Rate one-half convolutional coding is applied prior to interleaving to a depth of 384 bits. The modulation method is filtered OQPSK at a rate of 21.0 kbit/s. The data services may operate at a rate of 2.4, 4.8 or 9.6 kbit/s.

To ensure an acceptable spectral efficiency, it is no longer possible to ignore the Doppler shifts of about 1.5 kHz introduced by the aircraft motion, hence a Doppler compensation scheme is specified. As with the INMARSAT-B and M systems, power control and voice activation are used to ensure the maximum utilisation of spacecraft power. Power control is particularly necessary in the aeronautical case since the proposed aircraft antenna subsystems are anticipated to exhibit gain variations of several dB over their specified coverage region.

Great care has been taken in the design of the INMARSAT Aero system to ensure that it is capable of expansion and enhancement, as later generations of space segment are introduced.

Following a period of precommercial activity, initial commercial services were introduced during 1990 and 1991 by three service-provision consortia through fourteen GESs. All three, by carefully planned strategic alliances, provide four-ocean-region coverage.

Initially, the avionics equipment only provided a two-channel capability, however a Racal/Honeywell joint venture introduced three-channel and six-channel equipment. By June 1996 there were about 2600 voice channels and 600 data channels in service.

The ICAO future air navigation systems (FANS) committee has been studying new methods of air navigation throughout much of the 1980s. At the tenth air navigation conference, held during September 1991, the 85 states attending endorsed the FANS concept based upon the use of GPS/INMARSAT systems for air-traffic management in both enroute and terminal sectors. The system, known as automatic dependent surveillance (ADS), enables the ATC centre to poll the aircraft, which will then respond with a message containing the aircraft's position, heading etc. derived from GPS and other onboard sensors, all communications being via the INMARSAT aeronautical system. Such an ADS system is now operational in the Pacific Oceanic region and many experiments have been carried out in Europe and elsewhere. It is planned that, by the year 2000, the ADS system will be fully adopted alongside existing systems. By 2015 it is planned that only the new system will remain in use.

As with the INMARSAT-M system, INMARSAT is introducing a spot-beam-only derivative to be known as Aero-I. Aero-I will use an antenna gain of 6 dBi and provide 4.8 kbit/s vocoder voice and 2.4 kbit/s data over a C-channel. The use of 2/3 rate coding and OQPSK allows a transmission rate of 8.4 kbit/s in a 5 kHz spacing channel. Similar changes are being made to the original high gain standard, now known as Aero-H, to form the evolved Aero-H standard.

19.3.8 INMARSAT-D

A messaging service known as INMARSAT-D was launched during 1996. This service, which grew out of early trials carried out in the UK during the mid-1980s[2], provides a very low data rate bearer to a palm-sized message terminal. A degree of building penetration is claimed. The initial system is one way, but a

D + system which provides an acknowledgement capability is planned. Unlike for earlier INMARSAT systems, standard details of the INMARSAT-D- and D + standards have not been published.

19.4 Regional systems

19.4.1 OPTUS

In the early 1990s, the Australian government created a telecommunications duopoly. As part of this process, the government sold its wholly-owned satellite operator, AUSSAT, to OPTUS which won the licence to become the second public telecommunications operator. AUSSAT was originally formed to provide domestic Ku-band capacity. In 1987 it was announced that the operator's second generation of satellites would also accommodate a pair of L/Ku-band transponders to allow the provision of domestic mobile services. These services were named Mobilesat. The AUSSAT satellites have now been renamed as OPTUS A1, OPTUS B1, etc.

The contract for the OPTUS B satellites was placed with Hughes Aerospace in 1988 and the first, B1, was launched by a Long March rocket in August 1992. This satellite is now operational and is supporting Ku-band services. The second launch, B2, took place in December 1992, but unfortunately it was unsuccessful. A replacement satellite was launched in 1995. The Mobilesat service was finally introduced into service in early 1994. This postponement has resulted not only from the space-segment delays, but also from more significant delays in the establishment of the ground infrastructure and the development of mobile terminals.

The OPTUS Mobilesat system is designed primarily for use by land mobiles. By virtue of Australia's geographical characteristics, the OPTUS system benefits from being able to use a relatively high-gain antenna on the spacecraft, so providing more L-band EIRP (47 dBW) for a given DC power, and from reduced link margins resulting from the high elevation angles available.

The system is configured, as shown in Figure 19.4, around two L/Ku-band transponders on OPTUS B1 and B2 located at 160° E and 156° E, respectively. A control station and gateway stations are planned in a similar configuration to that used by the INMARSAT system. In addition, smaller private base stations are envisaged which may be used by one or more private closed user groups. Three types of mobile terminal were planned, a telephony terminal which was to provide full duplex telephony and data services, a mobile radio terminal which was to provide half duplex telephony and an ancillary messaging service and a messaging terminal to provide a low-capacity messaging service for SCADA and other fixed or transportable applications in remote areas.

The mobile terminals may be truly mobile or transportable, each with a G/T of −18 dB/K and EIRP of about 10 dBW.

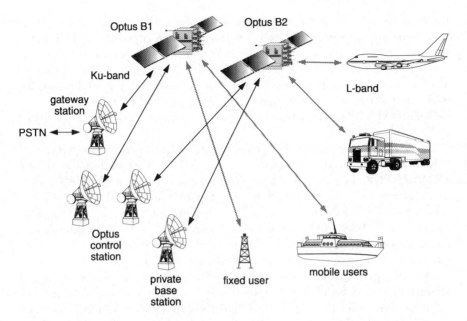

Figure 19.4   OPTUS Mobilesat system configuration

The telephony channel for interconnection with the public network will use vocoding and QPSK modulation operating at a transmission rate of 6.6 kbit/s. The vocoding algorithm is that selected in cooperation with INMARSAT. A channel spacing of 5 or 7.5 kHz is used to make best use of the limited spectrum available. The main data channels provide an information rate of 2.4 kbit/s and half-rate convolutional coding is applied, resulting in a transmission rate of 5.6 kbit/s following the addition of some subband data. The fixed-to-mobile signalling channel rate is 9.6 kbit/s, and BPSK modulation with rate three-quarters convolutional coding is used. This coding, coupled with dual repetition of signalling packets, is designed to provide a highly robust signalling and ancillary messaging channel. The mobile-to-fixed signalling channel is similar to that in the fixed-to-mobile direction, except that it uses a transmission rate of 2.4 kbit/s. The signalling channels operate in a TDM/TDMA mode, similar to that used by INMARSAT.

19.4.2 The North American mobile system

Following the introduction of terrestrial cellular radio systems in the USA during the 1970s, it was rapidly realised that a complementary satellite system would be needed to extend mobile services to the vast areas of rural North America. To this end, both the Canadian and US governments initiated R&D programmes which have eventually led to the licensing of the American Mobile Satellite Corporation (AMSC) and Telesat Mobile Inc (TMI) as service providers in the USA and Canada, respectively. These two operators continued the lead set by their respective governments by continuing to cooperate on space-segment procurement and operation. Each operator will initially purchase one of two similar satellites and enter into an agreement to provide capacity to the other party should its satellite fail. After a lengthy bidding phase, a contract was eventually let to a Hughes/Spar consortium in 1990. The first M-SAT satellite was launched in April 1995 into an orbital location of 101° W on behalf of AMSC. The second satellite was launched about a year later on behalf of TMI.

Interim low-rate data services (known as MDS in Canada) were introduced by both operators in late 1990. Both used the MARISAT F1 satellite leased from INMARSAT and located at 106° W. The system is similar to INMARSAT-C and the terminals are understood to be compatible with INMARSAT-C. The start up of the Canadian service has benefited from the procurement of 3000 mobile terminals by TMI. AMSC has also purchased quantities of mobile data terminals to provide INMARSAT-C-type services; these services have been transferred to the new space segment following its launch.

The full system configuration will be similar to, and will support similar services to, the AUSSAT mobile system with the use of both gateway FESs for public services and base FESs for closed user group services. Although the main market opportunities lie within the land mobile sector, both operators will also provide aeronautical services conforming to ICAO standards. The issue of spectrum coordination has been a major difficulty for both operators; their current approach is to advocate a dynamic partitioning of the available spectrum between the various services in a single pair of bands. This may make good sense to these two new operators, but it presents difficulties to the regulators, ICAO and existing space-segment providers that have structured their systems and policies around the fragmentation philosophy. However, WARC 92 adopted the principle of the generic MSS.

Details of the system design remain vague in the published literature. The Canadians are advocating an FDMA system, with 5 kHz spacing for voice channels, and are still proposing an analogue and a digital transmission format. The analogue format employs amplitude-companded single sideband (ACSSB) and the digital format employs a 4.8 kbit/s code-book-excited linear predictive voice coding (CELP) and trellis coded 16-QAM modulation. For the mobile radio service, a nominal unfaded $C/N_0 + I_0$ of 51 dBHz is specified.

With a mobile terminal G/T of -15 dB/K, an L-band EIRP of 30 dBW per channel is used. The satellite can therefore support up to about 700 channels, which is in turn estimated to cater for the needs of a population of up to sixty thousand users when voice activation is used. Mobile terminals have been developed, under contract from AMSC, for land, air and maritime use. The land mobile terminals are capable of dual mode (satellite/terrestrial cellular) and are either vehicle mounted or briefcase portable.

The take up of the AMSC services has been slower than anticipated and consequently the corporation has had to raise extra capital to continue operations. The next few years will be crucial for both AMSC and TMI.

19.4.3 European land mobile satellite system (ELMSS)

The European land mobile satellite system (ELMSS) is being promoted by the European Space Agency (ESA). The main objective is to provide a regional MSS for Europe in a similar way to that for Australia or Canada. Currently, there is substantial scepticism in northern Europe concerning the feasibility of satellite telephony services, mainly owing to the anticipated high penetration of cellular radio. However, ESA hold the view that, particularly in southern Europe where population densities (outside the major cities) tend to be lower, cellular radio will not be deployed for at least ten years and that satellite-based services can, in the short term, provide an economical alternative. Fortunately, it should be technically easier to provide adequate link budgets in southern Europe, since it benefits from reduced link margins owing to higher elevation angles and a more open environment.

The current programme of space-segment development is in three phases. The first phase consists of the development of a L/Ku band transponder which will be launched as a piggy-back payload on ITALSAT-F2. The payload will provide a single L-band beam with an EIRP of at least 42 dBW over the majority of Europe. The L-band G/T will exceed -2 dB/K. ITALSAT-F2 was finally launched in October 1996 and some small leases providing differential GPS corrections were activated in December 1996.

The second phase currently consists of a L/Ku-band payload known as the land mobile mission (LMM) to be flown on the ARTEMIS satellite which is due to be launched in 2000[3]. The LMM payload will provide European coverage at L-band with a Eurobeam and three spot beams. The total L-band EIRP is expected to be 45.5 dBW when only the spot beams are used. It is anticipated that the use of spot beams will also allow significant frequency reuse over Europe. To maximise the amount of frequency reuse achieved, the use of opposite hands of circular polarisation in adjacent beams has been considered. The third phase is known as Archimedes and forms part of the advanced orbital test satellite (AOTS) programme. This advanced programme will use a highly elliptical orbit in order to provide higher elevation angles over Europe.

Study work is currently in progress concerning the system architectures to be used with the EMS and LMM payloads. For telephony services, a public system based on an INMARSAT-M-like system is one possibility, however most of the study work is directed towards systems supporting closed user groups, each using one or more preassigned channels and PMR-type trunking protocols. This work goes under the name of mobile satellite network for business services. The use of CDMA schemes and their effects of frequency reuse is also under study and a technique known as quasisynchronous QS-CDMA has been developed.

Low-rate data services have been proposed, and it is likely that these will use a system known as PRODAT[4] which was developed as part of phase II of the PROSAT programme. The PRODAT system, in its original form, consists of a single gateway station which provides access between the public data network and a population of land, maritime or aeronautical terminals. The transmission format was designed to provide a high-integrity service with the best possible throughput. This is achieved with a novel two-dimensional Reed–Solomon coding scheme which enables the ARQ scheme to adapt to the prevailing propagation conditions so maximising the throughput under a given set of fading conditions. The second novel feature concerns the use of a CDMA scheme for the return link. CDMA was used as it is claimed that, at low data rate where message bursts are relatively long, a random-access channel can provide significantly more throughput than can the conventional approach. The advantage is strongly dependent upon the number of CDMA demodulators available at the fixed earth station. Extensive trials have demonstrated that the novel bidimensional coding and ARQ scheme provide a better throughput than the more conventional approaches adopted by INMARSAT.

More recently, ESA has embarked on a second phase of the PRODAT programme which involves the production of land mobile terminals and the replacement of the original network management system (NMS). The new NMS incorporates an X400 e-mail interface to the terrestrial network.

19.4.4 Qualcomm OmniTRACS and EUTELSAT EUTELTRACS systems

The Qualcomm OmniTRACS[5], and the closely related EUTELTRACS, systems are believed to be the only operational land mobile satellite systems using Ku-band. The system configuration, as shown in Figure 19.5, bears much similarity to a VSAT network, with a single hub earth station, two transponders of a domestic Ku-band satellite and many mobiles connected in a star configuration. The use of Ku-band offers two potential economic benefits, that is, the availability of space segment compared with L-band space segment and the advantage of using low-cost Ku-band components that have been developed for the large VSAT and DBS markets. Against these plus points, the use of Ku-band implies a secondary frequency allocation and so the system must be designed from the outset so as not to cause interference and not to be interfered with by the existing Ku-band systems. This implies the use of spread-spectrum techniques to compensate for the reduction in antenna directivity implicit in the use of an ultra small (in Ku-band terms) antenna.

The OmniTRACS system entered commercial service in June 1989. One year later there were about 9000 mobile terminals in use in the USA and a further 6000 terminals were on order. A single hub network management facility and two transponders on the GTE Spacenet Corporation's GSTAR-1 satellite provide two-way messaging, position reporting and fleet broadcasting services to a closed user group. Typical users include the public utilities and transportation, resource extraction, construction and agricultural industries.

Following trials with EUTELSAT space segment in Europe in 1989[6], EUTELSAT in cooperation with a joint venture company ALCATEL/ Qualcomm, introduced commercial service in early 1991 via a network management facility at Rambouillet near Paris and EUTELSAT I satellites. By the end

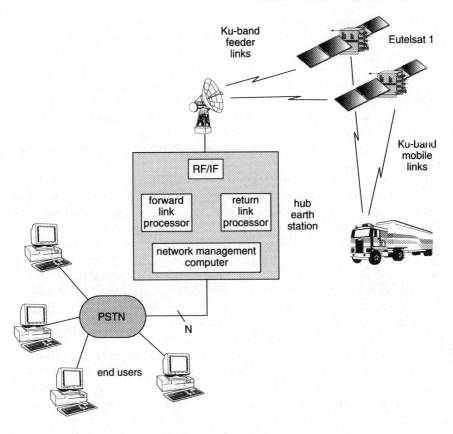

Figure 19.5   EUTELTRACS system configuration

of 1991, the service was believed to have service providers in at least three European countries.

The system supports a TDM with a transmission rate of between 5 and 15 kbit/s in the fixed-to-mobile direction and 55 to 155 bit/s in the mobile-to-fixed direction. It allows the provision of two-way message and polled-position reporting services, vehicle position being determined initially using the LORAN-C system. Each pair of transponders is claimed to support up to 60 000 users.

The mobile terminal is small enough to be readily acceptable to transport companies. The outdoor unit employs a 40 × 6 degree beamwidth antenna so that steering is only required in the azimuth plane. This antenna, coupled with a 1 W HPA, provides an EIRP of 19 dBW. The spread-spectrum technique is different in each link. On the forward link, after rate one-half coding, BPSK modulation is used prior to the application of a triangular-wave FM energy dispersal waveform. In the return direction, rate one-third convolutional coding is used. The resulting data stream is applied to a 32FSK modulator with a symbol duration of 15.1 ms. Each tone is then direct sequence spread at a rate of 1 Megachip per second. As each tone is separated by about 1.5 MHz, the transmit signal is spread over a bandwidth of 48 MHz.

The initial system provided a position-location function by using the LORAN-C navigation system and add on navigation receivers. In 1990 Qualcomm developed its own automatic satellite position reporting (QASPR) system making use of the TDMA timing signal formats of the basic OmniTRACS system and an additional satellite to obtain a second surface of position. Position determination is completed by the use of a terrain altitude database.

The system is an active navigation system in that the position of the mobile is determined at the central network management centre. The NMC estimates one surface of position from the primary satellite using a measurement of the round-trip delay during the normal message transfer process. The second ranging satellite transmits a similar forward link signal, but with no traffic. The mobile monitors this ranging signal on an occasional basis and measures the time difference between the primary and ranging signals. This time difference is transmitted, in the normal way, back to the NMC where it is used to determine the second surface of position. The final position is then determined by finding the point along the derived line of position which intersects with the surface of the earth. The accuracy of this position is enhanced by the use of a terrain database.

Trials of the system, in the USA, indicate that an accuracy of the order of 400 m is obtainable with a satellite spacing of 22°. Similar trials with EUTELSAT space segment in Europe using a satellite spacing of 9° indicates an accuracy of the order of 1000 m. This order of accuracy is perceived by Qualcomm as being adequate for vehicle position location and reporting applications.

19.5 Propagation

With the highly directive antennas employed in fixed-service satellite systems, multipath caused by signal reflections from near by objects is effectively nonexistent. Unfortunately, this is no longer the case when very low-gain antennas, such as those associated with INMARSAT-C type terminals, are considered. The effect of multipath propagation is to introduce an uncertainty into the instantaneous received signal amplitude and phase. When the multipath is caused by reflections, from objects other than the vehicle itself, the instantaneous received signal level clearly varies as a function of time as the antenna passes through an interference pattern. The effects of multipath are traditionally mitigated by additional link margins. In most cases this is not economically viable in a satellite service and so more advanced coding, cancellation or ARQ techniques are required. This Section will consider the basic fading process and review the findings of the experiments performed to characterise the various channels for the maritime, aeronautical and land mobile cases.

19.5.1 A basic fading model

Consider Figure 19.6. A satellite transmits a carrier, the direct ray or the wanted signal (C) and it enters the antenna at an angle of elevation ε. A second signal, the multipath signal (M), is reflected from the earth and enters the antenna at an angle of −ε.

The reflection coefficient of the earth is a variable which, for a given location, tends to be polarisation and elevation-angle sensitive. If the surface is smooth, then specular multipath occurs, as shown, giving rise to a classic interference pattern.

In practice, at L-band, the surface roughness is such that many reflected rays, at slightly different angles, arrive at the antenna. This is known as diffuse reflection and can be modelled as the sum of two quadrature-independent gaussian fading processes. The total received signal amplitude probability can then be modelled by the Rician distribution. The probability that a signal has a momentary received power greater than a certain level is given by:

$$P(s) = C.\exp\{-C(s+1)\}.I_0(2C\sqrt{s})$$

where s is the signal power relative to the mean, C is the received carrier-to-multipath ratio (Rice factor) and I_0 is the modified Bessel function of zeroth order. The mean received power is $(1 + C)^{-1}$. This is shown diagramatically in Figure 19.7 (from Reference 7). For example, a Rice factor (C/M) of 10 dB requires a margin of about 5.5 dB to ensure that the instantaneous received signal exceeds a given level for 99 per cent of the time. If, for some reason, the direct ray is obscured, then the Rice factor $= -\infty$ and the probability distribution reduces to a Rayleigh distribution.

Multipath fading is a multiplicative process and causes the received signal amplitude and phase to vary as a function of position and time. This noise-like amplitude and phase modulation therefore, causes the spectrum of the received signal to broaden. A measure of this effect is known as the fading bandwidth and is mainly influenced by the Rice factor and the vehicle's velocity. Clearly, any carrier tracking loops should be sufficiently wide to track the spectrum of the fading signal.

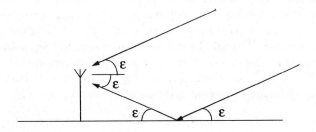

Figure 19.6   A simple multipath model

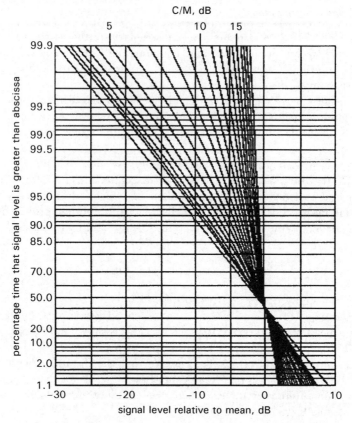

Figure 19.7   Rician cumulative probability distribution function for various values of Rice factor (C/M)

The additional path length of the reflected signal causes a differential delay between the direct and reflected rays. When this delay approaches a symbol duration, intersymbol interference results. So, unless adaptive equalisation is used, a maximum data-rate limit is imposed. The reciprocal of the differential delay is sometimes referred to as the coherence or correlation bandwidth.

The above discussion has indicated that the amount of fading experienced in a particular situation is very dependent upon the antenna radiation pattern, the elevation angle, the physical characteristics of the reflecting surface and the vehicle dynamics. This large number of variables clearly makes predictions difficult, hence a number of experiments have been carried out in order to establish realistic models for use by system engineers.

19.5.2 Maritime experiments

There have been a great number of maritime experiments, probably the most comprehensive being carried out by DLR for ESA. A recent study, funded by INMARSAT and carried out by COMSAT[7], has reviewed the results of many experiments and has produced some systems-oriented results. Figure 19.8 shows the effect of antenna gain and elevation angle upon Rice factor (C/M). These curves attempt to represent a typical case and the Rice factor should be lowered by at least 1 dB at low ($< 10°$) elevations and 2 dB for higher elevations for a more conservative estimate. It should be noted from this Figure that, at very low elevation angles, even high-directivity antenna systems experience Rice factors of 6 to 8 dB. However, at higher elevation angles, say 10°, an INMARSAT-A system would experience a C/M of around 20 dB. Conversely a INMARSAT-C system, at the same elevation would experience a Rice factor of about 9 dB.

The fading bandwidth is virtually impossible to characterise accurately since it depends upon factors such as the size of vessel and the sea state as well as velocity etc. Typically, the -3 dB bandwidth falls in the range of 0.1 to 1 Hz and the -20 dB bandwidth is usually within the 3 to 8 Hz range. These small bandwidths translate into long fade durations. These are briefly summarised in Table 19.2. It can be seen that with a low threshold (i.e. low margins) maximum fade durations of up to several seconds occur.

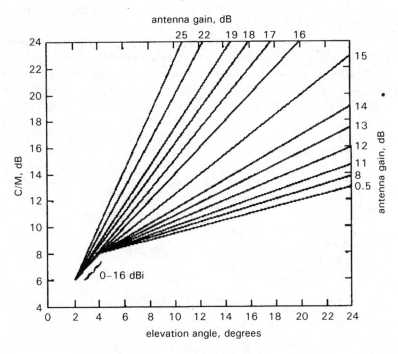

Figure 19.8   Effect of elevation angle and gain upon C/M

Table 19.2   Summary of measured fade durations for the maritime case

Fade threshold (dB)	Mean fade duration (s)	Maximum fade duration (s)
−2	0.02–0.17	0.3–3.5
−5	< 0.01–0.12	< 0.01–1.6
−8	< 0.01–0.08	< 0.01–0.9

Estimates of the coherence bandwidth indicate a −3 dB (e.g. 0.02–0.17) bandwidth of about 5 MHz. Such bandwidths are not generally the limiting factor for small terminals.

It should be noted that, particularly for small vessels, high Doppler rates of about 25–50 Hz/s can be experienced. In practice, it is this effect which sets the minimum carrier tracking loop bandwidth, and not the fading bandwidth.

19.5.3 Aeronautical experiments

Unlike the marine case, few serious experiments have been carried out on board aircraft. In Europe, those carried out by DLR[8] and ESA[9] are of interest.

The DLR experiment was more sophisticated as it attempted to measure direct and reflected rays independently using high gain (10 dB) antennas. Data was also collected from lower-gain antennas in the normal way.

All flights were carried out over the sea on courses set to provide a constant satellite elevation angle. Figure 19.9 shows a synthetic plot of Rice factor against elevation angle for an ideal, circularly polarised, isotropic antenna. It can be seen that the worst case Rice factor of 6 dB occurs at an elevation angle of about 4°. A curve derived directly from measured data for a 3 dB crossed dipole

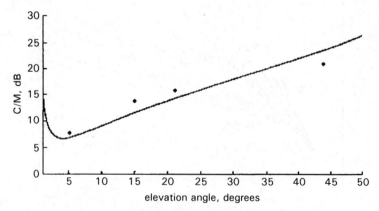

Figure 19.9   C/M as a function of elevation angle for a synthetic omnidirectional antenna for the aeronautical case

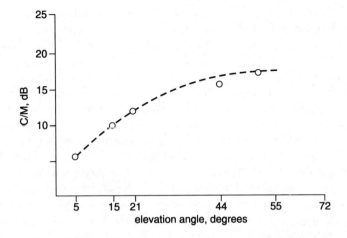

Figure 19.10   C/M as a function of elevation angle for a 3 dB gain antenna for the aero-nautical case

antenna is shown in Figure 19.10. Again, Rice factors of 5 dB and better can be seen.

The fade durations experienced are summarised in Table 19.3, and it can be seen that they are generally an order of magnitude less than for the marine channel. These fade durations equate to fading bandwidths of between 50 and 300 Hz. The main parameters affecting the bandwidth are the flight level and the elevation angle, in general, the fading bandwidth increases with flight level and elevation angle. The coherence bandwidth is also highly dependent on elevation angle and flight level. The DLR paper does not quote coherence bandwidths, but does present calculated differential delay times from the link geometry. This indicates delays of about 40 μs at enroute altitudes and with elevation angles of 30°.

Doppler rates of up to about 30 Hz/s can be anticipated in L band civil applications.

19.5.4 Land mobile experiments

As with the aeronautical case, most work in Europe on land systems has been carried out by ESA[9] and DLR[10]. In the USA, similar work has been undertaken,

Table 19.3   Summary of measured fade durations for the aeronautical case

Fade threshold (dB)	Mean fade duration (ms)	Maximum fade duration (ms)
0	11	100
−5	3	45

much of it being reviewed or published in the proceedings of the mobile satellite conference sponsored by NASA/JPL held in May 1988, and subsequent conferences in the same series. A useful, but not complete bibliography, is included in Reference 11. In general, the land mobile channel is more complex since signal shadowing occurs when the vehicle passes behind a building, tree etc.; when this occurs a more complex two-state model is appropriate. In the shadowed state, the model consists of a Rayleigh term to model multipath signals and an attenuation term to model increased loss due to absorption or imperfect reflection. The statistics of this attenuation term tend to be log normal. When the vehicle is unshadowed a direct ray is available and the Rician model is applicable; under these conditions the Rice factor is around 10 dB or better. The time share between unshadowed and shadowed events depends upon the elevation angle and topography of the environment, it typically ranges from 0.8 under highway conditions to 0.2 for inner city conditions. This time share as a function of elevation angle is shown in Figure 19.11. Figures 19.12 and 19.13 show a cumulative probability distribution of received power plotted on Rayleigh paper for two measurement sets carried out in a highway and city environment, respectively. This clearly shows the various elements of the model.

Fade durations are, again, strongly dependent upon the velocity of the vehicle and its environment. Typically, fade durations of up to ten seconds can be anticipated when travelling on a highway.

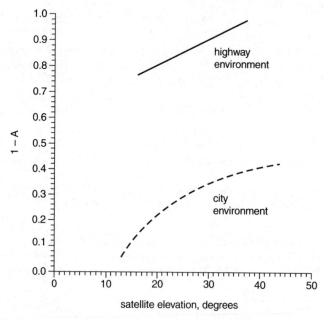

Figure 19.11   Time share of not shadowed events for the land mobile case (highway, city)

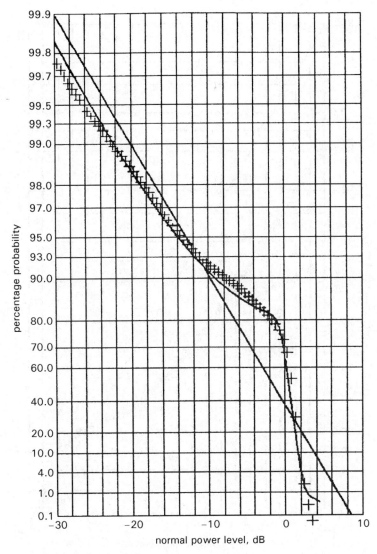

Figure 19.12   *Cumulative probability distribution function for a land mobile located on a highway*

The above comments relate to elevation angles below about 40°. The Australians have recently reported the results of measurements[12] at elevation angles of up to 60°, and a study has been carried out for ESA to design an experiment to gather propagation data in Europe at high elevation angles. A measurement campaign has also been carried out by the University of Bradford at elevation angles above 60°. Some results of L-band measurements are

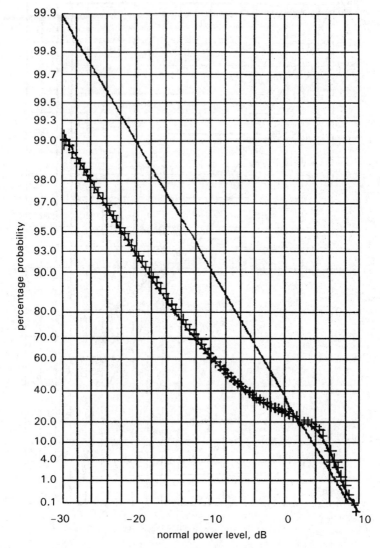

Figure 19.13   Cumulative probability distribution function for a land mobile located in the city

published in Reference 13 and some later results at S-band are published in Reference 14. More recently, the University of Surrey has carried out similar campaigns and ESA has sponsored a wideband sounding measurement campaign at 1820 MHz. This work found that the delay spreads in the various land mobile environments range from 500 ns to 5 s. As to be expected, the delay spread is inversely proportional to elevation angle[15].

19.5.5 Fading-channel mitigation techniques

For low-rate data services two basic schemes have been proposed, one by INMARSAT and the other by ESA; the ESA PRODAT scheme uses a bidimensional ARQ scheme which has proved to be robust under real live conditions. Other ARQ schemes have been proposed under ESA funding by ANT[16] and DLR[17]. The INMARSAT-C scheme employs bit interleaving and convolutional encoding with soft-decision Viterbi decoding. A further outer ARQ protection is applied at the packet level using a cyclic redundancy code (CRC) parity check. The interleaving period is about 8.64 s. The Japanese have produced schemes to reject multipath on the basis of polarisation discrimination[18]. References 2 and 19 show that various novel time-diversity schemes can effectively combat signal shadowing in the land mobile environment.

For telephony services normal ARQ is not feasible and only in the aeronautical case is useful bit interleaving/convolutional encoding acceptable from a delay point of view. The use of adaptive equalisation techniques for multipath mitigation has yet to be attempted in civil satellite mobile experiments, however it has been proposed for terrestrial land mobile systems.

19.6 Mobile terminals

19.6.1 Introduction

The system descriptions contained in Sections 19.3 and 19.4 indicate that currently there are three basic classes of terminal which can be differentiated by their antenna gain, they are:

(i)   Low gain (≈ 0 dBi).
(ii) Medium gain (≈ 12 dBi).
(iii) High gain (≈ 20 dBi).

This classification is significant because the required antenna gain determines the size of the terminal and its eventual range of applications. The tendency in system design is towards a common system infrastructure for both land, maritime and aeronautical mobiles. Unfortunately, this commonality cannot be extended to the mobile terminals since each type of mobile platform offers a significantly different environment and requires different interfaces. Further, the variation in size and cost of the various mobile platforms influences the design of a terminal intended for a particular market segment. The regulatory requirements of a particular sector of the transport industry also have a major impact on the design of the equipment; this is particularly significant where the terminal may be used for safety-related communications, for example ATC communications.

The build standard of the terminal is therefore determined by the type of vehicle in which the mobile terminal will be installed. This is most noticeable in the case of aeronautical mobile terminals, where equipment intended for airline

Figure 19.14   The Racal–Honeywell MCS6000

use must conform to rigorous environmental and detailed mechanical and electrical interface requirements. Figure 19.14 shows the Racal–Honeywell MCS6000 six channel INMARSAT-Aero avionics equipment which conforms to the ARINC 600 standard as applied within the ARINC 741 characteristic for satellite communication avionics equipment. The ARINC 741 characteristic also specifies the functionality of each box (line replaceable unit) so that airlines may source different units from different manufacturers.

As the capacity of the various systems increases, the commercial viability depends more and more on the selling price of the mobile terminals. This is because larger amounts of traffic are needed to provide sufficient service revenue to justify the investment. To generate this increased traffic, more mobile terminals must be deployed which must be suitable for use in larger application areas. The two major factors influencing the acceptability of a terminal to a potential user are its cost and size. The cost factor is particularly significant in the land and maritime sectors of the business. It was predicted in the early 1990s that the combined INMARSAT-C and INMARSAT-M user-terminal market will approach one million by the year 2000. When this total build is apportioned between the four or more designs which are needed to satisfy the market demand and amortised over the number of likely manufacturers, the annual production rate of each design of each manufacturer falls between 1000 and 10 000 terminals. Such rates are not sufficient to offer the economics of scale associated with consumer items.

Historically, the mobile terminal can be divided into two parts, the above-deck-equipment (ADE) and the below-deck-equipment (BDE). The ADE contains the antenna and, in the land and maritime case, the L-band RF electronics. The BDE contains the remainder. Often the tendency is to minimise the amount of electronics in the ADE since it must withstand worse environmental conditions and it adds to the size of the ADE, which for the case of low and medium-gain terminals is a key design criterion. However in the mid-1990s terminal sizes reduced significantly such that lap-top-size satellite mobile phones are a reality and handheld satellite phones are being developed for the satellite PCS.

The remainder of this Section will discuss the mobile terminal antenna and the relevant parts of the mobile terminal electronics.

19.6.2 Antennas for mobile terminals

19.6.2.1 Low-gain antennas

Low-gain antennas are used in low-rate data terminals such as INMARSAT-C. For deployment anywhere within the coverage area of global beam, geostationary satellites, such antennas are required to provide useful gain over the upper hemisphere. Typically, such an antenna will provide a minimum gain of 0 dBi for elevations in excess of 5°.

The most popular design is known as a quadrifilar helix, and this approach is capable of producing well controlled radiation patterns and acceptable circularity; a typical L-band quadrifilar helix is about 100 mm tall. The antenna element is enclosed in a suitably shaped radome. For the aeronautical case, the radome is blade shaped to reduce drag. In the case of a land or maritime data-only terminal a portion of the RF electronics is included within the radome to minimise feeder losses.

Care must be taken in the design of these antennas to ensure that, when mounted on the host vehicle, the antenna radiation pattern is not unduly distorted by the presence of reflections from, for example, the vehicle's roof. This can be particularly troublesome with low-gain antennas because their radiation pattern does not fall off steeply below the horizon.

For regional systems, where the geographical operating region is limited to the size of a continent or less, the radiation pattern of the antenna may be tailored to suit the range of elevation angles which will be encountered. Such radiation patterns are referred to as torroidal since they must still remain omnidirectional in the azimuth plane. Depending on the extent of elevation angle range, gains of 2 dBi or more can be achieved. A particularly popular vertical linear array antenna has been developed with a manually adjustable elevation angle. This array provides a gain of about 6 dBi.

19.6.2.2 Medium-gain antennas

There are three distinct applications for such medium-gain antennas: the high-gain antenna for terminals operating within the INMARSAT-Aero system, antennas aimed at land mobile telephony and maritime mobile low-cost telephony terminals.

The design of the high-gain antenna for the INMARSAT-Aero system has proved to be a significant technological challenge. The coverage requirement is shown in Figure 19.15, which shows the region of the super hemisphere where gains in excess of 12 dBi and 9 dBi are provided. The region in the forward direction over which less than 9 dBi of gain provided is known as the key hole. A

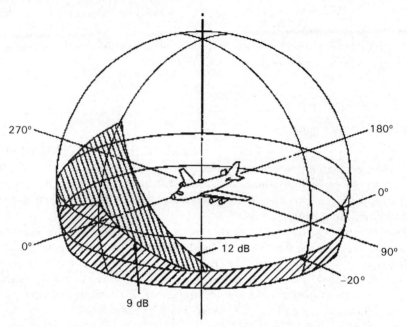

Figure 19.15   Coverage requirements for an aeronautical high-gain antenna (rear key hole not shown for clarity)

similar key hole exists in the rearward direction, but is not shown for the sake of clarity. The maximum gain of such an antenna is perhaps as high as 15 dBi; such gain variations are inevitable and make the use of link-power control highly desirable.

Four main solutions have been proposed, these are:

(i)   Conformal phased array.
(ii)   Low-profile phased array.
(iii)   Blade phase array.
(iv)   Mechanically-steered array of helices.

The conformal array requires the use of a port and starboard antenna array which is curved to suit the profile of the fuselage. The advantage of this solution is that it offers minimum drag since it is virtually conformal with the surface of the fuselage. The middle two solutions are mounted on the top centre of the fuselage. The low-profile array consists of an array of individual elements enclosed within a tear-drop-shaped radome. An example of the final approach is shown in Figure 19.16 and is intended to be mounted within a radome located on the top of the vertical stabiliser.

The design of antennas for the land mobile telephony terminals was also the subject of considerable work during the late 1980s[20]. The most common design concept consisted of a $1 \times n$ array of elements, where n ranges from four to eight

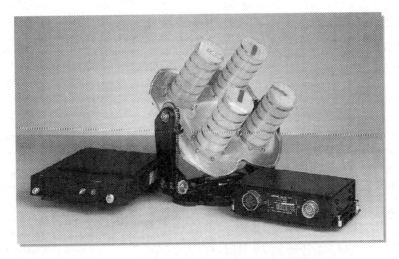

Figure 19.16   Aeronautical antenna of the mechanically-steered-array type

depending on the gain required. This array produces a fan beam so that, when the array is mounted horizontally, only azimuth steering is required. This eases the design of the steering mechanism and tracking circuits. Tracking methods used include open-loop steering using a compass and closed-loop methods such as step track or electronic beam scanning. The major problem with the mechanically-scanned antenna is its height, the most aesthetically sensitive dimension. With this in mind, JPL has funded work on an electronically-steered phased array which is virtually conformal with the roof of the vehicle. The major drawback of such design is its complexity and limited scan range, with minimum practical elevation angles of between 20 and 30 degrees.

The marine version of the medium-gain antenna has not been the subject of so much development. As a result of the greater amount of roll and pitch encountered on ships compared with a road vehicle stabilisation is required in two planes, thus making low-profile antennas very difficult. The most popular solution consists of a circular array of patches providing 16 dBi of gain. Such an array is mounted on an actively stabilised elevation-over-azimuth mount within a radome some 600 mm high and 500 mm in diameter.

19.6.2.3 High-gain antennas

The high-gain antennas used in applications such as INMARSAT-A or B terminals are very similar to other parabolic reflector antennas of a similar size. They are approximately 0.8 m in diameter and their major complexity lies in the associated stabilisation and tracking subsystems.

The favoured design solution employs a four-axis mount. The first pair of axes are maintained in the horizontal plane, independent of the ship's altitude, by two orthogonal gyroscopes. A conventional elevation-over-azimuth axis arrangement then allows the antenna to be pointed towards the satellite.

Tracking is normally accomplished using the step track method; an input from the ship's compass is used to assist with azimuth tracking. The complete antenna and ADE electronics are contained within a dome-shaped radome of about 1.8 m in diameter.

19.6.3 Mobile terminal electronics

At a simple block-diagram level, there is little difference between a mobile terminal and a VSAT terminal. The block diagram in Figure 19.17 shows the major functional elements of a low-rate data terminal; the major differences lie in the detailed design and the packaging techniques needed to suit the various operational and environmental requirements addressed above.

After the antenna subsystem, the high-power amplifier (HPA) is normally the most expensive item. The system design dictates the operating power and the degree of linearity that is required. For example, the data-only terminals for systems employing global-beam geostationary satellites employ low power (10–20 W) class-C designs since there is no linearity requirement. Such HPAs can use readily available (but expensive) silicon bipolar transistors. On the other hand, the INMARSAT-Aero HPA is required to operate in a multicarrier mode, with a 40 W multicarrier rating. This requires the use of class-A or AB designs. Currently, the highest rated silicon bipolar devices are rated at 9 W (at −1 dB gain compression). After an appropriate back off is applied to ensure the required degree of linearity, some eight or more devices must be operated in parallel to provide the required output power. Such an amplifier is clearly complex, large and expensive. The newer spot-beam geostationary systems and the LEO/MEO systems generally require lower HPA powers and are therefore cheaper and consume less DC power. The advent of PCS in the 1.8 GHz band has provided a significant market for suitable RF transistors and so devices are now much cheaper and more readily available compared to the situation in the mid 1980s. Clearly, battery-life considerations alone limit the power rating of the HPA in a hand portable to less than one watt.

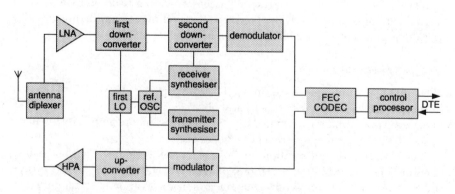

Figure 19.17   Simplified block diagram of a data-only terminal

The use of FDMA schemes with channel spacing of around 10 kHz and the high spectral purity needed to support low data rates have had a significant impact on the design of the frequency synthesiser and its associated frequency reference. In the aeronautical case, where Doppler compensation is used, it is necessary to provide step sizes of about 10 Hz to ensure accurate compensation. In all cases, the presence of vibration further complicates the design owing to degradation of the spectral purity caused by microphony. All of these constraints tend to make the design and manufacture of suitable frequency synthesisers a challenge.

All modern systems depend upon the use of digital-signal-processor (DSP) based RF modems and FEC codecs. Fortunately, the processing capability of the latest DSP chips is significantly greater than that of those first used in the early 1980s, and so it is now possible to employ very complex DSP algorithms at the data rates employed in the system discussed above to implement the demodulation and link-layer protocols now commonly used. Such algorithms lead to very low implementation loss, repeatable and stable designs. Furthermore, the ease of reprogramability facilitates multimode or multirate operation. Similarly, the availability of powerful general-purpose microprocessors allows the efficient implementation of the higher layers in the protocol stacks and the management of the terminal and its associated man–machine interface.

19.7 Satellite transponders for mobile systems

19.7.1 Introduction

The early 1990s marked a period of transition in the evolution of L-band payloads for the mobile satellite service. During the 1980s, INMARSAT had a monopoly of the L-band payloads; this monopoly was broken early in the 1990s with the launch of the satellites serving the domestic and regional such as OPTUS B-1 and B-2. As the 1990s have progressed, spot-beam satellites of the AMSC and INMARSAT-3 class have become a reality. The use of spot beams and the general shift to larger payloads combine to allow vastly increased L-band EIRPs to be provided; the histogram shown in Figure 19.18 illustrates this trend. Satellites currently being manufactured are even more ambitious with the incorporation of long-talked-about technologies such as digital onboard processing and large spot-beam antennas providing substantial numbers of spot beams. The following paragraphs are designed to provide an outline description of the payloads of the key MSS satellites which have been launched in recent years.

19.7.2 INMARSAT-2

Lifetime and traffic predictions made by INMARSAT in the early 1980s indicated the need for the launch of a second generation of INMARSAT satellites by the late 1980s. The new satellites have been designed not only to provide

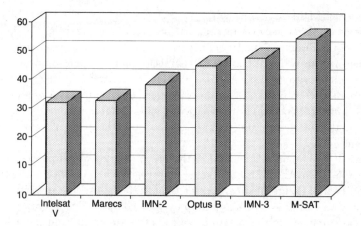

Figure 19.18   Satellite L-band EIRP (dBW) capability

more L-band EIRP but also increased bandwidth (18 MHz and 21.5 MHz in shore-to-ship and ship-to-shore directions at L-band, respectively). The four satellites manufactured by British Aerospace Limited were successfully launched in 1991/2. These satellites, like the first generation, employ global-coverage antennas. The development of further system standards (i.e. B, C, M and aeronautical) demands increased flexibility in the return direction as each link budget ideally requires a different value of satellite gain. This has led to the requirement for a four-channel transponder on the return link. The channel frequencies at L-band are:

- channel 1   1626.5–1631.0 MHz;
- channel 2   1631.5–1636.0 MHz;
- channel 3   1636.5–1643.8 MHz;
- channel 4   1644.3–1649.5 MHz.

The channel-separation filters are implemented using surface-acoustic-wave (SAW) filters. The L-band transmit antenna employs an array of 61-element cup dipoles to form a shaped global beam antenna pattern; a nine-element cup dipole array antenna is used in the ship-to-shore direction. Travelling-wave-tube (TWT) amplifiers are used in both transponders to provide an L-band EIRP of 39 dBW and a C-band EIRP of 24 dBW and the predistortion technique is used to linearise the transfer characteristics of the TWTAs to improve their efficiency under multicarrier conditions.

19.7.3 INMARSAT-3

The INMARSAT-3 series of satellites was designed to provide INMARSAT with a space segment with a similar capability to that developed by the new regional operators. These requirements imply an almost ten-fold increase in capacity per satellite and a capability to provide frequency reuse between the high-traffic regions of a given ocean region[21].

Such a requirement can only be satisfied by the use of spot beams; however a global beam is still required to support the large existing population of INMARSAT-A terminals which have network access protocols that cannot function in a multibeam environment. An analysis of projected traffic distribution in each of the ocean regions revealed that the majority of traffic is likely to be concentrated around the periphery of the northern hemisphere. This led to the specification of spot-beam coverage, in terms of coverage polygons, as shown in Figure 19.19.

To enable frequency reuse between high-traffic regions coverage isolations were also specified, for example, in the AOR the beam covering Europe must be isolated from that covering North America and vice versa.

The specification of EIRP for such a satellite becomes a more complicated issue. The payload needs to be specified in such a way that power can be allocated between the beams in a flexible manner to take account of changing traffic requirements. The payload is required to provide at least 48 dBW when all the power is allocated to the spot beams and at least 39 dBW when all the power is allocated to the global beam. Furthermore, up to 60 per cent of the total power must be able to be allocated to a given beam.

The payload is designed to cover the entire 1.5/1.6 GHz bands as allocated up to and including WARC 92. As a minimum, access to each beam (i.e. four or five spots plus global) by each of the three services must be provided. This must be achieved without using excessive C-band spectrum and excessive IF processor complexity.

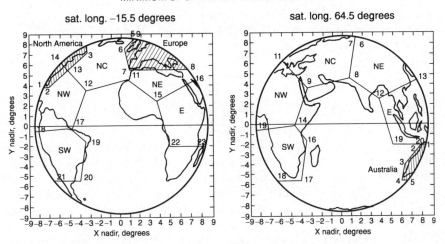

Figure 19.19   Example coverage polygons from the INMARSAT-3 specification

The payload also includes a navigation payload with a 2 MHz bandwidth centred on 1575.5 MHz. This payload option is intended to enable GPS-like signals to be transmitted, augmenting those already available and so enhancing the availability of navigation services. More importantly, the transmission will also contain GPS and possibly GLONASS integrity and wide-area differential correction data, thus providing the level of system integrity demanded by civil aviation which is not provided by the military systems alone[22]. This concept is primarily based upon the work of the ESA NAVSAT programme.

The INMARSAT-3 payloads were manufactured by Matra Marconi Space and some details of the implementation of the antenna and HPA subsystem have been published[23]. In essence, it consists of an offset feed imaging reflector antenna with low-level beamforming. This approach ensures that each SSPA operates at the array-element level and so its loading is more or less independent of beam EIRP distribution.

The transmit reflector diameter is 2.18 m and the feed consists of a 22-element array. Each element is realised as a short cupped helix. The output network which feeds the array consists of a stripline 4 × 4 Butler matrix and a cavity filter/tubular low-pass-filter combination on each input. A test directional coupler is provided in each array-element output. The Butler matrix is required since only five array elements are used to form a particular beam and the power must be shared between the 22 SSPAs. Each SSPA is required to handle about 20 W. The SSPAs are in turn driven by the beamforming network, the function of which is to provide the 22 outputs with the correct amplitude and phase distribution necessary to form the eight beams required by the combined coverage requirements. The complete transmit antenna subsystem is claimed to be capable of providing an EIRP of at least 47.4 dBW in the spot beams in addition to 39.0 dBW in the global beam with a total DC power consumption of 1500 W. The architecture of the system ensures an inherently good passive intermodulation performance owing to its distributed nature, so circumventing one of the major design difficulties of previous INMARSAT payloads. Much of the technology employed in this approach has been developed by Marconi during the ESA ARAMIS programme.

19.7.4 M-SAT satellite

The Americans (AMSC) and Canadians (TMI) cooperated in the provision of space segment for their respective mobile satellite services. Each country has launched one satellite with sufficient coverage to provide limited back up for the other should a satellite fail in orbit. Hence, the antenna coverage patterns for the two satellites are essentially identical.

The M-SAT satellite payload is described in Reference 24. The satellite uses two 5.75 × 5.25 m elliptical reflector antennas to provide six-beam coverage as shown in Figure 19.20. Note that the Alaskan beam is combined with the Hawaiian beams to form a single beam. The L-band antenna gain is around 31 dBi at edge of coverage. The resultant total L-band EIRP for the satellite is

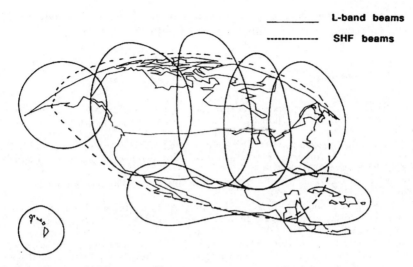

Figure 19.20   AMSC coverage diagram

56.6 dBW and the L-band G/T is 3.2 dB/K at edge of coverage. The use of a matrix HPA allows flexibility in the distribution of the L-band EIRP between all of the beams. Each matrix HPA consists of a bank of ten SSPAs, eight of which are operational at any one time; each SSPA is operated at a maximum level of 36.5 W. The transponders cover the top 29 MHz of each of the 1.5/1.6 bands and they are channelised into eight subbands. Each subband provides a bandwidth of either 3.5 or 4.5 MHz and the channelisation is performed using SAW filtering at IF. The feeder links operate in Ku-bands via a single-beam antenna covering the entire L-band coverage region. The US satellite is currently located at 101° W and the Canadian satellite is located at 106.5° W. The design lifetime of the satellites is ten years.

19.7.5 Future satellites

The satellites that are currently being constructed to support the various L- and S-band satellite PCSs range from the relatively simple to the complex. Several are using digital onboard processing[25] for the first time in the mobile arena and many will employ complex multibeam antennas. For example, the ACeS satellite is understood to be using an unfurlable antenna (about 12 m diameter) similar to that used by the TDRSS satellites, and the Iridium satellites are using digital onboard processing to demodulate the TDD/FDMA mobile link signals, and to switch and remodulate them onto the TDM of correct inter-satellite link signal. These techniques, when combined, will allow practical systems to be implemented using small and cost-effective handheld phones for the first time (see Chapter 20).

The exploitation of the 20/30 GHz bands has already started, both in the USA and Europe. For example, the NASA-funded team at JPL which worked on precompetitive research for L-band mobile systems (M-SAT X) is now developing mobile and personal terminals for use with the NASA ACTS satellite[26].

19.8 References

1 GODDARD, M.: 'Frequency allocations for mobile satellite systems and satellite navigation'. IEE 4th international conf. on *Satellite systems for mobile communications and navigation*, London, 1988, IEEE Conf. Pub. 294

2 CASEWELL, I.E. *et al.*: 'A satellite paging system for land mobile users'. IEE 4th international conf. on *Satellite systems for mobile communications and navigation*, London, 1988, IEEE Conf. Pub. 294

3 BENEDICTO, J. *et al.*: 'An L-band multi-beam land mobile payload for Europe'. Proc. IMSC-90, JPL Pub. 90-7

4 ROGARD, R., and PINELLE, M.: 'PRODAT aeronautical communication system: overall architecture and preliminary test results'. Presented at IEEE ICC-87, Seattle, USA

5 JACOBS, I.M.: 'An overview of the OmniTRACS: the first operational two-way mobile Ku-band satellite communications system'. Proc. of the international conf. on *Mobile satellite communications*, London, 1989

6 DUTRONC, J., and COLCY, J.N.: 'Land mobile communications in Ku-band. Results of a test campaign on Eutelsat I F1', *Int. J. Satell. Commun.*, **8**, pp. 43–63

7 SANDRIN, W., and FANG, D.J.: 'Multipath fading characterisation of L-band maritime mobile satellite links', *COMSAT Tech. Rev.*, 1986, **16**, (2)

8 NEUL, A., HAGENAUR, J., PAPKE, W., and EDBAUER, F.: 'Propagation measurements for the aero-nautical satellite channel'. IEEE VTC-87 conference

9 JONGEJONS, A.: 'PROSAT Phase I Report'. ESA STR-216, May 1986

10 LUTZ, E.: 'Land mobile satellite channel-recording and modelling'. 4th int. conf. on *Satellite systems for mobile communications and navigation*, London, 1988, IEE Conf. Pub. 294

11 STUTZMAN, W.L. (Ed.): 'Mobile satellite propagation measurements and modelling: a review of results for systems engineers'. Proc. of the *Mobile satellite* conf., JPL, Pasadena, USA, 1988

12 BUNDROCK, A., and HARVEY, R.: 'Propagation measurements for an Australian land mobile satellite system'. Proc. of the *Mobile satellite* conf., JPL, Pasadena, USA, 1988

13 SMITH, H., *et al.*: 'Service provision to land mobile units in northern latitude by a satellite in a non-geostationary orbit—propagation studies'. 5th international conf. on *Mobile radio and personal communications*, Warwick, UK, 1989 p. 15

14 SMITH, H., *et al.*: 'Propagation measurements for S-Band land mobile satellite systems using highly elliptical orbits'. Proc. ECSC-2, Oct 1991, ESA SP-332, p. 517

15 JAHN, A., *et al.*: 'Narrow and wide band channel characterisation for land mobile satellite systems: experimental results at L-band'. Proc. 4th international *Mobile satellite* conference 1995, Ottawa, Canada, JPL 95-12

16 SCHREITMULLER, W.: 'Novel ARQ techniques for land mobile satellite systems'. Proc. IDSCC-7, Munich 1986

17 LUTZ, E., DOLAINSKY, F., and PAPKE, W.: 'Land mobile satellite channel-model and error control' in 'satellite integrated communications networks' Eds. E. DEL RE, P. BARTHOLOM and R. P. NUSPL (Elsevier Science Publishers B.V., 1988)

18 MIURA, S., *et al.*: 'Aeronautical, maritime satellite communications experiments program using ETS V'. IEE 3rd international conf. on *Satellite systems for mobile communications and navigation*, London, 1983

19 IRISH, D., *et al.*: 'A robust signalling system for land mobile satellite services'. *Mobile satellite system architectures and multiple access techniques* workshop, JPL, 7–8 March 1989

20 RAFFERTY, W., *et al.*: 'Current development in NASA's mobile satellite programme'. IEE 4th international conf. on *Satellite systems for mobile communications and navigation*, London, 1988, IEE Conf. Pub. 294

21 MULLINS, D.R., *et al.*: 'INMARSAT-3 communications system requirements'. IEE Colloquium Digest, 1991/175, Nov. 1991

22 KINAL, G., and NAGLE, J.: 'Geostationary augmentation of global satellite navigation—1991 update'. Proc. of NAV 91, London, Nov 1991

23 GREENWOOD, D.A., *et al.*: 'Multi-matrix beamforming for semi-active antennas at L-Band'. IEE colloquium digest, 1991/175, Nov 1991

24 BLANGER, R., *et al.*: 'MSAT communication payload system overview'. Proc. AIAA, 1994, pp. 532–541

25 CASEWELL, I.E., *et al.*: 'A digital processing payload for a future, multibeam, land mobile communication system'. AIAA 12th international *Communication satellite systems* conf., Arlington, USA, 1988

26 SUE, M.K., *et al.*: 'A satellite-based personal communications system for the 21st century'. Proc. IMSC-90, JPL Pub. 90-7

Chapter 20

Satellite personal communication networks

L. Ghedia

20.1 Introduction and overview

The previous chapter described the history of mobile satellite communications based around the INMARSAT system and its recent privatisation activities. This chapter extends into the new satellite personal communication systems networks using hand-held terminals.

The marketplace which INMARSAT serves today is varied and includes, for example, very large terminals on ships providing primarily compressed voice and so called high-speed data services (64 kbit/s). More recently, with the proliferation of terrestrial cellular systems, the mobile satellite communications marketplace has been demanding smaller terminals with reduced tariffs. This has led to the development of smaller terminals which range from vehicles-mounted (INMARSAT-A analogue, M and B digital) to small hand-portable terminals the size of a laptop computer weighing a few kilograms (INMARSAT mini-M telephone).

The new challenges to mobile satellite communications have really come from the terrestrial cellular industry where small terminals have found very widespread use. The growth in the deployment of cellular telephony systems in the 1980s and 1990s has been phenomenal and this has led to massive reductions in the cost of terminals and call charges together with the offering of a wide range of services. With cellular telephony, a ground-based technology, it became apparent that industry was willing to deploy these systems only where there were sufficient numbers of subscribers, i.e. the densely populated areas — urban and suburban only. This would therefore leave out large geographical areas where the offering of cellular telephony would not be economical owing to coverage and mobility issues. This is the niche market which mobile satellite communications has come to fill. Therefore, mobile satellite communication systems tend to complement cellular systems by providing services in difficult areas. It is also possible that satellite communications may even eventually compete with cellular systems if the costs can be reduced significantly — at this stage a daunting task. The trend for the next-generation mobile systems — those referred to as the third-generation systems — is to provide seamless communication services to users wherever they are with a single multimode terminal and a single user number. This is the domain of personal communication services/networks (PCS/PCN) and satellite PCN for the systems being planned will be an inherent component of global PCN.

In this chapter details are provided on some of the systems proposed for the satellite component of PCN and in particular the satellite component of IMT 2000 (International Mobile Telephony 2000 — the successor to FPLMTS (Future Public Mobile Telecommunication Systems)). The various systems vying to represent the SPCN component are described and compared.

20.2 Wireless mobile radio systems

Growth in telecommunications in the past two decades has been primarily in the terrestrial cellular area. On a worldwide basis, the number of cellular subscribers by the year 2000 is expected to grow to about 200 million users. The projections for mobile growth indicate that the extent of coverage of cellular systems (predicted to the year 2000) is likely to be 20 per cent of the land surface area covered by cellular systems encompassing 80 per cent of the world's population. This leaves approximately 80 per cent of the world's surface area without cellular systems — a potential market for satellite communications. Historically, the systems first deployed were mainly analogue and are referred to as the first-generation cellular systems, examples of which include the advance mobile phone system (AMPS) primarily in the USA, Nordic mobile telecommunication systems — NMT, total access communication systems (TACS) primarily in Europe, E-TACS etc. These systems are now giving way to all digital systems of which the global system for mobile (GSM) — an ETSI-defined system — has seen prolific deployment globally. The counterpart of this is the digital AMPS system (DAMPS) being deployed in the USA and the personal digital cellular (PDC) in Japan. Variations of the GSM system are currently being deployed in Europe and the USA — the so called DCS1800, DCS1900, PCS1800 etc. Table 20.1 provides a summary of the salient parameters of the various systems.

20.3 Second-generation systems

The second-generation mobile cellular systems include primarily GSM, DAMPS and PDC, and of these systems GSM has been deployed globally at a frantic pace. The system arose in Europe owing to the lack of a cellular standard

Table 20.1 *Cellular mobile radio systems*

Parameter	AMPS	MCS-L1 / MCS-L2	NMT	C450	TACS	GSM	PCN	IS-54
Tx. freq.								
Base station	869–894	870–885	935–960	461–466	935–960	890–915	1710–1785	869–894
Subsystem (BSS)								
Mobile	824–849	925–940	890–915	451–456	890–915	935–960	1805–1880	824–849
Multiple access	FDMA	FDMA	FDMA	FDMA	FDMA	TDMA	TDMA	TDMA
Duplexing method	FDD	FDD	FDD	FDD	FDD	FDD	FDD	FDD
Channel bandwidth	30	25 12.5	12.5	20 10	25	200	200	30
Traffic/RF channel	1	1	1	1	1	8	16	3
Total traffic channels	832	600 1200	1999	222 444	1000	125*8	375*16	832*3
Voice	analogue	analogue	analogue	analogue	analogue	RELP	RELP	VSELP
Syllabic companding	2:1	2:1	2:1	2:1	2:1	–	–	–
Speech rate (kbit/s)	–	–	–	–	–	13.0	6.7	8.0
Modulation	PM	PM	PM	PM	PM	GMSK	GMSK	PI/4
Peak freq deviation (kHz)	+/–12	+/–5	+/–5	+/–4	+/–9.5	–	–	–
Channel rate (kbit/s)	–	–	–	–	–	207.8	207.8	48.6
Control	digital	digital	digital	digital	digital	digital	digital	digital
Modulation	FSK	FSK	FSK	FSK	FSK	GMSK	GMSK	PI/4
Baseband waveform	Manchester	NRZ	NRZ	Manchester	Manchester	NRZ	NRZ	NRZ
Peak deviation (kHz)	+/–8	+/–4.5	+/–3.5	+/–2.5	+/–6.4	–	–	–
Channel rate (kbit/s)	10.0	0.3	1.2	5.3	8.0	270.8	270.8	48.6
Channel coding	BCH	BCH	BCH	BCH	BCH	RS	RS	Conv
BSS–mobile	40,28	43,31	burst	15,7	40,28	12,8	12,8	1/2
Mobile–BSS	48,36	43,31 11,07	burst	15,7	48,36	12,8	12,8	1/2

between countries. The GSM system is configured around 128 channels, each 200 kHz wide, in the 900 MHz band (890–915 MHz mobile transmit and 935–960 MHz base transmit). This GSM band is referred to as GSM900. The wide channels allow high-speed digital transmissions which help to reduce the effects of fading and multipath. The access technique used in GSM is time division multiplex (TDMA) with eight time slots per channel and 217 frequency hops per second. Therefore eight users can be accommodated simultaneously per channel. Each mobile unit, as well as supporting digital cellular telephony, also supports synchronous and asynchronous data transmission and reception at 9600, 4800 and 2400 baud. The data portion interfaces with standard audio modems (V.22bis, V.32bis) and with the digital telephone system (ISDN). GSM also supports a feature which allows a two-way confirmed messaging service—a higher level of function than that achieved by pagers. In addition, paging messages can be sent and received invisibly during a telephone conversation.

GSM also makes use of a subscriber identity module (SIM) card which is issued by the service provider with the subscription. A user's portfolio of services is embedded in the SIM and, by using this card, a person can use their subscription to the service with any available GSM unit. GSM conforms to a set of European Telecommunications Standards Institute (ETSI) specifications, and all the GSM operators are signatories to the GSM memorandum of understanding (GSM MoU) which covers all technical aspects of GSM networks. A variation of GSM developed at the time of the GSM standard generation is the DCS1800 system. The major difference between GSM900 and DCS1800 is the lower transmitting levels required with the latter, designed to promote smaller cell sizes. GSM900 is similar to the digital cellular network in the USA, and the DCS1800 more closely resembles the USA PCS services in the 1.9 GHz band.

With reference to mobile satellite communication systems, the major constraints imposed on the cellular systems (handset size, transmit powers, safety considerations etc.) are also applicable to satellite services.

20.4 Third-generation systems

The third-generation systems are currently being worked at in the standardisation committees at the ITU, as IMT2000, and at ETSI, as UMTS (universal mobile telecommunication system).

Mobile communication via satellite is an integral component of IMT2000 and UMTS and the bandwidth allocated is 30 MHz in the S-band. This is referred to as SPCN. However, there are proponents that intend to use the original RDSS band (reclassified as generic MSS band at ITU-WARC 92) to provide SPCN services—most of them via low-earth orbit, LEO, satellite systems. The frequency bands defined for the IMT2000 band are as in Table 20.2.

Table 20.3 lists some of the cellular systems for the various generations.

Table 20.2   Frequency bands for SPCN

Return link (mobile to satellite)	2170–2200 MHz
Forward link (satellite to mobile)	1980–2010 MHz

Table 20.3   Mobile systems — cellular

1st generation	2nd generation	3rd generation
analogue	digital	digital
NMT	GSM	FPLMTS/IMT 2000
AMPS	DAMPS	UMTS
TACS	PDC	SPCN
	DCS1800	
	DCS1900	

20.5 Personal communications systems/networks — PCS/PCN

It is expected that the wireless systems of today will lead to the personal communications systems of the future which are characterised by much smaller handsets with lower powers and cells of very small size. At the network layers the following features are deemed necessary:

- calls to and from persons, not a location;
- calls dialled to a person, not a piece of equipment;
- calls characterised by a personally-defined set of features and services typically by way of a subscriber identity module (SIM) card.

PCS can be defined as follows:

'A personal communication system is one that provides universal accessibility to a wide range of services, such as voice, data, video, audio, messaging, positioning and navigation services, to individuals who may be in any one of many locations — home, work, remote location, in transit and within either, local, national and international boundaries'

The provision of such services is currently under intensive study at the ITU by way of UPT — universal personal telecommunications — which can be defined as:

'UPT is a telecommunications service which enables the provision of services in a total mobile environment. It enables a UPT subscriber to participate in a user-defined set of subscribed services and to initiate and receive calls on the basis of a unique personal, network-transparent telecommunications number (PTN) across multiple networks, public and private, at any terminal — fixed, mobile, movable — irrespective of geographic location, limited only by terminal and network capabilities and restrictions by the network service provider'

Some of the attributes that a PCS service and network must provide include the following:

- mobility;
- locatability;
- affordability;
- safety;
- quality;
- capacity;
- security;
- versatility;
- interoperability;
- availability.

20.6 Satellite PCN

The major characteristics of SPCN systems are defined as follows:

(i)   *Mobility*: the user location is not fixed.
(ii) *Services*: the user has overall control over the service profile to which they subscribe.
(iii) *Services*: global service accessible from any terminal (satellite, cellular) via an SIM card.
(iv) *Security*: services should be secure from eavesdropping and misuse.
(v)   *Access*: over many networks i.e. roaming via agreements rather than technology.

20.6.1 Functionalities of SPCN systems

An SPCN system must support the following functionalities.

20.6.1.1 Mobility management

The system must have the capability of providing call connection to mobiles which may roam globally. In order to support both mobile-originated and mobile-terminated calls, several procedures must be incorporated, some of which are as follows:

- location management (including updating);
- registration;

- authentication;
- number recognition and translation.

20.6.1.2 Service availability

The system must support the provision of services across multiple networks which may be nationally or globally distributed. This implies that all the networks must support a set of core services and additional services which may or may not be supported by all networks, i.e. the services subscribed for, as defined in the SIM, may not be available globally. Therefore, services must be available across:

- multiple networks;
- multiple satellites.

20.6.1.3 Call management

Call management must be supported and includes managing the subscriber call privileges. Some of these are as follows:

- caller identification;
- service type;
- message service.

20.6.1.4 Charge management

One of the major functions that needs supporting is that associated with the administrative data centre (ADC) — that related to billing. This includes requirements for:

- split billing;
- customised billing.

20.6.1.5 Customer control

The major attribute of modern telecommunications systems, and in particular mobile telecommunications systems, is that related to the provision of a wide portfolio of services from which the user may select and pay for a subset, i.e. customer-driven service provision and control rather then network driven. The attributes which support customer control are as follows:

- service profile management;
- terminal configuration.

20.6.1.6 Security

A successful telecommunications service must support the requirements of caller security including access to the network, eavesdropping, fraud control etc. This can be achieved by having the system support the vital function of user:

- authentication;

which may be done on a call-by-call basis. This introduces the concepts of equipment identity register (EIR), the authentication centre (AuC) and the associated encryption algorithms.

20.7 Satellite PCN — challenges

The provision of satellite-based PCN services faces a multitude of challenges — those related to user terminals, the space segment, the ground segment, regulatory challenges etc. In this section some of the major challenges and design drivers are considered. For PCN via satellite, several orbit options are available with their own specific attributes. The choice of a mobile system is then governed by the following:

- constellation — orbit, number and satellites (planes and satellite per plane);
- satellite and payload capability;
- system capacity as a design driver;
- access techniques for user links;
- application of link diversity;
- the quality of signal desired (QoS) and the associated grade of service (GoS);
- overall system cost for implementation (nonrecurring);
- recurring cost — operations.

20.7.1 Key technical areas

20.7.1.1 Frequency sharing and management

It is expected that the band specified for SPCN will be used by a multiple of system operators. For example, in the case of the use of the radio determination satellite services (RDSS) band for services via LEO satellites, the FCC in the USA authorised the segmentation of the band for TDMA and CDMA systems, with the CDMA proponents having to share a common band. The issue of such sharing is very important since for a global system even if a virgin band were available it would have to be shared by multiple operators. Such sharing is currently being done by the GEO operators, but the task gets complex when the space and user ground segments are not fixed. Another added issue is related to the availability of the SPCN band globally in all countries which brings in regulatory, commercial and political issues.

20.7.1.2 Access techniques

The choice of frequency-division multiple access (FDMA), time-division multiple access (TDMA) or code-division multiple access (CDMA) arises when wireless communication is concerned and the specifics may be quite different from those relevant to its application in terrestrial mobile. For example, some of the SPCN proponents have specified CDMA as the access techniques which allows full frequency reuse for every beam by having different spreading codes. As long as a sufficient number of mutually-orthogonal codes can be found, the same frequency can be reused in every beam. However, the very nature of CDMA calls for very precise power control and the time delay owing to larger satellite distances may impact the signal quality in a fast fading channel. The use of TDMA techniques, on the other hand, may limit the reuse capabilities and the peak transmit powers can also impact safety and battery requirements.

In addition, there are two frequency assignment techniques available for mobile applications:

1 *Region-oriented frequency assignment (ROFA)*: geographic regions on the ground are assigned fixed frequencies and the moving satellite has to dynamically change frequencies on the satellite for calls to a particular region. As different beams cross a given area the calls are handed off and the frequencies matched to those assigned to that region.
2 *Satellite-oriented frequency assignment (SOFA)*: the satellite beams are assigned fixed frequencies and as the satellite moves, the mobile terminals and the gateways switch frequencies to match the assignment on the beam. This simplifies the satellite payload, but the complexity migrates to the ground which now has to manage the frequency assignments on a dynamic basis.

20.7.1.3 Radiation and user safety

The major factor which drives the satellite system design and more importantly the handheld terminal is user safety. Radiation from the terminal held close to the head must satisfy strict requirements and this is achieved by having low radiated powers on the return link. This, rather than the battery power, terminal size etc. drives the overall design. These challenges are captured under Section 20.7.2.

20.7.1.4 Physical-layer issues—margins, modulation and coding

The physical layer is the most critical layer, since failure at this level means that support of upper-layer functions becomes impossible. The design of this layer needs to account for the following:

- link margin;
- shadowing—environment;

- Doppler;
- body effects;
- multipath effects;
- data rates;
- service types.

The link-margin trade off impacts the ground and satellite designs and to some extent the choice of orbit height. Shadowing arises owing to user movement and satellite motion, where a direct line-of-sight link may not be sustainable and the call has to be handed off to other beams and/or satellite or may even terminate. The support of higher data rates is also not possible for a satellite system owing to limited power margins and rates exceeding 32/64 kbit/s are typically referred to as high-speed data. The exception to this is the wideband satellite communication systems proposed (CyberStar, SPACEWAY, Teledesic etc.) which use the Ka-band where higher powers are possible but the losses are also high (see Chapter 25).

20.7.1.5 Channel modelling — propagation issues

All systems have to work in a specified channel and a mobile channel is typically a difficult channel because of motion, shadowing, multipath etc. effects. In the case of the cellular systems the major effects are avoided by having a very large margin (40 to 50 dB) and therefore the system works using multipath components in channels which typically suffers from Raleigh fading. The satellite channels can only be sustained in a Rican channel — a direct line-of-site link with minimal margins (see Chapter 19). The worst-case margins specified for satellite systems range from 16 dB for some of the LEO constellations to 9 dB for the MEO systems. Enhanced margins may be available via satellite path diversity which primarily avoids shadowing at the expense of a more complex ground infrastructure.

20.7.1.6 Satellite — cellular integration

The integration of satellite/cellular components implies that it should be possible for subscribers to roam from a cellular network into a satellite network and *vice versa*. This assumes the availability of mobility-management databases which support multiple standards e.g. DAMPS and satellite, GSM and satellite, PDC and satellite. An extreme case of integration relates to in-call handovers between the satellite and cellular components — implying translation between different MAP (mobile application part) protocols.

20.7.1.7 Mobility management

Mobility management implies the ability to roam freely globally and make and receive calls from all places. This is achieved by the use of databases referred to as the home location register (HLR — where the user is registered for services and his or her service profile and privileges are kept) and the visitor location register (VLR—databases where visitor profiles are stored). Typically, the numbering adopted specifies the ID of the user's HLR. The ability to roam between different networks (e.g. satellite and GSM etc.) implies the design of databases which support multiple MAP protocols.

20.7.1.8 Call alerting—high-penetration paging

Almost all satellite PCN systems proposed will not support services to users located in difficult environments (depths of a building, so called urban canyons). However, it is possible to alert a user to an incoming call so that the user may move to a favourable reception position. This is achieved via a high-power short message (only a few bytes long).

20.7.2 Terminal challenges

The terminal challenges are driven by the rapid developments occurring in the cellular arena where terminals are becoming more powerful and also smaller with enhanced features. Historically, mobile satellite terminals have been fairly large as evidenced by INMARSAT M/B and mini-M terminals. These terminals are not truly mobile as they cannot be placed next to the ear. Rather, they would be classified as small portable terminals.

20.7.2.1 Handheld antenna

In the case of a fully mobile handheld terminal, the radiation must be effectively omnidirectional, i.e. antenna gain of approximately 0 dB. For the case of satellite communications the antenna pattern desired is generally referred to as cardoidal, i.e. radiation in the upper half of the hemisphere which gives an antenna gain of approximately 1 to 2 dB. In addition, the terminal cannot track the polarisation from the satellite, and linear polarisation needs to be avoided. More generally, RHCP or LHPC polarisation is utilised and the antenna construction is of the quadrifilar helix (QFH) type. The constraints on the antenna are as follows:

- dimensions;
- radiation pattern and polarisation purity;
- transmit and receive gains;
- operating frequencies and bandwidths;
- cellular and satellite band support.

20.7.2.2 Battery technology

Although the electronics in a mobile telephone has advanced, battery technology has not seen any revolutionary breakthroughs and to date remains an important issue in terms of talk time and standby time supported. Most users would prefer a talk time of 24 hours and standby time of one week, if possible.

Battery power is conserved by use of dynamic power control, support of idle mode etc. The batteries most often used are of the nickel-metal hydride types (NMH).

20.7.2.3 *Radiation exposure — safety*

A mobile telephone must be a safe telephone and this puts severe restrictions on the transmit power. The limits are generally set by the various regulatory bodies and to date there does not seem to be any consensus on whether mobile telephone radiation poses any health hazards. However, interference from mobile telephones into other electronics has been noticed (pacemakers, hearing aids etc.)

20.7.2.4 *Multimode operation (satellite and cellular)*

Almost all proposed SPCN operators will support dual-mode telephones. The issues of integration, therefore, relate to the sharing of the electronics in the telephone (vocoder, antenna, power supply, plastics etc.) to in-call handovers between satellite and cellular. It is unlikely that the latter will be supported. Other relevant issues are those related to registration in one system, numbering etc. At this stage only dual mode and single mode (satellite) telephones have been proposed.

20.7.2.5 *Receiver (single mode and diversity)*

The receiver must have the capability of supporting multiple standards and in some cases also diversity modes—specifically those operating in systems employing CDMA access. Even in the TDMA case, if diversity is implemented then the receiver must accept bursts from different satellites, synchronise them and produce the best output by burst-by-burst selection, coherent combining etc.

20.7.2.6 *Low-rate vocoder*

From capacity and power constraints (power and bandwidth) all SPCN systems can only support low-bit-rate voice—typically 2.4 kbit/s to 9.6 kbit/s with a mean opinion score of around 3.5 (i.e. communications-quality voice only). In addition, the cellular component typically uses a higher-rate voice codec and therefore there is no compatibility in vocoder usage. In the case of INMARSAT and other proposed mobile systems, the vocoder from DVSI referred to as the AMBE (advanced multiband excited) codec has been investigated. Some of the SPCN systems are planning to deploy either a dual-rate codec (2.4 kbit/s and 4.8 kbit/s) or even a variable-rate codec dynamically matched to the operating channel. The employment of error-correcting codes (e.g. FEC) also needs to be considered in the selection of the codec (e.g. is the error control embedded in the codec or is coding applied outside the codec).

Table 20.4   Terminal types

Parameter	Handheld	Portable	Vehicle
Size	pocket	laptop	various
Antenna gain	0–2 dBi	7–9 dBi	3 dBi
Transmit power	<500 mW	1 W	1 W
EIRP (dBW)	−3.0–0.0	+7–9	+7
Receiver G/T (dB/K)	−24 to −22	−17	−20

20.7.2.7 Size—MMIC/VLSI integration

The size of the handheld terminal depends on the level of integration possible and this depends on the application. Table 20.4 presents examples of three different types of terminal and their main attributes.

20.7.3 Satellite challenges

There are basically two orbit options available for the provision of SPCN service, geostationary orbits (GSO) and nonGSO orbits.

20.7.3.1 GSO orbits

The major challenges to the GSO orbit option include the following:

- large reflector antennas (diameter typically greater than 10 m);
- large number of beams (typically greater than 200 spot beams).

These challenges are then related to the design of the feed arrays, the reflector size, design and deployment, the payload channelisation and bandwidth flexibility desired. It should be noted that the NASA first-generation mobile satellites (now AMSC/TMI) use a five metre diameter antenna, the second generation proposed in the 1980s having a 20 metre diameter antenna and the third generation having antenna diameter greater than 60 metres. The design and deployment complexity, in addition to station keeping, are high-risk areas.

20.7.3.2 GSO and non-GSO orbits

The following are some of the main complexities and risk areas associated with both GSO and nonGSO systems:

- onboard processing;
- digital payload/repeater;
- mass and power constraints;
- intersatellite links;
- launcher availability and number of launches, satellites per launch.

20.7.4 SPCN design drivers

Some of the key drivers are as follows:

- satellite design compatibility with handheld terminal transmit and receive;
- speech-quality issues with low rate;
- QOS—link availability both forward and return;
- commercial issues—mobile pricing and tariffs issues, billing;
- regulatory issues (i.e. permission to provide services in a country);
- roaming and integration issues;
- coverage—regional and global requirements.

Table 20.5 shows the progression of global mobile satellite communication systems from the earlier generation to systems for the 1990s and beyond, the latter being referred to as the SPCN system.

20.7.5 SPCN—space-segment orbit options

The following orbit options are possible for the provision of satellite PCN services:

- geostationary—GSO/GEO (orbits at an altitude of 36 000 km);
- medium-earth orbit—MEO (orbits at an altitude of 10 000 km);
- intermediate circular orbit ICO (orbits at an altitude between LEO and GEO orbit heights);
- low-earth orbit—LEO (orbits up to an altitude of 2000 km);
- highly-elliptical orbit—HEO (example apogee of 40 000 km and perigee of 2000 km).

20.7.6 Comparisons of orbit options

The major attributes of the three basic orbit options are listed in Table 20.6.

Table 20.5   Progression—satellite systems

1970/1980	1990	1996	Late 1990s–2000
INMARSAT A	INMARSAT M/B AMSC OPTUS Regionals	Mini-M	ICO MEO/IMT2000 Iridium LEO/RDSS band Globalstar/LEO/RDSS bands APMT Super-GEO ASC Super-GEO Aces Super-GEO Thuraya Super-GEO

Table 20.6 Comparisons of SPCN space-segment orbit options

Parameter	LEO	ICO/MEO	GEO
Delay (milliseconds)	low (15)	medium (70)	high (280)
Link diversity	possible	possible	limited
Satellite power	low	medium	high
Satellite mass	low	medium-high	high
Doppler	high	medium	very low
Number of satellites/global coverage	40	10–18	3
Hand-off rate	high	medium	none
Ground-segment complexity (no ISL)*	high	high	low
Ground-segment complexity (ISL)*	low-medium	low-medium	low
System cost (relative)**	high	medium-high	medium

*ISL: intersatellite link (typically for LEO and MEO)
** relative to current operational systems (nonhandheld)

20.7.6.1 GEO mobile systems

There are several proposals and systems under construction for the provision of SPCN via a GEO space segment. A typical GSO satellite is capable of providing a capacity equivalent of 5000 to 15 000 duplex voice channels (typically 4800 bit/s). For global service a total of three to six satellites is needed. Since the handheld terminal EIRP is limited by safety considerations, the propagation loss has to be made up by having a large antenna diameter and a large number of spot beams. An antenna diameter of 10–12 metres is possible with the fairing sizes possible with the current launch vehicles. Typical examples of GSO mobile systems include ASC (Agrani Satellite Corporation), AceS (Asian satellite cellular system) and APMT (Asia pacific mobile telecommunications system). Besides these there are the regional systems—AMSC and OPTUS.

The ASC system

ASC has awarded a contract to Lockheed Martin for the delivery of a mobile satellite system that is centred on an LM A2100 satellite and an uncertain number of gateways. The satellite will be a hybrid design, relying on Ku-band and L-band frequencies to link users and gateways to the satellite. The choice of the system was largely owing to the success of the similar ACeS system from Lockheed Martin. The system could be operational in two and a half years. The hybrid satellite for ASC has the capability to provide L-band MSS services to the mobile segment and DSS service via the Ku-band for digital video distribution.

The ASC system is capable of supporting three modes of communication:

(i)   Mobile originated (mobile to PSTN—easiest to implement).
(ii)   Mobile terminated (PSTN to mobile).
(iii) Mobile to mobile (most difficult to implement).

The payload proposed uses analogue techniques for beamforming, frequency-band-to-beam mapping and return link band to feeder-link spectrum compaction. In general, owing to isolation problems with a single antenna, two antennas are deployed—one for transmit and one for receive. (This is in contrast to the 1980s Hughes-proposed GEO mobile tritium system which had a single reflector antenna for transmit and receive.)

For the 12 metre antenna proposed, the beamwidth is approximately 0.9 degrees, and the scan angle is limited to ± 6 degrees. This results in a total of 245 beams which may be generated by either a focussed antenna, a reflector array antenna or a defocussed reflector antenna (proposed for the ASC system). The defocussed antenna requires approximately 200 antenna elements, each with its own amplifier, and the beamforming being accomplished digitally.

One of the major advantages of a GSO option is the ease with which beam-to-beam hand offs and satellite hand offs can be accomplished since the motion of the satellites, and hence the Doppler, is negligible. Unless there is a large displacement in position (for example with an aeronautical service), the question of satellite-to-satellite hand off does not arise. This therefore results in a very significant reduction on the signalling load on the air interface for location updates.

The Thuraya system

The Thuraya-system contract was placed with Hughes Space and Communications and Hughes Network Systems. The first of Thuraya's two satellites will be launched around May 2000, the second will be launched in 2002. The space system is designed to accommodate changes in traffic by means of a reprogramable payload in the satellite. The payload proposed is an all digital construction which permits the use of digital processors for channelisation, digital beamforming, frequency-band-to-beam mapping and return-link beam band to feeder-link spectrum compaction. In general, owing to isolation problems with a single antenna, two antennas are deployed—one for transmit and one for receive. The digital payload will allow modifications to the coverage area after launch and optimisation of performance over areas where demand is highest.

Hughes Space and Communications has awarded Spar subsidiary company Astro Aerospace a contract for two 12.25 m diameter AstroMesh unfurlable antennas. The antenna opens in space to become a large parabolic-shaped dish.

The user terminals are to be provided by Ascom which has been chosen as the second supplier for Thuraya telephones. The Ascom telephone will be a dual-mode—satellite and GSM—handheld terminal. Hughes and Ascom will provide 235 000 terminals.

Access into PSTN networks is via the primary gateway which will be set up in Sharjah, and four to five additional national gateways will be built in the coverage area to provide national access to the space segment.

The AceS system

Lockheed Martin (LM) has been selected for the development of the AceS system and Ericsson has the prime responsibility for providing the handheld terminals.

The AceS system will offer a low data rate at 2.4 kbit/s as well as the core voice service at approximately 4.8 kbit/s. Other services are similar to those found in the GSM system, with the exception of high-rate services which the GEO mobile air interface bearer cannot support to handheld terminal users.

In most aspects the AceS system is very similar to the ASC system, also being developed by LM. The first ACeS satellite is due for launch in 1999/2000 and will begin commercial service six months later.

Access into the PSTN is via the Thailand gateway which is nearing completion. The gateway will be operated by ACeS Regional Services Co. Ltd.

As far as service sharges are concerned, the ACeS retail rate will be US$1.00/minute of airtime, plus applicable last-mile cost. The subscription fee will vary from one national service provider to another, however, such fees will not be substantially different from current GSM network subscription fees.

The ACeS dual-mode handset will cost in the range US$1000–1500. The handset size will be 13.5 × 6.7 × 3.1 cm, with a weight of 200 gram with the antenna and standard battery with 24-hour battery stand-by time.

20.7.6.2 MEO systems

The proposed MEO systems are listed in Table 20.7. Of these, only the ICO system is currently under construction (TRW—the parent of Odyssey system joined with ICO in January 1998 and disbanded the Odyssey program). The proposed AMSC system is the second-generation system for which an application has been filed with the FCC, but to date there have been no real developments since AMSC itself faces problems in terms of markets and revenues.

20.7.6.3 ICO Global Communications

During the late 1980s and early 1990s INMARSAT initiated a program to find a replacement for the INMARSAT-M system—a system which incorporated

Table 20.7   MEO satellite systems for mobile services

System	AMSC (proposal)	Odyssey (disbanded)	ICO
Operator	AMSC	Odyssey	ICO
Satellites	12	12	10 + 2
Orbit inclination	47.5	55	45
Mobile frequency—Tx (GHz)	1610–1626.5	1610–1626.5	1985–2015
Mobile frequency—Rx (GHz)	2483.6–2500	2483.6–2500	2170–2200
Approximate cost (US$)	3.10 billion	N/A	4.60 billion

services mainly to handheld terminals. The program was labelled INMARSAT-P, and Project 21 was envisaged to include such terminals and other offerings of service typical of the terrestrial cellular market. In January 1995, INMARSAT spun off the INMARSAT-P program into an independent commercial company called ICO Global Communications. Investment in the new company was restricted to the investors of INMARSAT only, with INMARSAT itself taking the largest share, of the order of 15 per cent, and two seats on the ICO board. Together with another 49 investors, ICO Global communications was capitalised to the tune of 1.4 billion US dollars. Soon after, Hughes Space and Communications of the US was awarded the contract to construct and deploy the 12 satellites, and Hughes itself invested approximately 100 million dollars into ICO—the first commercial company to do so. Under contract, Hughes will construct and launch the satellites on a multitude of launch vehicles including Sea-Launch.

Unlike all other nonGSO systems, the ICO system uses the FPLMTS/ IMT2000 frequency bands allocated to satellite services. In addition, the gateway stations, referred to as the satellite access nodes (SANs), access the satellite via the C-band (the commercial C-band operated in reverse mode). A total of 12 satellites will be launched with one satellite per plane acting as an in-orbit spare.

Table 20.8 presents the salient parameters for the ICO global system.

ICO system architecture
The ICO system consists of $10 + 2$ satellites in intermediate circular orbit (10 355 km). The ground infrastructure consists of 12 satellite access nodes

Table 20.8   ICO Global Communications system overview

Parameter	Value	Note
Service uplink	1985–2015 MHz	SPCN band
Service downlink	2170–2200 MHz	SPCN band
Feeder	7/5 GHz	C-band (reverse mode operation)
Service link bandwidth	10 MHz	
Access technique	TDMA	offset transmit and receive
Mobile spot beams/ satellite	163	
Constellation	2/10	2 planes of 5 satellites each
Orbit height	10 390 km	
Inclination	45 degrees	
Orbit period	6 hours	
Earth stations—satellite access nodes	12	5 antennas per SAN
Handovers	beam and satellite	seamless

(SANs) to be provisionally located in India, Australia, China, South Korea, Indonesia, USA, Germany, UAE, South Africa, Mexico, Chile and Brazil. The SANs are interconnected to form the ICONET. All calls are placed via the SANs, which comprise five C-band tracking antennas, the satellite base station equipment (channel units and channel managers), the switch (MSSC) and the associated databases for mobility management (VLR, HLR etc.). The SANs typically connect into service-provider gateways (ISCs). The management of the ground infrastructure is under the network management system and NMC is the highest point at which frequency and power planning for the satellites and the SANs is performed.

The satellites are monitored and controlled via the satellite control centre (SCC).

Figure 20.1 shows the ICO system architecture.

ICO network architecture
As pointed out, the SANs are all interconnected into a ring-type architecture referred to as the ICONET. Figure 20.2 depicts the network architecture of the ICO global system.

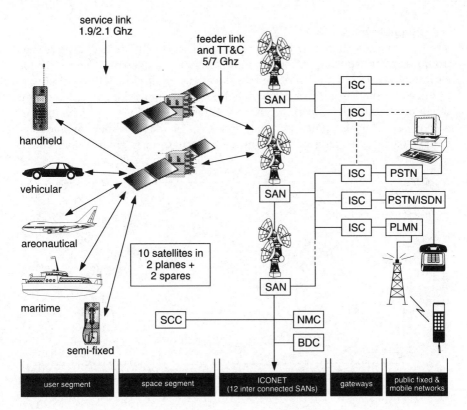

Figure 20.1   ICO system architecture

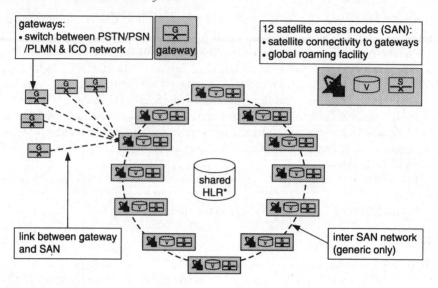

Figure 20.2   ICONET architecture

ICO satellite coverage

The constellation chosen by ICO is one from a set of optimised Walker configurations and the coverage at a given instant in time is depicted in Figure 20.3 for elevations down to ten degrees. In most locations, multiple-satellite visibility is possible and therefore the user has a choice of communicating via two or more satellites, i.e. diversity of path. With the system designed for a minimum margin of 8 dB, the diversity path effectively improves the performance, by avoiding shadowing and blocking effects—the effective margin is greater than 16 dB.

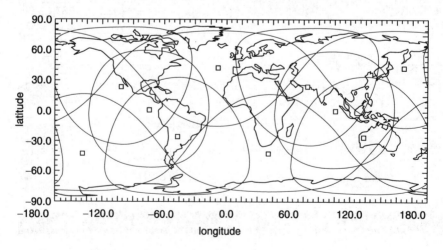

Figure 20.3   ICO satellite coverage for ten-degree elevation

Since all mobile satellite systems work only with line-of-sight to satellite, diversity offers much improved performance. However, this is at the expense of higher bandwidth requirements compared with the LEO systems which offer no diversity but provide a 16–17 dB or greater margin (power expense).

ICO satellite payload

The ICO satellite is designed around a digital payload offering variable bandwidth to the 163 beams. From a set of 30 MHz, bandwidth granularity of 150 kHz can be achieved and blocks of 150 kHz assigned to beams on an as-needed basis. This is accomplished via channelisation plans formulated at the NMC and transmitted to the SANs, for onward routing to the satellite via the payload command subsystem. The constellation uses no intersatellite links and this, therefore, imposes a greater complexity on the ground instead of on the satellites. The variability of bandwidth allocation to the beams also means that there is an inherent flexibility for the offering of current and future services which may need bandwidths that are undefined at satellite construction time. Figure 20.4 shows the channelisation and the corresponding frequency routing.

Figure 20.5 shows the architecture of the fully digital payload on the ICO satellite. It allows the assignment of bandwidth to particular beams on an as-needed basis.

ICO ground segment

The ground segment of the ICO global mobile personal communications system consists of twelve SANs which are interconnected to form the ICONET, the terrestrial backbone of the ICO system. The minimum number of SANs required to provide global services is seven, but it has been decided to install twelve stations to achieve better coverage and reliability. The system design is such that any number of SANs can easily be accommodated. Figure 20.6 shows

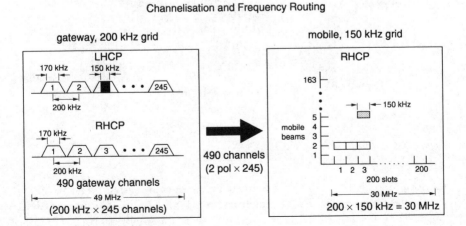

Figure 20.4   C-band to S-band mapping

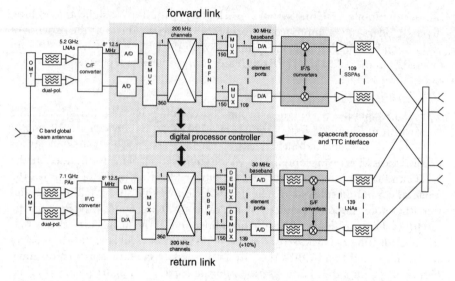

Figure 20.5   ICO satellite payload architecture

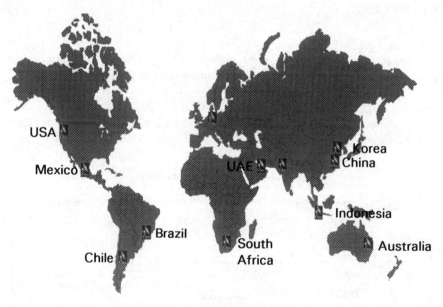

Figure 20.6   The 12 ICONET SAN sites

the locations of the 12 SANs baselined in the current system. The six SANs which have colocated TT&C facilities are USA, India, Germany, Chile, Australia and South Africa. The in-orbit tests for the satellites (IOT) are to be conducted from the SAN in India.

Figure 20.7 shows the block diagram of a typical SAN. The RFT comprises the antennas and the RF subsystems, which are being designed by NEC. The satellite base station, which is being designed by Hughes Network Systems, is similar to a cellular base-station system (BSS). The SBS comprises the channel units (similar to BTS) and channel managers (similar to BSC). An additional component of the SBS is involved with the management of the SAN resources and this is the local real-time resource management system which interfaces with the SRMC at the centralised network management centre (NMC). The SBS interfaces with the mobile satellite services switching centre (MSSC), a modified GSM cellular switch (the AXE10 from Ericsson). Service providers connect to the MSSC via their respective point of interconnect (POI).

The commercial element of the ICO system consists of the business operations support system which has the responsibility of service provisioning, customer care and billing. The customers for the ICO system are the service providers and not the end users.

The ICO system avoids bypassing existing terrestrial networks. Instead, it integrates them, connecting customers through the existing local cellular service providers and fixed network operators and offering the ICONET as a means of interconnection to/from the satellite segments anywhere, whenever local infrastructure is not available. The ICONET also includes a network management

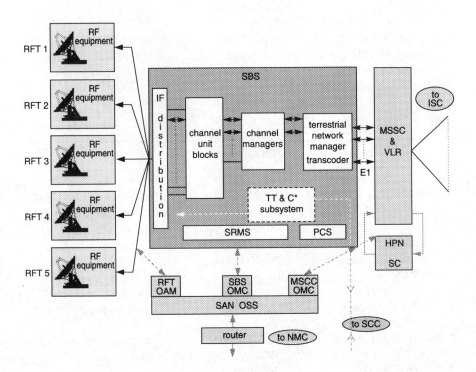

Figure 20.7   SAN architecture

centre (NMC), a satellite control centre (SCC) and their respective back-ups. The SANs, the NMC and the SCC are connected via terrestrial links for carrying network management, signalling and communications traffic.

The six SANs which have colocated TT&C facilities guarantee seamless telemetry, tracking and commands of the satellite fleet on a real-time basis. Five high-availability radio-frequency terminals are installed at each SAN site, four active and the fifth as a hot stand by. Communications and TT&C carriers at the TT&C sites share the RFT equipment, with the TT&C having the capability to override the communications traffic in case of satellite emergencies.

The satellites transmit telemetry and payload data, on a continuous basis, which is initially analysed at the SANs. The overall control of the satellite resides at the SCC, but the payload configuration is maintained by the NMC.

Examples of ICO services

The services offered by the proposed SPCN systems follow very closely those offered by the second-generation cellular mobile systems. However, owing to the restrictions on powers, only the lower-bit-rate services can be offered. For example, data services to handheld terminals are typically restricted to 9.6 kbit/s or lower. Higher rates may be achieved via nonhandheld terminals, where the sacrifice in margin is avoided by terminals with a higher antenna gain. The classification of services also follows very closely the definition used in the global system for mobile (GSM) system.

Table 20.9 shows the services typically being offered by the ICO system. The data service with a handheld terminal is restricted to 2.4 kbit/s, however, services to 64 kbit/s are planned for higher-rate modulation schemes and non-handheld terminals.

Table 20.9   ICO teleservices

	Tele-Service	Attribute	Note
1	basic telephony	4800 bit/s MOS 3.5	global, full roaming, mobile originated (MO), mobile terminated (MT), full duplex (FDX)
2	emergency call	as above	MO
3	short message service (SMS)	≤ 140 bytes	MO, MT, unidirectional > 15 dB margin, multi-alphabet, encrypted
4	high penetration notification (HPN)	<20 bytes	priority levels, delivery <3 seconds
5	asynchronous data	300–2400 bit/s	circuit mode
6	synchronous data	1200 and 2400 bit/s	circuit mode
7	group 3 fax	2400 bit/s	MO, MT

Table 20.10   ICO supplementary services

1	CLIR	calling line identification restriction
2	CoLP	connected line identification presentation
3	CoLR	connected line identification restriction
4	CFU	call forwarding unconditional
5	CFB	call forwarding on mobile subscriber busy
6	CFNRy	call forwarding on no reply
7	CFNRc	call forwarding on mobile subscriber not reachable
8	CW	call waiting
9	HOLD	call hold
10	MPTY	multiparty service
11	CUG	closed user group
12	BAOC	barring of all outgoing calls
13	BAIC	barring of all incoming calls

Table 20.10 shows the supplementary services planned for the ICO system. These services are typical of those found in the GSM network and may include the offering of intelligent network services via IN platforms and nodes.

20.7.7 LEO systems

The proposed LEO system comprises offerings from Iridium, Globalstar, Ellipsat, Constellation, Vitasat, Leosat, Teledesic, Orbcom etc. However, only the major systems are considered in this Section.

The following two Tables (20.11 and 20.12) provide the salient features for the big and little LEO systems.

20.7.7.1 LEO system — OrbComm

This system was proposed by Orbital Sciences Corporation and the FCC authorised it for operations in 1997. It features LEO satellites at an orbital height of 970 km with an inclination of 50 and 90 degrees, with satellites distributed in three planes. Initially, there are six satellites per plane.

The proposed services and their attributes are listed in Table 20.13.

20.7.7.2 LEO system — Iridium

Perhaps the most famous system and the one which has received the most publicity is the Iridium system — initially proposed by Motorola. The Iridium system comprises a constellation of 66 satellites in low-earth orbit arranged as six polar planes of 11 satellites per plane. In addition, there are six spare satellites — one per plane — giving a total of 72 satellites. The salient parameters of the Iridium system are given in Table 20.14. The system operates in a dedicated, FCC-approved band of approximately 5 MHz. The access technique utilised is

Table 20.11   Big LEO satellite systems for mobile services

System	Globalstar	Iridium	Aries	ECCO	Ellipsat	Teledesic
Operator	Globalstar LP	Iridium Inc.	Aries	ECCO	Ellipsat	Teledesic
Satellites	48	66	8 + 2		12 (24)	288
Orbit inclination	52	86.4	equatorial		63	82
Mobile frequency —tx GHz	1.610– 1.6265	1.610– 1.6265	L/S		1.610– 1.6265	20/30
Mobile frequency —rx GHz	2.4836– 2.500	1.610– 1.6265	L/S		2.4836– 2.500	20/30
Approximate cost (US$)	1.90 billion	3.80 billion	1.70 billion		560 million	9.0 billion
Services	low data/voice	low data/voice			low data/ voice	Internet

Table 20.12   Little LEO satellite systems for mobile services

System	OrbComm	Starsys	Vitasat
Operator	Orbital Communications Corp.	Starsys Global Positioning Inc.	Volunteers in Technical Assistance
Satellites	26 (36)	24	2
Orbit inclination	45 + 2 polar	50–60	88
Mobile frequency —tx MHz	148–150.5	148–150.05	149.8–149.9
Mobile frequency —rx MHz	400.15–401	400.15–401	400.15–401
Approximate cost (US$)	320 million	320 million	10 million
Services	paging messaging	paging messaging	paging messaging

Table 20.13   OrbComm proposed services

Emergency services SecurNet	Data acquisition DataNet	Tracking services MapNet	Messaging
road—towing search and rescue medical	environment industrial utility remote asset SCADA	boxcar containers property animal customs	business personal handicapped

TDMA with time-division duplex transmission—the mobile transmitting or receiving only at any time (this avoids the use of a diplexer in the receiver and provides an additional 2 dB for the receive G/T). Access to the L-band is via gateways operating in the Ka-band. Each satellite generates spot beams at Ka-band which are pointed at and track the gateways until hand off to another

Table 20.14   Iridium system parameters

Parameter	Value	Note
Service uplink	1610.0–1621.35 MHz	original RDSS band
Service downlink	1610.0–1621.35 MHz	original RDSS band
Feeder	18.8–20.2/27.5–30.0 GHz	Ka-band—4 spot beams (4 gateways/satellite)
Intersatellite links (ISL)	22.55–23.55 MHz	
Service-link bandwidth	10.5 MHz	
Access technique	TDMA	
Mobile spot beams/satellite	48	scanning and optimised for land and coastal coverage
Constellation	6/11	6 planes of 11 satellites each
Satellite mass/power	1000 kg/1000 W	approximate values
Satellite antenna	phased array	tx and rx arrays
Orbit height	780 km	
Inclination	86 degrees	
Orbit period	1.7 hours	
Ground segment: gateways	3 antennas	site diversity
Number of gateways	TBD	for rain
System/satellite operation and network management centre	1	fades TBD
Handovers	frequent	beam and satellite
Ground segment: user		
Dual mode	GSM, AMPS, PDC	
Voice	4800 bit/s	MOS 3.5
Data	4800 bit/s	
SIM card		
Battery talk/standby	60 min/24 hrs	
RF power average/peak	−4.4 dBW/6.0 dBW	
Antenna	cardoidal	QFH

gateway is needed. The Iridium system features a single line-of-sight link to the satellites of 16 dB minimum margin which allows the link to close in most line-of-sight environments (i.e. multipath-dominated channels). The system also allows the dynamic allocation of bandwidth thereby avoiding bandwidth wastage, for example, over ocean areas.

Each satellite projects a cluster of 48 beams onto the surface of the earth and as the satellites approach the poles the overlap increases. In order to avoid interference, selective beams are switched off. In fact, at a given time, only 2100 out of the 3168 beams are simultaneously active. Unlike other proponents, the Iridium system uses intersatellite links (ISL); in-plane links (to the satellites in front and behind) and cross-plane links (to two satellites in planes on either side of the subject satellite). Hence, a total of six ISLs per satellite are incorporated. This allows calls to be routed in the sky and in the limit only one gateway is needed for operations to begin (unlike other global systems where the gateways must all be connected for global service to be provided). Although the ground infrastructure cost is reduced, the complexity is migrated to space where the ISLs must be carefully managed and the calls assigned and maintained for their durations despite the constellation moving, the earth precessing and the users also not being stationary.

Iridium system architecture

Figure 20.8 shows the Iridium system architecture. It comprises the space segment (satellites referred to as the space vehicle, SV), the ground stations working at Ka-band and multiple terminal types.

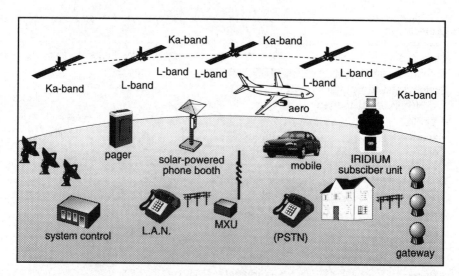

Figure 20.8   Iridium system architecture

Iridium satellite coverage

Iridium constellation

The planned Iridium constellation is depicted in Figure 20.10. The coverage from the Iridium constellation is depicted in Figure 20.9, which shows the overlap from all the beams from all the satellites at a given instant in time.

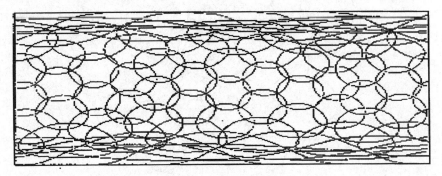

Figure 20.9   Iridium constellation footprint

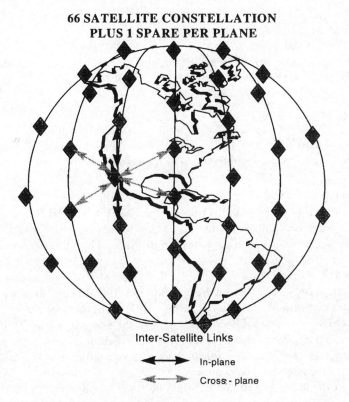

Figure 20.10   Iridium satellite constellation

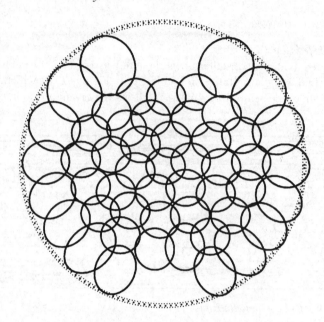

Figure 20.11   Iridium satellite mobile beam organisation

Iridium mobile beam organisation

Each of the Iridium satellites projects a cluster of 48 beams onto the earth. Figure 20.11 shows the beam organisation.

Iridium space-vehicle configuration

Figure 20.12 shows the satellite in a deployed configuration. Phased-array antennas mounted on three body panels are used to generate the 48 beams. In addition open-ended waveguides act as antennas for the crosslines.

Iridium gateway

Figure 20.13 shows the architecture of the Iridium gateway. The gateway consists of three Ka-band terminals (Scientific Atlanta) together with a terminal controller acting as the satellite base station (SBS). The SBS is connected to a GSM switch (Siemens) and thereon to the public networks. For mobility management the visitor location register (VLR) and the home location register (HLR) are also colocated at the gateway. Authentication and security is achieved via the authentication centre (AuC) and the equipment identity register (EIR) databases. The management of the gateway is under the operations and management centre (OMC) of which there are three—the OMC-R for the radio equipment, the OMC-S for the switching component and the OMC-P for the paging services nodes.

Some of the major parameters of the two types of gateway envisaged in the Iridium system are listed in Table 20.15.

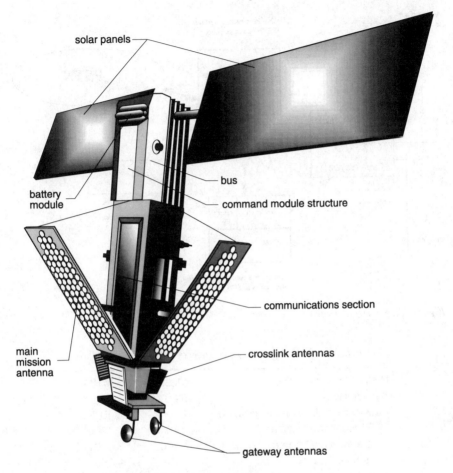

solar panels

bus

battery
module

command module structure

communications section

main
mission
antenna

crosslink antennas

gateway antennas

Figure 20.12   Iridium satellite deployed configuration

Iridium operation and management centre
Figure 20.14 shows the architecture of the OMC in the Iridium network. The
satellite control centre is located in Virginia.

20.7.7.3 LEO system—Globalstar

Globalstar is another big-LEO systems scheduled to go into service in 1999
with a constellation of 48 satellites. Unlike Iridium—which uses (the RDSS
band) TDMA-TDD (time-division multiple access, time-division duplex)
access, Globalstar uses CDMA access.

Satellite description
The constellation is arranged as eight planes with six satellites each at an
orbital height of 414 km (750 nautical miles). The inclination of the planes is 52°.

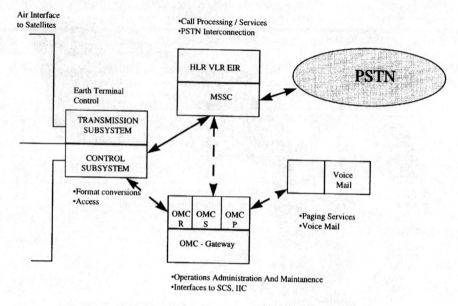

Figure 20.13 Iridium gateway overview

Table 20.15 Iridium gateway parameters

	M030	M120
Performance/capacity	M030	M120
Availability	99.8%	99.8%
Busy-hour call attempts (BCHA)	20 000	40 000
HLR subscriber sizing	30 000	120 000
VLR subscriber sizing	30 000	120 000
Terrestrial channels	240	750
Paging subscriber database	25 000	100 000
Paging input circuits	60	240
Radio channels per link	480	480
Radio links	1	1
PSTN signal control	SS7	SS7
Transmission facilities	PCM 30	PCM30

With the LEO orbit proposed the orbit period is 113 minutes. The selected constellation provides optimum coverage between 70°S to 70°N latitude with the possibility of multiple satellite coverage in temperate zones which can result in higher signal quality and availability.

The Globalstar satellite is a simple low-cost satellite designed to minimise both satellite costs and launch costs. Globalstar satellites do not directly connect one Globalstar user to another. Rather, they relay communications between the user and a gateway. The party being called is connected with the gateway

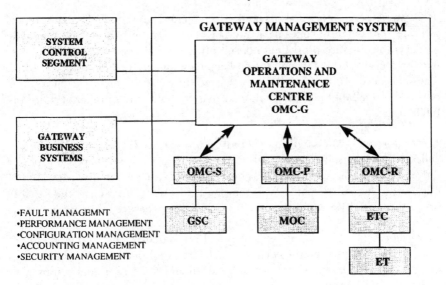

Figure 20.14   Iridium NMC architecture

through the public switched telephone network (thus maximising the use of existing, low cost communications services) or back through a satellite if the party is another Globalstar user.

A total of 56 Globalstar satellites will be placed into low-earth orbit, 48 of which will be operational, with eight on-orbit spares. The satellites will be placed in eight orbital planes of six satellites each with a 1414 kilometer circular orbit inclined at 52 degrees. The satellite is three-axis stabilised and consists of a trapezoidal main body (configured to facilitate the launching of multiple satellites on one space launch vehicle), two deployable solar arrays and a deployable magnetometer on a boom. Unlike many geosynchronous satellites, the antennas are not deployable. The satellite mass is approximately 450 kilograms and requires some 1100 W of power for normal operations. The satellites in the first-generation constellation are designed to operate at full performance for a minimum of seven and a half years.

Body: The satellite's trapezoidal shape, fabricated from a rigid aluminum honeycomb, is designed to conserve volume and facilitate the mounting of multiple satellites within the fairing of a space launch vehicle. Separation of the satellite is radially away from a central core dispenser on the launch vehicle. When the satellite is mounted within the fairing of the launch vehicle, the earth face is oriented outward and the antiearth face is mated to the control and communications subsystems.

Stabilisation: The satellite operates in a body-stabilised, three-axis attitude control mode. It is unique in that it is among the first of its kind to utilise the global positioning system to keep track of its orbital location and attitude. It also uses sun sensors, earth sensors and a magnetic sensor to help maintain its attitude. The satellite contains a number of momentum wheels and magnetic torquers to minimise propellant consumption for attitude control, and utilises five thrusters for orbit-raising, station-keeping manoeuvres and attitude control. The spacecraft's monopropellant hydrazine thrusters are fuelled from a single onboard propellant tank, sufficient to keep the satellite in its proper orbit for the full lifetime.

Electrical power: The satellite's primary power source is the energy of the sun collected by two solar arrays, although it utilises batteries to provide power during eclipses. The sun-tracking solar panels provide the satellite with 1100 W of power. These solar panels will automatically track the sun as the satellite orbits the earth, providing the maximum possible exposure to the sun's energy.

Communications: The heart of a Globalstar satellite, its communications systems, are mounted on the earth deck, the larger of the two rectangular faces on the satellite's body. There are C-band antennas for communications with gateways, and L- and S-band antennas for communications with user terminals. These antennas are of a phased-array design which projects a pattern of 16 spot beams onto the earth's surface, covering a service area or footprint of several thousand kilometers in diameter.

Globalstar frequency plan

The terminal-to-satellite link uses the old RDSS-band, L-band (1610–1626.5 MHz), and the satellite-to-terminal link also uses the RDSS-band, S-band (2483.5–2500 MHz). The gateway-to-satellite link uses the C-band (5091–5250 MHz) and the satellite-to-gateway link uses the C-band (6875–7055 MHz).

20.8 SPCN system comparisons

The previous sections have provided details on the various systems proposed for SPCN. This Section provides a summary comparison of the major characteristics of the SPCN systems. Table 20.16 lists the general characteristics of the SPCN system.

20.8.1 SPCN system services

The services expected of the SPCN network are shown in Figure 20.15 and comprise teleservices and bearer services. The supplementary services are a feature of the switch and support network and cannot be offered on their own. Tele and bearer services in an end-to-end connection are shown and defined below (Figure 20.15). Table 20.17 lists the services on offer from the various SPCN systems.

Table 20.16 SPCN, general characteristics

	Iridium	Globalstar	ICO	Aries	Teledesic
Main partner	Motorola	Loral Qualcomm	P11s	Constellation	AT&T, Motorola Microsoft
# satellites initial	66	4	6	48	288 + spares
# satellites final	66	48	10 + 2	48	288 + spares
Orbit height	789 km	1414 km	10 390 km	1020 km	700 km
sat/planes	11S/6P	I = 52, 6S/8P	6S/2P	12S/4P	
# ground stations			12		
Terminal class	P/M/F	P/M/F	P/M/F	M/PT	F/M
Technology status	new	mature	mature	new	conceptual

P: personal   M: mobile   F: fixed   PT: portable

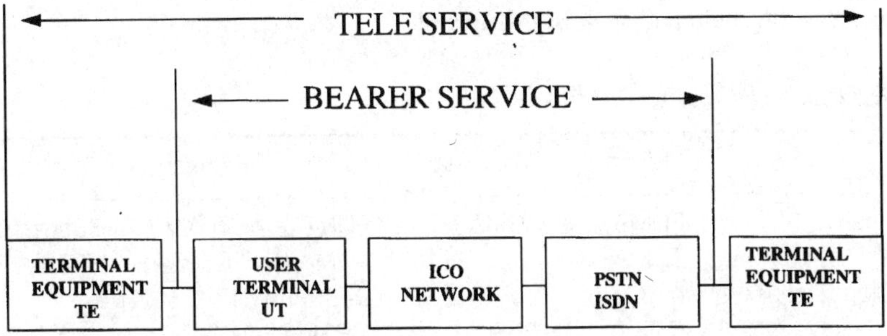

Figure 20.15   ICO tele and bearer connection

Bearer Services: Enable the transmission of signals between mobile users and the PSTN/ISDN user. Transport is defined by the lower three layers of the ISO model. Bearer supports transparent (no error correction) and nontransparent services and synchronous and asynchronous data transfer.

Teleservices: Enable application-type communications using standardised protocols. Includes all seven layers of the ISO model.

Supplementary services: Additional services, typically network provided and cannot be offered standalone.

20.8.2 SPCN access

Table 20.18 summarises the salient features of the proposed access techniques associated with the SPCN systems.

Table 20.17   Services on offer from SPCN systems

	Iridium	Globalstar	ICO	Aries	Teledesic
Service start	1998	1999	2000	1997	2001
Service offered	V/D/R/P	V/D/R	V/D/R/P	V/D/R	M/B
Service area	Global	+/−70 lat	global	global	global
Ch/satellite	2100	5600	4500	500	N/A
Min. elevation	10	10	20	5–10	40
Voice rate, bit/s	4800	600–9600	4800	4800	4800
Data rate, kbit/s	2.4 – 9.6	2.4	1.2 – 64.0	2.4	16 – 2048
BER voice	10^{-2}	10^{-2}	10^{-3}	10^{-3}	
BER data	10^{-3}	10^{-5}	10^{-5}	10^{-5}	
Delay ms, RT	10–33	43	70	25	6
Position accuracy, km	1.5	0.3–2.0	5		
Availability, %	90–95	90–96	90–95		high
Security	y	y	y		

V: voice   RT: round trip   %: user link   D: data   R: RDSS
P: paging   M: multimedia   B: broadband

Table 20.18   SPCN system access

	Iridium	Globalstar	ICO	Aries	Teledesic
Access	FDMA/ TDMA	CDMA	FDMA/ TDMA	TDM/ CDMA	Fast packet FDMA
Frequency reuse	4–5	4–6	3–7		
Modulation	QPSK	QPSK	QPSK	QPSK	
Coding data	2/3	1/3–1/2		3/4	

20.8.3 SPCN space-segment

Table 20.19 provides a summary of the features of the various space-segment options in the SPCN scenario.

20.8.4 SPCN satellite and payload

Table 20.20 lists the features of the SPCN satellites and the associated payloads. Note that some of these numbers are approximations as the exact details are not made public by the proponents.

20.8.5 Global mobile comparisons

A brief comparison of the major mobile systems is presented in Table 20.21.

Table 20.19   SPCN space segment

	Iridium	Globalstar	ICO	Aries	Teledesic
Satellite power, W	1429	875	7000	250	
Launch vehicle	CGWI, Proton	Pegasus/Ariane		Delta/Atlas	
Launches/yr	66	4/5	6/7	48	

* Power: bus and payload

Table 20.20   SPCN satellite and payload

	Iridium	Globalstar	ICO	Aries	Teledesic
Mass		60			
Power		800	2000		
Antenna type	phased array	phased array	phased array*2		phased array
Antenna coverage	48	16	163	7	576
OBP (on-board processing)	y	n	y	n	y
ISL	y	n	n	n	y

20.9 Conclusion

The growth in mobile cellular communication systems in the 1980s and 1990s has given rise to expectations of seamless global wireless communications and mobile satellite communication systems can easily fulfil that role. This can easily be noted from the multitude of systems which have been proposed, some of which will be operational as early as 1999. This Chapter has provided an overview of some of these systems. It is expected that these and other multimedia satellite communication systems (for fixed and mobile communications), requiring investments in the several tens of billions of dollars, will herald a revolution in the field of communications, especially that related to global mobility.

Table 20.21   SPCN system technical comparison

	Iridium	Globalstar	ICO	GEO-Mobile
# Satellites	66	48	10	1
Space segment approximate cost	high	low	medium	low
Satellite lifetime years	5	8	12	12
Terrestrial access, earth-station cost	high	high	high	medium
Handset cost	high	low	medium	medium
Per-minute cost	low	low	medium	medium
Propagation delay	low	low	medium	high
Voice quality	good	good	good	good
Service area	global	global	global	regional
Elevation-angle range	high	high, limited	high	fair, limited
Operations	complex	complex	complex	medium
Call hand offs	frequent	frequent	occasional	none
User corporation	unlikely	likely	likely	required
User convenience	low	low	medium	medium
Channel power flux	high	high	medium	medium
Onboard processing	yes	no	yes	yes
Visibility range	poor	poor	good	best
Global connectivity	ISL	PSTN	PSTN	PSTN
Mobile–mobile connectivity	ISL	PSTN	PSTN	OBP
Phased start up	no	no	limited	yes
Development time	medium	short	medium	short
Deployment time	long	long	medium	short
Technology risk	high	medium	medium	low
Feeder link	Ka	C	C	Ku
Mobile-link availability	high	high	medium	satisfactory
Feeder-link availability	high	high	high	high
Feeder-link bandwidth	small	large	small	n/medium
Global-frequency reuse	264	770	TBD	N/A
Frequency availability	yes	yes	pending	yes

20.10 References

1 FCC filing of Globalstar
2 FCC filing of Iridium
3 Globalstar FCC filing
4 ITU documentation on FPLMTS and IMT2000—TG8/1
5 ETSI documentation on UMTS, SMG5
6 Websites: http://www.Iridium.com, http://www.ico.com
  http://www.Globalstar.com
7 MOULEY and PUTET: 'The GSM system for mobile communications'
8 'GSM engineering' (Merhotra, Artech House)

Chapter 21

Satellite navigation: a brief introduction

I. E. Casewell

(with A. Batchelor, Racal Research Ltd.)

21.1 Introduction

The use of satellites as an aid to navigation can be traced back to early in 1959 when development of the Navy navigation satellite system was started in the United States. This system is based on the Doppler principle and became known as the TRANSIT system. It was introduced into service in 1964 and remains operational to this day. Around the same time, the USSR introduced a very similar system known as TSIKADA. The main limitation of Doppler systems is that they are only applicable to platforms with low dynamics, such as ships, because the time to obtain a fix is of the order of tens of minutes. For these reasons, systems based on transmission time-delay measurements have been developed. The US system is known as NAVSTAR/GPS and the Russian system is known as GLONASS. Both of these systems have been under development for fifteen or twenty years and have now been declared operational. The massive cost of developing a global, high-accuracy satellite navigation system has meant that such systems can only be afforded by the militaries of the USA and USSR.

Over recent years attempts by GEOSTAR and LOCSTAR to build a commercial system have only ended in bankruptcy. One regional and highly specialised system has been operated successfully in the Gulf of Mexico, for the benefit of the offshore oil industry, by John Chance and Associates. This system is known as STARFIX and uses capacity leased on three conventional C-band DOMSATs. Pseudorandom codes are used to modulate the carrier in a manner similar to GPS and a quasidifferential mode is used to provide an accuracy of

5 m (95%). The Qualcomm OmniTRACS system, and its European counter-part EUTELTRACS, also provides a crude position-reporting capability using its integral ranging subsystem. This is an active two-way ranging system using the TDM/TDMA communications carriers. The use of only two satellites and height aiding provides a horizontal positioning accuracy of about ±1 km.

During the 1980s, the European Space Agency carried out a number of studies under the name of NAVSAT. Initially, NAVSAT was intended to be a civil replacement for GPS. However, the high costs of developing a GPS-like system became prohibitive. As time passed, the NAVSAT system changed into a regional system using a combination of geostationary and highly-inclined satellite orbits and then eventually into a geostationary overlay to GPS and GLONASS. This overlay concept has been taken up by INMARSAT and a piggy-back navigation transponder has been procured as part of the INMARSAT III programme. The geostationary overlay provides a means of broadcasting integrity messages and additional ranging signals on a frequency very close to the GPS L1 frequency. This approach has now been adopted by the aviation community and operational GPS augmentation systems are currently being developed.

This brief introduction to the rapidly expanding field of satellite navigation must be limited to the explanation of the basic principles of, and a short description of, the NAVSTAR/GPS system. This will be preceded by a discussion of some of the ever increasing range of applications of GPS.

There are a number of books and specialist publications covering the field of GPS. The three books found most useful by the author are cited in the references section at the end of this Chapter. They contain extensive bibliographies and so provide an excellent starting point for new readers in the field.

21.2 Applications

Not surprisingly, the military sponsors of GPS envisaged that the major applications would be mainly military. Some twenty years after the initial plans were laid, GPS had found even wider application in a range of military and civil applications.

The military applications envisaged included general navigation, weapons targeting and mine countermeasures. The general navigation of a wide range of vehicles, from tanks to ships to jet fighters, clearly provides the accurate position information vital for efficient command, control, communications and intelligence (C^3I) aspects of military operations. This was amply illustrated during the Gulf War; the availability of a common, highly accurate navigation system to all sections of the allied forces proved to be a significant factor in achieving their objectives in the largely featureless terrain found in the Persian Gulf. The GPS system is also a useful aid to weapons targeting. For example, the position of artillery pieces must be known accurately for precision targeting to be achieved.

More recently, attitude-determination techniques using GPS signals have been developed to provide accurate aiming information for artillery. GPS receivers are also used, along with other sensors, to guide missiles to their targets. The high accuracy capabilities of GPS are invaluable during mine countermeasures operations and during other cleaning up operations once the war is over.

Civil applications also include general navigation for the whole range of civil vehicles, from the luxury car to the super tanker. The use of GPS is rapidly establishing itself in the field of air navigation; initially as a sensor in automatic dependant surveillance (ADS) systems for aiding air traffic control over oceanic or other remote regions. More recently GPS has become a candidate for the future precision landing aid. The FAA, and more recently Eurocontrol, are studying and developing GPS-based solutions for all phases of flight including using them as a CAT III landing aid with the objective of replacing the microwave landing system (MLS). The requirements of accuracy, integrity and availability demanded by an air navigation system are the key issues currently being tackled by many researchers in this area. Many technical solutions are being studied. The extent to which GPS will encroach upon the preserve of ILS and MLS will be fiercely contested by the various lobbies concerned over the next few years.

The use of GPS as a sensor in automatic dependent surveillance systems for aircraft has its parallels in both the maritime and land sectors. In the maritime case, automatic position reporting for fleet and vessel traffic monitoring is becoming commonplace. Similar systems for land use are also now being introduced. At first they were used entirely to track dangerous or valuable cargoes, but now they are also being introduced into fleet management and customer information systems. Several such systems are being introduced by public transport authorities in a number of cities in both the USA and Europe.

The offshore oil industry has many requirements for accurate position determination such as seismic surveying, rig positioning and pipe laying. In each of these applications GPS is rapidly becoming accepted. In order to provide the real-time accuracy demanded by the oil industry, it is commonly operated in a differential mode. Many companies now compete to provide a differential corrections service in all of the major regions of offshore activity. Differential GPS (DGPS) is also used as a positioning aid for navigation in coastal areas, harbour approach and dredging operations. With the introduction of digital maps, DGPS is starting to find application in the creation of data for use in geographical information systems (GIS).

Land surveyors and geodesists were quick to realise the potential of GPS for precision surveying. Following many years of development, the use of the carrier phase observable has been perfected to such an extent that an accuracy of one part in 10^7 over baselines of 1000 km is now routinely obtained. State-of-the-art techniques now allow intercontinental baselines to be measured with an accuracy of one part in 10^8. Normally these techniques operate in post-processing mode, using double difference observables, but much research is currently being directed at extending these techniques to allow them to be operated in real time on mobile platforms. Such techniques require ambiguity resolution to be carried out on the fly.

In addition to the ever increasing range of navigation and surveying applications of GPS, the other envisaged application is time transfer. In its basic form GPS allows the distribution of time accurate to within $1\,\mu s$ relative to UTC. Although this may be more than adequate for normal scientific and engineering applications, the international time-keeping community needs to measure the time difference between national time standards to within 10 ns. This can be achieved using the common-view-technique provided accurate satellite positions, antenna positions and estimates of propagation delays are available.

21.3 Fundamentals of time ranging

21.3.1 Basic concept

GPS is a passive time-ranging system in which the time taken for a signal to travel between a satellite and the receiver is measured. This transit time is then easily related to the distance between the satellite and the receiver by the velocity of light. The three-dimensional position of the receiver (x, y, z) can be calculated if the absolute transit time, and hence distance from three satellites, at known positions is measured.

In practice, the measurement of absolute transit time is not possible in a passive system, hence it is necessary to measure the transit time relative to a local time reference at the receiver. This local time reference will be offset from GPS time by an unknown amount and is known as the receiver clock offset. The product of the velocity of light and the transit time relative to the receiver clock is known as the pseudorange. As the receiver clock offset is common to all transit-time measurements, by taking a total of four pseudorange measurements between four GPS satellites and the receiver it becomes possible to solve for the position (x, y, z) and the receiver clock offset. Thus, this simple concept has the capability of both allowing the three-dimensional position of an unknown point to be determined and the dissemination of GPS time. This is only possible provided the receiver has an accurate knowledge of the position of the four satellites and the time offsets between each satellite and GPS time. The estimation of these parameters is the key function of the GPS control segment. Once this information has been determined it is uploaded to the satellites and then broadcast to the users in the navigation message.

21.3.2 The navigation solution

The preceding qualitative discussion introduces the concept of pseudorange and the solution of the receiver's position. More formally, the pseudorange between the ith satellite and jth receiver (p_j^i) can be written as:

$$p_j^i = \rho_j^i + (t - t_j)c$$

where ρ_j^i is the true range, t is GPS time, t_j is the receiver time and c is the velocity of light. The true range is given by:

$$\rho_j^i = \sqrt{\left(x^i - x_j\right)^2 + \left(y^i - y_j\right)^2 + \left(z^i - z_j\right)^2}$$

where x^i, y^i and the z^i are the Cartesian coordinates of the ith satellite and x_j, y_j and z_j are the Cartesian coordinates of the jth receiver in an earth-centred earth-fixed (ECEF) coordinate reference system.

Hence, when there are four satellites, $i = 1$, 2, 3 and 4, in view from receiver j, a series of four simultaneous equations can be written:

$$\rho_j^1 = \sqrt{\left(x^1 - x_j\right)^2 + \left(y^1 - y_j\right)^2 + \left(z^1 - z_j\right)^2} + \left(t - t_j\right)c$$

$$\rho_j^2 = \sqrt{\left(x^2 - x_j\right)^2 + \left(y^2 - y_j\right)^2 + \left(z^2 - z_j\right)^2} + \left(t - t_j\right)c$$

$$\rho_j^3 = \sqrt{\left(y^3 - x_j\right)^2 + \left(y^3 - y_j\right)^2 + \left(z^3 - z_j\right)^2} + \left(t - t_j\right)c$$

$$\rho_j^4 = \sqrt{\left(x^4 - x_j\right)^2 + \left(y^4 - y_j\right)^2 + \left(z^4 - z_j\right)^2} + \left(t - t_j\right)c$$

These four equations may be solved iteratively, or algebraically, for x_j, y_j, z_j and $t - t_j$. A commonly-used iterative method is the variation of coordinates technique. As in practice, the time of each satellite is slightly offset from GPS time, the observed pseudorange is corrected with this offset to reference the pseudorange back to GPS time. This satellite clock offset is determined by the control segment and broadcast to the users.

In reality, the ionosphere and the troposphere cause additional small delays. The effects of these delays are minimised by compensation of the pseudoranges using values predicted by suitable models of the atmosphere. In the case of the ionospheric delay, parameters of the model used to predict the zenith delay are broadcast in the navigation message. For the tropospheric delay, standard values of temperature and pressure can be used to estimate the zenith. Finally, the use of obliquity models allows the estimation of each delay for the specific elevation angle.

21.3.3 *Spread-spectrum ranging*

Spread-spectrum ranging is the name given to the method by which the pseudorange is measured in the GPS. The basic principle revolves around the properties of the idealised autocorrelation function of a pseudorandom sequence, shown in Figure 21.1. It can be seen that when the offset time $\tau = 0$ chip periods, i.e. when the codes are perfectly aligned, the value of the autocorrelation function is one. The term chip period refers to the duration of a single bit in the pseudorandom sequence. The magnitude of the autocorrelation function linearly decreases to zero as the offset time approaches $\tau = \pm 1$ chip periods. At

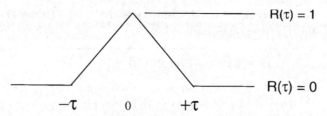

Figure 21.1   Idealised autocorrelation function of a pseudorandom sequence

all other values of τ the autocorrelation is approximately zero. Thus it is possible to determine when a received pseudorandom sequence is synchronised with a locally-generated replica to within a small fraction of a chip duration. Having performed the synchronisation of a local code with the received code, it is simple matter to determine the pseudorange by measuring the time at which the replica code starts relative to the receiver clock. The precision of the pseudorange measurement is dependent on the signal-to-noise ratio in the code tracking loop. The number of chips in the sequence determines the largest time difference which can be unambiguously measured, and so long codes are desirable in a navigation system.

The crosscorrelation function between two codes of the same family generally approaches zero and so signals sharing the same frequency band, but using different codes, can be separated. This is the basis behind the concept of code-division multiple access (CDMA) scheme used by GPS.

21.4 System description of GPS

This Section will present an overview of GPS. The system consists of three key segments, the space segment, the control segment and the user segment. Each of these will be addressed below. These descriptions will be followed by some details of the navigation message and a discussion on system accuracy.

21.4.1 Space segment

The space segment consists of a constellation of 21 operational plus three active spare satellites. The orbits are circular and are inclined, in the case of the operational Block II satellites, at an angle of 55°. The developmental Block I satellites were placed in orbits with an inclination of 63°. The orbits are sun synchronous, having a period of 11 hours 58 minutes which corresponds to a height of 20 200 km. The satellites are placed into six planes, each separated by 60°, and each plane will eventually contain four satellites, as illustrated in Figure 21.2. During December 1993, the US DoD declared that an initial operational constellation (IOC) of 24 satellites had been established. This constellation consisted of a few remaining developmental Block I satellites and a balance of

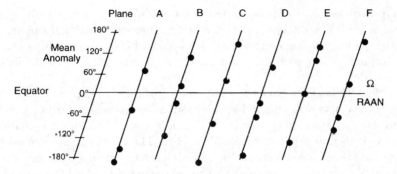

Figure 21.2 The 21 optimal satellite constellation

satellites making a total of 24 satellites. Today, all of the old Block I satellites have been replaced with Block II satellites.

The Block II satellites each weigh about 1667 kg and are designed to last at least seven and a half years. They are three-axis stabilised and the solar arrays provide about 600 W of power. The navigation signals are transmitted on two frequencies in L-band via an array of helical antennas. The satellite TT&C is carried out at S-band. The heart of the navigation payload is a group of two caesium and two rubidium clocks running at 10.23 MHz.

All of the key frequencies in the transmitted navigation signals are derived from this master frequency, as shown in Figure 21.3. The L1 signal is transmitted at 1575.42 MHz and is modulated with a 1.023 Mchip C/A-code signal and a 10.23 Mchip P-code signal in phase quadrature. The C/A code is 1023 chips long and therefore repeats every millisecond, and the P code is 2.35×10^{14} chips long and is restarted each week. A 50 bit/s BPSK inner modulation is used to convey the navigation messages. Each satellite transmits on the same frequencies, but each is assigned a different PRN sequence; the C/A-code EIRP is 14.25 dBW

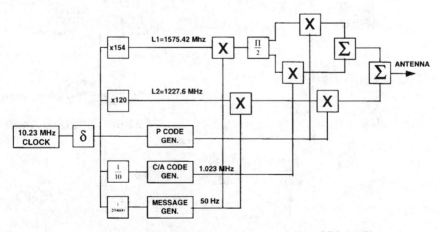

Figure 21.3 Block diagram of signal-generation system within a GPS satellite

which provides a received signal power of $-160\,\text{dBW}$ when a $0\,\text{dBi}$ gain antenna is used. The L2 signal, at 1227.6 MHz, is normally modulated by the P code and the inner navigation message. It enables a P-code receiver to estimate the ionospheric delay with greater accuracy than that provided by the ionospheric model.

The Block II satellites have the additional capability to phase modulate the output of the clocks with a noise-like jitter so that the accuracy of the C/A-code derived standard positioning service (SPS) can be controlled. This degradation process is known as selective availability (SA). The P-code signal is used to provide the precise positioning service (PPS). Through the optional use of the encrypted Y code, the use of the PPS can be restricted to allied military and other government agencies. These agencies have the means to counter the effects of SA.

21.4.2 Control segment

The control segment consists of the master control station located in Colorado Springs and a number of monitor stations and uplink stations (see Figure 21.4). The control segment carries out all of the necessary tasks needed to manage the operation of the GPS; the main tasks of interest here are the determination of the satellite orbits and the individual satellite clock offsets from the master clock at Colorado Springs.

The monitor stations are located at Colorado Springs, Kwajalein, Diego Garcia, Ascension Island and Hawaii, and they track each of the satellites in

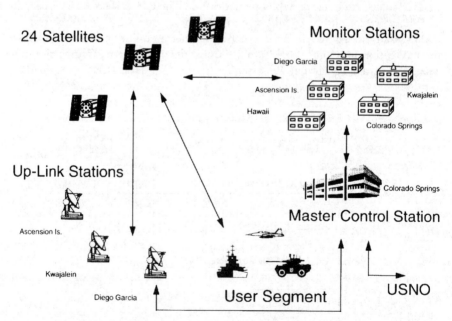

Figure 21.4   GPS configuration

view and take pseudorange measurements which are transmitted back to the MCS. At the MCS, these observations are used to compute the satellite orbit and clock offsets which, along with other information, is formatted into the navigation message. The navigation message is then periodically up loaded to the satellite via one of the S-band uplink stations located in Kwajalein, Diego Garcia and Ascension Island.

The satellite orbits are determined relative to the WGS-84 coordinate system. The offset between GPS time is maintained within ± 180 ns (95%) of USNO UTC. This offset is included within the navigation message to enable the dissemination of time relative to UTC to take place.

21.4.3 User segment

The user segment consists of the various types of user receiver needed to satisfy the wide range of applications addressed in Section 21.2 of this Chapter. In the civil community, a wide range of receivers is available from small handheld C/A-code receivers retailing for under £200 to complex receivers intended for geodetic surveying applications costing upwards of £20 000. Today's receivers employ as much digital processing as possible. They generally downcovert the spread-spectrum signals to complex baseband or very low IF where they are sampled and converted to the digital domain. All the remaining processes are carried out digitally. Even the smallest receivers now process five or six C/A-code signals in parallel. Position estimates are normally output at one-second intervals.

21.4.4 Navigation message

The navigation message contains a large amount of information which is needed to compute an accurate position and to define the status of the system. The data is transmitted at 50 bit/s and has a total length of 37 500 bits. The data is divided into 25 frames which are in turn divided into five subframes of 300 bits each. Each subframe consists of ten 30-bit words containing up to 24 information bits and six parity bits derived using a systematic BCH code. The whole message takes 12.5 minutes to transmit. The data is split into three blocks, Blocks I and II are transmitted at 30-second intervals and Block III messages at 12.5-minute intervals.

The Block I data mainly contains the clock corrections relating to the particular satellite and the ionospheric corrections transmitted in the form of the Klobuchar model. Block II data contains the particular satellite orbit parameters (ephemeris) expressed in the form of an extended set of Keplerian parameters. The final block, Block III, contains coarse ephemerides and clock correction parameters of all the satellites; this data is known as the almanac and is used by the receiver to assist with the faster acquisition of subsequent satellites after the first has been acquired. The almanac data may also be used in satellite visibility calculations for scheduling observations. This data is also used to steer a P-code receiver to the correct part of the very long PRN code.

21.4.5 *System accuracy*

The US DoD declared system accuracy (95% or 2d r.m.s.) provided by GPS is:

$$\begin{array}{lll} \text{SPS}: & 100\,\text{m} & \text{horizontal} \\ & 140\,\text{m} & \text{vertical} \\ \\ \text{PPS}: & 18\,\text{m} & \text{horizontal} \\ & 27\,\text{m} & \text{vertical} \end{array}$$

As, in practice, the undegraded C/A code has been found to provide a horizontal accuracy of about 36 m (95%), the DoD has introduced the capability to apply deliberate degradation known as selective availability (SA). The accuracy of the SPS when SA is employed can be varied, but is generally set to provide the declared accuracy. The users of the PPS have the necessary information to remove the effects of SA and remove the Y-code encryption. Selective availability can take the form of clock dither, as described in Section 21.4.1, known as δ. In addition, further degradation known as ε can be applied. The ε degradation results from the ability to degrade the ephemeris and clock offset information transmitted in the navigation message to provide very low-frequency variations in position.

The accuracy of the system depends on a number of factors; an accuracy budget is shown in Table 21.1 (after Forssell). The position accuracy is related to the receiver pseudorange error (UERE) and the dilution of precision (DOP), which takes account of the effect of the satellite geometry as seen by the receiver. There are several types of DOP depending on the context. For example, GDOP relates to x, y, z and t, and HDOP just relates to horizontal position, i.e. x and y. When the full constellation is operational, the worst-case HDOP will be about 1.5 in the majority of receiver locations. The lowest DOP is obtained when one satellite is at a high elevation, and the remaining three are at low elevations and evenly spaced in azimuth. A modern receiver with more than four channels will observe as many satellites as possible to determine its position using a weighted least-squares process; this has the effect of reducing the DOP further. In some applications where the height is known by another means, the process known as height aiding may be used to further improve the position solution.

The effects of SA can be overcome by the application of differential navigation techniques where a reference station at a known position observes the GPS satellites and transmits pseudorange corrections to users in the vicinity. Provided the separation between a user and the reference stations is not too great, SA and other errors cancel allowing an accuracy of about 5 m (2d r.m.s.) to be obtained. The use of differential GPS is now common in the offshore oil industry.

Even greater accuracies are possible if the C/A code observable is replaced by the carrier phase observable. The carrier phase observable provides noise levels of the order of 1 cm r.m.s. and is significantly less susceptible to multipath distortion.

Table 21.1   Static error budget for GPS (no SA)

Error source	Expected measurement range error (m r.m.s.)	
	P-code	C/A-code
Satellite clock errors	1–3	1–3
Ephemeris errors	2.5–7	2.5–7
Ionospheric delay	0.4–2	2–15
Tropospheric delay	0.4–2	0.4–2
Receiver noise and quantisation errors	0.1–0.3	1–2
Channel biases	0–0.1	0–0.2
Multipath distortion	1–2	2–4
Resulting UERE	3–8	4–18
Resulting position error		
horizontally (HDOP = 1.5)	4.5–12	6–27
vertically (VDOP = 2.5)	7.5–20	10–45

The major difficulty is that the ambiguity distance is about 18 cm at L1 frequencies. The surveying community has developed complex double-difference methods to resolve these ambiguities. Static relative positioning can be performed with an accuracy of about 1×10^7 of the baseline length. The research community is now developing these carrier-phase techniques to enable the measurements to be made on a moving platform. These techniques are known as kinematic carrier-phase GPS.

21.5 The future of global positioning systems

21.5.1 The twentieth century

Both GPS and GLONASS were developed, and have been maintained, under military control. However, in the early 1990s, the number of civilian users of these systems, particularly GPS, fast began to overtake the military users. As we approach the turn of the century civil markets are beginning to take precedence. Certain civilian applications require better accuracy, integrity, continuity and availability than is achievable with the standalone GPS and GLONASS systems. An obvious example is the international civil-aviation community, which stipulates stringent precision-approach specifications for all phases of flight. Without augmenting the GPS and GLONASS ground infrastructure and applying differential techniques, the most demanding integrity and accuracy requirements for category III precision-approach landing cannot be met.

There is a clear need for a global navigation satellite system that satisfies commercial requirements in terms of performance and safety considerations. If such a system is proven viable, it could lead to a second-generation system. Indeed, the European Space Agency (ESA) is already actively pursuing research and development programmes for such a system, GNSS2. Ideally, this system will be seamless, international and under civilian control (it would also need to be available to the military and to be deniable to some users in times of conflict).

21.5.2 The twenty-first century

GPS and GLONASS work exceedingly well in the rural regions for positioning activities such as surveying. Recently, the development of in-car navigation facilities has enabled better than ten-metre accuracy to be obtained using local-area differential augmentations. Future markets will require even greater accuracy and more raid updating of the position. In essence, the emphasis of applications will shift from global positioning to global navigation. Assuming ten-metre accuracy in the 1980s and an order of magnitude improvement in accuracy every decade, it is anticipated that requirements for navigation will reach the ten-millimetre level by the year 2010. Of key importance will be the safety-critical markets for applications involving aircraft, rail and automated highways. A revolutionary approach to navigation is expected for science and space applications.

In the 1980s it was possible to disseminate time to system users at the 20 ns level. We might, therefore, expect that it should be possible to disseminate time to system users at the 20 ps level in 2010. It is clear that the need for better time synchronisation will become increasingly pressing for applications such as global time and frequency dissemination for digital network synchronisation and global mobile communications. Given the reliability, a global synchronisation network in space may prove to be more accurate and more economical to maintain than one on the ground.

Let us assume that the next-generation GNSS is to be launched around the year 2010. Then the system architecture would need to be able to deliver a navigational accuracy of around ten millimetres and a time-transfer capability of approximately 30 ps, if it is to stand a good chance of meeting industry's future requirements. Already, theoretical and experimental studies are beginning to indicate that this type of performance may well be achievable[4].

However, if such systems are to be implemented, there will have to be significant markets, either extant or emerging, to support the investment needed to establish the satellite network and operate such a global service with sufficient integrity and accuracy to meet all requirements. More work is needed on understanding how system accuracy and reliability are affected by propagation and terrain effects before many potential market applications can be confirmed.

Three are currently a number of augmentations to GPS and GLONASS under development. For example, the US has its wide area augmentation system (WAAS) and in Europe there is the European geostationary navigation overlay system (EGNOS). These systems are aimed directly at supporting current navigation performance requirements and are driven primarily by aviation requirements. However, the long-term goal is towards sole-means category III precision approach for landing and in the terminal area. This requires a vertical accuracy of between 0.6 m and 0.8 m and, more importantly, an integrity which measures the probability of losing the system once an approach manoeuvre has started as 3.3×10^{-7}.

Further evidence for increased accuracy requirements comes from intelligent vehicle highway systems, which require position accuracy of the order of five metres. However, ultimately automated highway systems may require navigational accuracies of better than 0.05 m with a 0.1 s update rate[5]. Requirements for railway and maritime communities are also projected to be high; rail services might require an accuracy of 0.1 m at junctions and terminal areas[6]. For the maritime community, harbour-approach manoeuvres could require accuracies of 0.05 m. The surveying community is already asking for 0.001 m, with much less postprocessing of data. There are many other examples.

The bandwidth for the transport of navigation data to a system user is very small compared to that for communications data. For instance, in Japan, digital telecommunications networks using fibre-optic cables and semiconductor diode lasers for signal detection and amplification are synchronised at the nanosecond level. It is predicted that the next generation systems will require time synchronisation at the ten picosecond level! The rapid development of global mobile digital communications networks will lead to ever increasing demands for greater bandwidth and synchronisation performance.

Ultimately, the demand for a truly global time scale disseminated from space will drive the need for synchronisation to the ten picosecond level. Today time is distributed through national standards laboratories from ground-to-space-to-ground. Common-view and two-way satellite time transfer are used. However, eventually it could be strictly from space-to-ground. There will be a need to disseminate time accurately on a global basis. The ability to monitor the earth's surface at the ten millimetre level using Doppler ranging could become a reality.

21.6 References

1 FORSSELL, B.: 'Radionavigation systems' (Prentice Hall International (UK) Ltd., 1991)
2 LEICK, A.: 'GPS satellite surveying' (John Wiley & Sons Inc.)
3 KAPLAN, E. et al: 'Understanding GPS, principles and applications' (Artech House Inc, 1996)
4 ALLEN, D.W., and ASHBY, N.: 'Navigation and timing accuracy at 30 centimetre and sub-nanosecond level'. Proceedings of ION GPS-96, September 17–20, 1996
5 SINKO, J.W.: 'An evolutionary automated highway concept based on GPS'. Proceedings of ION GPS-96, September 17–20, 1996
6 DE LA FUENTE, C., and PRESTON, K.: 'The use of GPS on British Rail'. Proceedings of ION GPS-93, Salt Lake City, Utah, USA, September 22–24 1993, pp. 1631–1636

Chapter 22

VSATs for business systems

R. Heron

22.1 VSATs for business systems

VSATs, very small aperture terminals, are small earth stations capable of receiving from and sometimes transmitting to satellites. They represent an important addition to the telecommunications world because they can provide a service directly to the user at virtually any geographic location covered by a suitable satellite beam. They do not require any support from a local terrestrial communications network and can even be run from portable or alternative power supplies.

The term VSAT appears to have been coined around 1979 and was first used in connection with the systems offered by Equatorial in the USA. These early systems operated at C-band and represented a radical departure from the normal usage of satellite communications links.

VSATs can be used individually or, more usually, in networks of related users and they are often used in conjunction with a hub station. A hub station is usually a larger earth station at the centre of a star network. The small size of VSATs brings with it special system and equipment design problems, both for the VSAT itself and for the hub to which it operates. These problems are discussed later in this Chapter. The deregulated telecommunications policy in the USA meant that VSATs could be used at an early stage. In Europe, regulation prevented their use for a number of years. However, there are now many VSAT systems in Europe and the UK lottery is a well known example of a network which uses approximately 2500 VSATs.

Although the term VSAT is generally used in connection with small, fixed-location terminals for business use, there are comparable developments in related fields. For example, low-cost TVROs could offer VSAT-type services at higher data rates and lower cost than traditional designs. The various INMARSAT systems can provide different types of voice and data service anywhere in the world and are widely used by people who need to stay in touch.

The communications scenario has been changing in Europe since the initial UK Green Paper on the development of the common market for telecommunication services and equipment of June 1987. Most countries have liberalised their internal markets and have taken steps towards opening up their telecommunications sectors to foreign competition. These policies have had a major effect on the use of VSATs, SNG terminals and other satellite-related services.

The small size of a VSAT is the key to understanding its importance. Before the advent of VSATs, satellite earth stations were large, expensive facilities run by national PTTs and large corporations. Antennas ranged in size from 30 metres down to around ten metres. Such facilities are, to this day, rather expensive to procure and expensive to run.

A VSAT, in contrast, is relatively cheap to procure, can be installed more or less anywhere and has low, predictable running costs.

Thus, over the last ten years, VSATs have become a tool for business and a means of bypassing existing terrestrial infrastructure. This has been particularly true in the USA where the liberalised regulatory environment allowed people to use VSATs at a very early stage. In the rest of the world, the situation was quite different. In the 1970s, most countries still regarded telecommunications as a state monopoly and very little use of VSATs was allowed.

Things have moved on. We are now moving to a phase where VSATs can almost be considered as commodities. VSAT vendors are now marketing products suitable for any business and we are on the verge of seeing these products in use by the private individual.

22.2 Applications of VSAT systems

22.2.1 One-way and two-way VSATs

VSATs are frequently classified as either one-way or two-way devices. One-way VSATs can receive but not transmit; two-way VSATs can receive and transmit. In the broadest sense, the term one-way VSAT includes the data terminals and TVROs of different types.

There is no inherent limit on the transmit capability of a VSAT but two-way VSATs usually transmit at rates up to a couple of hundred kbit/s. The low directivity of VSAT antennas is a practical limit on the power, and hence the boresight EIRP, which may be transmitted. The transmit equipment is usually completely solid state and highly integrated.

22.2.2 Data distribution

There are many kinds of system where data originates consistently at one source and must be transmitted to many sinks. One example is a library, which has a store of documents (an archive), some of which are copied to users spread

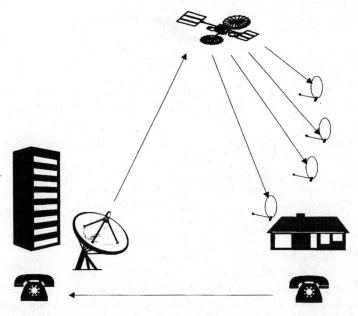

Figure 22.1   Data distribution

over a wide area. The data flow is primarily one way. The user can request a copy of a document using a few characters (e.g. a British Library request form) but the document itself may be very large. This is illustrated in Figure 22.1.

Conventionally, paper copies are posted to the person making the request; however they could be transmitted by satellite to a VSAT. Of course they could also be transmitted by facsimile or by telex, and often are, but the inherently-wide bandwidth of a satellite transponder is ideally suited to this kind of bulk data transfer. Another common situation concerns the transfer of data between computers. For many years some research establishments have transferred computer tapes by road on a regular basis. The transmission of such data by telephone link using modems is certainly possible, but at a transmission rate of a few kilobits per second the transfer time is unacceptably long. However, on a satellite link, many megabits of data can be transferred in a few seconds, with a quality of transmission and a tariff which are almost independent of location.

Another example of this type of traffic is seen on the Internet. A search request requires only a few bytes of data, but the response can be thousands of document references. Even to download one or two of the documents can require very consid-erable data transfer to the user. The Hughes DirecPC^{TM} system and other systems can be applied to this situation and provide the user with satellite-based Internet and much higher data rates than dial up customers would normally receive.

22.2.3 Rural applications

In some countries the terrestrial infrastructure is still underdeveloped and access to telephone lines is scare. Some estimates suggest that over half the

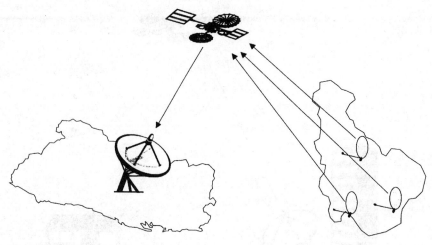

Figure 22.2   Rural applications

world's population has still never made a single telephone call! VSATs offer a means of providing the necessary communications capability without the massive financial investment which is normally required for a terrestrial infrastructure. In India, a network called **AGROMET** has been used to collect and distribute both agricultural and meteorological data. This system uses VSATs with the capability to transmit information from rural areas to a central location (Figure 22.2) where the data could be processed; this may bring significant benefits to a large, widely dispersed, rural community.

22.2.4 Disaster monitoring

There is sometimes a need to provide reliable communications to remote sites for the emergency services; this can be readily achieved using VSATs almost regardless of the terrain and location. In Florida, USA, Hughes has provided an emergency management system of 106 sites. The availability of the remote terminals is quoted as 99.887% for September 1995 to December 1996.[1]

22.2.5 Business applications

In the business world VSATs are of interest because of their potential for providing companies with a competitive edge. International agencies, such as INTELSAT, EUTELSAT, INMARSAT and ESA, have all undertaken extensive research in connection with VSATs and offer services or technology. Businesses can use two-way VSATs to provide a range of services both internal and external. The major motor companies, such as GM, Renault and VW, all have VSAT systems. Such systems are used for parts ordering, stock control and many other purposes.

A study by Logica for the Commission of the European Communities (CEC) forecast an annual market size of US$351 million by the year 2000 for two-way VSATs. Although at present there are many thousands of one-way VSATs in operation in Europe, there are only a few thousand two-way VSATs. In general, businesses want to establish private networks so that they can link their locations and move their information about in a safe manner and for the lowest possible cost. In the USA there are many such networks, some of which are extended beyond national boundaries and span the globe. The applications are numerous, some examples are:

- supermarkets;
- airline reservations;
- car rental companies;
- conference facilities;
- insurance companies;
- newsgathering;
- law firms;
- motor manufacturers.

22.3 Regulatory issues

22.3.1 Licencing

Despite deregulation of the markets, it is still necessary and important that VSATs are operated in a manner which does not cause excessive interference to other users. To this end, VSATs in Europe must be licensed with the appropriate national radiocommunication agency.

There are two main issues:

(i)   The VSAT must not interfere with other users.
(ii) The VSAT must comply with the satellite operators' requirements.

The broad conditions under which all satellite equipment operate are laid down by the ITU. The ITU is a body of the United Nations and it operates through the World Radio Conferences which take place every two years. In the UK, the Radiocommunications Agency (RA) is responsible for interpreting and implementing the regulations. The RA also uses the telecommunications acts which embody the relevant UK legislation.

European directives now require member states to deregulate all telecommunications services, including satellite services. Telephony was excluded from some of the legislation but member states, with the exception of Spain, Ireland, Greece and Portugal, had liberalised voice in January 1998. However, all states are required to be fully open to competition by 2003.[2]

22.3.2 *Frequency*

The electromagnetic spectrum is a precious resource and must be used wisely. One of the ways in which this is done is by splitting the spectrum into bands which can be used for similar purposes. The main split of spectrum is:

(i)   L-band is for mobile applications (primarily the INMARSAT system) and also for forthcoming satellite cellular systems such as Iridium, GlobalStar etc.

(ii)   C-band is in principle for large and small fixed services. It is very crowded but still important where propagation losses are high. Some terrestrial systems plan to use C-band for cellular telephony; an example of this is the Freedom system.

(iii)   X-band is used for military systems.

(iv)   Ku-band has been used for most VSAT systems to date.

(v)   Ka-band is emerging as a likely band for future systems, but high rain losses have always made it less attractive.

22.3.3 *Performance specifications*

The European Union has attempted to produce a set of standard specifications for VSATs. The standards are produced by ETSI.

Each satellite operator produces its own standards for both receive and transmit performance and it is necessary to comply with these standards to be allowed to use a satellite system. VSAT vendors seek to have their equipment type approved so that they can install and operate a VSAT with the minimum of formality.

22.4 Service provision

22.4.1 *Procuring a VSAT system*

Most people, and companies, like to deal with a one-stop shop when procuring satellite services. The customer wants a ready-made solution which includes: equipment supply and installation, approvals (satellite operator, local planning authority, RA), safety legislation (EMC, low voltage directive etc.), network services (routers, interface standards, management, disaster recovery etc.).

In exactly the same way as we expect to procure a telephone, plug it in and find that it works (in Europe at least), customers want to be able to do that with VSAT systems as well. The fact that it's a satellite system is irrelevant to the user; he or she simply wants a system that works, with well defined performance and price parameters.

This sort of service is available from several different vendors.

22.4.2 Vendor-provided services

Most service providers are actually equipment manufacturers as well. After a period of development and growth, the main equipment providers are now almost all large corporations with vertical integration of all the resources necessary to provide a VSAT system. Some of them build and launch the satellites, build the VSATs and hubs (where appropriate) and provide all of the network equipment and application software to run on them.

They have hub stations which are shared so that a user does not have to provide any of the infrastructure and only pays for the part of it that he uses.

Each vendor has built equipment to its own standards and tried to make it universally acceptable. Much of this equipment is highly versatile and can support many different types of interface.

22.4.3 Private services

Some large companies have built their own networks for intracompany use and started to sell spare capacity to other companies. Shared hubs have evolved in this way.

22.4.4 Service features

A service provider has to produce a service with desirable features. Some of the features which are required are:

- a single point of contact for the user;
- the provision of suitable equipment;
- the procurement of space segment;
- installation, commissioning and maintenance services;
- the integration of the system with customer-interface equipment;
- network-management services;
- local regulatory approval;
- disaster-recovery services.

22.5 Voice and data

22.5.1 Speech

Speech can be transmitted via VSAT if it is suitably encoded to reduce the data rate required. High-quality speech can be carried on satellite channels with a data rate of 9.6 kbit/s using modern voice encoders and modems capable of operating at low C/N values. As work on speech encoding progresses the data rate required for high-quality speech will continue to fall. Systems operating at 4.8 kbit/s are now in use. It is important that the VSAT data protocol is designed for such services as otherwise its data-framing mechanism may prevent the voice encoders from functioning correctly. Modern VSAT systems are designed to integrate voice and data.

A speech circuit set up between two VSATs via a hub will include a double hop and accordingly an end-to-end delay of at least 0.5 s. Outlying areas want to connect to the cities, both for speech and data, and a hub located in a city can provide both services to many villages. The provision of speech circuits to a rural community can strongly affect its economic development. Although the delay can exceed the ITU recommendations for speech circuits, the provision of good quality echo compensation can render the service acceptable. An ITU report on the subject estimated the ratio of benefit to cost in third-world countries to be between 85 : 1 and 200 : 1.

22.5.2 Bulk data

Bulk-data transfer may require large files, databases for example, to be transmitted with very low error rates. Some applications will tolerate no errors at all. The transfer time for such data is often not critical so that it can be transmitted with low priority or in the quiet hours.

At the same time, it is always desirable to minimise transmission times.

22.5.3 Transactional data

Some activities are transactional in nature. A classic example of this is credit-card verification and associated activities: cash balance, cash withdrawal etc.

The volume of data in both directions is small, the data must be absolutely error free and the transaction must be as quick as possible. Nobody wants to stand by a cash machine for ages waiting for a response.

Other types of data in this category include airline reservation systems and booking systems in general. In these systems also, errors cannot be tolerated.

22.5.4 Store and forward

Other data may fall into an intermediate category where the message size is not large and time is not critical; nevertheless delivery within a few tens of minutes is desirable. E-mail could be considered in this category.

22.6 Example systems

22.6.1 Hughes Olivetti Telecom

Hughes, like the other major players in this field, has a range of systems for different applications. See the company's web site for the latest data. The DirecPC system provides Internet access at rates of up to 2 Mbit/s; Turbo Internet access service is said to cost from $16 to $40 per month; the HOT (Hughes Olivetti Telecom) system can transmit a 675 Mbyte CDROM over a satellite in under 30 minutes and provide a constant data stream to an individual at 400 kbit/s.[3]

Market data indicates that Hughes has approximately 70 per cent of the world market for VSATs.

22.6.2 *Matra-Marconi Space*

Matra-Marconi supplies the Skydata VSAT system in Europe; three photographs of Matra's equipment are included in this Chapter (Figures 22.3, 22.4 and 22.5). The Skydata system can use frame relay to provide a very flexible, high data rate solution. Frame relay is capable of handling disparate types of source material such as voice and data (which have very different system requirements).

Some of Matra's equipment is shown in the following photographs. Figure 22.3 shows a system with a 2.4 metre antenna; this is used for videoconferencing. Figure 22.4 shows a small terminal developed in collaboration with POLYCOM for data broadcasting. More than 3000 reception terminals are in operation.

Figure 22.3   A 2.4 metre VSAT for videoconferencing

Figure 22.4   A POLYCOM receive-only VSAT

Figure 22.5   A 1.8 metre VSAT

Figure 22.5 shows another videoconferencing terminal with a smaller antenna, illustrating the integration of the feed, LNA and SSPA at the prime focus of the antenna.

22.6.3 GE Spacenet

GE Spacenet provides a wide range of VSAT and hub equipment;[4] its Skystar Advantage private hub system can operate with networks of 50 to 5000 remote sites.

In 1997, GE Spacenet acquired AT&T Tridom, and it now has over 70 000 VSAT sites around the world.

22.6.4 AT&T (now part of GE Spacenet)

The AT&T Clearlink system incorporates TCP/IP, SNA/SDLC and X25, and data broadcast, packetised voice and digital voice overlay are offered. Prices for Ku-band systems start at $6000 and for C-band systems $9000.[5]

22.6.5 Scientific Atlanta

The Skyrelay VSAT network can support voice, data and video. Prices are said to start from $4900 per remote terminal.

There are now many companies involved in the provision of equipment and services for VSAT. Some others are:

- SIS;
- Multipoint;
- Orion;
- GlobeCast;

22.6.6 DVB and satellite interactive terminals

DVB will have a major effect on VSATs and VSAT standards. An ad-hoc group hosted by ESA-ESTEC has prepared specifications for SITs (satellite interactive terminals) taking into account the use of DVB. The SIT could be thought of as a VSAT or as a TV receiver only terminal but with a satellite return channel. Clearly, the aims are to provide workable standards and to drive down costs.

DVB can be used for data broadcasting as defined in the draft ETSI Standard EN 301 192. Data broadcasting includes the transmission of IP datagrams as well as interactive TV and software downloading. These facilities may be some of the most important aspects of DVB in the middle future. The transmission of ATM and IP over DVB are now major issues of research around the world. DVB carries MPEG2 video and audio signals as well as program information signals in a prescribed but flexible format. It is an international standard which reflects many years of research and development by all of the world's leading players in the broadcasting business. DVB has been defined for all current transmission media.

SITs could be used for many purposes including Internet connections. The astonishing growth of the Internet and multimedia services provide new opportunities for business and domestic users. See, for example, the SES web site, for information about their plans for SIT usage on the ASTRA satellite system.[7] The growth of use of multimedia-capable personal computers to access the Internet and the World Wide Web (www) for the transfer of image, video and audio information has resulted in a higher demand for Internet bandwidth. The acceptance of DVB has resulted in the development of digital platforms to provide television, radio and data services over satellite and terrestrial networks. There is a convergence of technology which will lead to lower costs and higher performance for small terminal satellite users.

22.7 Design considerations

22.7.1 Antennas

Hub stations generally have antennas in the range 5 to 8 m diameter; although some systems have hubs as small as 2.4 m. The size required depends on the type of VSAT system, the satellite being used, the location and all the other parameters which are involved in a link budget.

The VSAT antenna usually has a diameter of 60 cm to 90 cm for Ku-band. However, antennas as small as 45 cm would be adequate close to beam centre for some services. Antennas as large as 2.4 m could be required for other services. A diameter of 1.8 m would be used beyond the normal beam edge of 4 dB; this is the case with television transmissions from current EUTELSAT transponders. The efficiency of small receive-only antennas can be high (about 70 per cent); they are light and inexpensive.

Despite its small size and low cost, the VSAT antenna must meet demanding specifications. The relevant parameters in Europe are in the specifications produced by the ETSI technical committee on satellite earth stations (TC-SES).

Standard DE/SES-2002, draft pr ETS 300 159, May 1991 covers this area. It deals with two types of specification:

(i) Type approval requirements such as:

- off-axis EIRP;
- tx antenna polarisation discrimination;
- antenna pointing accuracy capability;
- polarisation plane alignment capability.

(ii) Recommendation on receiving quality.

The most common type of VSAT antenna is the front-fed paraboloid (i.e. a section of a parabola). Two forms are used, the offset and the axisymetric.

The offset antenna has no blockage and its efficiency can be high at between 60 and 70 per cent. It also has low sidelobes owing to the absence of blocking (and diffracting) material in the wave front.

Hub antennas can also be of offset or axisymetric design. Some hubs will use Cassegrain antennas.

For Ka-band, which is a likely candidate for some future VSAT systems, the antenna could be even smaller. Some flat, phased-array-type antennas have been developed for this application and they could become more common.

22.7.2 Receive-only VSATs

Typically, an offset antenna will be used for receive-only VSATs. This can have an RF front end, comprising an LNA and a downconverter, mounted at the prime focus behind the feed. The Cassegrain has the same items mounted on the feed behind the main reflector. In both cases, it is common to use an integrated front end containing the LNA and downconverter. The front end housing on some units also incorporates the feed (i.e. the feed is part of the casting or moulding).

The received signal must be processed to extract the user data, and the stages of processing depend on the design. The first stage is demodulation, whereby a signal is extracted from the carrier. The demodulator will most probably be either BPSK or QPSK and may be followed by a deinterleaver, a decoder and a descrambler. The INMARSAT-C system, which provides a service to very small vehicle-mounted terminals, uses all of these processes to compensate for the very poor channel conditions under which it has to operate.

The user data can then be passed directly to a device (such as a PC or a fax machine), or it can be passed to a demultiplexer which will split it into several streams and supply them to different users or applications.

22.7.3 Two-way VSATs

A block diagram of a two-way VSAT is shown in Figure 22.6.

The receive side is essentially the same as for the one-way VSAT but the antenna performance may be changed. In order to provide optimum performance at both receive and transmit frequencies the antenna efficiency may be lower, a value of 60 per cent would be normal. If the receive efficiency reduces from 75 to 60 per cent the received signal level drops by $10 \log(60/75) = 0.97 \text{ dB}$. The carrier-to-noise ratio will also drop by approximately the same value.

An ortho-mode transducer (OMT) is necessary to separate the receive and transmit paths and to ensure that they are on orthogonal polarisations. The power amplifier produces about 1 W of output power at saturation and is a compact solid-state device; these are now readily available at low cost. The transmit chain can be realised in several ways.

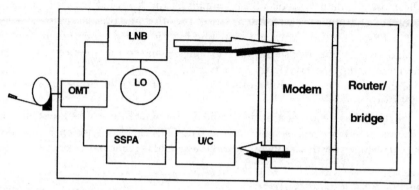

Figure 22.6   A two-way VSAT

22.7.4 Size constraints

Traditionally, the earth stations in any particular network have all been of approximately the same size or, more importantly, the same specification. This approach balances the requirements of the transmitting and receiving facilities and ensures that all participants in the network share the costs on an equitable basis. However, in a VSAT network, the distribution of facilities is highly asymmetrical with many VSATs and only one (or very few) hubs. In such a network the main requirement is to minimise the cost of the VSAT terminal and to transfer, as far as is possible, the technical complexity and cost to the hub station. Because the VSAT has a small aperture antenna, it must receive a correspondingly higher power from the satellite in order to provide an adequate carrier-to-noise ratio for error-free demodulation. If, however, the signal level from the satellite is arbitrarily increased to provide the necessary carrier-to-noise ratio for a VSAT, it is most likely that the limit on allowable flux density per kilohertz at the earth's surface will be exceeded. At present there is a limit for downlink power spectral density of 6 dBW/kHz in the USA and this may also be adopted in Europe.

22.7.5 Phase noise and carrier recovery

The local oscillators in earth-station frequency converters are usually generated using phase-locked oscillators. The cost of these components may be significant compared with the overall price of a VSAT and it is preferable to use a dielectric resonator oscillator (DRO) if possible. However, this may affect operation with low data rates, where close to carrier phase noise is most important. A DRO, which has poor phase noise compared with a PLO, can be used, but the effects on carrier and clock recovery and on the overall stability must be considered. Phase demodulation can be achieved by regenerating a carrier or by using a pilot signal. Carrier regeneration can be accomplished by one of the following methods:

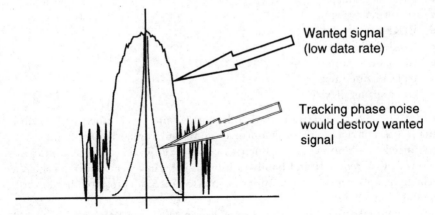

Wanted signal
(low data rate)

Tracking phase noise
would destroy wanted
signal

Figure 22.7   Phase noise and low data rates

- squaring loop;
- Costas loop;
- decision feedback.

In a system operating with a low C/N (such as a VSAT), a Costas loop has advantages.

If the phase-noise bandwidth is similar to that of the signal it may be impossible to demodulate the signal. This is because any circuit which tracks out the phase noise will also track out the modulation on the signal and the information content will then be lost. Even if the information on the signal is not lost, this effect will increase the probability of an error in the demodulator.

22.7.6 VSAT advantages

The small size of a VSAT has several inherent design advantages. As satellites are generally maintained on station to an accuracy of $+/-0.05°$ in both north/south and east/west directions, there is no need to track the antenna. However, it is worth noting that satellite operators, including INTELSAT, have considered ways of tracking small antennas. This is due to the shortage of transponders and the need to continue operating satellites even when there is effectively no fuel left for north/south station keeping. Another advantage of very small antennas is their lack of sun outages. The beamwidth of a VSAT is such that the increase in system noise temperature when pointed directly towards the sun is negligible.

22.8 Modulation and coding

22.8.1 Modulation and coding methods

There are many trade offs in achieving a desired level of performance. The modulation and coding techniques used can be traded against parameters such as:

- antenna diameter;
- received power;
- EIRP;
- bandwidth;
- transmission time;
- transmission rate;
- cost and complexity.

Unfortunately, most of these trade offs work against each other. For example, since it is not necessary to conform to common standards one could use an advanced modulation scheme such as coded 8PSK. This has advantages in terms of C/N required and bandwidth occupied. However, modems using this scheme are expensive and sensitive to phase noise (as are most higher-order modulation schemes).

A commonly-used forward-error-correction (FEC) technique, such as half-rate convolutional encoding with Viterbi decoding, will improve the user BER by several orders of magnitude but it will also double the occupied bandwidth. BPSK and QPSK with some form of coding or spectrum spreading are often used for transmissions to the VSAT.

The use of BPSK enables a particularly low-cost demodulator to be used, even if it is combined with spectrum spreading. BPSK is resistant to phase noise since phase shifts of up to $\pm 90°$ will not (in the absence of thermal noise) cause an error. BPSK has only one decision threshold, which means that the distance to that threshold is maximised; it is therefore also resistant to thermal noise (see Chapter 9 for details). Transmissions from VSATs to the hub also often use FSK since this can be easily implemented by, for example, direct modulation of an oscillator. There is no requirement for spectrum spreading, since the EIRP of the signal is low. It may, however, be necessary to use coding so that the data can be accurately recovered.

There are several types of code in common use which are suitable for implementation in a VSAT system; some of these are:

- linear block coding;
- convolutional coding with Viterbi or sequential decoding;
- concatenated coding e.g. Reed–Solomon/biorthogonal or Reed–Solomon/convolutional.

Reed–Solomon codes are the best known codes in the family of linear block codes. The BER achieved is proportional to the length of the code for a given C/N; a long code can correct a larger number of errors. The number of correctable errors is given by:

$$t = (n - k)/2 = n(1 - k/n)/2$$

where t is the number of correctable symbols and k/n is the code rate (see Chapter 10 for more details).

Convolutional encoders, being constructed from shift registers, are simple to implement. The simplest decoders can also be made from shift registers but they are some 3 dB less efficient than Viterbi or sequential decoders. Many current VSAT systems and conventional earth stations use rate 1/2, constraint length seven Viterbi decoders. These can provide a BER of 1×10^6 with an E_b/N_o of 5.5 dB and are available as integrated circuits for inclusion in OEM equipment.

Sequential decoders are powerful and their complexity is not dependent on constraint length as is the case with Viterbi decoders. They suffer because the processing time is variable for each output bit and a large buffer memory is needed when the C/N is poor.

A very powerful code can be realised by concatenating two codes. With this technique, an effective code can be achieved with relatively simple implementation because the decoder is constructed from two simpler decoders. Many modems can use two concatenated codes. The result is a data stream which is almost literally error free as long as the C/N of the link is above a threshold. However, if the C/N falls below the threshold the performance of such systems will degrade rapidly.

22.8.2 Modulation techniques

In general, the same considerations apply to VSATs as to all other satellite communications systems. However, there are some differences. VSATs systems must be reasonably priced and this has limited the complexity of the modem used. Equipment space is limited by the need to have a compact installation and this also limits what can be used.

Modems have become highly integrated and the use of DSP-based designs means that modem performance can be several dB better than in the past.

INMARSAT used OQPSK (offset QPSK) for its systems. OQPSK is now being used for other systems which seek to exploit the advantages of a constant-envelope modulation scheme.

Applying a filter to PSK makes it nonconstant envelope; new schemes like MSK are constant envelope and can have better spectral efficiency than PSK.

22.8.3 Coherent and noncoherent demodulation

Coherent demodulation gives the best performance and is based on the regeneration of a local oscillator in the demodulator by phase locking to the received signal. However, noncoherent demodulation can also be used. In the case of PSK, this is also called differential demodulation since it uses the phase of the last symbol as the phase reference to demodulate the current symbol. Noncoherent demodulation can be used in situations where the C/N is low and a low-cost circuit is required. It has worse performance than coherent demodulation, as one would expect.

22.8.4 Noise

Since satellite links are between earth and space, there is little scope for man-made noise to be introduced. Most of the noise on a satellite link is thermal in origin; thus it is spectrally flat and wideband. However, flat, wideband noise can cause bursts of errors in a decoder; an interleaver can spread such error bursts and minimise their impact.

22.8.5 Coding

Most codes are designed for use on an AWGN channel and Shannon's information theory is relevant to this type of channel. Modern coding schemes are getting close to the Shannon limit of C/N which in fact limits the coding gain which it is possible to achieve.

BCH codes can be decoded by simple decoders using shift registers with logic gates, but soft-decision decoding cannot readily be applied to block codes.

22.8.6 Spread spectrum

In order to achieve an adequate received C/N at a VSAT (with its very small antenna), it may be necessary to increase the power transmitted by the satellite to such an extent that the power flux density (PFD) at the earth's surface is unacceptable high. One way of overcoming this problem is by spreading the transmitted signal over a wider frequency range and thus reducing the PFD. The spread-spectrum technique can be used for both digital and analogue modulation methods, although the term spread spectrum is usually used only in connection with digital schemes.

With a spread-spectrum signal, the PFD at the earth's surface is held within allowable limits, but the total power, equal to PFD multiplied by the occupied bandwidth, can be as high as required. In this way the carrier power received by the VSAT can be increased to the level required to provide adequate signal quality. In the case of a digital signal, the signal quality is measured in terms of BER and the required BER is usually determined by the user's application. The effects of spectrum spreading are illustrated in Figure 22.8.

A spread-spectrum system spreads the energy of the data signal over a bandwidth which is 100 to 1000 times higher than would normally be used. The term processing gain is used to describe the degree of spreading. If the occupied spectrum is increased by 1000 times, the processing gain is 30 dB (i.e. 10 log 1000). However, there is no real gain or loss in the system. Apart from a small implementation loss, a few tenths of a dB, the overall end-to-end link budget in terms of received C/N is the same whether spectrum spreading is used or not.

Because a VSAT is small, its beamwidth is high which means that it is capable of receiving excessive interference from adjacent satellites. Also, the satellite may have to radiate high power levels so that the VSAT has a sufficiently high C/N. Spread spectrum can avoid both of these problems. Interfering signals received by the VSAT are themselves spread in the demodulation process and their effect is made negligible.

Two common methods of producing a spread-spectrum signal are direct-sequence (DS) modulation and frequency hopping (FH). Civilian systems use DS but FH is used in military systems.

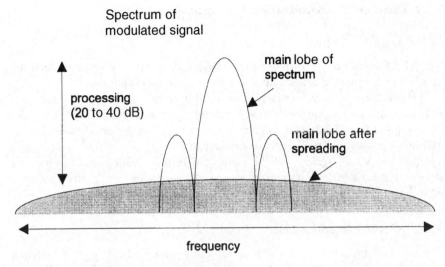

Figure 22.8   The effect of spectrum spreading

Spread spectrum has been used primarily for C-band applications where spectrum crowding is most severe. C-band systems have lower-gain antennas and thus adjacent satellite interference is worse. Spread spectrum makes a system relatively immune to multipath, ASI, ACI and sun outages.

In a spread-spectrum system each bit of information is replaced by a pseudo-random (PN) sequence; this is shown in Figure 22.9. Conventionally, the bits of the spreading sequence are referred to as chips. The length of the sequence is system dependent but would typically be 1000 chips in a commercial system.

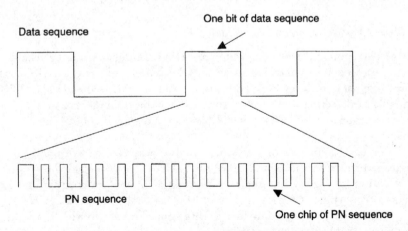

Figure 22.9   The PN sequence

22.9 Data transmission and protocols

22.9.1 Quality of service

The VSAT system must be designed to provide the quality of service which the application requires in order to function in a satisfactory manner.

Quality of service involves factors such as bit error rate or packet error rate, link availability, throughput delay and call set-up time. The VSAT system could be optimised for a particular application or it could be designed to be flexible and support many different types of application.

Often the VSAT system will be connected to one or more data-processing devices and is essentially part of a network. Therefore, network optimisation is an important part of VSAT design.

22.9.2 Circuit and packet switching

Circuits can be packet switched or circuit switched. In a packet-switched system each packet of data is individually routed from one end of the application to the other, possible over different routes. In a circuit-switched system, circuits are allocated for the duration of a call, at least, and all data travels along the same route from end to end.

Each type of switching has advantages for different types of traffic. The most flexible VSAT system can actually switch between the two modes of operation depending on the traffic profile at a given time. It can be very important to know what type of traffic is flowing at each part of the day in order to optimise a system.

There are usually many parameters that can be altered to improve a network's performance. Parameters such as buffer size, number of buffers and various time constants depending on the network protocol.

22.9.3 Error detection and correction

Error detection and error correction are two different but related techniques. Errors can be easily detected using a cyclic redundancy check (CRC). Most CRCs consist of a 16-bit word added to the end of a data sequence. The CRC is capable of detecting almost all errors. In fact the residual error rate, i.e. the error rate due to errors which the CRC does not detect, is about 1 in 10^{14} bits or less.

Error correction can be performed in two ways. Either through forward error correction, or through ARQ. With FEC, the data is encoded on transmission with many parity bits and this information is used on reception to detect and correct errors. A common form of FEC uses half rate, in which case there are as many parity bits as data bits, or three-quarters rate, in which case there is one parity bit to every three data bits.

An FEC system alone can supply data which has an arbitrarily low error rate, but it does not guarantee that the data is error free.

An ARQ system uses a mechanism such as a CRC to determine that something is wrong within a packet or frame of data and arranges for that packet to be transmitted again.

If a system requires data which is error free then an ARQ system must be used.

22.9.4 *ACKs and NACKS*

End-to-end protocols usually function by sending acknowledgments (ACKS) and negative acknowledgments (NACKS). An ACK indicates that a particular packet or frame has been received. A NACK indicates that a packet or frame was not received within a particular time frame.

A protocol can function with ACKS alone or with NACKS to speed things up but it cannot operate with NACKs alone.

22.9.5 *Frame formats*

The ISO model refers to packets and frames and these terms are often used interchangeably. In principle, each system vendor could use its own frame format but there is merit in using a standard format.

HDLC, in particular, is widely used as a model for the frame even if the data is not being used in a proper HDLC environment.

22.9.6 *VSATs and interface standards*

The value of using standard interfaces between communicating devices, whether part of a network or not, has long been recognised. Standards such as RS-232, X21 and X25 are widely used and organisations around the world are developing their systems with the OSI model in mind. One-way VSATs do not easily fit into such systems since they cannot directly (i.e. on the satellite channel) provide acknowledgments of data received. This problem can be overcome by spoofing at the hub or tunneling through a terrestrial link.

Two-way VSATs are able to provide acknowledgments but often, especially at Ku-band, cannot receive their own transmission owing to their small size. This is typically a requirement of distributed control systems, such as IEEE-802.3, where each device is responsible for ensuring that it does not disrupt the communications of others. This problem can be overcome in systems with a central control entity such as a hub station. A practical example of a distributed control system is the operation of SATNET, which is part of the well known global network ARPANET. SATNET is used by C-band INTELSAT earth stations of 30 metre diameter and cannot easily be used by VSATs.

The way in which VSATs fit into the OSI model is illustrated by Figure 22.10.

When implementation of a VSAT system is being considered, the cost of ownership must be compared with the cost of using existing communications services. The cost of ownership includes the purchase price, which will normally

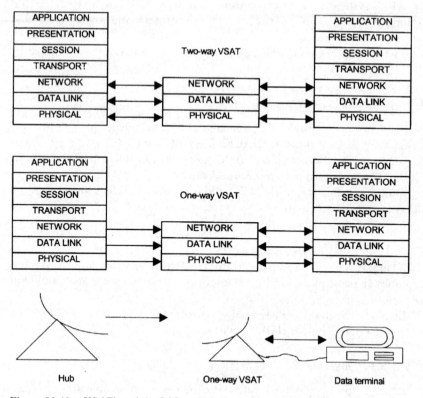

Figure 22.10   VSATs and the OSI model

be amortised over several years, plus running and maintenance costs. It is impossible to give definitive prices for VSAT equipment because the functions and services offered by different suppliers vary so much. However, the take up of VSATs will depend strongly on cost of ownership since, from the user's point of view, they are simply a means to an end, and the lowest-cost system with comparable performance will win.

In the USA, low-speed C-band two-way VSATs cost about $6000 when purchased in quantities of several hundred. Two-way Ku-band VSATs may cost two or three times as much. The cost of ownership of a $6000 VSAT, amortised over three years, would be about $400 per month (the total cost of ownership is about two or three times the equipment purchase price).

The hub is an expensive item, the costs of which are generally shared by the network users either directly or through an element of their connection charges. According to Morgan[6], the monthly cost of a hub station, including operating and administrative overheads, is approximately $108 000. Therefore, in a network of 1000 users, the total cost per user would be $108 per month. In this case the total running costs per user would be $580 per month. Whether or not this cost is attractive depends on the service required and the value placed by the user on having a degree of control over their own communications system. There are strong indications that many users value this aspect of the system very highly.

Consider that a forward link data rate of 2 Mbit/s is available and that each user has 1/1000 of the total channel time in each month. The volume of data which each could transmit or receive is:

$$\frac{30 \text{ days} \times 24 \text{ hours} \times 60 \text{ minutes} \times 60 \text{ seconds} \times 2 \text{ Mbit}}{1000} = 24.9 \text{ Mbit}$$

The cost of this data is then 5184 Mbit/$580 which is 9 Mbit/$. This is a highly simplified example but it demonstrates that the cost of using a VSAT can be comparable to the cost of using existing terrestrial services

22.9.7 DVB

DBV is associated with television but is now being used to carry Internet traffic. One of the main advantages of this is the ready availability of equipment such as data inserters and PC cards. This equipment can be used as part of a system which can deliver high data rates to the home or office.

Possibly the greatest advantage of DVB is that it is an international standard which is clearly going to be around for some time.

DBV signals can carry data which uses other protocols, and many organisations are examining the use of ATM and TCP/IP over DVB.

22.9.8 ATM and frame relay

The terms ATM (asynchronous transfer mode) and frame relay refer to systems which route packets of data over a network. The difference between them is essentially that ATM operates at the OSI model level 2, the packet level, and frame relay operates at level 3, the frame level.

They are very popular at present not least because they appear to be able to handle both voice and data. Routing devices (switches) are available that can be integrated into a VSAT system.

22.10 Satellite access techniques

22.10.1 Sharing the satellite

A satellite is a relatively expensive and scarce resource. In traditional systems, the satellite transponder was used in a very well defined and stable manner which did not change from year to year. However, when VSAT systems are used, a single satellite transponder may be used by many different VSAT networks. Each network may carry different traffic at different times of the day. The environment is essentially dynamic and can change over a period of hours.

The VSATs must access the transponder in an efficient manner to avoid wasting the satellite capacity and to reduce the cost of that capacity by allowing as many users as possible to access it.

Satellite access concerns the means of utilising, and sharing, the satellite's resources.

The use of many VSATs means that terminal cost rather than satellite cost may be a limiting factor, therefore, the most efficient use of the transponder may not be of greatest relevance.

22.10.2 *Addressing and channel-access techniques*

In a receive-only VSAT network channel access is a minor problem which only concerns the hub or hubs. The hub typically has exclusive use of a fixed-frequency channel and is able to transmit as and when required. However, it may still be necessary to address information to each individual VSAT or to groups of VSATs. Each VSAT receives all of the information on the channel to which it is tuned; it only uses, however, the information which is addressed to it. The address can be unique to a particular VSAT or it can be a group address which covers many VSATs.

Each VSAT could have an individual address and, in addition, one or more group addresses. A sophisticated system may also transmit, to the VSAT, information which will allow it to decrypt the data and prevent others from doing so.

In a two-way system the situation is clearly quite different. There may be several hundreds or even thousands of VSATs which wish to transmit to the hub. It is not feasible to consider having a separate channel for each VSAT so they must share a channel or channels. Many schemes have been proposed to deal with this problem but there is no clear favourite since system requirements differ so much. However, there are many systems which need to share a single channel for communication at essentially random times. A common technique for dealing with this problem is to use ALOHA or a token-passing technique. Both of these methods are now used as part of the IEEE-802 specification for local area networks and the principles also apply well to a VSAT system. A possible channel frequency plan is shown in Figure 22.11.

The satellite's capacity can be shared in one of two domains, namely, the time domain and the frequency domain. The time domain can be shared by several VSATs in time-division multiple access (TDMA) such that data from a given VSAT is compressed in time and transmitted so as to interleave at the satellite with data from other VSATs. The transmission of bursts such that they do not overlap and interfere with each other usually requires a mechanism for acquisition and synchronisation. For this mechanism, the hub can usually provide a master timing reference onto which all VSATs can lock. Although such a system may appear simple, there are problems caused by satellite movement and VSAT geographic location to take into account.

In a frequency-division multiple access (FDMA) system, each signal accessing the satellite is assigned its own carrier frequency. The frequency allocated is available all of the time and the access method can be simple.

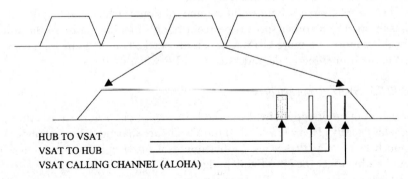

Figure 22.11 Two-way frequency plan

However, FDMA would be very inefficient in terms of channel utilisation for only one VSAT. The efficiency is increased by using time-division multiplexing (TDM). In this case many VSATs can use the channel but the controlling mechanism is very much simpler than with TDMA.

22.10.3 Access methods

Satellite access methods include:

22.10.3.1 FDMA (frequency-division multiple access)

This is the simplest type of system. Each user has a different frequency allocated and transmits a single carrier on that frequency.

22.10.3.2 TDMA (time-division multiple access)

Although generally considered to be very complicated, this method is nevertheless suitable for VSAT systems. A reference burst can include a control and data channel to inform each VSAT when and on what frequency it should transmit. A separate channel can be used by the VSATs to request capacity. Usually there is a master station in such systems which is responsible for capacity allocation.

22.10.3.3 FM2 (FM squared)

This system uses a number of FM signals, each at different subcarrier frequencies, which are modulated onto a single, TV like, carrier; this allows low-cost FM receivers to be used. Because each VSAT receives the composite signal, which is high power and wideband, the carrier-to-noise ratio can be high and demodulation relatively straightforward.

22.10.3.4 CDMA (code-division multiple access)

Several spread-spectrum signals can use the same frequency at the same time as long as each uses a different spreading code. This is referred to as CDMA. As long as the different codes used by each signal have low crosscorrelation coefficients they will not interfere with each other.

There is, of course, a limit to the number of signals which can be piled on top of each other. The other spread-spectrum signals look like noise to the wanted signal. As more and more CDMA signals are stacked on the same frequency, the C/N of the wanted signal degrades and the BER will also degrade.

22.10.3.5 Sharing transponders

In the inbound direction, i.e. from the VSAT to the hub (if there is a hub), the signal is usually of very low level. If the VSATs are sharing the transponder with a much larger signal there is the possibility of small-signal suppression which could badly affect the VSAT link budget (a single small carrier in the presence of a large carrier suffers 6 dB suppression).

22.10.4 Multiple access strategies

In addition to deciding how the satellite transponder will be shared (frequency, time, code domains), the system must handle the allocation of capacity to users (VSATs) when they require it. Capacity can be assigned in several ways. Some common techniques are:

22.10.4.1 Fixed assigned (or preassigned)

The channels or time slots are preassigned and remain so indefinitely or for a significant time (hours to days).

22.10.4.2 Demand assigned

The channels or time slots are dynamically assigned as the VSATs request them.

22.10.4.3 Voice activation

Each speech channel is used for about 50% of the time during conversation and it makes sense for VSATs carrying voice traffic to switch off their transmitters during gaps in the speech and thus save on carrier power. Some systems use the gaps in speech to insert data channels rendering a very efficient speech and data system.

22.11 Network configurations and availability

22.11.1 Star network

If the hub is affected by atmospheric attenuation then all of the VSATs in a network will be affected. The availability of the hub must be very high, typically 99.995% (i.e. 26 minutes outage per year) in order to prevent this.

22.11.2 Mesh network

In a mesh network, atmospheric attenuation will affect only a part of the network. The availability of each VSAT is determined by the service being provided.

22.12 Link budgets

22.12.1 One-way systems

To illustrate the way in which VSATs are used in real systems, the link budgets in Tables 22.1 and 22.2 give examples of the system performance which can be achieved for one-way and two-way cases using satellites of the ECS type.

Table 22.1   One-way link budget

Hub to VSAT	No FEC	FEC
Uplink at 14.25 GHz		
Earth station EIRP	63.6 dBW	58.4 dBW
Spreading loss	163.6 dB	163.6 dB
Path loss	208.0 dBW	208.0 dB
Atmospheric loss (clear sky)	0.2 dB	0.2 dB
IPFD for saturation	-76.4 dBW/m^2	-76.4 dBW/m^2
IPBO at the satellite	23.8 dB	29.0 dB
IPFD at the satellite	-100.2 dBW/m^2	-105.4 dBW/m^2
Satellite G/T	-5.3 dB/K	-5.3 dB/K
Uplink C/N_0	78.7 dBHz	73.5 dBHz
Downlink at 10.95 GHz		
Saturated EIRP (-2 dB contour)	43.0 dBW	43.0 dBW
OPBO	17.8 dB	23.0 dB
Satellite EIRP	25.2 dBW	20.0 dBW
Path loss	205.8 dB	205.8 dB
Atmospheric loss (99.9% of the year)	4.0 dB	4.0 dB
Antenna gain	38.5 dBi	38.5 dBi
Receiver G/T	12.1 dB/K	12.1 dB/K
C/N_0	56.1 dBHz	50.9 dBHz
Overall link budget		
Overall C/N_0	56.1 dBHz	50.9 dBHz
Overall C/N	12.9 dB	7.8 dB
Implementation margin	2.0 dB	2.0 dB
E_b/N_0	10.9 dB	5.8 dB
BER	1.0×10^{-7}	1.0×10^{-7}

These link budgets were calculated to show how the system must be set up in order to obtain, in each case, a user (as opposed to channel) BER of 1×10^{-7}. This is adequate for many applications. The cases with and without half-rate FEC are illustrated and the benefit of having FEC is clearly seen in the reduced EIRP required from the hub and from the satellite. In the FEC case the output back off is 23.0 dB, thus the VSAT link is using only 0.5% of the transponder's saturated power output. Assuming that the maximum power available from the transponder under multicarrier conditions is 41.0 dBW, i.e. 2 dB down on saturation, then a further 63 such VSAT links could be accommodated in the same transponder. Since each carrier can serve a large number of VSATs on a time-shared basis, it is clear that a single transponder could actually accommodate communications to many thousands of VSATs. This fact is important for the system economics.

Table 22.2   Two-way link budget

Uplink at 14.25 GHz	
VSAT EIRP dBW	40.5 dBW
Spreading loss	163.6 dB
Path loss	208.0 dB
Atmospheric loss (clear sky)	0.2 dB
IPFD for saturation	-76.4 dBW/m^2
IPBO at the satellite	46.9 dB
IPFD at the satellite	-123.3 dBW/m^2
Satellite G/T (-4 dB contour)	-5.3 dB/K
C/N_0	55.6 dBHz
Downlink at 10.95 GHz	
Saturated EIRP (-2 dB contour)	43.0 dBW
OPBO	40.9 dB
Allowance for small carrier suppression	2.2 dB
Satellite EIRP	2.1 dBW
Path loss	205.8 dB
Atmospheric loss (99.9% of year)	4.0 dB
Receiver G/T	29.0 dB/K
C/N_0	47.7 dBHz
Overall link budget	
Overall C/N_0	47.1 dBHz
Overall C/N	7.3 dB
Implementation margin	1.5 dB
E_b/N_0	5.8 dB
BER	1.0×10^{-7}

22.12.2   Two-way link budget

The EIRP of the VSAT is quite low. This is advantageous for reasons of cost, safety and reliability. The link budget assumes the use of rate half FEC and a user data rate of 9.6 kbit/s. The link uses only 0.01% of the transponder power but is still able to provide a BER at the hub of 1.0×10^{-7}.

22.13 References

1 'Network performance summary'. Hughes VSAT August 1996 report.
2 'Satellite services: the European regulatory framework' (http://www2.echo.lu/legal/en/converge/satellite.html)
3 EUTELSAT Press Release, EUTELSAT at CEBIT 97 (http://www.eutelsat.org/press/release/press14.html)
4 GE Spacenet has a very comprehensive web site packed with VSAT information. Start at: http://www.ge.com/capital/spacenet/prodserv/ssadat1.htm
5 Equipment show review
6 MORGAN, W.L., and ROUFFET, D.: 'Business earth stations for telecommunications' (ISBN: 0 471 635561)
7 SES Website: www.aia.lu/recept/arcs/index.html

Chapter 23

Military satellite communications

D. Bertram

23.1 Background

Military satellite communications (milsatcoms) have been well established for a number of years, and are major features of the communications systems of the USA, UK, France, NATO and nations of the former Soviet Union. They are significant in terms of both current systems and associated research and development activities: there is little doubt, at least in the UK, that military programmes have helped to maintain the commercial industrial base, and in many ways military systems lead the field in terms of technology development and sophistication. Recent developments in commercial civil programmes, such as Iridium, Teledesic and GlobalStar, have reversed the balance in the spending on satellite communications development between the military and civil sectors, as the commercial investment currently dwarfs the military spending on communications.

This Chapter attempts to outline, from a UK perspective, the features which distinguish milsatcoms from civil systems. Clearly, there are many technical and operational aspects which it is not possible to describe in detail, and the views expressed do not necessarily represent official policy.

A full decade before the advent of the UK milsatcom era, experiments were underway in the UK and US to investigate the feasibility of using the moon to bounce radio signals off for military communications traffic. Using the Moon as a reflector had three major drawbacks:

(i)   The extended shape of the moon spread the reflected signals out temporally.
(ii)   The topography of the moon's surface induced deep multipath fading.
(iii) It was only available 12 hours per day.

The first two problems were soluble, even in the early 1960s, but the last was (and still is) insurmountable. To fill the availability gap, experiments using orbiting space junk were performed from 1964 onwards. The most successful trials involved using the second and third stages of the launch vehicle, as these had a large radar cross-section and a relatively useful orbit. The geostationary orbit was known about, of course, but its use did not attract as much support at the time and, although not exactly regarded as heresy, not many would have predicted the ubiquitous use of the geostationary arc we see today.

The principal characteristic of milsatcoms is its varied nature: it needs the ability to cope with a wide range of users, traffic and scenarios. Thus it encompasses both high-capacity fixed links and low-data-rate mobile traffic (often called the strategic and tactical elements). Also, in contrast to its civil counterparts, many parts of the system may be required to operate in hostile environments, and will be designed to withstand threats such as jamming, intercept and indirect nuclear weapon effects such as ionospheric scintillation and EMP.

Following early experiments with the US nongeostationary interim defence communications satellite programme (IDCSP), the UK military satellite communications programme began in 1969 with SKYNET 1 — the world's first geostationary defence communications satellite. This was followed in 1973/4 with the launch of two upgraded satellites called SKYNET 2. The SKYNET 1 and 2 satellites operated at SHF (8/7 GHz) and utilised a spin-stabilised design which provided one 2 MHz and one 20 MHz channel sharing a single travelling-wave-tube amplifier. A simple despun horn and plane reflector provided earth coverage for both the uplink and downlink. One SKYNET 2 satellite was still in use more than 20 years after launch, despite only a 95 per cent expectancy of a three-year lifetime estimated from the design life of three years for the TWTA. This is quite exceptional, but how it was achieved is unknown, as TWTAs made since then have not achieved nearly such a long life. In order to celebrate its majority, Marconi and DERA Defford managed to set up a communications link through SKYNET 2 in 1994.

The major defence review in 1974, following the associated policy of withdrawal from East of Suez, cancelled the SKYNET 3 programme. The UK milsatcom programme resumed in 1981 with plans for SKYNET 4. This programme called for an initial three satellites (stage 1) with the first launch scheduled for June 1986, followed by a further two satellites (stage 2) to be launched in the mid-1990s. The launch programme slipped a little, entirely normal in space programmes apart from those run by Japan!

As of January 1997, three SKYNET 4 satellites have been successfully launched—the first in December 1988 (on board an ARIANE 4 rocket), the second in January 1990 (on board a TITAN 3D rocket) and the third in August 1990 (on board an ARIANE rocket). SKYNET 4B was actually launched first, as SKYNET 4A was originally due to be launched on the space shuttle, but the *Challenger* disaster in January 1986 meant that the launch plans were revised. A further satellite was launched as NATO IVA in January 1991 (on board a Delta rocket) and NATO IVB followed in late 1993. The first phase of the SKYNET 4 programme has orbital slots at 34°W, 53°E and 1°W for SKYNETs 4A, B and C, respectively. The slot at 6°E, currently occupied by NATO IVB, is to be used by SKYNET 4D, at least for the in-orbit check out, verification and test phase of the mission.

The three stage 1 SKYNET 4 satellites are fully in service and plans are well developed for the two stage 2 satellites, to the extent that the launch of SKYNET 4D is anticipated to be at the end of 1997. Major planning efforts are now underway for the next generation of UK milsatcoms—SKYNET 5. This next generation will be based on completely revised and updated user requirements and the satellites are scheduled for launch from the middle of the next decade.

As with civil systems, complete milsatcoms systems include both the space and ground segments with a range of associated ground terminals and other ground facilities for telemetry tracking and command (TT&C).

The demand for milsatcoms is continually increasing. This is due partly to increased communication requirements in the face of enemy threats and partly to increasingly sophisticated end-user requirements (e.g. computers and sensors exchanging digital information as part of command, control communications, computers and intelligence, C^4I, networks).

23.2 Military applications

Satellite communication is particularly attractive for military applications in that it provides a highly reliable, high-capacity service over a wide coverage area. This service is available at short notice in virtually any part of the world without reliance on any national communications infrastructure. Communications can thus be established in an environment with little or no communications infrastructure or in a conditions where use of the local infrastucture is denied or deemed unreliable. In comparison, the traditional carriers of HF and VHF/UHF offer, in the first case, limited availability through erratic propagation and, in the second case, limited coverage. Furthermore, capacity is often very limited.

The wide-area coverage available with satellite communication can in some circumstances be a disadvantage in that the satellite may be equally visible to an enemy and the links may be subject to interception and disruption. Because of the crucial requirement for survivability of communications at various levels of conflict, military networks generally aim to use diverse modes of communications for which satellite communications form one part.

However, satellite communications are playing an increasingly important military role with the current preoccupations with out-of-area operations (e.g. The Falkland Islands, Persian Gulf) where satellite communications may represent the only viable means of communication. Satellite communications are used extensively by UK forces on peacekeeping operations in Bosnia.

Milsatcoms prove attractive for both large fixed links and small portable or mobile links, and are employed by all three armed services. Consequently, there is a variety of data rates, modulation schemes, terminals and modems in use, all of which may need to be accommodated over a common satellite. This leads to considerable complexity in planning and allocation of accesses (i.e. terminal frequency and power). Bearing in mind the added difficulties of coordinating users who may be operating in adverse environments (perhaps under enemy threat), we see a varied community of users who are unlikely to be as well controlled as, say, a PTT civil system. Interoperability between different networks and systems, and flexible and responsive access planning represent major demands in milsatcoms.

23.3 Frequency bands

Milsatcoms operate currently in the UHF and SHF frequency bands, an experimental EHF package is aboard SKYNET 4 and the operational use of EHF is supported for US forces via Milstar. British operational use of EHF is planned for SKYNET 5 in the next decade. Optical and infrared frequencies are becoming of increasing scientific interest. These radio frequencies are not universally recognised as the exclusive prerogative of the military, although the SHF and EHF bands are generally accepted in this way within NATO. Circular polarisation is commonly used in all bands (frequency reuse through polarisation diversity is not commonly employed in milsatcoms). The band characteristics may be summarised as follows:

23.3.1 UHF

Broad range 225–400 MHz. The major user attraction lies in the simplicity and cheapness of the ground equipment. UHF is popular with mobile terminals (e.g. submarines which can deploy a simple nondirectional antenna just above the sea surface while remaining submerged) and often allows direct communication between mobile terminals without the need for a large anchor station.

Antennas are physically large even for moderate directivity, this is due to the large wavelength at UHF. As a direct consequence an earth-cover antenna deployed on a satellite represents major size and mass penalties.

The limited bandwidth available for milsatcoms at UHF, the sharing of these bands with terrestrial systems and the generally limited directivity of radiators, combine to produce a particular vulnerability to unintentional interference from other systems and to deliberate jamming. Furthermore, frequency reuse between satellites with overlapping coverage areas on the earth is not normally possible, and transponders normally use nonoverlapping channels with individual bandwidths of typically tens of kHz. An important consequence of the narrow channels is that low-power terminals are able to capture most of the satellite downlink effective isotropic radiated power (EIRP) making direct terminal-to-terminal communication possible (e.g. between mobiles).

23.3.2 SHF

Uplink 7.9–8.4 GHz, downlink 7.25–7.75 GHz. Part of the band is allocated exclusively to satellite communications (but not to milsatcoms). However, interference from other users is not generally a problem since practical ground antennas possess reasonable directivity (e.g. a 1.7 m dish antenna has a beamwidth of around 1°). This directivity also permits frequency reuse by other satellites within the geostationary arc, although the limits of capacity are now being reached. Transponders are generally transparent and wideband (up to around 100 MHz BW), carrying a number of channels simultaneously. A translation frequency of 725 MHz is traditionally used for interoperability, although this does not optimally match the up- and downlink bands.

23.3.3 EHF

Uplink 43.5–45.5 GHz, downlink 20.2–21.2 GHz. This is quite a recent development for operational use, but considerable future application may be anticipated in milsatcoms particularly for highly survivable communications requirements. The use of EHF offers better antijam performance, lower probability of exploitation and intercept and lower vulnerability to geolocation than SHF.

23.4 Satellites

The majority of military communication satellites are in geostationary orbit. Although obviously convenient, this configuration provides limited coverage of extreme latitudes, which is a weakness (regions such as northern Norway are of military interest). For many years, the Russians have used highly-inclined elliptical orbits (Molnya) for polar-region coverage, where three satellites may each provide eight hours coverage per day with high elevation. Figure 23.1 shows views of the earth from a geostationary satellite at 0° longitude (Figure 23.1a) and a Molnya orbit satellite at apogee at 0° longitude, (Figure 23.1b) making apparent the coverage advantages. The disadvantages of Molnya orbits are: the need for three satellites, increased orbit decay (and hence shorter satellite life), increased satellite fuel requirements and greater environmental radiation levels.

The Molnya approach has not hitherto been adopted for western milsatcoms although Molnya and Tundra (24-hour period but otherwise similar to Molnya), inclined geosynchronous and low-earth orbits have been considered to improve northern latitude coverage.

Geostationary military satellites tend to have less stringent station-keeping requirements than do civil satellites; east–west station keeping is typically to

a

Figure 23.1   Satellite's view of the earth

   a From geostationary orbit

0.1°, and active north–south station keeping may not be used in order to increase payload mass at the expense of station-keeping fuel. The natural precession of the orbit plane can, with the correct initial setting, keep the inclination within ±3.5 degrees over the expected lifetime of a spacecraft.

   Military communication satellites are similar to their civilian counterparts in using transparent transponders, but they also tend to incorporate features to provide communications under jamming, such as special antennas and spread-spectrum processors. Additionally, they may be hardened against nuclear effects, and employ secure encrypt/decrypt coding for telemetry, tracking and command (TT&C). This feature is necessary to prevent unauthorised commanding of the spacecraft.

   Earth-coverage antennas are generally used, where possible, to provide a service over the widest area (military deployments tend to be unpredictable in their location). In order to permit operation with small terminals, high-gain spot-beam antennas are also needed, and this represents a conflict with the wide coverage requirements. Some advanced satellites may employ steerable spot

Figure 23.1   cont.

b From Molnya orbit (apogee)

beams with mechanical steering, selectable beams or phased-array steering: such solutions, however, are neither straightforward nor cheap.

Transparent transponders tend to be used for the bulk of traffic. Greatest flexibility is achieved in this way, allowing usage by a wide range of terminals. Design features include amplifier stages which have good small signal linearity and stable outputs under extreme overload (jamming) conditions, very low AM-to-PM conversion performance and stable, switchable gain over the complete dynamic range. Figure 23.2 shows a functional outline of a transparent transponder, where the uplink passband is selected (by filtering), amplified, downconverted (single conversion) and transmitted via a traveling-wave-tube amplifier (TWTA), together with further filtering (to remove images, intermodulation products etc.). The channel amplifiers can have commandable gain steps, allowing the system operator to choose required back off from limiting (i.e. saturation) mode, and transponder gain may be sufficient to permit saturation on front-end thermal noise if required.

A typical military satellite is illustrated by SKYNET 4. The SKYNET 4 system[1] aims to provide flexible communication for maritime and land forces, together with fixed strategic services. Among its major design features are

COMMUNICATIONS SATELLITE

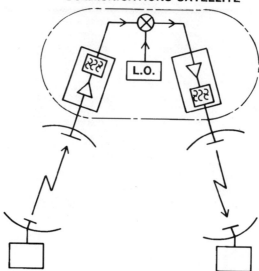

Figure 23.2   Transparent transponder—simplified functional outline

- multifrequency capability, i.e. operational UHF and SHF service;
- survivability, i.e. hardening against direct and indirect nuclear effects; also electronic counter-countermeasure (ECCM) features are incorporated to give protection against jamming;
- operational flexibility, i.e. selectable antennas, channels and gain steps to optimally support a varying user community;
- long service life: the operational design lifetime is seven years.

The SKYNET 4 satellite is three-axis stabilised, and is based upon the successful ECS design, with a service module and a payload module. British Aerospace was the prime contractor, and Marconi Space Systems the major subcontractor for phase 1 with the successor company, Matra Marconi Space (UK) being the prime contractor for SKYNET 4 phase 2.

The design has been engineered for launch either by the space shuttle or by any of the principal expendable launch vehicles including ARIANE 3/5, Titan, Delta and Atlas Centaur. Injection into geostationary transfer orbit is by payload assist module D-2 (except for ARIANE) and final injection into geosynchronous earth orbit (geostationary) is achieved with an apogee kick motor which is integrated into the main body of the satellite. The physical configuration of the spacecraft on station is shown in Figure 23.3

In orbit, control is based on a momentum-wheel bias system coupled with sun and earth sensors. Coarse attitude control, momentum wheel dumping and east–west station keeping are achieved using a system of hydrazine thrusters. Lightweight carbon-fibre materials are used for many structural parts. The power supply is regulated in sunlight, and unregulated in eclipse.

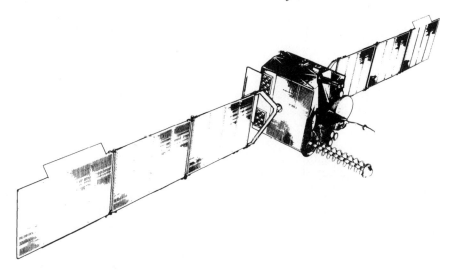

Figure 23.3 SKYNET 4: on-station perspective

An outline of the SHF payload is shown in Figure 23.4. The SHF amplifier technology is based on gallium-arsenide FETs with a receive noise temperature of 700 K (1000 K in orbit including the earth's contribution). The SHF portion of the payload comprises four channels with bandwidths ranging from 60 to 135 MHz, and single conversion is employed with a translation frequency of, 725 MHz. The uplink antenna is an earth-cover HE_{11} mode corrugated horn in all cases, except channel 4 where a spot beam is selectable to give a higher gain for small terminals.

There are a variety of transmit antenna options, with separate higher-gain offset-fed reflector antennas serving to concentrate the transmit power over smaller regions of the earth, to provide increased capacity to small terminals. In addition to earth cover, the widebeam antenna serves interests in the North Atlantic region, the narrowbeam serves the European area and the spot beam serves central Europe (specifically for small manpack and other tactical army terminals). The specified coverage regions are shown in Figure 23.5.

The SKYNET 4 antenna plan is shown in Figure 23.4 and the channel parameters are summarised below, illustrating the increased downlink EIRP as the coverage area is reduced in size.

A beacon transmission is provided at SHF from a solid-state transmitter. The transmitter feeds an earth-cover antenna, and may be used by terminals for acquisition and tracking. The beacon also carriers full satellite telemetry and timing signals. The SHF payload incorporates several filter units: these include bandpass filters to prevent intermodulation products from multicarrier signals in one channel falling into adjacent channels; bandstop filters to reject interfering signals at the beacon frequency; bandstop filters to reject transmitted

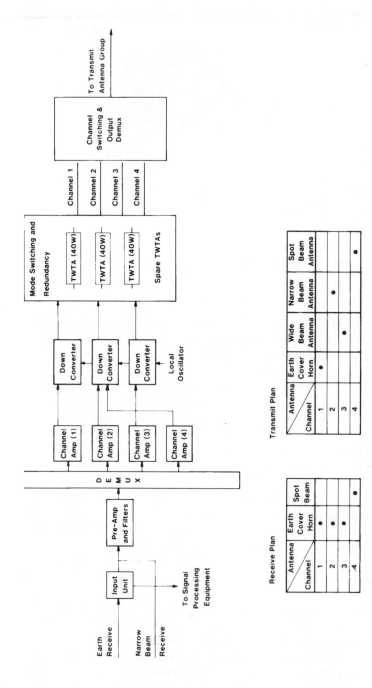

Receive Plan

Antenna Channel	Earth Cover Horn	Spot Beam
1	●	
2	●	
3	●	
4		●

Transmit Plan

Antenna Channel	Earth Cover Horn	Wide Beam Antenna	Narrow Beam Antenna	Spot Beam Antenna
1	●			
2			●	
3		●		
4				●

Figure 23.4   SKYNET 4: SHF payload simplified outline, and antenna plan

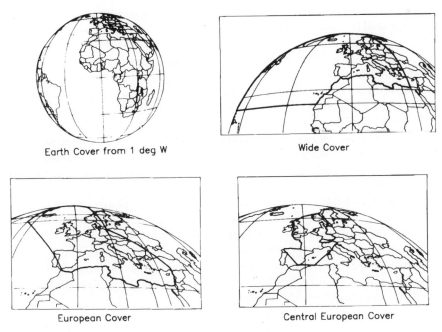

Earth Cover from 1 deg W

Wide Cover

European Cover

Central European Cover

Figure 23.5 SKYNET 4: specified coverage areas

Table 23.1 SKYNET 4 channel parameters

Channel	BW (MHz)	Transmit coverage	EIRP (dBW)
1	135	earth beam	31
2	85	narrowbeam	34
3	60	widebeam	35
4	60	spotbeam	39

noise and intermodulation products at the receiver frequency; low-pass filters to reject TWTA harmonics.

Two UHF channels are also provided, each with 25 kHz bandwidth, and operate within the band 305–315 MHz (uplink) and 250–260 MHz (downlink). A common helix antenna is used for both transmit and receive, with a multiplexer unit separating the two. This antenna is relatively large (2.4 m length) yet it provides only earth coverage. The UHF transponder is all solid state, with each channel delivering 40 W of RF power and an EIRP of 26 dBW. Receive G/T is −18 dB/K.

A self-contained spread-spectrum onboard receiver provides communication facilities at SHF with protection against jamming. For additional protection, a nulling antenna is also provided for use with the SHF payload.

An experimental EHF payload is incorporated into each stage 1 satellite. This payload is an advanced research and development package and is aimed at providing an in-orbit capability to assist in the investigation of the future exploitation of the EHF bands.

The SKYNET 4 payload may be reconfigured by telecommand to change the SHF antennas, for example. Furthermore, there is considerable redundancy built into the payload, as reliability is a major consideration. This redundancy may be switched in through telecommand from dedicated computer facilities on the ground.

Because SKYNET 4 provides facilities for all three UK armed services, it is necessarily complex: if a greater number of satellites were provided, each dedicated to a particular user community (as is more common in the US), they might be somewhat simpler. The close proximity of a number of antennas on the satellite gives particular problems of electromagnetic compatibility (EMC); the use of multiple satellites is currently very expensive and the single satellite represents the most cost-effective solution.

23.5 Traffic and terminals

23.5.1 Traffic

Military communications traffic may be categorised as strategic or tactical; the former generally relates to large and fixed terminals, and the latter to small and mobile terminals. This terminology is usually extended to the terminals themselves, even though both types of traffic may be handled by all terminals. Strategic traffic (at SHF) is handled in a similar way to civil traffic. The traffic patterns and routing requirements may, however, be less predictable than in a civil scenario.

Tactical terminals are of major importance to the military, and their usage differs from most existing civil applications. The number of small terminals deployed may be very much greater than the number of large terminals, and each is essentially independent with single-channel-per-carrier (SCPC) transmission. Their capacity is limited, and traffic may be a combination of a few data and speech channels. Data are often telegraphed at 50 or 75 baud, several channels of which may be multiplexed, prior to transmission, into a higher data rate. Other forms of data may include transfers between battlefield computers and sensing devices, or slow-scan television.

Speech in military systems is generally in digital form to allow the use of encryption devices. Two forms of speech traffic are common: 2.4 kbit/s vocoded speech and 16 kbit/s continuously-variable slope-delta modulation (CVSD), which is a simple form of adaptive delta modulation, akin to one-bit differential pulse-code modulation (PCM). CVSD has the merit of tolerance to relatively high bit-error rates, BER of around 10^{-2} may still be intelligible, although the use of encryption devices may worsen the situation through error extension. Furthermore, voice recognition and tolerance to background noise are superior to that found in vocoded systems. CVSD at 16 kbit/s is common in army communications (being applied mainly to UHF portable radios where simple lightweight encoders have been mass produced).

Vocoded speech at 2.4 kbit/s can be provided by a range of encoders which with modern IC technology can be acceptably small. There are several schemes currently employed, including those using some form of linear predictive coding (LPC). Vocoders of this type produce a model of the vocal tract which is periodically updated; the parameters of this model are transmitted over the channel and used to drive a voice synthesiser, which is essentially a transversal filter fed by a pulse stream at the pitch rate. There are other forms of channel vocoder, which essentially split the speech spectrum into narrow frequency bands and transmit their amplitudes.

Vocoded speech systems, however, may not be very tolerant of errors (because of the high information content per bit, and the structured format), and a channel BER of not worse than 10^{-2} is usually required: if cryptographic encoding is also used then, again, the requirements may be even more stringent. Vocoded speech at 2.4 kbit/s has been used traditionally by the navy, where its narrow BW suits HF radio channels.

There are high bandwidth uses of milsatcoms too, as a trunk communications pipe into theatre and as part of the research programme destined for future operational use of milsatcoms. As an example, the author has run 2 Mbit/s trials using SKYNET 4 and NATO IV satellites as part of UK-based research and international collaboration with allied nations. In the civil satellite communications arena much higher bandwidths are available, as a highly sophisticated example will show. The NASA advanced communications-technology satellite (ACTS) has run 620 Mbit/s trials.

23.5.2 *Modulation*

A variety of carrier-modulation schemes are used in milsatcoms. The choice of these may be partially dictated by the need for operation in a suitable multiple-access scheme, or in conjunction with some form of spread-spectrum technique. It is generally important that signals are constant envelope within a transponder, suggesting the use of PSK or FSK rather than ASK. Binary PSK (BPSK) and four-phase PSK (QPSK) are commonly used, together with variants of these which marginally improve upon BW and envelope characteristics. Phase coherence is usually lost between hops in a frequency-hopped spread-spectrum system (FHSS) which makes FSK the preferred modulation scheme for this type of spread spectrum.

23.5.3 Coding

Error-control coding is widely used and is a feature of most modem designs, as it helps to maintain high integrity for data communications and speech links, and may also provide link power benefits. However, given that speech links may be working at relatively high error rates in the presence of fading, the overall benefits in terms of power budgets need careful consideration.

Coding is particularly important for frequency-hopped spread spectrum where a combination of bit interleaving, coding and symbol diversity maintain the data integrity in the presence of jammed hops. Advanced spread-spectrum modems can employ combinations of coding schemes (e.g. convolutional and Reed–Solomon codes).

23.5.4 Terminals

Milsatcom systems employ a variety of terminals, ranging from large fixed stations to smaller portable or mobile terminals. Each milsatcom system will include one or more main base or anchor stations. These serve as an interface between satellite traffic and fixed networks, as switching and routing centres, and can often be colocated with TT&C facilities and the system control and operational staff.

Figure 23.6   UK Military Satellite Communication system main anchor station at RAF Oakhanger

The feature distinguishing military anchor stations from their civilian counterparts is the fact that they generally handle many different types of traffic, and use a variety of types of modem and multiple-access scheme. In other respects they may be very similar to civil stations, with antennas of diameter 20 m or more and with transmit power of tens of kilowatts at SHF. Figure 23.6 shows the main UK military satellite communications system main anchor station at RAF Oakhanger.

Anchor stations provide local loopback facilities, to permit communication between small satellite communication terminals, which would otherwise not be possible directly because of power constraints. Anchor stations also provide patching and switching connections to a variety of terrestrial networks, using cable, fibre-optic and microwave links. Anchor stations are generally major facilities requiring major investment; the procurement of an anchor station can be as costly as the procurement of the space segment itself. RAF Colerne is the main backup anchor station with DERA Defford providing further cover for operational users.

Besides a few large stations, a range of smaller fixed terminals may be employed to handle traffic at military centres. The majority of their links will be to or from an anchor station, but some terminals may also communicate directly with one another or with small tactical terminals.

Land tactical terminals come in a variety of sizes, being transportable by a vehicle, by air drop or by a soldier. A typical terminal is the UK VSC 501 (Figure 23.7), which is transported by a Land Rover, and may be erected within a few minutes. This has a 1.9 m dish and a TWTA giving 60 W to provide one or two speech channels and telegraph channels; multiplexed data channels may also be used. A similar terminal is the UK TSC 502 (Figure 23.8), which is also intended for rapid deployment, and may be transported by air.

A smaller terminal is the UK man-portable PSC 505 Manpack (Figure 23.9), with a 45 cm dish, digital speech (if operating in a satellite spot beam),

Figure 23.7   UK portable SHF Satcom Terminal VSC 501

Figure 23.8   UK portable SHF Satcom Terminal TSC 502

Figure 23.9   'Manpack' SHF Satcom Terminal PSC 505

Figure 23.10 '*Manpack*' *UHF Satcom Terminal*

analogue speech and 50-baud telegraph. Highly portable terminals of this type are extremely useful to land mobile forces, but have the disadvantage that a high satellite downlink EIRP is required to support each transmission. The terminals perform optimally in regions covered by uplink and particularly downlink spot-beam antennas. A description of the development of the prototype Manpack is given in Reference 2.

The UHF Manpack offers particular convenience of use for communication to and from medium-sized terminals and is shown in Figure 23.10. Although these UHF manpacks are convenient to use, the extreme limitation on the availability of UHF satellite capacity and the well established vulnerabilities of the UHF band mitigate against their general use for milsatcom.

Land mobile terminals are also of importance and may be used on tanks and other military vehicles. Their application is restricted primarily by constraints on antenna mounting.

Ships are prime users of milsatcoms and have traditionally used both SHF and UHF. The exploitation of EHF is also of current interest, largely as it can provide better antijam (AJ) and low-probability-of-exploitation (LPE) performance.

At frequencies above UHF, ship motion makes antenna pointing demanding; installation space is also at a premium and parts of the superstructure may block the field of view (known as wooding).

Figure 23.11   'SCOT' Navy Satcom Terminal

The SCOT is the principal example of a UK milsatcom ship terminal. This terminal uses two antennas, one either side of the mast, to alleviate wooding. These are contained in weather-protecting radomes, as shown in Figure 23.11. The SHF dish size is in the range of 1–2 m.

Submarines need to maintain a low profile, and to remain submerged whenever possible. Their use of milsatcoms is currently confined to UHF, with a simple antenna deployed above the surface for short periods. Future developments in small SHF and EHF antennas may be foreseen and there is also interest in optical frequencies where blue-green lasers offer communications to submerged submarines. To minimise detection, message durations are kept as short as possible.

Aircraft milsatcoms are less well developed. There are increased problems of antenna pointing and stabilisation, and Doppler shift. For high-speed fighter aircraft these problems are even more severe. Potential outages may also occur owing to path obstruction while manoeuvring. Antenna size is severely constrained on an aircraft, and values of receive G/T of 0 dB/K may be typical. The UK Marconi airborne satellite communication terminal (MASTER) demonstrator programme[3] has proven the practicaly of SHF satellite communications for Nimrod maritime reconnaissance aircraft and it is intended to start installing satellite communications for tactical aircraft in the next decade.

23.6 Link budgets and multiple access

23.6.1 Link budgets

The performance and capacity of any satellite communications system is largely dependent upon the link budgets. The wide variety of simultaneous links through a shared transponder, together with the fact that the majority of military links are power (rather than bandwidth) limited, distinguishes most military from civil systems. It is generally downlink EIRP which is at a premium, and for each user there is a potential trade off between power, antenna size and achievable data rate. An understanding of link budgets is thus crucial to milsatcom designers, planners and operators.

Transponders may be operated in a backed-off mode, or close to saturation. In the former case, the transponder output power for an access will be linearly related to its input power. In the latter case, the fixed available output power will be apportioned between accesses and noise according to their proportion of the total input power to the transponder together with any small signal suppression effects.

Small-signal suppression (SSS) arises when more than one signal shares a common limiting (i.e. saturating) amplifier. The analysis of this nonlinear situation can be very complex, especially where several signals are concerned. There are two simple extreme cases however:

(i)   A single carrier signal with power which is far below accompanying wideband gaussian noise: the carrier suffers additional SSS of 1 dB over and above the simple power apportionment between the signal and the noise.
(ii) A single carrier signal in the presence of another much higher power carrier (e.g. a CW jammer): the carrier suffers additional SSS of up to 6 dB.

Front-end thermal noise must be included among the input power levels: some uplink accesses may be operating well below the power levels of others, or even below the total transponder noise power, i.e. an access with an overall negative (in dB) signal-to-noise ratio (SNR) in the transponder BW. Thus, with uplinks from the smallest terminals alone, we find that the satellite downlink power comprises mainly thermal noise! In practice a transponder will often carry many signals with a wide dynamic range and, in the presence of jamming, great care must be taken to calculate the effects of saturation, including small-signal suppression and intermodulation.

It can easily be seen that the capacity of small terminals is relatively limited, and that they may require a significant proportion of available satellite EIRP in order to operate.

23.6.2 Link margins

Suitable margins must be included in link equations to provide for the effects of propagation loss, antenna pointing and engineering implementation. These will depend upon the frequency band, area of operation (including the weather and the elevation angle to the satellite) and type of terminal. In general, military systems are not as controllable or as predictable as their civil counterparts, and larger margins are taken. One need only envisage a soldier operating a small terminal under hostile conditions in a battlefield environment to appreciate that niceties of fractions of a decibel are academic.

Some typical practical single-path (i.e. uplink and downlink) margins, for moderate elevation angles, might be:

UHF:	1 dB (considerably more if multipath fading)
SHF:	3 dB
EHF (20 GHz):	3 dB (for 95% availability)
	8 dB (for 99% availability)
EHF (44 GHz):	10 dB (for 95% availability)
	20 dB (for 99% availability)

Actual link performance in practice will depend on detailed implementation and weather. At EHF particularly, the elevation angle and effect of precipitation, clouds and gaseous absorption will be highly significant, and required margins may vary considerably. Margins are significantly affected by operating region. For example, a link to a satellite from a terminal in the arctic will suffer less attenuation than a link from a terminal situated in an equatorial rain forest but having the same elevation angle to the satellite. At the higher operating frequencies, i.e. SHF and EHF, line-of-sight propagation between the terminal and satellite is assumed. If a reflecting surface intersects the terminal beam, the multipath propagation will need a significant extra margin. This consideration also applies to anomalous propagation at low elevation angles.

23.6.3 Multiple-access techniques

A milsatcom transponder may be required to handle simultaneously a large number of links, operating with different forms of traffic and protocols from a number of terminals. It is characteristic of the military scenario that accesses come and go unpredictably, and requirements may vary rapidly. The system controller will need to respond to this, and in periods of high demand also take into account user priority. In practice, a combination of manual assignment (with computer back up) and automatic demand assignment schemes may be in operation.

The multiple-access problem is that of allocating and implementing the sharing of transponder capacity between a number of terminals, most of which are accessing the channel independently. Individual users may access a channel by prior arrangement, or in conjunction with a polling or request channel. A group of users (a net) may employ their own protocol over an allocated channel (e.g. time-slot allocation or random-access contention operation). A link may also be multiplexed by a user; for example, several telegraph channels may be combined in time division multiplex (TDM) over a 2.4 kbit/s speech circuit.

Although traffic may feasibly be routed directly between medium-sized terminals (e.g. 5 m diameter or more), the link budgets seldom permit direct small-terminal-to-terminal working. This would be achieved via an anchor station with the penalty that a double hop is needed, and the user of the anchor node represents added vulnerability. The additional propagation delay may also be of significance.

One important military requirement is that of the broadcast, which is a transmission from a large station to a number of users, including small terminals. Such a service may be required to operate under the worst-case threat conditions, and may employ protection against jamming.

At UHF, transponders provide only a narrowband capability, representing perhaps a single terminal access at any one time, and control is accordingly fairly straightforward.

At SHF, a number of users may share a wideband transponder and there are often several accesses simultaneously present. If the combined power of these accesses is sufficient to drive the transponder towards saturation, then access and channel noise interact, resulting in power sharing, intermodulation and small-signal suppression. Nevertheless, despite these complications, military transponders are often operated further into saturation than most civil systems, as this maximises the utilisation of downlink EIRP. Indeed, under jamming, operation hard into saturation may be inevitable. Operation near saturation calls for careful planning (with appropriate frequency and power allocations to each user) together with good user power control.

Time-division multiple access (TDMA) may be used by high data rate (strategic) users in a similar manner to civil systems.

Frequency-division multiple access (FDMA) is commonly used in milsatcoms. A frequency slot and power allocation is given to each link, and these are placed at appropriate intervals over the transponder BW, having regard to the precise position of intermodulation products. Usually it will be EIRP rather than available BW which limits the capacity of FDMA systems.

The advantages are:

- users may be independent in terms of power, traffic and modulation scheme (e.g. analogue, PSK, FSK etc.);
- no overall time synchronisation is required;
- implementation is relatively simple and cheap.

The disadvantages of FDMA are:

- significant intermodulation products owing to mixing of frequencies in the transponder; these can interfere with other links, especially where strong and weak (i.e. high and low data rate) links are mixed; intermodulation power also wastes satellite EIRP;
- signal suppression of weak carriers by strong carriers may be important, especially with mix of link types.

The use of a large back off from saturation of the TWTA will significantly reduce the above disadvantages of FDMA; however, this will be at the expense of an overall reduction in transponder EIRP.

Code-division multiple access (CDMA) is an alternative technique in wide military use. It is a form of spread-spectrum (SS) communication and is also known as spread-spectrum multiple access (SSMA). Most CDMA systems use the direct sequence (DS) form of SS, but frequency-hopping (FH) versions are being used increasingly. Somewhat different implementations of SS are widely employed for antijam purposes.

In CDMA, each user modulates their transmit signal (typically at a few kbit/s) onto a wideband spreading waveform, which for DS-CDMA would be a pseudorandom code at a chip rate of a few Mchip/s. Thus the transmission occupies a wide bandwidth. The wanted receiver correlates its input with the same spreading function, suitably synchronised, to recover the signal. Each transmit-receive pair employs a unique code, which is uncorrelated with codes in use by other transmit-receive pairs. In this way transmissions from unwanted users, and also any interference (including intermodulation products), are rejected by the receiver despreading process, and may be modelled simply as gaussian noise.

Additionally, the noise-peak levels of discrete intermodulation products generated in a transponder from the CDMA accesses are considerably smaller than would be the case with narrowband signals, permitting operation closer to saturation.

CDMA users in a channel may overlap in frequency totally, partially or not at all. As the number of overlapping accesses increases, the effective noise level rises and performance degrades.

Although DS-CDMA is theoretically not as spectrally efficient as FDMA or TDMA, in practice the reduction in intermodulation levels permits a greater number of users to share a power-limited transponder than would be the case with FDMA. A penalty is that the terminal equipment is a good deal more complex, and costs more. An advantage, however, is that less stringent control and planning of CDMA accesses is required than in either FDMA or TDMA schemes. Figure 23.12 depicts the outline of a DS-CDMA transmitter and

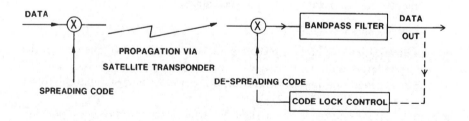

TERMINAL TRANSMITTER   TERMINAL RECEIVER

Figure 23.12   Direct Sequence CDMA: outline implementation

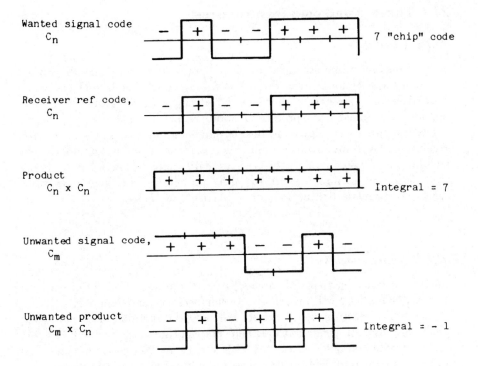

C_n and C_m are "Quasi Orthogonal".

Code periods are typically hundreds or thousands of chips.

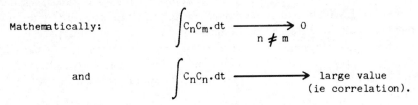

Figure 23.13   Illustration of code correlation in CDMA

receiver (the difficult problem of code synchronisation at the receiver is not addressed here). Figure 23.13 illustrates the concept of correlated and uncorrelated codes in a multiple access scheme.[4]

CDMA is currently being developed for civil small-terminal application. The principal merit of this application is that the relative immunity of CDMA to narrowband interference (and *vice versa*) permits operation where small terminal antenna beamwidths are so broad as to illuminate more than one satellite.[5]

23.7 Threats and countermeasures

23.7.1 General

Milsatcoms are distinguished from civil systems by the requirement to provide communications under threat. Protection against threats tends to be costly and restrictive, and may not be affordable for all users: thus a system is likely to provide different degrees of protection for different classes of user.

Physical threats are fairly obvious. Ground stations are clearly vulnerable, suggesting that some diversification of large anchor stations (for both traffic handling and TT&C) is desirable. Geostationary satellites would require considerable resources to threaten directly, although one can speculate about high-energy lasers and particle beam or directed energy weapons.

23.7.2 Nuclear threats

Nuclear weapons pose special threats.[6] Apart from any effects of direct blast there are radiation, EMP and ionosphere enhancement considerations.

Spaceborne electronics are subject to the natural radiation environment which calls for appropriate shielding and choice of component technologies (e.g. TTL or CMOS-on-SOS rather than NMOS devices). Nuclear-radiation threat levels may be considerably higher than the natural environment and require further protection (hardening). Nuclear hardening is commonly employed in many military equipments and adds to costs.

A nuclear detonation also produces an electromagnetic pulse (EMP). Whereas the primary EMP may be insignificant at great distance, significant secondary EMP effects can be induced in a spacecraft structure by the initial incident radiation pulse. This produces unwanted currents in a spacecraft, and careful engineering and shielding are required to protect electronic circuitry. Such measures are similar to those required on all spacecraft against electrostatic discharge (ESD), which can follow the build up of charge on external insulating surfaces.

High-altitude nuclear explosions (HANE) can cause significant disruption to earth-space propagation over a very wide region of the earth and over a wide range of operating frequencies. This disruption is due to an enhancement of the ionospheric electron density caused by the ionising radiation from a nuclear detonation. A high initial electron density decays with time after the event, as recombination takes place with positive ions. The high initial electron density causes severe attenuation (owing to collisions with atmospheric molecules) and later scintillation. These effects are strongly frequency dependent and may disrupt UHF communications for many hours, SHF communications for a few hours and EHF momentarily. Where there is a requirement to operate in the presence of HANE, the use of EHF may be mandatory. Coding and interleaving may also be necessary for extra survivability.

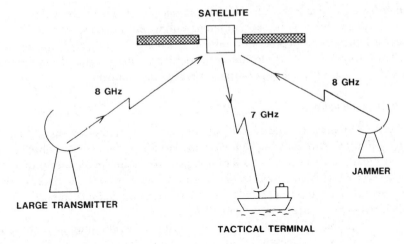

Figure 23.14   Uplink jamming concept

23.7.3 Jamming

Jamming is an attempt by an enemy to prevent communication by swamping a system with radiated power; it may be either uplink or downlink. Uplink jamming of a satellite transponder can be a serious threat, especially as most satellite receive antennas view hostile territory. It may be assumed that whatever high power can be radiated by a large ground station, a somewhat larger power can be radiated by a jammer of similar scale. The use of very-high-power gyrotron tubes at higher microwave frequencies may permit extremely high jammer EIRPs.

Figure 23.14 depicts uplink jamming. This will affect the signal directly, resulting in reduced signal-to-(jammer) noise ratio and, as the transponder is power limiting, downlink power will be captured. This capture results in an absolute reduction in the wanted downlink EIRP, plus additional small-signal suppression of up to 6 dB. The effect on communications is that the transponder throughput is greatly reduced, and normal traffic rates may become virtually impossible to maintain.

Downlink jamming from an aircraft or other platforms is also a threat, but one which may be physically removed (e.g. shot down). Both forms of jamming come within the area of electronic countermeasures (ECM).

23.7.4 ECCM

Electronic counter-countermeasures (ECCM), are the steps taken to alleviate ECM. In order to maximise communications capacity under jamming, it is necessary to remove as much jammer power as possible by techniques on board the satellite and/or at ground terminals. Although antijam (AJ) protection may be usefully applied at terminals, it may also be worthwhile to prevent an uplink jammer from capturing the satellite transponder power. There are two principal AJ techniques which may be used: antenna nulling and spread spectrum. Each of these can achieve some degree of protection, at a penalty of added complexity and cost. The introduction of such techniques generally results in reduced communications throughput, even when operating in a benign environment.

23.7.5 Antenna nulling

Antenna techniques on board the satellite can help alleviate the effects of uplink jamming. A reduced uplink coverage area may be used to enhance the wanted signal at the expense of a jammer, provided that the two are physically well separated. This is a simple and obvious measure, but requires a large aperture antenna and provides only limited discrimination; more importantly, it conflicts with the requirement for global coverage.

The concept may be extended, however, to the provision of an array of spot-beam antennas, with selection of the appropriate coverage region. This would be integrated as a multiple beam antenna (MBA), employing a number of feeds having a common dish reflector or waveguide lens structure. An MBA has the merit of providing flexible coverage with high gain, but in its simplest form gives only limited jammer rejection.

Improved jammer rejection of specific interference sources may be achieved by combining the signals from two or more elements of an MBA. One simple example realisation of nulling is shown in Figure 23.15, where the attenuated output from a spot-beam antenna, with a narrow beamwidth, is subtracted from that of an earth-cover antenna with a wide beamwidth. On a purely amplitude basis, it can be seen that a narrow (and unique) null will be produced.

Conceptually, this technique is similar to that of the interferometer, which is depicted in Figure 23.16. Here a signal is received by two identical antennas, the

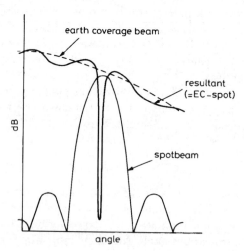

Figure 23.15   Example of nulling antenna implementation

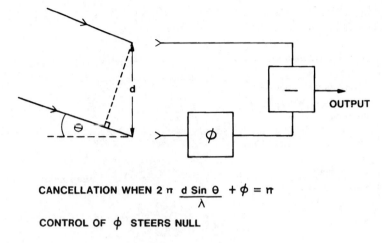

CANCELLATION WHEN $2\pi \dfrac{d \text{ Sin } \theta}{\lambda} + \phi = \pi$

CONTROL OF ϕ STEERS NULL

Figure 23.16   Interferometer principle for nulling

outputs of which are subtracted after imposition of phase shift. The relative phase is a function of the path difference and depends upon the angle of incidence; thus cancellation can be produced in a particular direction. This direction can be varied by control of the phase shifter.

The use of several antenna elements, together with phase and amplitude control and combination, may permit considerable flexibility as a nulling antenna, and may allow simultaneous nulling of several interference sources. It might also be possible to synthesise area nulls, for example over hostile territory, without knowledge of specific jammer locations.

Such sophisticated antennas would represent considerable complexity on a spacecraft. Constraints include the insertion loss which inevitably affects even wanted users, and there are demands for maintenance of performance over wide bandwidths and environmental temperature ranges. Efficient control is a major aspect, and this may be either by remote telecommand or locally through onboard adaptive algorithms; if the latter can be achieved, this helps to alleviate some of the above difficulties.

Jammer nulling on a satellite is a far greater problem than for most land-based applications, as the narrow field of view may contain both wanted users and a number of interference sources with very small angular separation between them. There are clearly difficulties in distinguishing jammers from wanted signals, and future systems may be expected to rely upon distinctive signal coding, leading to integration of antenna subsystems with spread-spectrum processors.

23.7.5.1 Spread-spectrum techniques

Spread spectrum (SS) is an AJ technique which relies on the wanted user spreading the signal with a spreading function which cannot be replicated by an

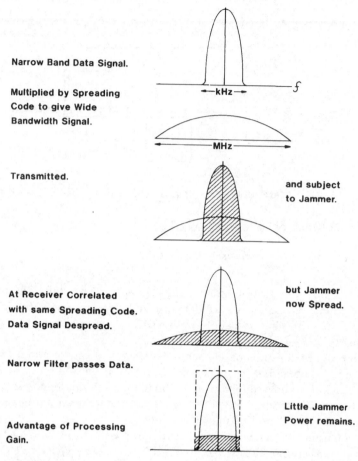

Narrow Band Data Signal.

Multiplied by Spreading
Code to give Wide
Bandwidth Signal.

Transmitted.

and subject
to Jammer.

At Receiver Correlated
with same Spreading Code.
Data Signal Despread.

but Jammer
now Spread.

Narrow Filter passes Data.

Advantage of Processing
Gain.

Little Jammer
Power remains.

Figure 23.17   Jamming protection through spread spectrum

enemy. The receiver performs the inverse despreading operation, and the original signal is recovered through a narrow bandpass filter. This process spreads any uncorrelated interference, such as jamming, and the bulk of this is removed by the filter. The advantage given to the wanted signal over the interference is called the processing gain (PG), and is given broadly by the ratio of the spread bandwidth to the signal bandwidth. Figure 23.17 illustrates the concept.

There are two basic spread-spectrum techniques: direct sequence (DS) also called pseudonoise (PN) and frequency hopping (FH).

Direct sequence involves the linear modulation of the signal with a pseudorandom biphase code, typically at several Mchip/s; thus the effective spectral width might be increased from (say) 10 kHz to 20 MHz. At the despreading receiver, the identical operation is performed with the same spreading code, suitably synchronised in time and code phase. With these example parameters, a PG of 33 dB could be achieved—a typical figure.

Technology tends to limit the code chip rate to a few tens of MHz, restricting the PG. The impact of nuclear scintillation is to further reduce the available PG by limiting the coherence BW of the propagation path.

DS has been established in military communications for many years and has the advantage that it may be used in a transparent fashion with virtually any form of constant-envelope modulation.

Frequency hopping requires the carrier frequency to jump in discrete hops over a wide BW; the receiver recovers the signal by hopping its local oscillator in synchronism. Outline implementation is shown in Figure 23.18. A narrowband jammer will statistically affect only a small proportion of the hops, and error-correction coding with interleaving at the receiver will deal with this. If the jammer adopts a strategy of spreading power over the hopping bandwidth, processing gain is still obtained because the jammer power will affect the received, dehopped signal which will be reduced by roughly the ratio of the per hop BW to the total spreading BW. With optimum interleaving and coding techniques, it is currently possible to achieve arbitrarily low output bit error rate with up to 30 per cent of hops being jammed. Hopping rates may range from 10 hop/s to 20 khop/s. The processing gain achievable is set principally by the satellite transponder BW and is not essentially technology limited.

The price paid for spread spectrum is the complexity and the difficulty of achieving synchronisation at the distant receiver; e.g. a 10 Mchip/s DS system requires time synchronisation (sync) to a few nanoseconds. FH systems are more robust, and may maintain sync for long periods with conventional crystal clocks set up on the time of day. The acquisition and tracking circuitry itself has to function under conditions of jamming. Initial acquisition time may be a significant parameter of a receiver, and will itself be a function of the required maximum operating jammer-to-signal (J/S) ratio and of initial timing or code-state uncertainty.

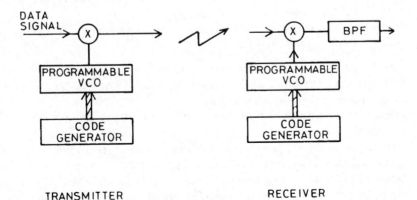

TRANSMITTER   RECEIVER

Figure 23.18   Frequency hopping: outline implementation

In milsatcom, spread spectrum may be used end-to-end with some advantage over a transparent wideband channel. Significant performance can only be maintained under heavy jamming if the despreading receiver itself is on board the satellite. The synchronisation problems here are evident, given uncertain propagation delays and Doppler shift; the engineering of a self-contained spaceborne SS receiver represents a major technological challenge. The cost and complexity leads to such a facility being realistic for only a relatively small community of users.

An essential feature of spread spectrum for AJ purposes is that the spreading code must not be capable of replication by an unauthorsised user. This means that no amount of knowledge of past transmission will enable prediction of future code states, and for this reason AJ systems will tend to use not simple repetitive codes (e.g. *m*-sequences, or Gold codes) but nonlinear spreading codes which do not repeat during the lifetime of the system.

With increasing threat levels and requirements to work from small terminals, greater spread-spectrum processing gains are being called for: this is leading to the exploitation of EHF, where there is up to 2 GHz of allocated BW. The processing gain requirements will be met by FH, and hybrid SS schemes, where a DS signal is frequency hopped, are also feasible. General and readable descriptions of SS techniques may be found in References 7 and 8, and References 9 and 4 provide full and detailed analyses for the system designer.

23.7.6 Low probability of exploitation (LPE)

Reception of milsatcom signals by an enemy represents a significant exploitation threat, especially for tactical terminals. It may permit location of the terminal, together with identification of its type and its activity (the traffic itself is usually protected through encryption).

Although the downlink from a satellite may be readily received and yield some information, the reception of the uplink often presents a more direct threat as terminal location may be determined directly by direction finding (DF).

The exploitability of a signal is determined, in the limit of low signal strength, by its detectability against (enemy) receiver noise.

Among important measures which a system designer may incorporate to reduce exploitation vulnerability are:

- spread spectrum, which reduces the signal power spectral density and forces an interceptor to look over a greater bandwidth, with a correspondingly smaller probability of detection;
- operating procedures which minimise transmission power level and duration;
- terminal antenna design to reduce sidelobe radiation, as a guide, it may be taken that the radiation performance of a tactical terminal can lie within the range −10 to 0 dBi in any direction away from the main beam; the −10 dBi figure might apply to a land-based terminal well clear of surrounding clutter, whereas the 0 dBi figure might apply to a ship where the antenna may be cluttered by superstructure.

Bearing in mind the significant potential range advantage of an enemy receiver compared with a geostationary satellite, it can be appreciated that exploitation can be a very significant threat. For further reading on satellite communication LPE, see References 10, 9 and 7.

23.7.7 Geolocation

For many commercial satellite operators unauthorised accesses on their satellites is a growing concern, and these unauthorised accesses from interferers are likely to increase with the growth of traffic worldwide. Most interference is transient in nature, quickly located by administrative means, but some interference is long term and may cause considerable disruption, requiring the reallocation of resources. The technique is not a conventional DF technique and therefore does not require a direct detection of the uplink. The concept is outlined in Figure 23.19 and the results depicted in Figure 23.20.

The DRA has operated a transmitter location system (TLS) since the mid 1980s for military satellites. In 1993 a commercial organisation requested proposals for the provision of a study aimed at identifying the most suitable cost-effective solution for the location of interferers with their satellite communication services and demonstrating its performance. DRA won the contract for the study and measurement campaigns, which progressed to a fully operational service contract which was awarded in 1995 and continues today.

The military implications are that when either jamming an enemy satellite or communicating via satellite communications on your own (or allied owned)

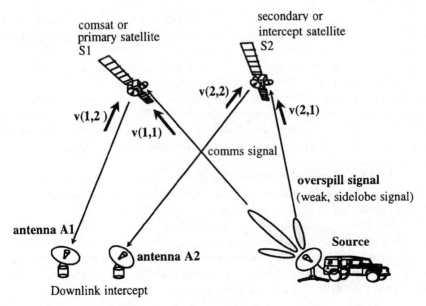

Figure 23.19   Interference source location

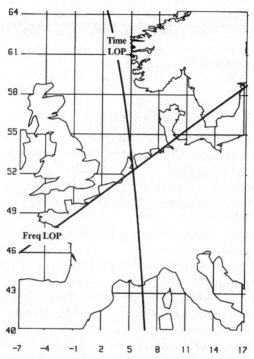

Figure 23.20   Transmitter location from intersection of time and frequency LOPs

system, the user is potentially vulnerable to geolocation, with potentially undesirable consequences.

The technique is based on the simultaneous determination of the differential
slant range (DSR) and its time derivative, the DSR rate (DSRR) that pertains
to the interferer's location and to two satellites.[11] The DSR is determined by a
knowledge of the satellite's position and by measuring the path-length difference
between the interferer and the target satellite and the interferer and an adjacent
satellite which, owing to the nature of the radiation pattern from the antenna, is
receiving some overspill signal from the interferer. The measurement is taken by
determining the time difference of arrival (TDOA) at two colocated ground
terminals, visible to the target satellites. For this technique to be viable, the interferer's signal must be modulated.

The DSRR is determined by knowing the velocities of the satellite and determining the frequency difference of arrival (FDOA) of the signals at the
measuring stations.

For any given satellite geometry, lines of position (LOP) can be drawn on
the earth which are the loci corresponding to the FDOA and TDOA. Where
they intersect is the estimate of the location of the transmitter. For satellites with
small eccentricities and small inclinations (the case for most commercial service
providers), the LOPs for the DSR look like straight lines and do not vary much

with time, however, in contrast, the DSRR LOPs depend on the difference in the satellites velocities, which is strongly dependent on time and has components both in the direction of earth and perpendicular to it. This behaviour has a useful property, inasmuch as when it is impossible to calculate the DSR (owing to the signal being unmodulated) then two measurements of the DSRR at different times may exploit the variability of the LOPs with time and provide a location.

The uncertainty in determination of the DSR and DSRR depends on the uncertainties in the TDOA and FDOA measurements, the uncertainty in the positions and velocities of the satellites and the systematic error introduced by any propagation effects. The location measurement error can be greatly reduced by employing reference signals on both satellites from a known transmitter; it is preferable that the known transmitter data and the interferer signals should be coprocessed with the same bandwidth, integration time and on the same satellite transponders, thus allowing the systematic errors to be reduced to a minimum. The effects of systematic errors in the ephemeris of the satellites is minimised by having the known reference transmitter close to the interferer, allowing an iterative search to be employed, where the initial TDOA /FDOA measurement is used as the basis for providing the reference transmitter location.

The study demonstrated that the TLS can provide useful information on the interferer in most cases, but that sometimes the information may not be sufficient to unambiguously identify the interferer, depending on operational conditions. The sensitivity of the TLS system is dependent on the signal characteristics, the SNR in both the primary and secondary satellite, the integration time of the sample taken to process the antenna size of the interferer, the amount of traffic on the intercept satellite and the sensitivity of the equipment used in the system.

23.8 Future trends in milsatcom

23.8.1 Political and military context

It is now believed that the majority of future conflicts in which British military forces will be deployed operationally will be in the context of joint or coalition forces, under the auspices of the UN, NATO or the WEU. The use and especially the management of milsatcoms will become increasingly important, as will the use of commercial satellite communications in the support of deployments of British forces in peacekeeping and humanitarian missions, which have been increasing over the last few years.

23.8.2 Use of LEOs/MEOs — military use of commercial satellite communications

Procurement of bespoke military satellite communications is expensive. Serious investigations have been started into the possible military use of the LEO/MEO constellations that are about to be deployed for the purpose of personal mobile communications. The vulnerability issues of such usage are being studied, as is the military use of commercial geostationary systems. The two major drawbacks are:

(i)   Commercial organisations may be reluctant, as they have stated in the past, to take on potentially life-threatening military communications in time of crisis.

(ii)   The operation of the satellites would no longer be under national control.

23.8.3 Enhanced onboard processing

To extend protection against jamming to greater numbers of users, especially tactical terminals, greater use may be anticipated of SS onboard processing. This could, for example, handle FH uplinks from small terminals in multiple-access fashion (perhaps with FDMA overlay), and could take advantage of the wide available BW at EHF to give very considerable PG. Other features such as demodulation to baseband, and data routing on board, may be envisaged to be akin to those proposed for civil systems using small terminals.[12] Downlinks may be TDM for efficiency (which eases the intermodulation problem). Some scenarios may call for scanning downlink narrow spot beams to provide a high-peak EIRP on a time-shared basis. Developments may be anticipated in phased-array antennas for spacecraft, where a reflecting dish or lens aperture is replaced by an array of elements, each with suitably controlled amplitude and phase combining. Onboard adaptive algorithms will discriminate against jamming or interference. Onboard processing also allows the up- and downlink budgets to be decoupled, with the advantage of denial of power capture to a jammer.

23.8.4 EHF and optical frequencies

Major exploitation of EHF may be envisaged in future milsatcom systems.[13] The principal uplink band is 43.5–45.5 GHz, with downlinks at 20.2–21.2 GHz. The 2 GHz of uplink BW permits considerable SS PG (e.g. 2 GHz/ 10 kHz = 53 dB), giving realistic AJ protection with onboard despreading. The use of a new band suggests relative freedom from regulatory constraints; orbital spacing of satellites could also be reduced, as even small tactical antennas will be capable of $1°$ beamwidths. EHF promises yet further advantages in ECCM, as the EIRP of a small terminal increases with frequency, and that of very large terminals (e.g. a jammer) tends to reach practical limits.

It has been proposed that SKYNET 5 has an EHF package comprising of (at least) an earth cover, a theatre spot, a steerable spot and an agile beam overlay, with full interconnectivity between them. The military requirements are to provide robust acquisition to disadvantaged users, a highly survivable TT&C system and a highly robust and survivable low data rate capability. The waveform to be employed will be a commonly agreed international standard and the signal will be hopped across the entire band, providing an improved AJ performance. The size of the constellation and the cost to the country depends on the options taken for deployment. This could range from an entirely national UK-only programme to a collaborative effort (with cost-sharing advantages) between the UK and a number of allied nations.

LPE may be considerably enhanced at EHF: apart from potential for wider BW SS (reducing intercept detectability), small terminal antennas may permit narrower beamwidths and improved sidelobe performance, and local rain attenuation may be more likely to disadvantage an interceptor than the user.

Other benefits of EHF lie in the small size and mass of hardware, especially satellite antennas. This may encourage the application of sophisticated onboard null-steering or adaptive antennas, which would otherwise be a problem at SHF. If operation in a post-nuclear scenario is required, the use of EHF may be almost mandatory to overcome nuclear scintillation propagation effects.

The disadvantages of EHF may be summarised as follows:

(i) *Heavy attenuation during rain:* This is a real problem, and can to some extent be overcome by including large link margins (e.g. typically 10 dB 95% link availability in the UK, compared with only about 3 dB at SHF). However, one needs to examine the detailed statistics of rain outages. For military application, the duration of outages may be of more importance than the average availability (which is often the commercial criterion), and it may be more realistic to try and accept short duration outages than to aim for significantly increased link margins.

(ii) *Antenna pointing:* The narrow beamwidths achievable at 44 GHz may demand more accurate pointing mechanisms for terminals e.g. closed loop rather than open loop for 1.7 m terminals. This could be a constraint for smaller man-portable terminals, where the operator is unlikely to be in a position to engage in accurate pointing exercises.

(iii) *High cost and advanced technology:* Component and system costs are high but have fallen as the MILSTAR[14] programme has matured. The ultimate cost impact on satellites themselves may be advantageous owing to size and mass savings.

There is interest in optical laser communication for milsatcom both for inter-satellite links and for downlinks. Calculations suggest that the power and data rate parameters need not be significantly different from those of EHF links, although clearly clouds and rain present special problems.

The optical link could operate at around a 1.3 micron wavelength most likely with Nd: YAG sources, and using pulse-position modulation (PPM).[15] The benefits of optical satellite communications include very good LPE, mainly through very narrow beamwidths, small antenna (i.e. optical telescopes) apertures, wide bandwidth capability and potentially good jamming resistance. Current technology is largely based upon direct detection methods, but coherent detection systems, where the optical signal is heterodyned down to RF, offer considerable potential.

It has also been suggested that the use of blue-green light may permit communication to submarines below the sea surface.[16] This might involve a one-way broadcast from a low orbiting satellite using a modulated scanning spot beam, an optical wavelength to transmission in sea water, and a very narrow receiver optical filter to reject background noise.

23.8.5 System management and networking

As the use of milsatcom and the resource base widens, including the diversification of anchor-station facilities, there is an increasing need to manage these resources to maximise efficiency. Not only does the resource management need to be efficient, but it also needs to have the flexibility to respond effectively and rapidly to system stress, however caused. It is paramount that military communication systems degrade gracefully under stress and do not collapse precipitously.

Concomitant with increasing maturity is the need to integrate milsatcom with other communications systems. Networking concepts that are currently applied to terrestrial systems need application to and harmonisation with the particular characteristics and constraints of milsatcoms. For example, 'robust protocols' need to be further developed that can tolerate the delay and possible jamming experienced on milsatcom links. Interesting work has been done on X.25, IP and is now starting on ATM to this end. The requirement for milsatcom to be seamlessly integrated into terrestrial WANs is being addressed as are issues concerning network architecture and traffic management.

23.8.6 Cost reduction

It has long been recognised that, for a wide variety of reasons, procurement of military equipment is expensive, often to the point of prohibition. Milsatcoms are no exception to this phenomenon. Methods of cost reduction are being investigated. Some areas where cost reductions have been suggested are simpler satellites with less expensive launches,[17] and standardisation of equipment allied with international collaboration in procurement. Another important way of achieving cost reductions is to use as much commercial off-the-shelf equipment as possible (COTS) and to do as little as possible to the COTS product to integrate it into the military systems. An important approach is to consider the degree of ruggedisation as is appropriate to the role of the equipment. With some equipment in certain roles it is cheaper to apply no ruggedisation whatsoever to the COTS equipment and to simply have some spare units available.

23.8.7 Extended coverage

A satellite in geostationary orbit provides coverage to a region bounded by longitudes roughly plus and minus 70 degrees from the satellite longitude and between latitudes 70 degrees north and 70 degrees south. For coverage outside this region, alternative orbits need to be considered. Promising orbits include Molnya (highly-elliptical inclined at 63.4 degrees to the equatorial plane with a 12-hour period), Tundra (as Molnya but with a 24-hour period) and, of special interest currently, low earth orbits. There is considerable interest in the use of the civil commercial services that will be available in the near future, employing constellations of satellites, examples of such programmes include Iridium, Globalstar and Teledesic.

Each orbit has its advantages and disadvantage, dependent on the application, but the relative merits will not be pursued here. A feature of the nongeostationary orbit (of whatever type) is that the coverage is not fixed; consequently store-and-forward techniques and intersatellite and interorbit links are becoming of increasing interest.

In store and forward, data is uploaded from a ground terminal to a passing satellite and stored in memory. At a later time the data is downloaded when the satellite passes over the appropriate receiver. Such a system could be of use in data communication where a significant delay between transmission and reception can be tolerated. Clearly, speech transmission by this means will be of limited application.

Intersatellite and interorbit links may be used to extend the coverage area of a geostationary system, eliminating the need for intermediate anchor stations (and reducing delay), or as links between low-orbiting and geostationary satellites. With reduced dependence on vulnerable intermediate ground anchor stations, overall physical survivability may be enhanced. For optimum technical performance, the link could be either EHF at 60 GHz (which is in the oxygen absorption band, greatly reducing probability of intercept on the ground) or at optical frequencies. Excessive transmit powers are not required, and the main problems lie in the antenna/aperture acquisition and tracking. With present technology, 60 GHz systems appear to have an advantage over optical systems as the final payload mass and power consumption seems to be significantly less. A convenient solution, however, would be to use UHF, where the problems of acquisition and tracking are minimised.

23.9 Brief conclusions

Milsatcoms are distinguished from civil systems by the requirement to handle a variety of different terminals deployed over a wide area, and in the face of potential threats. Most traffic is carried at SHF, using shared wideband transparent transponders, with earth-cover satellite antennas. Spot beams having smaller areas of coverage are also particularly important, however, especially for use with small tactical terminals, which may nevertheless be constrained to operate at low data rates.

Survivability is a prime requirement, against which jamming represents the principal threat. This threat may be alleviated by antenna nulling and by spread-spectrum techniques. Such facilities should ideally be provided on board the satellite itself; they are costly in this form and such protection may only be afforded to a limited community of users.

Future milsatcom developments are likely to include the exploitation of the EHF band, together with enhanced onboard processing, nulling antennas, improved management, cost reduction, higher bandwidths and extended coverage.

23.10 Acknowledgments

Particular acknowledgment is due to Mr. R. L. Harris of the Strategic Communications and Networks Department, DERA Defford, who provided, from his long experience, many detailed contributions towards both the first and second revisions of this Chapter. Many thanks are also due to Dr. D. P. Haworth, DERA Defford, who wrote the first revision to this Chapter and may I also express my gratitude to Mr. T. C. Tozer, University of York (formerly DRA Defford), who wrote the original version. Finally, many thanks to my colleagues within the department and outside who have contributed directly or indirectly.

23.11 References

1 DAWSON, T.W.G. *et al.*: 'United Kingdom SKYNET 4 defence satellite communications'. Proceedings AIAA 9th *Communications satellite systems conference*, San Diego, California, 7–11 March 1982

2 JONES, C.H.: 'A manpack satellite communications earth station', *Radio Electron. Eng.*, 1981, **51**, pp. 259–271

3 CUMMINGS, D., and WILDEY, C.G.: 'Military aeronautical satellite communications', *IEE Proc. F*, 1986, **133**, pp. 411–419

4 HARRIS, R.L.: 'Introduction to spread-spectrum techniques' *in* 'Spread-spectrum communications'. AGARD lecture series 58

5 VITERBI, A.J.: 'When not to spread spectrum—a sequel'. *IEEE Commun. Mag.*, 1985, **23**, pp. 12–17

6 SMITH, E.: 'Nuclear effects on RF propagation', *Defence Electron.*, October 1984

7 TORRIERI, D.J.: 'Principles of secure communication systems' (Artech House, 1985, 2nd edn.)

8 DIXON, R.C.: 'Spread-spectrum systems' (Wiley & Sons, 1984, 2nd edn)

9 SIMON, M.K., OMURA, J.K., SCHOLTZ, R.A., and LEVITT, B.K.: 'Spread-spectrum communications', (Computer Science Press, 1985, 3 vols.)

10 GLENN, A.B.: 'Low probability of intercept', *IEEE Commun. Mag.*, July, 1983, pp. 26–33

11 BARDELLI, R., HAWORTH, D.P., and SMITH, N.G.: 'Interference Localisation for the Eutelsat satellite system'. Globecom 95, Singapore, 13–17 November 1995

12 EVANS, B.G. *et al.*: 'Baseband switches and transmultiplexers for use in an on-board processing mobile/business satellite system', *IEE Proc. F*, 1986, **133**, pp 356–363

13 HAWORTH, D.P. *et al.*: ' Rationale for the future use of EHF for military satcoms', *Military microwaves 88*, London, 5–7 July 1988

14 SHULTZ, J.B.: 'Milstar to close dangerous C^{3}I gap', *Defense Electron.*, 1983, pp. 46–59

15 SVOREC, R.W., and GERARDI, F.R.: 'Space laser communications', *Photonics Spectra*, April 1984, pp. 78–81

16 WIENER, T.W., and KARP, S.: 'The role of blue/green laser systems in strategic submarine communications', *IEE Trans.*, 1980, **COM 28**, pp. 1602–1607

17 RAWLES, J.W.: 'LIGHTSAT: All systems are go', *Defense Electron*, May 1988, pp. 64–70

Chapter 24

Microsatellites and minisatellites for affordable access to space

M.N. Sweeting

24.1 Introduction

Satellites have become increasingly large and expensive, generally taking many years to mature from concept to useful orbital operation, and this process has limited access to space to only a relatively few nations or international agencies. Furthermore, new ideas, technologies and scientific experiments find it difficult to gain timely access to space, and space technology lags considerably that now taken for granted on earth. The advance of increasingly low-power, miniature electronic components, combined with growing financial pressures, have focused attention on the use of small satellites to complement large satellite systems for many space applications. Indeed, in the field of satellite communications, numerous constellations of small satellites have been proposed to provide a range of global services either real-time voice or nonreal-time data.

However, whether a particular satellite is large or small depends somewhat upon viewpoint. For instance, the small satellites proposed for the Iridium system weigh in at around 600 kg each whereas those for HealthNet are a mere 50 kg! In view of this, the classification in Table 24.1 has become widely adopted.

Although there have been many examples of large, small and even minisatellites, it is only relatively recently that capable microsatellites have shown that it is possible to execute both civil and military missions very effectively, rapidly and at low cost and risk, for the following applications:

Table 24.1   *Size classification for satellites*

Class	Mass	Cost
Large satellite	> 1000 kg	£ >100 million
Small satellite	500–1000 kg	£30–100 million
Minisatellite	100–500 kg	£7–20 million
Microsatellite	10–00 kg	£2–4 million
Nanosatellite	< 10 kg	£0.5–1 million

- specialised communications;
- earth observation;
- small-scale space science;
- technology demonstration/verification;
- education and training

A sustained, commercial microsatellite programme requires predictable and regular access to orbit through formal launch service contracts—since the ability to construct sophisticated yet inexpensive microsatellites must also be matched by a correspondingly inexpensive method of launch into orbit. In 1988, Arianespace developed the Ariane structure for auxiliary payloads (ASAP) specifically to provide, for the first time, regular and affordable launch opportunities for 50 kg microsatellites into both LEO and GTO on a commercial basis. Although the ASAP has been key to catalysing microsatellites, it cannot now provide the number of launch opportunities needed to meet the burgeoning growth of small satellites and so alternative inexpensive launch options from the CIS on Tsyklon, Zenit, Cosmos and START are now increasingly being used for micro/minisatellites. Satellites in this category have been called little LEOs to distinguish them from their larger brothers. Constellations of such satellites, often with store-and-forward technology, and operating in the 140/150 MHz allocated frequency bands, have recently been proposed. OrbKomm is an example of such a system.

24.2 Surrey microsatellites

The University of Surrey has pioneered microsatellite technologies since beginning its UoSAT programme in 1978, and its space-related research, teaching and commercial activities are now housed within a purpose-built Centre for Satellite Engineering Research—with over 100 staff and postgraduate research students. The objectives of the Centre's programmes are:

- to research cost-effective small-satellite techniques;
- to demonstrate the capabilities of micro/minisatellites;
- to catalyse commercial use of micro/minisatellites;
- to promote space education and training.

Figure 24.1   The Surrey Space Centre and SSTL facilities at the University of Surrey

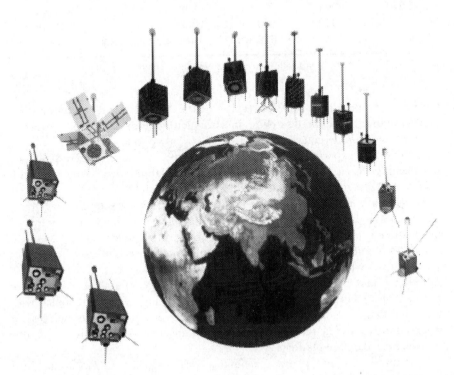

Figure 24.2   SSTL microsatellite missions in LEO

Table 24.2   University of Surrey microsatellite missions

Microsatellite	Launch	Orbit	Customer	Payloads
UoSAT-1	1984-D	560 km	UoS	research
UoSAT-2	1984-D	700 km	UoS	S&F,EO,rad
UoSAT-3	1990-A	900 km	UoS	S&F
UoSAT-4	1990-A	900 km	UoS/ESA	technology
UoSAT-5	1991-A	900 km	SatelLife	S&F,EO,rad
S80/T	1992-A	1330 km	CNES	LEO comms
KitSat-1	1992-A	1330 km	KAIST	S&F,EO,rad
KitSat-2	1993-A	900 km	KAIST	S&F,EO,rad
PoSAT-1	1993-A	900 km	Portugal	S&F,EO,rad
HealthSat-2	1993-A	900 km	SatelLife	S&F
Cerise	1995-A	735 km	CNES	military
FASat–Alfa	1995-T	873 km	Chile	S&F,EO
FASat-Bravo	1998-Z	835 km	Chile	S&F,EO
TMSAT	1998-Z	835 km	Thailand	S&F,EO
UoSAT-12	1999-Z	650 km	SSTL & Singapore	EO,comms
Clementine	1999-A	735 km	CNES	military
PICOSAT	1999	800 km	USAF	military
TIUNGSAT	1999	800 km	Malaysia	S&F,EO
SNAP-1	1999	650 km	UoS	EO,comms
Tsinghua-1		650 km	Tsinghua University	EO,comms

D = Delta; A = Ariane; T = Tsyklon; Z = Zenit; S = SS18/Dnepr

Over the last decade, CSER has established an international reputation as a pioneer of innovative small satellites in a uniquely combined academic research, teaching and commercial environment.

Surreys' first experimental microsatellites (UoSAT-1 and 2) were launched free of charge as piggy-back payloads through a collaborative arrangement with NASA on DELTA rockets in 1981 and 1984, respectively[1,2].

Since then, a further ten low-cost yet sophisticated microsatellites have been placed in low earth orbit using Ariane and Tsyklon launchers for a variety of international customers and carrying a wide range of payloads.

UoSAT-1 and 2 both used a rather conventional structure—a framework skeleton onto which were mounted modules containing the various electronic subsystems and payloads. However, the need to accommodate a variety of payload customers within a standard (ASAP) launcher envelope, coupled with increased demands on packing density, economy of manufacture and ease of integration, catalysed the development of a novel modular design of multi-mission microsatellite platform.

This novel structure is based around a series of standard CNC-machined module trays which house the electronic circuits and themselves form the mechanical structure onto which solar arrays are mounted. Each module box

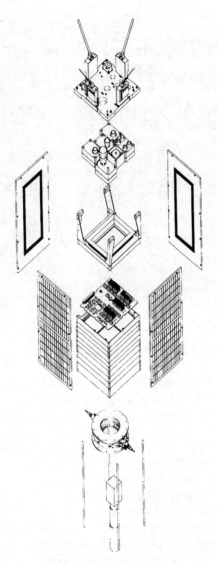

Figure 24.3   Exploded view of microsatellite structure

houses the various microsatellite subsystems—eg. batteries, power conditioning, onboard data handling, communications and attitude control. Payloads are housed either in similar modules or on top of the platform alongside antennas and attitude sensors as appropriate.

Electronically, the microsatellite uses modern, sophisticated, but not necessarily space-proven, electronic circuits to provide a high degree of capability. These are underpinned by space-proven subsystems—resulting in a layered architecture which achieves redundancy by using alternative technologies rather than by simple duplication.

Figure 24.4   Typical microsatellite CNC module

Communications and earth-observation payloads require an earth-pointing platform and so the microsatellite is maintained to within two degrees of nadir by employing a combination of gravity-gradient stabilisation (using a 6 m boom) and closed-loop active damping using electromagnets operated by the onboard computer. Attitude determination is provided by sun, geomagnetic field sensors, and star-field cameras, and orbital position is determined autonomously to within ± 50 m by an onboard GPS receiver.

Electrical power is generated by four body-mounted GaAs solar-array panels, each generating around 35 W, and is stored in a 7 Ah NiCd rechargeable battery. Communications are supported by VHF uplinks and UHF downlinks, using fully error-protected AX.25 packet link protocols operating at 9.6 to 76.8 kbit/s, and are capable of transferring several hundred kbytes of data to briefcase-sized communications terminals. This modular microsatellite platform has been used successfully on twelve missions, each with different payload requirements.

It is the onboard data handling (OBDH) system that is the key to the sophisticated capability of the microsatellite. At the heart of the OBDH system is a 80C386 onboard computer, which runs a 500 kbyte real-time multitasking operating system with a 128 Mbyte solid-state CMOS RAM. In addition, there is a secondary 80C186 OBC with 16 Mbyte SRAM, two 20 MHz T805 transputers with four Mbytes SRAM and a dozen other microcontrollers. A primary feature of the OBDH philosophy is that all the software onboard the microsatel-

Figure 24.5   Integrated microsatellite structure

lite is loaded after launch and can be upgraded and reloaded by the control ground station at will thereafter. Normally, the satellite is operated via the primary OBC-386 and the real-time multitasking operating system. All telecommand instructions are formulated into a diary at the ground station and then transferred to the satellite OBC for execution either immediately or, more usually, at some future time. Telemetry from onboard platform systems and payloads is similarly gathered by the OBC-386 and either transmitted immediately and/or stored in the RAMDISK while the satellite is out of range of the control station. The OBCs also operate the attitude-control systems according to control algorithms which take input from the various attitude sensors and then act accordingly. Thus it is this OBDH environment that allows such a tiny microsatellite to operate in a highly complex, flexible and sophisticated manner, enabling fully automatic and autonomous control of the satellite systems and payloads.

The latest SSTL microsatellite platforms have enhanced subsystems supporting the following major new features for greatly increased performance and payload capacity. The first version of the enhanced microsatellite platform was used on the FASat-A mission for Chile, launched into LEO by a Ukrainian

Figure 24.6   Microsatellite orbital configuration

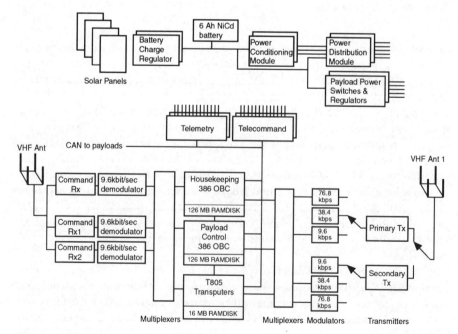

Figure 24.7   SSTL microsatelllite platform systems

Tsyklon launcher in 1995, and has since been further developed and used on the new FASat-Bravo, TMSAT and TIUNGSAT, PICOSAT and Clementine missions. The very latest microsatellite platforms can provide:

- distributed telemetry/telecommand for easy expansion;
- 5 MIPS (386) OBC + 256 MB RAMDISK;
- 1 Mbit/s onboard LAN;
- autonomous GPS navigation (50 m);
- attitude determination (0.001°);
- 1° nadir pointing (three-axis).

24.3 Applications of micro/minisatellites

The early UoSAT missions demonstrated the potential capabilities of micro-satellites and generated considerable interest in applications such as digital store-and-forward file transfer and in-orbit technology verification, however wider applications such as space science and remote sensing were slow to develop. Emerging space nations were quick to recognise the benefits of entering space with an affordable first step via an extremely inexpensive yet realistic microsatellite programme. The need to handle this growing interest, to catalyse wider industrial and commercial applications and to generate regular income to sustain a research activity in satellite engineering at the University of Surrey brought about the formation of a University company—Surrey Satellite Technology Ltd (SSTL). SSTL provides a formal mechanism for handling the transfer of small satellite technologies from the university's academic research laboratories into industry in a professional manner via commercial contracts. The income generated by SSTL is then reinvested to support the academic activities of the Surrey Space Centre, which has now become the largest European centre of excellence in satellite engineering research, teaching and applications.

Since UoSAT-5, all the Surrey microsatellites have been designed and built for individual commercial customers and the range of applications can be demonstrated by reviewing recent examples of payloads carried by the UoSAT/SSTL microsatellite platform.

24.3.1 LEO communications

Various constellations of small satellites in LEO have been proposed to provide worldwide communications using only handheld portable terminals; these broadly fall into two main categories:

(i) Real-time voice and data services.
(ii) Nonreal-time data transfer.

The close proximity of the satellites in LEO to the user and the consequent reduction in transmission loss and delay time appear attractive when compared to traditional communications satellites in a distant geostationary orbit—holding out the promise of less expensive ground terminals and regional frequency reuse. The communications characteristics associated with a LEO constellation pose, however, quite different and demanding problems, such as varying communications path and links, high Doppler shifts, and handover from satellite to satellite.

Currently, there is only one small LEO satellite service in operation (HealthNet) which employs a constellation of just two microsatellites, HealthSat-1 and 2, both built by SSTL for the network operator, SatelLife (USA). HealthNet, and the services waiting to be implemented (e.g. TemiSat, OrbComm, VITASat, GEMStar), uses narrowband VHF/UHF frequencies recently allocated to the little-LEO services to provide digital data store-and-forward capabilities for use with small, low-power user ground terminals which can be located in remote regions where existing the telecommunications infrastructure is inadequate or nonexistent[3].

These VHF/UHF frequency allocations exhibit such effects as multipath and, in particular, man-made cochannel interference which can very significantly reduce the performance that can be achieved in practice by the satellite communications system compared to that expected from a simple theoretical model. A thorough understanding of the real LEO communications environment at VHF and UHF is therefore necessary in order to select optimum modulation and coding schemes.

The KITSAT-1 and PoSAT-1 microsatellites carry a digital signal-processing experiment (DSPE) which has been designed to provide a sophisticated in-orbit test bed for research into optimising communications links with satellites in LEO. The DSPE comprises a TMS320C25 and a TMS320C30 with supporting PROM, RAM and data interfaces to the spacecraft communications subsystems—enabling it to replace the hardware onboard modems with an in-orbit, reprogrammable modem. The DSPE is being used in a research programme to

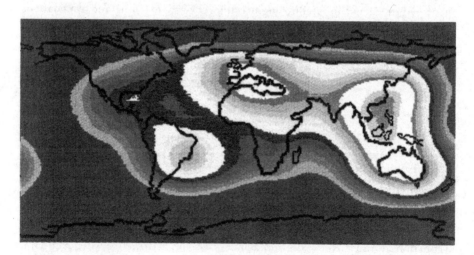

Figure 24.8   LEO VHF interference environment

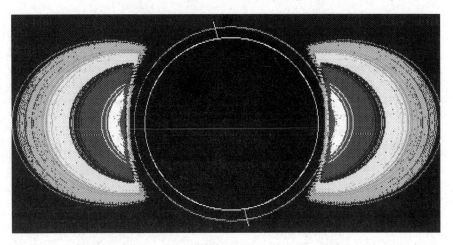

Figure 24.9   LEO radiation environment from PoSAT-1

evaluate adaptive communications links—continuously optimising modulation/demodulation techniques, data rates and coding schemes in response to traffic characteristics during the microsatellite's transit of the ground station.

The interference characteristics of the VHF LEO frequency allocations have been measured using an experimental communications repeater on the S80/T and HealthSat-2 microsatellite missions. In conjunction with a mobile ground-station, S80/T has measured the VHF spectrum noise and interfering signals to evaluate the use of VHF frequencies for a full-scale LEO communications service (S80). S80/T was completed by SSTL for the Centre National d'Etudes Spatiales (France), from proposal to launch, within 12 months!

24.3.2 Space science

Microsatellites can offer a very quick turnaround and inexpensive means of exploring well focused, small-scale science objectives (e.g. monitoring the space radiation environment, updating the international geomagnetic reference field etc.) or providing an early proof of concept prior to the development of large-scale instrumentation in a fully complementary manner to expensive, long-gestation, large-scale space science missions. The UoSAT missions have demonstrated that it is possible to progress from concept through to launch and orbital operation within 12 months and within a budget of £2 million. This not only yields early scientific data but also provides opportunities for young scientists and engineers to gain real-life experience of satellite and payload engineering (an invaluable experience for later large-scale missions) and to be able to initiate a programme of research, propose and build an instrument and retrieve orbital data for analysis and presentation for a thesis within a normal period of post-graduate study.

UoSAT-3 and 5, KITSAT-1 and 2 and PoSAT-1 provide examples of the use of a microsatellite platform for collaborative space science research between the University of Surrey, the UK Defence Research Agency, UK AEA, the UK Science and Engineering Research Council, KAIST and Portugal. A cosmic ray effects and dosimetry (CREDO) payload monitors the near-earth radiation environment and provides an important opportunity to validate ground-based numerical models with flight data yielding simultaneous measurements of the radiation environment and its effects upon onboard systems (especially SEUs in VLSI devices)[4].

24.3.3 Earth observation

Conventional earth-observation and remote-sensing satellite missions are extremely costly—typically £100–500 million each—and thus there are relatively few such missions and the resulting data, although impressive, is correspondingly expensive. The development of high-density two-dimensional semiconductor charge-coupled device optical detectors, coupled with low-power consumption yet computationally powerful microprocessors, presents a new opportunity for remote sensing using inexpensive small satellites.

UoSAT-1 and 2 both carried the first experimental 2D-CCD earth imaging cameras which led to the development of the CCD earth imaging system on board UoSAT-5, intended to demonstrate the potential of inexpensive, rapid-response microsatellite missions to support remote sensing applications. Clearly, the limited mass, volume, attitude stability and optics that can be achieved with a tiny microsatellite mean that a different approach must be taken to produce worthwhile earth observation. For these reasons, SSTL employs electronic cameras with two-dimensional CCD arrays to gather imagery from its microsatellites. Because cameras capture whole images in a single snap shot, they preserve scene geometry and are therefore immune to the residual attitude drift experienced on microsatellites. The latest SSTL microsatellites will use these techniques to provide 75 m resolution, multispectral imaging which approaches conventional (large) satellites but at a fraction of the cost! This is attractive to developing nations in particular which are interested in possessing an independent remote-sensing facility that can be under their direct copntrol and at a very modest cost.

The earth imaging systems (EIS) on board the UoSAT-5, KITSat and PoSAT microsatellites comprise an EEV (UK) 576×578 pixel area CCD digitised to 256 levels of grey. The latest TMSAT uses 1024×1024 Kodak CCDs with three special bands. The digitised data is stored within a 2 Mbyte CMOS RAM which is accessed by two T-800 transputers to allow the image data to be processed to enhance quality and compressed to reduce storage and transmission requirements.

The EIS data is transferred via the microsatellite's local area network to the 80C186/386 onboard computers and stored as files within the 32 Mbyte RAM DISK for later transmission to ground—some 60 images can be stored within

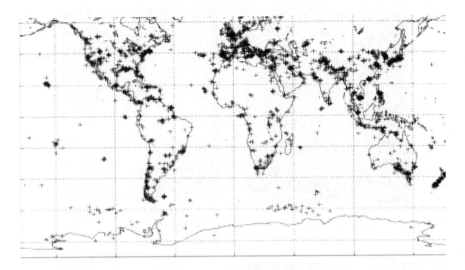

Figure 24.10   Daily, worldwide imaging by SSTL microsatellites

the RAM at any one time. The EIS is commanded to collect an image of a particular area of the earth's surface by the onboard computer which operates a multitasking, real-time operating system responsible for the automatic (and in some cases, autonomous) operation of the microsatellite mission.

Ground controllers select a sequence of images of areas of interest anywhere on the earth's surface and, checking the time and position of the microsatellite using an onboard GPS receiver and orbital model, instruct the onboard computer to collect the images according to a diary that is loaded periodically in advance to the microsatellite.

 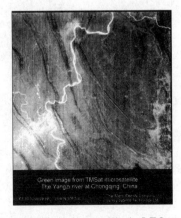

Figure 24.11   Multispectral Earth images from TMSat microsatellite in LEO

Figure 24.12   Multispectral Earth images from TMSat microsatellite in LEO

Figure 24.13   Meterological images from PoSAT-1

The TMSAT microsatellite carries four independent cameras providing a wide-field ground resolution of 1 km for meteorological imaging and a narrow-field ground resolution of 90 metres for environmental monitoring, with three band optical filters providing multispectral images comparable to LANDSAT-TM data sent at a very small fraction of the spacecraft and launch cost[5].

Figure 24.14   FASat-Bravo microsatellite

The results from current UoSAT EIS in orbit have been remarkable, however the next generation of small-earth observation satellites using the SSTL platform, such as TSINGHUA-1, will support EIS cameras yielding better than 50-m resolution with three spectral bands.

24.3.4 In-orbit technology verification

Microsatellites also provide an attractive and low-cost means of demonstrating, verifying and evaluating new technologies or services rapidly in a realistic orbital environment and within acceptable risks prior to a commitment to a full-scale, expensive mission.

UoSAT-based microsatellites have supported a wide range of such in-orbit technology demonstrations, covering:

- new solar cell technologies;
- modern VLSI devices in space radiation;
- demonstration of advanced communications;
- pilot demonstrations of new communications.

For example, satellites depend upon the performance of solar-cell arrays for the production of primary power to support onboard housekeeping systems and payloads. Knowledge of the long-term behaviour of different types of cells in the radiation environment experienced in orbit is, therefore, essential. The continuing development of solar-cell technology, based upon a variety of materials and different process techniques, yields a range of candidate cells potentially suitable for satellite missions.

Unfortunately, ground-based, short-term radiation susceptance testing does not necessarily yield accurate data on the eventual in-orbit performance of the different cells and hence there is a real need for evaluation in an extended realistic orbital environment. UoSAT-5 carries a solar cell technology experiment (SCTE) designed to evaluate the performance of a range of 27 samples of GaAs, Si and InP solar cells from a variety of manufacturers. When the sun passes directly overhead of the panel mounted on the body of the microsatellite, the monitoring electronics are triggered automatically and measure typically 100 current/voltage points for each cell sample. These data are then sent in a burst to the microsatellite's onboard computer, together with associated temperature and radiation dose data, for storage prior to transmission later to ground. SCTE measurements are taken repeatedly immediately after launch, when the radiation degradation is most rapid, and then at increasing intervals thereafter.

24.3.5 Military applications

The demands of a military-style satellite procurement and the cost-effective approach to microsatellite engineering might, at first sight, appear incompatible! However, retaining the essential characteristics of low cost and rapid response, a military version of the SSTL microsatellite platform with deployable solar panels has been developed to support various military payloads. The main differences between the commercial and military versions of the platform is in the specification of components and, particularly, in the amount of paperwork which traces hardware and procedures. An optimum tradeoff between the constraints of a military programme and economy has been sought which results in an increase factor for cost and timescale of approximately 1.5 when compared to the commercial procurement of the platform.

The first use of the SSTL military microsatellite platform was on the CERISE mission designed and built for the French MoD and launched into 700 km low-earth orbit by Ariane in July 1995.

A second mission for the French MoD has now commenced (Clementine) for launch into LEO in 1999.

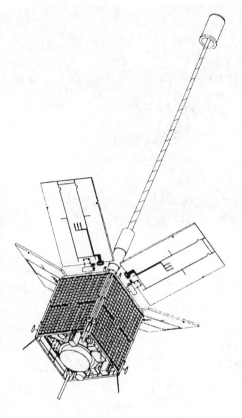

Figure 24.15   CERISE orbital configuration

The geostationary transfer orbit (GTO) provides a good opportunity to study the effects of a severe radiation environment on satellite components, especially solar cells and VLSI components. Surrey has provided platform subsystems and payloads to the UK Defence Evaluation and Research Agency (DERA, Farnborough) for its two space technology research vehicle (STRV-1) microsatellites which were launched into GTO in early 1994. STRV-1a and b carry a range of in-orbit technology demonstration payloads—particularly to study the effects of the space radiation environment on military satellite components.

24.3.6 Education and training using microsatellites

Although microsatellites are physically very small, they are nevertheless complex and exhibit virtually all the characteristics of a large satellite—but in a microcosm. This makes them particularly suitable as a focus for the education and training of scientists and engineers by providing a means for direct, hands-on experience of all stages and aspects (both technical and managerial) of a real satellite mission—from design, construction, test and launch through to orbital operation.

Figure 24.16   Technology transfer with FASat-Alpha (Chile)

The very low cost, rapid timescale and manageable proportions makes this approach very attractive to emerging space nations that wish to develop and establish a national expertise in space technology through an affordable small-satellite programme.

Each technology transfer and training (TTT) programme is carefully structured according to the specific requirements or circumstances of the country or organisation concerned, but the first phase typically comprises:

- academic education (MSc, PhD degrees);
- technology training (seconded to SSTL);
- ground station (installed in country);
- microsatellites (first at SSTL, second in country);
- technology transfer (satellite design licence).

Five highly successful international technology-transfer programmes have been completed by SSC and SSTL at Surrey and three new programmes are now underway.

A total of 70 engineers have been trained through these in-depth TTT programmes—a further 320 students from countries worldwide have graduated

Table 24.3 SSTL technology transfer and training programmes

Country	Dates	Satellites
Pakistan	1985–89	BADR-1
South Africa	1989–91	UoSAT-3/4/5
S.Korea	1990–94	KITSat-1/2
Portugal	1993–94	PoSAT-1
Chile	1995–97	FASat–Alfa/Bravo
Thailand	1995–98	TMSAT-1
Singapore	1995–99	Merlion payload
Malaysia	1996–98	TiungSAT-1
China	1998–99	TSINGHUA-1

from the MSc course in satellite communications engineering, unrelated to these TTT programmes.

24.4 Microsatellite ground stations

Compact and low-cost mission-control ground stations have been developed by SSTL to operate the microsatellites once in orbit. These ground stations are based on PCs and are highly automated—interacting autonomously with the microsatellite in orbit to reduce manpower requirements and to increase reliability.

Figure 24.17 SSTL mission-control ground station at Surrey Space Centre

Figure 24.18   SSTL ground station in Portugal for PoSAT-1

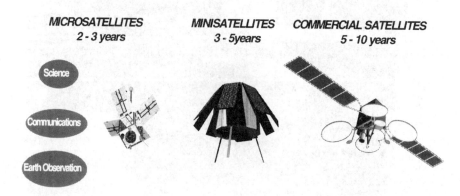

Figure 24.19   Small satellites provide evolutionary low-risk space development

The SSTL mission control ground station at CSER operates nine microsatellites in LEO with just a single operator.

24.5 SSTL minisatellites

In response to growing payload demands for power, volume and mass, but still within a small-scale financial budget, UoSAT and SSTL are developing an enhanced, modular, multimission minisatellite platform capable of supporting missions up to 400 kg.

Once developing space nations have mastered microsatellite technology, the minisatellite provides a logical next step in the development of an increasingly capable national space infrastructure.

The minisatellite platform has been designed to meet a variety of mission objectives and to be capable of operating in different orbits, its primary features are:

- 400 kg total mass;
- 150 kg payload capacity;
- 1.2 m diameter, 1 m height;
- three-axis, 0.1 degree, attitude control;
- GPS autonomous orbit and attitude determination;

Figure 24.20   SSTL Minisatellite configurations

- 1 Mbits L/S-band communications links;
- onboard propulsion for orbit manoeuvres;
- cold gas thrusters for attitude control;
- 300 W orbit average power, 1 kW peak power.

The SSTL minisatellite platform has been designed according to similar cost-effective principles which have proved so successful on the UoSAT/SSTL micro-satellites—resulting in a basic platform cost of ~ £5million. Considerable effort has been expended to ensure compatibility with a range of affordable launch options—on Ariane, CIS, Chinese and US (Pegasus) launchers.

24.5.1   UoSAT-12 minisatellite

The first SSTL minisatellite (UoSAT-12) will carry experimental earth observation and communications payloads and be launched by SS18-DNEPR into LEO in April 1999.

UoSAT-12 will carry 32 m resolution multispectral and 8 m monochromatic CCD earth cameras with powerful onboard image processing capabilities.

Sophisticated frequency-agile VHF/UHF and L/S-band DSP regenerative transponders will provide both real-time and store-and-forward communications to small terminals.

Three-axis control is provided by a combination of magnetorquers, momentum wheels and cold gas N_2 thrusters, and an experimental electric resisto-jet thruster will provide orbit trimming and maintenance demonstrations for future network constellations.

Figure 24.21   SSTL UoSAT-12 minisatellite in LEO

24.6 Project management

Solving the technical challenges associated with the design, construction, test and operation of a microsatellite is less than half of the story—in parallel with the technical considerations of the mission, effective project management is crucial to the realisation of a successful low-cost, sophisticated small-satellite project.

Affordable small satellites require a very different approach to management as well as technology if cost, performance and delivery targets are to be met. Several attempts at taking a traditional aerospace organisation to produce such satellites have failed because of the rigidity of management structure and mind-set. Small teams (25 persons), working in close proximity with good communications, with well informed and responsive management are essential.

These characteristics are best found in small companies or research teams rather than large aerospace organisations, which may find it difficult to adopt or modify procedures necessary to produce affordable small satellites using staff and structures that are designed for conventional aerospace projects. The main ingredients for a successful small-satellite project can be summarised as:

- highly innovative technical staff;
- small, motivated teams;
- personal responsibility for work rigour and quality;
- good team communications, close physical proximity;
- well defined mission objectives and constraints;
- knowledgeable use of volume, modern components ;
- layered, failure-resilient system architecture;
- thorough burn-in;
- technically-competent project management;
- short timescale (prevents escalation of objectives!).

24.7 Summary and conclusions

The University of Surrey embarked upon the design of its first experimental microsatellite in 1978 and UoSAT-1 was launched by NASA in 1981—since then a further 13 low-cost yet highly sophisticated microsatellites have been built and launched into low earth orbit. The UoSAT missions have demonstrated that microsatellites can play a useful role in supporting specialised communications, earth observation, small-scale space science and in-orbit technology verification missions.

Although obviously limited in payload mass, volume and power, but with very real and attractive advantages in terms of cost and response time, microsatellites offer a complementary role to traditional large satellites by providing an alternative gap filler for affordable quick-response or exploratory missions for both civil and military objectives. Developing space nations have used rapid and inexpensive microsatellite projects to act as the focus for effective technology transfer and an affordable first step into orbit. A further four new microsatellite missions will be launched by SSTL in 1999.

Based upon its success with microsatellites, Surrey is now building an afford-able, modular (400 kg) minisatellite capable of supporting more demanding payloads – but still within a cost of $\sim £5$ million – in order to stimulate further use of small satellites to complement large space programmes.

24.8 Acknowledgments

The author would like to recognise the contributions of colleagues at and support from Surrey Satellite Technology Ltd, UK Defence Evaluation Research Agency, Science & Engineering Research Council, Arianespace (France), Alcatel Espace (France), KAIST (Korea), PoSAT Concorcium (Portugal), Matra-Marconi Space (France), SatelLife (USA) and the University of Surrey. FASat-Alfa and FASat-Bravo are owned by the Chilean Air Force (FACH) and have been developed, constructed and will be operated under a technical collaboration program between FACH and Surrey Satellite Technology Ltd (SSTL) UK. TMSAT is a collaborative programme between Mahanakorn University of Technology (Thailand), the Thai Microsatellite Company (TMSC) and SSTL. The L/S-band communications payload on UoSAT-12 is supported by Nanyang Technological University, Singapore. TiungSAT-1 is supported by ASTB, Malaysia. PICOSat is supported by the USAF Foreign Comparative Test program.

24.9 References

1 SWEETING, M.N.: 'UoSAT – an investigation into cost-effective spacecraft engineering', *J. Inst. Elec. Radio Eng.*, 1982, **52**, (8/9), pp. 363–378

2 SWEETING, M.M.: 'The University of Surrey UoSAT-2 Spacecraft Mission', *J. Inst. Elec. Radio Eng.* 1987, **57**, (5) (supplement), pp. S99–S115

3 WARD, J.W.: 'Microsatellites for global electronic mail networks', *Electron. Commun. Eng. J.*, Dec. 1991, **3**, (6), pp. 267–272

4 UNDERWOOD C.I., WARD, J.W., DYER, C.S., SIMS, A.J., FARREN J., and STEPHEN, J.: 'Space science and microsatellites – a case study: observations of the near-earth radiation environment using cosmic-ray effects and dosimetry (CREDO) payload on-board UoSAT-3'. Fifth annual AIAA/USU Conference on *Small satellites*, Logan, Utah, August 1991

5 SWEETING, M.N.: 'UoSAT microsatellite missions', *Electron. Commun. Eng. J.*, 1992, **4**, (3), pp. 141–150

6 FOUQUET, M., and SWEETING, M.N. 'Earth observation using low cost SSTL microsatellites', IAF-96-B.3.P215, 47th IAF Congress, Oct 1996, Beijing, PR China

7 SUN, W., and SWEETING, M.N.: 'Global email system and internet access in remote regions via microsatellites', IAF-96-M.507, 47th IAF Congress, Oct 1996, Beijing, PR China

8 FOUQUET, M., and SWEETING, M.N.: 'UoSAT-12 minisatellite for high performance earth observation at low cost'. IAF-96-B.3.P215, 47th IAF Congress, Oct 1996, Beijing, PR China

9 SWEETING, M.N.: 'Technology transfer and training at Surrey', 47th IAF Congress, Oct 1996, Beijing, PR China

10 SWEETING, M.N.: 'TMSAT: Thailand's first microsatellite', 47th IAF congress, Oct 1996, Beijing, PR China

Chapter 25

Future trends in satellite communications

B. G. Evans

25.1 Introduction

As stated in the introductory Chapter, satellite communications in the late 1990s is in another interesting period of change and this is not the best time to, make predictions for the future. As this Section is being written the first of the GMPCS satellite constellations (Iridium) is about to launch its service. This represents a tremendous change from the thirty-odd years in which the GEO ruled supreme. Both of the major IGOs, INTELSAT and INMARSAT, are embarked on the process of becoming private organisations after having dominated innovation in satellite communications for the same period of time. Thus satellite communication, despite being a truly global business, is set to become a predominantly commercial one. We are also witnessing convergence of manufacturers, service providers and content providers to provide new systems in a different format than hitherto. There is a convergence of media data, voice and video and the beginnings of a demand for interactive/multimedia services on the back of the explosion in demand for Internet.

The commercial world is one of greater risks and we already see very large satellite projects of between US$1–10 billion well into implementation. Obviously, key to the risks is the ability to be able to predict markets and to design and engineer the satellite system efficiently for coverage and service types.

This, in itself, is a paradigm shift from the old days when one launched a general purpose satellite and waited for the markets to emerge, this is no longer a viable scenario.

The triggers of opportunity now lie in:

- global deregulation;
- complete integrated digital networking;
- global players and consortia;
- global information infrastructure (GII) and the spread of the Internet;
- move to multimedia services;
- requirements for greater mobility;
- more synergy between military and civil systems;
- low-cost processing and terminals;
- ability to handle complexity in control, (IN and software agents).

All of these opportunities will be provided in a new business climate where convergences of services, providers and technology development is providing new competition in an increasingly deregulated scenario. In this environment satellites are reaching further down the distribution chain right to the end user and are now real competitors even in the local loop. The new enterprise models are creating a value chain which has at the top the old bandwidth providers and progresses down via service providers and content providers. The share of the consumer revenues is weighted towards the lower half of the value chain and more towards the consumer. Thus, satellite systems have to move down the value chain if they are to survive.

With this explosion in demand for services and the new business regime it is becoming easier and more fashionable to attract funds for new satellite systems which would have erstwhile been considered too risky. Via the big LEOs we are seeing billion dollar funding packages which are a mixture of equity, debt from the banks, plus bonds, shares etc. on the public market.

It is estimated that the worldwide communications and navigation satellite market in the year 2000 will be as shown in Table 25.1

This is a huge market and yet it represents only three to four per cent of the global market for telecommunications. Figure 25.1 shows a breakdown of the market through to the year 2005 and Table 25.2 outlines the growth trends for the various sectors of the business. It will be the availability of cheap space segment capacity that opens the door to the new innovative services. However, it must be remembered that there are many competitors to satellite delivery

Table 25.1   Worldwide satellite market in year 2000 (source: ESYS Report, 1998)

	US$ billion
Satellites and launchers	3–4
Ground segment/operations	2–3
User equipment	8–10
Value-added (including software)	15–20

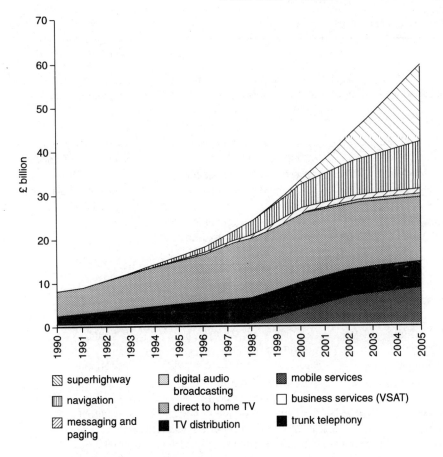

Figure 25.1   Worldwide satellite market sectors
Source: ESYS Report, 1998

(cable, terrestrial radio, cellular, stratospheric platforms etc.), and only if satellite can provide the service cheaper and with equivalent, or better, quality of service will it achieve business success in the climate of competition. As has always been the case, satellites have the advantage of rapid deployment and will always be an excellent early entrant to any new service.

New technologies that are already being integrated into satellite systems are:

● multispot-beam antennas—12 m deployables with around 200 spots;
● onboard digital signal processing—DSP functions but not with onboard regeneration and switching;
● intersatellite links—millimetre-wave links but not as yet optical;
● higher power platforms—up to 20 kW with miniature components.

These will continue to be areas of important innovation and development and will progress in the next five to ten years. The overall networking scenario

Table 25.2   The future opportunities

	Market status	Trends	Key issues	Window
Fixed service	Large	strong growth	• deregulation • launch costs • networking standards	continuous
Mobile service	small	very strong growth	• deregulation • standards • system costs • spectrum	speech–5 years low-rate multimedia: 2002–2006 high-rate multimedia: 2005–2010
Television services	very large	strong growth	• deregulation • digital technology • standards • two-way services	continuous
Audio broadcasting	very small	strong growth	• spectrum • system costs • terminal costs	5 years
Messaging/ paging	small	medium growth	• spectrum • Launch costs	5 years
Navigation	medium	strong growth	• GPS access • integrity • technology	continuous
Satellites/ launchers	large	medium growth	• production lines • launch costs reusables	continuous
Military	medium	medium growth	• change in threats • cost	continuous

and the ability to deal with increasing complexity in hybrid constellations, integrated fixed-network mobile and satellite systems, are areas in which more and more intelligence will be built into the network and more systems will rely on intercommunicating software agents.

Satellite systems will reflect the general trends in telecommunication systems and will need to keep pace with, and provide the same facilities as, fixed or mobile terrestrial networks.

25.2 Television and audio broadcasting

At the end of the millennium we are moving rapidly into the digital video age with the first set-top boxes being produced and first digital satellite TV broadcasts in 1998. The reduced satellite bandwidth needed for broadcast will enable hundreds of satellite channels and near video-on-demand-type services. The growth here is very rapid, in particular in Asia, the Pacific Rim and Latin America. The revenues from the digital broadcast market are set to rise rapidly (see Figure 25.2). The equipment and operators' revenues will be dwarfed by the content and programming side, with thousands of new production houses feeding the hundreds of broadcasters. This represents a greater added-value service and new businesses.

Analogue services have at most a ten-year lifetime. Digital satellite services to the home face competition from cable and terrestrial radio and much will depend on the nature and speed of deregulation as to which succeeds.

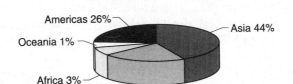

market distribution

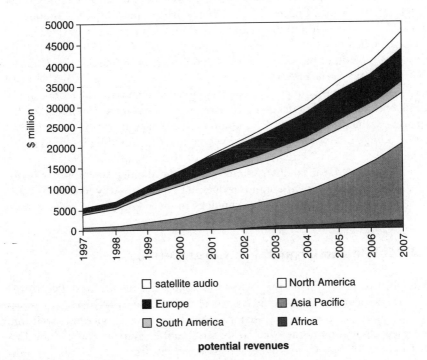

potential revenues

Figure 25.2   World digital satellite broadcasting markets
(Source: ESYS Report, 1998)

Convergence of standards will also play a part, with set-top boxes being one of the immediate areas of contention. Key in the satellite-system competitiveness is the reduction of launcher costs via use of reusable launchers.

In many parts of the world satellite will have the edge over its competitors owing to: difficult terrain, dated cable systems and the ability to provide the services quickly to a large area.

Standards play a role in the digital television area as well, with MPEG source coding standards and DVB transmission standards being the trail blazers. Proprietary systems such as DirecTV have done well initially but it will always be likely that international standards will take over eventually in an expanded market.

Once into digital format, interactive television systems and multimedia service provision become possible. We are then in the realms of the entrepreneurs and in the next five years it is expected that we will see a plethora of new services developed on the back of the digital technology with all sorts of user/audience participation schemes.

Audio broadcasting via satellite is in its infancy in 1998. Most DTH-TV service providers offer audio radio broadcast but these are very much as piggy backs to the television service. In Europe, ASTRA and EUTELSAT have subscription services for hundreds of digital programmes. There is very strong interest in the new satellite systems dedicated to providing digital audio broadcasting (DAB). World Space will provide services to portables and mobiles in Africa, the Caribbean and Asia, and CD Radio is to provide services in the Americas. Cheap mobile receivers will become available by 2000 and this will herald the take off of the service with global broadcasters such as the BBC and CNN showing a keen interest. Issues are: the limited spectrum available in L- and S-bands, the ability to provide small compact antennas for the mobile applications and the reduction in cost of the satellite systems, again dominated by the expensive launchers. Market deregulation is necessary in many countries but there is a huge potential for indirect revenues and for providing low-rate multimedia services. By 2002 satellite DAB is likely to be widespread.

Both DVB and DAB satellite systems are among the first to employ onboard digital multiplexing and although this does not go the full way to OBP, it is the first step in the use of the new DSP technology in space.

25.3 Mobile and personal communications

Arguably this area will be the fastest to expand during the next five to ten years. It is already, in 1998, in the process of major change. INMARSAT is privatising; there are three to five major GMPCS worldwide operators well into implementation and another four to six GEO mobile systems planned for Asia (see Chapter 20). Spectrum is a major issue with pressure on L-/S-bands, coordination difficulties and movements to bring about fixed mobile convergence. The generations of mobile satellite systems, as seen today, are shown in Table 25.3.

Table 25.3   Mobile satellite generations

Generation	Time period	Comments
1	1990s	INMARSAT-M ⇒ mini-M laptops, speech/LDR 10s kbit/s
2	1998–2000	GMPCS or SPCNs handhelds—proprietary standards, speech/LDR 10c kbit/s
3	2002–2006	IMT 2000—multimedia, laptops 64–144 kbit/s
4	2004–2010	MBS, Ka-band multimedia, portable palm tops—10s Mbit/s

INMARSAT has produced three generations of mobile satellites which currently provide LDR/speech communications to lap-top-sized terminals or up to 64 kbits to slightly larger dish versions. These satellites consist of only a few spot beams and thus cannot support higher bit rates or larger numbers of channels at low bit rate, making the service expensive.

The key to provision of higher-rate services is multispot-beam satellites which use large deployable antennas. The state-of-art in this technology is in the 10–15 m range (1998) but will rapidly move to 20–40 m in the next three to five years.

Starting in 1998, the dynamic satellite-constellation systems of Iridium, Globalstar and ICO will commence operation over a two-year period. These systems are aimed at handheld voice communications on LDR services and have projected markets shown in Table 25.4 which are only two to five per cent of the total mobile markets.

The major new technologies involved are those of the software control of large and complex distributed information systems, including constellations of 10–66 satellites and large numbers (up to hundreds) of GES. Much of the networking is based upon the GSM network standard with different air interfaces

Table 25.4   GMPCS market predictions (Source: average three studies (C.A. Ingley, L. Taylor Assoc., Lehman Bros.) in 1998)

Year	Market Size (million)	Market ($ billion)
2002	6	5
2004	12	13
2007	20	21–33

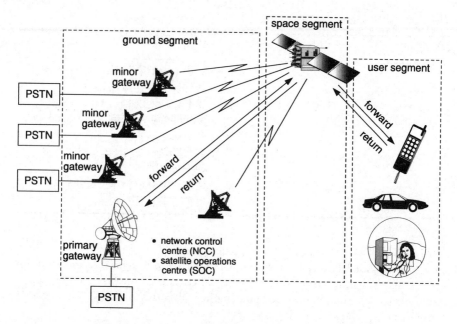

Figure 25.3   Architecture of GEO mobile satellite system

which are proprietary to each GMPCS. Multimode terminals developed by the consumer electronic markets to allow users multiaccess to satellite and cellular systems are major new innovations.

In addition to the GMPCS (as described in Chapter 20) GEO mobile systems are to operate from 1999/2000 in equatorial regions, with principal markets in Asia, China etc. It has been estimated that this region alone will be a US$3 billion market by the year 2000 with around 300 thousand channels required. China alone, it is estimated, can support at least one GEO mobile and the latent potential is huge. Satellites have great potential in allowing roaming and single billing in a region which has a vast range of different cellular standards and operators as well as a major need for rural telephony. The GEO mobiles are around US$1 billion projects involving two satellites with large deployable antennas (12–14 m) producing hundreds of spot beams–see Fig. 25.3.

Both the GMPCS and GEO mobiles are attempting to serve the market shown in Table 25.4 Acceptability will be judged on:

- coverage;
- QoS/availability;
- price;
- timing to market;
- service offerings;

with similar market issues to the terrestrial cellular. The spectrum and coordination aspects remain the largest barriers to profitability of such systems.

Table 25.5 GMPCS/mobile market positions

Market	End users Density	Sensitivity to QoS	Price sensitivity	
High	Global roamers/int. business travellers	low	high	low
Middle	Enhanced cellular coverage/landline	medium	medium	medium
Low	Rural users who have never made a call	high	low	high

In 1998 ITU/ETSI and other standards bodies made initial decisions for the specification of the third-generation mobile system. Both ETSI in Europe and Japan have embraced a wideband CDMA system (WCDMA) for terrestrial cellular. The USA has only recently established a second generation system based on narrowband CDMA and is proposing an extension of this, CRMA 2000, for the third generation. However, there does seem to be a move from TDMA towards CDMA, based upon the promise of increased capacity. The collection of third-generation standards is referred to as IMT 2000 by ITU and satellites are meant to be an integrated part of such systems. Thus, unlike the second generation, where GSM was very much a cellular standard and the satellite systems developed it using proprietary protocols, in the third generation, satellites are meant to be an integrated part so as to remove the need for multimode terminals and provide a single cheaper terminal.

At the time of writing the satellite part of IMT-2000 has still to be defined. However, the idea of third-generation is to embrace multimedia services at variable bit rates up to 2 Mbit/s in cellular and perhaps up to 144 kbit/s for larger satellite cells. Thus, second-generation mobile satellite is part of IMT-2000.

Satellite multimedia is predicted to be around five to ten per cent of the total mobile multimedia market by 2010. It will clearly enable extensions of the cellular IMT-2000 which are only economic in urban areas, to suburban and rural areas. It will facilitate GSM/IMT-2000 coexistence and could be the only way for an early entry to the new multimedia market.

The candidates for satellite IMT-2000 will be the second-generation GMPCS/SPCNs and extensions of the current INMARSAT system. The latter is called HORIZONS and is essentially a more advanced GEO mobile system targeting the multimedia market. Whether it is the latter or the second-generation dynamic satellite constellations which end up providing this service in the 64–144 kbit/s range, the networking will be very different from first-generation systems. To deal effectively with VBR traffic requires a packet-based system which can be a derivative of ATM or IP and hence a new approach to the network design. Such systems have a niche of around 2002–2006 to coincide with the start up of the cellular IMT-2000. Thereafter we will see the mobile broadband systems (MBS) emerging.

25.4 Broadband multimedia systems

Current fixed satellite services are dominated by television with 75 per cent of all capacity, with telephony and other thin route services accounting for ten to 20 per cent of current transponder revenues. The latter is declining as the penetration of fibre increases, but demand still grows.

An expanding market in FSS is for VSATs, although around 73 per cent of these are in the US. With deregulation the rest of the world is catching up and there is rapid expansion in Asia, South America, and Eastern Europe. It is estimated that there will be 2.5 million VSATs by 2004—an increase of ten to 15 times today's numbers. VSAT terminals are now very small (<1 m) and very cheap (<$10 k) which, together with cheaper space-segment costs, is providing the growth rates of twenty to thirty per cent.

Overall costs are still being dominated by satellite launching, which consumes 50 per cent of the space-segment costs. The first reusable launchers will be trialled by the end of 1998 and if successful will have a major impact on the reduction of future satellite system costs.

To date applications have been limited to constant bit rates which in general are below 2 Mbit/s and mostly in the 100s kbit/s region. There are applications now appearing which receive more and cheaper bandwidth on demand. The first of these is for Internet/www connections and several systems are now in operation (eg. DirecPC) at low rates. Broadband satellite interactive multimedia services are driven by the superhighway/GII concept where satellites can both extend fibre networks into areas where it is uneconomic for terrestrial roll out and more importantly provide an early availability of such services. The market is predominantly corporate, to fixed terminals, in the first instance but also includes significant domestic elements. There is also a demand for portable and even mobile applications but these are more difficult to serve (see Section 25.3).

The multimedia services (see Table 25.6) require considerably greater bandwidth and here operators are forced to look at higher frequency bands, Ka or V/Q or to reuse spectrum at Ku-band by using new technology/techniques. In 1997 WRC extended the spectrum available in Ka-band to 500 MHz which has caused a flurry of activity in new satellite-system filings with around 300 orbital slot bids for 800–900 satellites. Some of the major systems are shown in Table 25.7, but it must be recognised that not all of these systems will be realised. In addition the existing operators INTELSAT, INMARSAT, PANAMSAT, ORION, EUTELSAT, ASTRA etc. all have filings for GEO orbital slots. The eventual decision to proceed will depend on the market assessment and the business plans.

In a recent (1998) market study for Booz Allen and Hamilton it is shown (Figure 25.4) that a significant market exists, growing from around 50 to 330

Table 25.6   *Multimedia services*

Services	Rate	S/AS
E-mail/messaging	28.8–144 kbit/s	S
Remote LAN access	64–2 Mbit/s	AS
Remote database access	64–2 Mbit/s	AS
Document showing	64–512 kbit/s	S
One-way data/images	64–2 Mbit/s	S
Video telephony	32–144 kbit/s	S
Voice	2.4–4.8 kbit/s	S
Videoconferencing	256–2 Mbit/s	S
Broadcast video	2–8 Mbit/s	AS
Video on demand	20–100 Mbit/s	AS

Table 25.7   *Some major multimedia satellite systems*

Ka-band		
Teledesic/Celestri	LEO (288/63)	Motorola/McCaw Microsoft/Matra
SPACEWAY	GEO (15)	Hughes
Astrolink	GEO (9)	Lockheed Martin
Cyberstar	GEO (4)	Loral Space
VOICESTAR	GEO (12)	AT & T
Echo Star	GEO (2)	
WEST	GEO/MEO	Matra
Euroskyway	GEO (13)	Alenia
Ku-band		
Skybridge	LEO (64)	Alcatel
V-band		
Deneli Telecom LLC	HEO (9)	
Globalstar LP	LEO (80)	
Hughes Expressway	GEO (14)	
Spacecast	GEO(6)	
Star Lynx	GEO (4)+MEO (20)	
Lockheed Martin	GEO (9)	
Launch Space Com	GEO (10)	
LEO One	LEO (48)	
Orbital Sciences	MEO (7)	
TRW	GEO (4)+MEO (15)	
Teledesic/Motorola*	LEO (288–72)	

*Recent merger of Teledesic (288 LEO) with Motorola Celestri (72 LEO + 1 GEO) system

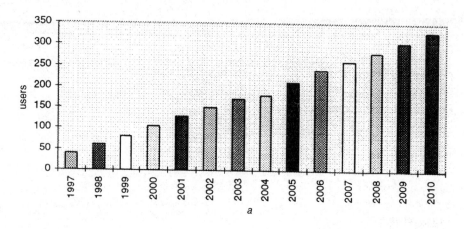

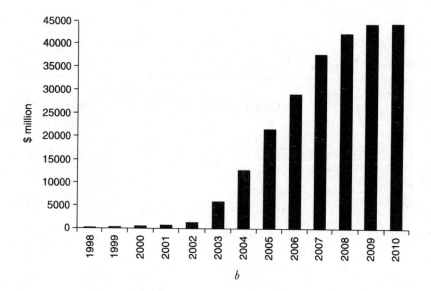

Figure 25.4   *Broadband satellite market predictions*
   a Size (Booz, Allen and Hamilton)
   b Potential revenues (Merill Lynch)

millions by 2010. The cumulative service revenues for satellite systems over the period 2000–2010 is estimated to be US$120 billion. Europe and the US represent about 30 per cent of the market each, the rest being distributed over other regions. It is estimated that around 16 per cent of the market is capturable by satellite.

At Ka-band and above rain attenuation becomes a significant factor in providing services for low percentages of the time. The situation is exacerbated by the low-power availabilities from current Ka-band amplifiers and the high cost of such devices. Thus fade countermeasure techniques using DSP need to be employed to enable closure of the link budgets.

The cheap equipment availability can be solved by the realisation of high market demands, although Ka-band technology has been around now for two decades without significant market applications to bring about equipment cost reductions. Link budget closure will require novelty in system design and technology for services which may demand BERs of as low as 10^{-10}.

Systems proposed in Table 25.7 fall mainly into three categories:

(i) GEO.
(ii) LEO.
(iii) GEO + MEO (hybrids).

The GEOs are offering variable bit-rate services from 8 kbit/s–10 Mbit/s to fixed corporate antennas ranging from 0.5 to 2 m diameter but can receive up to 100 Mbit/s. Target prices for terminals are less than \$10 k, with target transmit powers of less than 10 W.

The satellites are multibeam (30–60 spot) and incorporate full onboard processing as well as packet or cell switching—examples are shown in Figure 25.5.

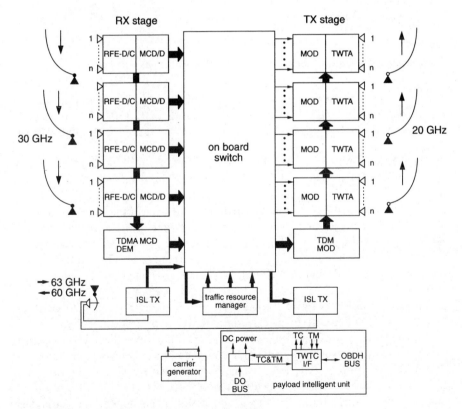

Figure 25.5   Typical architecture of OBP Ka-band satellite

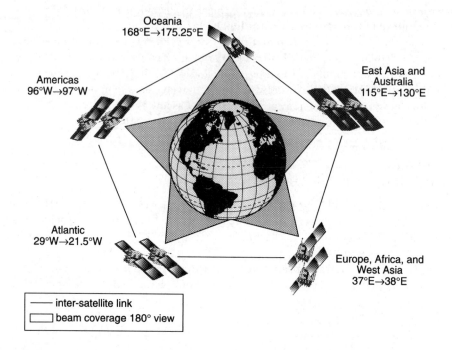

Oceania
168°E→175.25°E

Americas
96°W→97°W

East Asia and
Australia
115°E→130°E

Atlantic
29°W→21.5°W

Europe, Africa, and
West Asia
37°E→38°E

——— inter-satellite link
⬚ beam coverage 180° view

Figure 25.6   The Global Ring—inter-regional ISC connections

Several of the systems introduce the concept of commencing operations in one region and then expanding to a global service using a GEO ring (Figure 25.6) of satellites connected by optical intersatellite links operating in the 100s Mbit/s range.

The LEO grouping typified by Teledesic/Celestri and Skybridge use the concept of earth-fixed cells (EFC) rather than satellite-fixed cells (SFC) as used in all of the GMPCS (Chapter 20). In this system, as shown in Figure 25.7, the cells are maintained constant on the earth from the time of the satellite visibility within a supercell and then all switch at the same time to the subsequent visible satellite in the constellation. Thus, the handovers are reduced by eliminating the beam to beam, with only the satellite to satellite remaining. This would remove one source of degradation in the attempt to achieve the low BERs required for multimedia services.

The other problem is maintaining a high elevation angle to the LEO satellites as low angles will result in a degradation of the propagation channel and increased errors. This is why Teledesic first proposed 840 satellites. However, the overall cost of the system is very high and a compromise has to be made in slimming down the constellation between performance and cost. LEO constellations such as those of Teledesic/Celestri also involve ISLs between the satellites, onboard processing and switching and new packet networking protocols (discussed later).

Constellation Parameters

Total number of Satellites	288+spares
Number of Planes	12
Satellites per Plane	24
Altitude	1,375 km
Eccentricity	Circular
Inclination	84.7°
Inter-Plane Spacing	15.36°
Inter-Plane Satellite Spacing	Uniform
Inter-Plane Satellite Phasing	Random
Elevation Mark Angle	40°

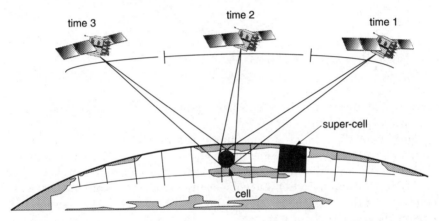

Figure 25.7   Original Teledesic constellation parameters and idea of earth-fixed cells

Skybridge is the only proposed LEO constellation to reuse the Ku-band allocations: It proposes to do this by switching off satellites as they fly through the geostationery orbit so that they do not interfere into the GEO arc. This system again is aimed at both residential and corporate users with an emphasis on the former, using 50 cm antennas. Domestic antenna tracking is an area for novel technology; each of the 64 satellites creates 45 spot beams (350 km radius) and EFCs are used. Skybridge is the only Ka-band system to propose CDMA on the air interface.

The hybrid constellation group (see Figure 25.8) have mainly gone for GEO-MEO combinations although GEOs and MEOs have also been suggested in some cases. In these systems the services are divided between the GEO, for nondelay sensitive lower quality (home shopping, online services etc) and the MEO, for higher quality and more delay sensitive (teleworking, telemedicine, videoconferencing etc). The MEOs give good coverage off the equator and offer the prospects of improved quality over the GEOs. ISLs are used to link the two elements of the constellation and to provide efficient connectivity between

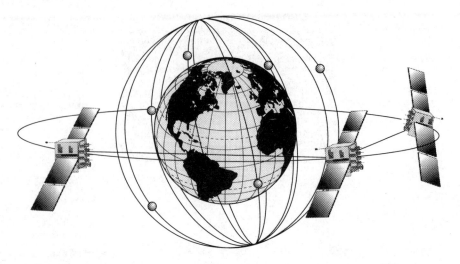

Figure 25.8   Hybrid orbit of GEO and MEO

services and users as well as enabling early start up in one region to be extended to global operations.

These new Ka/V-band systems produce some R&D challenges for the future. Phased-array antennas and millimetre front ends are the major terminal challenges, as already mentioned. On the air-interface most systems have gone for MF-TDMA but for GEO systems synchronous CDMA is also attractive. In either system there are onboard processing implications as well as synchronisation problems to be solved.

The choice of modulation and coding in order to achieve BERs of 10^{-10} is crucial and hence higher order QAM, TCM, concatenated or turbo codes, and perhaps most importantly adaptive modulation and coding schemes for the nonCBR service, will be crucial.

As far as payloads are concerned, the MBAs, ISLs and OBP/DSP are important elements as is the miniaturisation of components in order to reduce the payload mass and volume. The central OBP switch is a feature of the networking strategy and here we have two scenarios. The first is to use a fast packet-switching idea (at frame level) and this has advantages in that it is available and will allow interworking via the GES to a range of existing protocols. The second and more adventurous approach is to employ an ATM switch on board with service switching at cell rates. This produces a much more flexible approach in terms of connectivity but requires greater changes of the basic ATM structure. For efficient transmission over the satellite channel special adaptations are needed of the ATM headers forming a satellite ATM cell. This idea has already been incorporated in HORIZONS in a nonOBP satellite, but needs to be extended to the processing system where the switch is now on board the satellite. The two different approaches are shown in Figure 25.9 with the changes needed

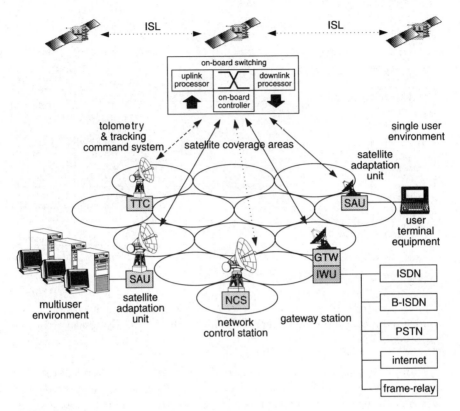

Figure 25.9a   Generic ATM satellite architecture

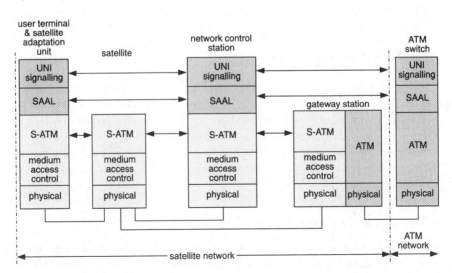

Figure 25.9b   ATM control plane protocol stack

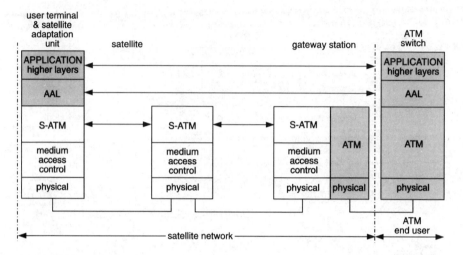

Figure 25.9c   ATM protocol user plane stack

in the ATM protocol structure. With an onboard switch, problems of congestion control, connection admission control and bandwidth on demand need to be addressed and this has knock-on implications for the satellite buffers as well as the partition of control between the satellite and the GES.

With more applications moving to IP, the transition of IP over ATM may in the longer term not prove to be efficient and the whole system may evolve to a native IP network with routers on board satellite and complex multicasting via the constellation. This latter idea is embraced by Teledesic but is at an early stage of development.

25.5 Satellite navigation

Commercial navigation services have been enabled by the free access to positional data from Navstar GPS satellites since 1995. Most GPS markets are still embryonic with many different field trials/prototypes in progress. Although GPS is the major system the exSoviet GLONAS system is also used but is suffering from lack of replenishment and under investment. The market is essentially the sale of user equipment and its integration into other products. The key growth areas are:

- in-car navigation;
- wireless tracking applications;

but with the 1998 cost of the GPS chip set approaching $10 there must be very many other applications and the only limitation is imagination (e.g. when will Nike produce their first trainers with GPS chips for monitoring children's movements?).

The total global market by 2005 is estimated to be US$50 billion, and with the recent US declarations of maintaining a free service well into the 2000s this looks secure. For safety-of-life markets, air transport, maritime etc. specialised augmented systems such as the European EGNOS are under development for 2002. Such systems will improve the integrity of the MEO GPS system by adding a GEO component. Eventually, new systems (GNSS2) will emerge from the US and possibly elsewhere that will use hybrid orbits to provide the coverage and integrity needed to extend into these safety-of-life markets. The real problem here is cost and securing a business.

These new markets rely on partnerships between equipment suppliers, information providers, telecommunications and Internet operators and the semiconductor producers. A good example is dynamic route guidance (in-car navigation) which guides drivers to a destination by voice prompts, requires metre-level accuracy, digital map provision and offers additional services of traffic information, diversions, tourism services, hotels, comparative fuel prices at service stations etc. GPS is now standard fit to all luxury cars and optional for most mid ranges. The full set of services will be available by 2003 and all new cars will be equipped. Other major markets are in intelligent transport systems, survey, aviation and maritime. These will all be realised as the navigation, IT and telecommunication/Internet technologies converge.

25.6 Messaging and monitoring

A rather specialised market which is sometimes very much allied to that of navigation is short messaging/monitoring. Currently QualComm offers two-way messaging services via GEO satellites and 150 000 OmniTRACS units have been sold in 30 or more countries. In the US around 500 trucking companies use OmniTRACS. The service in Europe, marketed by Alcatel QualComm and operated via EUTELSAT, was slow to start but is now growing rapidly. INMARSAT-C offers similar services.

New systems involving constellations of little LEO satellites are just emerging with the 26 satellite constellation of OrbComm being the first to operational status. These systems operate in the VHF bands and provide global coverage for utility meter reading, asset tracking, short messaging, e-mail etc. They are claimed to be five to ten times cheaper than OmniTRACS as they charge on a per byte basis. OrbComm claims to have more than 60 licences worldwide and an FCC licence to sell 200 000 terminals. Such systems are claimed to need only 50 000 subscribers to break even and so there is a niche market. The projected market for these services is shown in Figure 25.10.

The major downside has been the lack of spectrum with only 3–4 MHz being available (and then not in all countries) and this would only support three to four global systems. A sufficient worldwide allocation of bandwidth is crucial to the expansion of these little LEO systems. Again in this area launch costs dominate the economics and cheap and regular launchers for micro- and minisa-

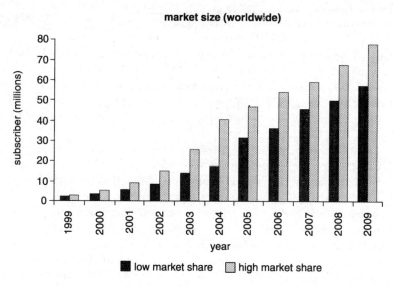

Figure 25.10   Market for little LEO messaging and monitoring services
(Source: ESYS Report, 1998)

tellites are key. The little LEOs have a niche which they must exploit before
the larger big LEOs start to compete.

25.7 Conclusions

One thing is certain, the pace of change is accelerating dramatically in satellite
communications. The satellite-communications markets as described in this
Chapter are buoyant worldwide across many business areas. Although the
satellite markets may only be five to ten per cent of the total, they are extremely
significant (up to $150 billion by 2010) but do need major capital investment in
equipment. There is a growing interest and investment in regional GEO systems
for equatorial regions. In this respect the markets in the Pacific Rim, Asia, South
America are very large and have much higher growth than in North America or
Europe. There will continue to be a larger scale investment in hybrid satellite
constellations for mobile and fixed service provision. There is a tendency for
fixed-mobile convergence and a move to produce a fully integrated system environment. Services are moving away from speech-only towards multimedia, and
there is a huge growth in asymmetrical interactive services to low-cost terminals
(both mobile and fixed). Service mobility is very important to the customer and
sometimes difficult to provide over a satellite link. The emphasis will move much
more to that of service provision and not just on technology development. New
vertically-structured enterprise models will emerge for network operators,
service providers, content providers and brokers, but at the end of the day the

customer wants a single bill and so convergent billing is key. We will see a high degree of integration of services (eg. communications/navigation/broadcasting etc.) with a larger range of network types—a network of networks.

The satellite communication system, until now treated in isolation, will become merely one part of this network structure, the management and control of which is increasing in complexity. Embedded intelligence and software agents will be the eventual key to the operation of what will become a software telecommunications environment.

25.8 Acknowledgments

This chapter has been based upon the contribution of two guest speakers in the 1996 and 1997 IEE vacation schools. I would like to credit both Mike Dillon of ESYS Ltd and Pat Norris of Logica Ltd for allowing use of some of their material. Some of the market material was taken from a satellite systems market report produced by ESYS Ltd. in 1995 and updated in 1998 for BNSC. Further details may be obtained from ESYS Ltd, Istoke Road, Guildford, GU1 4HW, UK. The author has, however, used this material to support his own ideas of what lies in the future.

Appendix A

Link budgets and planning: worked examples

B. G. Evans and T. C. Tozer

A.1 Introduction

An essential part of the planning of a satellite system is the link-budget calculation. In Chapter 11 we presented the equations to be used in this exercise, but anyone entering the satellite communications field will need familiarity with the use of these equations. In this Appendix we have collected together some examples of link budgets and their use in planning satellite systems. It is recommended that the reader work through these in detail and by so doing master the art of satellite design. The detailed methodology and terminology, and the degree of accuracy employed, vary according to the application and requirement. It should be stressed that there is no one correct way to present a link budget.

The examples have been chosen to illustrate currently operating systems with a range of network scenarios, modulation and multiple-access techniques. They are as follows:

(i)   A simplified link budget for a TVRO from the EUTELSAT satellite.
(ii) An FDMA system design, typical of that used in the INTELSAT system, together with an earth-station G/T calculation.
(iii) A high-data rate TDMA design, again typical of the rates used in the INTELSAT system. Also illustrated is the calculation of frame efficiency.
(iv) An SCPC design example typical of that on domestic systems using leased transponders from INTELSAT. Also illustrated here is the use of companded FM.
(v)   A mobile satellite system example based upon the INMARSAT standard A ships earth terminal to and from a coastal earth station.

(vi)   A military systems design example illustrating the use of mixed-size terminals.

(vii)   A Ka-band VSAT system example using digital modulation and unbalanced inbound/outbound transmissions.

A.2 Television transmission from medium power satellite to TVRO

In this link-budget example we illustrate the basic principles by considering the one-way transmission of a TV (PAL, NTSC, SECAM or MAC) signal from the ECS-FI satellite to a TVRO dish to be situated in London. The maximum available transponder EIRP on the London contour (available from EUTELSAT) is +40 dBW, and it is required to calculate the size of dish required at the TVRO to provide CCIR grade four picture quality, equivalent to an S/N of 42.3 dB (see Chapter 11). The full link budget is shown in Table A.1 for clear weather and for an annual availability of 99.5% of the year. The following are notes on the method of calculation:

(i)   Starting point is $S/N = 42.3$ dB
        From eqn. 11.30:

$$\frac{S}{N} = \frac{C}{N} + 20 \log\left(\sqrt{3}\left(r\frac{\Delta F_{pp}}{f_m}\right)\right) + 10 \log\left(\frac{B_{rf}}{f_m}\right) + W + P$$

ΔF_{pp} is the peak-to-peak frequency deviation $= 13.5$ MHz; r is the ratio of peak-to-peak amplitude of luminance to peak-to-peak amplitude of the video $= 0.7$ for 625/50 Hz systems; f_m is the top video base bandwidth, but this is taken as 5 MHz when the unified weighting is used; $W + P$ is the weighting (unified) plus emphasis gain $= 13.2$ dB; B_{rf} is the receiver bandwidth $= 27$ MHz. Whence:

$$42.3 = \frac{C}{N} + 20 \log\left(\sqrt{3}.\, 0.7.\frac{13.5}{5}\right) + 10 \log\left(\frac{27}{5}\right) + 13.2$$

$$\therefore C/N = 11.6 \text{ dB}$$

The downlink budget is:

$$\frac{C}{N} = \text{EIRP} - \text{FSL} - L_f - L_e + \frac{G}{T} - 10 \log(B) - K$$

$$\text{EIRP} = +40 \text{ dBW}$$

$$\text{FSL} = 20 \log\left(\frac{4\pi d}{\lambda}\right) \text{dB} \qquad \lambda = 3 \times 10^8 / 1.17 \times 10^{10}$$

$$= 205.65 \text{ dB} \qquad\qquad d = \text{distance from London to the ECS-FI satellite}$$

The distance may be found by inserting the latitude and longitude of London and the position of the satellite into the equation for slant path distance given in Section 11.3.

Table A.1   ECS-FI satellite TVRO link budget for London

Parameter	Value		Units
ECS-F1 satellite EIRP	40.0	40.0	dBW
Free-space loss (11.7 GHz)†	−205.65	−205.65	dB
Annual availability	clear weather	99.5%	
Downlink fade depth	–	0.7	dB
Received satellite EIRP	−165.6	−166.4	dBW
Antenna diameter	23.	2.9	metres
Antenna gain	46.6	48.8	dBi
Antenna pointing loss	−0.8	−1.4	dB
Misalignment/ageing losses	−0.2	−0.2	dB
Waveguide losses	−0.3	−0.3	dB
Received carrier power	−120.4	−119.5	dBW
Boltzmann constant	−228.6	−228.6	dBW/Hz/K
Received carrier power	108.2	109.1	dBW
System noise temperature	22.3	23.2	dBK
Receiver bandwidth	74.3	74.3	dBHz
Received C/N ratio	11.6	11.6	dB
Received weighted S/N ratio	42.3	42.3	dB
CCIR grade 4 required S/N ratio	42.3	42.3	dB
Link margin	0.0	0.0	dB

† *The uplink free-space loss includes a 0.11 dB clear-weather fade*

L_f are the fixed losses made up of antenna misalignment on the receiver using a fixed pointing antenna, waveguide losses and an allowance for setting up misalignment and ageing.

L_e are the rain losses which are calculated according to a 99.5% availability for London from the CCIR rain model given in Chapter 6.

Receiver bandwidth in logarithmic form is $10 \log (2.7 \times 10^7) = 74.3$ dB-Hz. Boltzmann's constant $k = 1.38 \times 10^{-23}$ joules/kelvin (J/K); in logarithmic form $K = -228.6$ dBW/Hz/K.

The latter enables G/T to be calculated in the clear-weather case as 24.3 dB/K.

Now:

$$T = \alpha T_a + (1 - \alpha)T_0 + T_1$$

where T_a is the sky temperature at 11.7 GHz at 30° elevation, 40 K; T_0 is ambient 290 K; T_1 is the LNA noise temperature $= 110$ K, and 0.3 dB $= 10$ log $(1/\alpha)$; therefore $\alpha = 0.93$, whence $T = 22.3$ dB/K.

Therefore:

$$G - T = 24.3$$
$$G = 24.3 + 22.3 = 46.6 \text{ dB}$$
$$= 20 \log\left(\frac{\sqrt{\eta}\pi D}{\lambda}\right) dB + 0.3 \text{ dB}$$

taking $\eta = 0.6$, $\lambda = 3 \times 10^8/1.17 \times 10^{10}$ produces $D = 2.3$ m.

The calculation for the 95.5% case is similar except that the rain fade 0.7 dB reduces the carrier level and the noise will be increased by $(1 - \alpha_R)T_R$, where 0.7 dB $= 10$ log $(1/\alpha_R)$ and T_R is taken as 273 K. This leads to a 1.3 dB increase in the G/T required to overcome the reduction in C/N, and the requirement for a 2.9 m dish.

Table A.1 shows how the link budget should be set out.

A.3 FDMA system design

(*a*) An earth-station configuration operating in the 4/6 GHz bands is illustrated below in Figure A.1. Calculate the G/T of the earth station if the effective antenna temperature at the operating elevation is 40 K.

(*b*) Eight earth stations of the type illustrated above operate in FM/FDMA and share equally the total transponder EIRP_{SAT} of $+28$ dBW, with the following parameters:

Downpath loss	$= 197$ dB
Propagation margin	$= 4$ dB
C/N_u (uppath)	$= 27$ dB
C/N_l (intermodulation)	$= 20$ dB
Output back off	$= 6$ dB
Peak frequency deviation of carrier	$= 260$ kHz

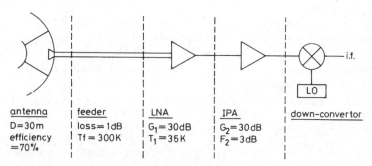

antenna
D=30m
efficiency
=70%

feeder
loss=1dB
Tf = 300K

LNA
G_1= 30dB
T_1=35K

IPA
G_2= 30dB
F_2=3dB

down-convertor

Figure A.1   Earth-station configuration

Ignoring equipment and interference noise, calculate the number of 4 kHz telephone channels which each earth station can transmit in order to meet an overall C/N_{TOT} of 16 dB at the receiver input.

If the same bandwidth transponder (36 MHz) were available for a four-phase PSK/TDMA carrier, how many additional channels would it be able to sustain?

A.3.1 Solution

(a)
$$\frac{G}{T} = \text{gain (dB)} - 10 \; \log(T_s)$$

$$G = 20 \; \log_{10} \frac{(\sqrt{\eta}\pi D)}{\lambda} \; \text{dBi}$$

$$\eta = 0.7, \; D = 30 \; \text{m}, \; \lambda = \frac{3 \times 10^8}{4 \times 10^9} = 7.5 \times 10^{-2} \; \text{m}$$

Therefore:

$$G = 60.4 \; \text{dBi}$$

$$T_s = \alpha T_a + (1 - \alpha) T_f + T_1 + \frac{T_2}{G_1}$$

$$T_a = 40 \; \text{K}, \; T_f = 300 \; \text{K}, \; T_1 = 35 \text{K}, \; \frac{T_2}{G_1} \; \text{negligible}$$

$$L = 1 \; \text{dB} = 10 \; \log(1/\alpha), \; \alpha = 0.79$$

$$T_s = 0.79 \times 40 + 0.21 \times 300 + 35$$

$$T_s = 130 \; \text{K}$$

$$\frac{G}{T} = (60.4 - 1) - 10 \; \log(130)$$

$$\frac{G}{T} = 38.3 \; \text{dB/K}$$

(b)
$$C/N_{TOT} = \frac{C}{(N_u + N_1 + N_D)} = \frac{1}{\left(\dfrac{N_u}{C} + \dfrac{N_1}{C} + \dfrac{N_D}{C} \right)}$$

Therefore:

$$\frac{N_D}{C} = \left[\frac{N_T}{C} - \frac{N_u}{C} - \frac{N_1}{C} \right]$$

All above must be in ratios:

$$C/N_T = 16 \; \text{dB. (39.8)}, \; C/N_u = 27 \; \text{dB (501.2)}, \; C/N_I = 20 \; \text{dB (100)}$$

Therefore:

$$C/N_D = 18.8 \text{ dB}$$

Downpath link equation

$$C/N_D = (\text{EIRP/carrier} - B_0) - L_d - M + \frac{G}{T} - K - 10 \ \log (B)$$

$$L_d = 197 \text{ dB}, \ M = 4 \text{ dB}, \ B = \text{bandwidth in Hz}$$

$$\text{EIRP/carrier} = +28 \text{ dBW} - 10 \ \log(8) = +19 \text{ dBW}$$

Therefore:

$$18.8 = (+19 - 6) - 197 - 4 + 38.8 + 228.6 - 10 \ \log (B)$$

$$10 \ \log (B) = 59.8 \text{ dB} - \text{Hz}$$

$$B \simeq 1 \text{ MHz}$$

Bandwidth needed to transmit FM carrier:

$$B_c = 2(\hat{F} + \text{FM})$$

$$\hat{F} = 260 \text{ kHz}$$

$$\text{FM} = \text{top modulation frequency}$$

Note: really the carrier bandwidth B_c is smaller than the RF bandwidth needed to transmit the signal by about ten per cent. Here we will assume that they are equal:

$$500 = 260 + \text{FM}$$

$$\text{FM} = 240 \text{kHz}$$

Therefore $240 \simeq N \times 4 \text{ kHz}$; i.e. $N \simeq 60$
In QPSK the channel bandwidth $= (1.2 \times 32) \text{kHz}$
Note: 1.2 is the assumed ratio of B/R_s for the system.

Number of channels in 36 MHz $= \dfrac{36\ 000}{1.2 \times 32} = 937$

This makes no allowance for guard bands, preamble etc. (see example A.4) and probably a frame efficiency of 95 per cent would be conservative. Number of channels $= 937 \times 0.95 = 890$.

Increase in capacity from $8 \times \text{FDMA}$ carrier $(8 \times 60) = 480$ channels to 89 TDMA channels is 510. The improvement is due to the fact that intermodulation does not restrict the carrier bandwidths as in FM/FDMA.

A.4 TDMA systems design (high bit rate)

Given that the uplink is BW limited, then:

$$R = \frac{WB}{Cw}\text{bit/s}$$

where

R = TDMA channel TX rate (bit/s)
W = bit-to-symbol-rate ratio of PSK modulation (1 for 2-phase, 2 for
4-phase, 3 for 8-phase)
B = transponder bandwidth (Hz)
Cw = modulation scheme bandwidth to symbol rate ratio ($\simeq 1.2$ for QPSK)
Therefore:

$$R = \frac{2 \times 36}{1.2} = 60 \text{ Mbit/s}$$

In order to improve the transmission rate through a 36 MHz transponder a higher phase modulation scheme is required. The downlink could be power-limited:

$$R = \text{EIRP} - L_d + G/T_E - K - \frac{E_b}{N_0} - M$$

Let us assume:

$\text{EIRP} = +22.5 \text{ dBW}$
$L_d = 197 \text{ dB}$
$G/T_E = 40.7 \text{ dB/K}$
$E_b/N_0 = 8.4 \text{ dB}$ (theoretical for QPSK for BER of 10^{-4})
M = practical margin to account for:

$$\left.\begin{array}{l} \text{—modem implementation } (\simeq 2\text{ dB}) \\ \text{—pointing errors } (\simeq 1\text{ dB}) \\ \text{—propagation margin } (\simeq 4\text{ dB}) \\ \text{—ISI degradations etc. } (\simeq 1\text{ dB}) \end{array}\right\} \simeq (8\text{ dB})$$

Therefore $R = 22.5 - 197 + 40.7 + 228.6 - 8.4 - 8 = 78 \text{ dB} \simeq 60 \text{ Mbit/s}$
Note to improve on this we must use FEC to give 2–5 dB coding gain advantage.

$$\text{Example of maximum voice channels} = \frac{R}{v} = \frac{60 \text{ Mbit/s}}{64 \text{ kbit/s}} = 937 \text{ channels}$$

$$\textit{Note}: \text{ frame efficiency } \eta_f = \frac{\text{no. traffic bits/frame}}{\text{total no. bits/frame}}$$

where:

total no. bits/frame $= R_{Tx}.T_F$
no. traffic bits/frame (assume QPSK)

$$= R_{TX}T_F - 2[NS_P + KS_{RB} + (N + K)S_G]$$

$\mathcal{N}$ = no. traffic bursts/frame
S_p = preamble (PA) length in symbols
K = number of RB/frame
S_{RB} = length of recovery bits (RB) in symbols
S_G = guard time in symbols

$$n_f = \frac{R_{TX}T_F - 2[\mathcal{N}S_p + KS_{RB} + (\mathcal{N} + K)S_G]}{R_{TX}T_F} \times 100\%$$

Number of telephone channels $= \dfrac{R}{v} \times n_f$

for 95% frame efficiency $= \dfrac{60 \times 10^6}{64 \times 10^3} \times 0.95 = 890$

Example (Telecom I with 36 E/S)

R_x = 24.576 Mbit/s, $T_F = 20$ ms, $\mathcal{N} = 36$

K = 2, CR.BTR = 112 symbol, UW = 16 symbol

$S_{RB} = 256$ symbol, $G_B = \pm 16$ symbol, assume QPSK

$$n_f = \frac{24.576 \text{ Mbit/s, } \times 20 \text{ ms} - 2[36 \times 128 + 2 \times 256 + 38 \times 32]}{24.576 \text{ Mbit/s} \times 20 \text{ ms}}$$

$n_f = 97\%$

A.5 Mobile system link budget

This link budget is courtesy of INMARSAT and for communications from ship-to-shore at 1.64 GHz transmit for the ship to satellite and 4.2 GHz from the satellite to coastal earth station as given in Table A.2. The ship's terminal is considered to be a standard-A installation. In Table A.3 is presented the reverse direction communications of shore-to-ship; 6.4 GHz for coastal earth station to satellite and 1.5 GHz for satellite to the ship's standard A $(G/T = -4 \text{ dB/K})$.

The up and downlink budget for C/N_0 are fairly standard except that remember:

$$C/N_0 = \text{EIRP} - \text{FSL} - L_f + G/T - K$$

The intermodulation on board the satellite is given as follows: the total intermodulation noise in the 7.5 MHz (68.8 dBHz) transponder is 24 dBW. Hence this equates to an intermodulation noise density of

$$24 \text{ dBW} - 68.8 \text{ dBHz} = -44.8 \text{ dBW/Hz}$$

whence the carrier to intermodulation noise is:

$$\left(\frac{C}{N_0}\right)_{IM} = 18 \text{ dBW}(-44.8 \text{ dBW/Hz})$$
$$= 62.8 \text{ dB} - \text{Hz}$$

The overall carrier-to-noise is calculated as per eqn. 11.27:

$$\left(\frac{C}{N_0}\right)_{TOT} = \left\{\left(\frac{C}{N_0}\right)_u^{-1} + \left(\frac{C}{N_0}\right)_0^{-1} + \left(\frac{C}{N_0}\right)_{IM}^{-1} + \left(\frac{C}{I_0}\right)^{-1}\right\}^{-1}$$

and the difference between the resultant (C/N_0) and the required (C/N_0) to support the service is the system margin. The required C/N_0 is obtained

Table A.2   Ship-to-shore for standard-A system

Ship-to-satellite link	
Ship-station EIRP/carrier	36.0 dBW
Free and absorption path losses (1.64 GHz, 5 deg. elev.)	189.4 dB
Satellite receive G/T	-13.0 dB/K
Uplink C/N_0	62.2 dBHz
Satellite intermodulation noise	
Total satellite EIRP	16.0 dBW
Intermod. noise power ratio	15.0 dB
Transponder bandwidth (7.5 MHz)	68.8 dBHz
Satellite EIRP/carrier	-5.0 dBW
Satellite C/N_{0IM}	62.8 dBHz
Satellite-to-shore link	
Satellite EIRP/carrier	-5.0 dBW
Free space and atmospheric losses (4.2 GHz, 5 deg. elev.)	197.6 dB
Shore earth-station receive G/T	32.0 dB/K
Downlink C/N_0	58.0 dBHz
Satellite link C/N_0 (uplink, intermod. and downlink)	55.7 dBHz
Intersystem interference C/I_0	64.4 dBHz
Overall C/N_0	55.2 dBHz
Required C/N_0	52.5 dBHz
Margin	2.7 dB

Table A.3   Shore-to-ship for standard-A system

Shore-to-satellite link	
Shore station EIRP/carrier	58.0 dBW
Free and absorption path losses (6.42 GHz, 5 deg. elev.)	201.3 dB
Satellite receive G/T	-14.0 dB/K
Uplink C/N_0	71.3 dBHz
Satellite intermodulation noise	
Total satellite EIRP	33.0 dBW
Intermod. noise power ratio	9.0 dB
Transponder bandwidth (7.5 MHz)	68.8 dBHz
Satellite EIRP/carrier	18.0 dBW
Satellite C/N_{oIM}	62.8 dBHz
Satellite to ship link	
Satellite EIRP/carrier	18.0 dBW
Free space and atmospheric losses (1.5 GHz, 5 deg. elev.)	188.9 dB
Ship earth station receive G/T	-4.0 dB/K
Downlink C/N_0	53.7 dBHz
Satellite link C/N_0 (uplink, intermod. and downlink)	53.1 dBHz
Intersystem interference C/I_0	61.8 dBHz
Overall C/N_0	52.6 dBHz
Required C/N_0	52.5 dBHz
Margin	0.1 dB

knowing the demodulation threshold, the margin and the receiver bandwidth, e.g.

$$\left(\frac{C}{N_0}\right)_{\text{Required}} = \left(\frac{C}{N}\right)_{\text{Threshold}} + M - 10 \log (\text{BW})$$
$$= 6.5 \text{ dB} + 2 \text{ dB} - 10 \log (30 \text{ kHz})$$
$$= 52.5 \text{ dB–Hz}$$

A.6 SCPC system design

A developing country wishes to lease a half transponder of an INTELSAT-IVA satellite for use in domestic telephony distribution.

It has been decided to use SCPC equipment with the following characteristics:

channel bandwidth $\qquad = 22.5$ kHz
demodulator threshold $C/N = 7.5$ dB

and standard-B earth stations are to be used $(G/T = 31.7$ dB/K$)$. Propagation conditions indicate a worst fade for 0.3% of the month, experienced on the worst link of 3 dB.

The INTELSAT satellite back offs for leased service operation are:

output $= 5$ dB
input $ = 10$ dB

Calculate the domestic telephone capacity with and without the use of voice activation.

An INTELSAT condition of lease is that the earth-station off-axis EIRP density should not exceed $42{-}25 \log (\theta)$ dBW/40 kHz, where θ is the off-axis angle.

Would the above service meet this requirement?

A.6.1 Solution SCPC

$$\text{uplink}\left[\frac{C}{N_0}\right]_u = (\text{EIRP} - Bo_i) + \frac{G}{T_s} - Lu + K$$

Note: flux $(S) = \text{EIRP} - 10 \log (4\pi S^2)$

$$Lu = 20 \log \frac{(4\pi S)}{\lambda}$$

Therefore:

$$(C/N_0)u = (S - Bo_i) + \frac{G}{T_s} - 37 + K \tag{A.1}$$

$$\left(37\text{dB} = 10 \log\left[\frac{4\pi}{\lambda^2}\right], \lambda = 5 \times 10^{-2} \text{ m at 6 GHz}\right)$$

Downpath

$$\left[\frac{C}{N_0}\right]_D = (\text{EIRP} - Bo_0) - L_d + \frac{G}{T_e} + K \tag{A.2}$$

Intermod

$$(C/N_0)_1 = 84.5 + 1.78 \times Bo_0 \tag{A.3}$$

(approximate formula)

Specification

$$10 \log (N) = \left[\frac{C}{N_0}\right]_{TOT} - \left[\frac{C}{N_0}\right] \text{ required} \qquad (A.4)$$

Solve above equations to find N.

(C/N_0) required:

$(C/N)_{th} = 7.5$ dB
Prop. margin $= 3$ dB
$(C/N)_w = 10.5$ dB
$B = 22.5$ kHz

Therefore:

$$\left[\frac{C}{N_0}\right]_{TOT} = 10.5 + 10 \log (2.25 \times 10^4)$$

$$= 54 \text{ dB-Hz}$$

For 1/2 transponder lease:

	EIRP	Flux
Full saturation	+22.5 dBW	−67.5 dBW/m²
Half saturation	+19.5 dBW	−70.6 dBW/m²
→ BO	+14.3 dBW	−80.5 dBW/m²

(1) $(C/N_o)_u = -80.5 - 18.6 - 37 + 228.6$
$\qquad = 92.4$ dB-Hz

(2) $(C/N_o)_D = +14.5 - 197.6 + 31.7 + 228.6$
$\qquad = 77.2$ dB-Hz

(3) $(C/N_o)_1 = 84.5 + 1.78 \times 5 = 93.4$ dB-Hz

(4) $(C/N_o)_{TOT} = \dfrac{C}{N_{ou} + N_{oD} + N_{o1}} = 77.5$ dB-Hz

Therefore:

$$10 \log (N) = 77.5 - 54$$

$$N = 223 \text{ channels}$$

Voice activation, 4 dB improvement due to activity:

$$10 \log (N) = 27.5 \text{ dB}$$

$$N = 562 \text{ channels}$$

Existing standard B meets $32 - 25 \log (\theta)$ antenna gain pattern
Note: off-axis EIRP $\leqslant 42 - 25 \log (\theta)$

Therefore, for any carrier spectrum in 40 kHz, the power $\leqslant +10$ dBW (10 W) on individual SCPC (22.5 kHz) $\leqslant 7.5$ dBW (5.6 W).

$$\text{EIRP/carrier} = -80.5 + 10 \log (4\pi S^2)$$
$$= -80.5 + 163.3$$
$$= 82.8 \text{ dBW}$$
$$\text{EIRP/Ch} = 82.8 \text{ dBW} - 10 \log (N).$$

(*a*) No. voice activation

$$\text{EIRP/Ch} = 82.8 - 23.5$$
$$= +50.3 \text{ dBW}$$

(*b*) with voice activation

$$\text{EIRP/Ch} = +55.3 \text{ dBW when averaged over a reasonable time.}$$

$$\text{Gain of standard B} \simeq 55 \text{ dB} \left[= 20 \log \frac{(\sqrt{\eta}\pi D)}{\lambda} \right]$$

Therefore:

$$\text{Power/channel} = +0.3 \text{ dBW with voice activation}$$
$$= +4.3 \text{ dBW without voice activation}$$

This is less than $+7.5$ dBW, so both cases will meet the requirement.

A.7 Military systems design

By considering a few simplified link budgets, we illustrate some typical parameters and in particular the constraints of small-terminal satellite communication operation. We are concerned with an SHF geostationary satellite, having a transparent transponder of 10 MHz BW. We look at two terminal types:

(*a*) *Large terminal*: this might be an anchor station, and has a 10 m dish antenna with a 1 kW transmitter and receive noise temperature of 200 K.

(*b*) *Small terminal*: this is based on the UK Manpack, and has a 45 cm dish with a 2 W transmit power and 100 K receive noise temperature. The actual terminal EIRPs are 88 dBW and 31 dBW, respectively; these figures are taken from the specification (and account for antenna efficiency).

The path loss is as determined by frequency (which is taken here as 8 GHz for both up- and downlinks, for simplicity), and by range. Range is taken as 37 000 km to a geostationary satellite. The satellite receive-antenna gain is that of an earth-cover antenna, which is about 17 dB at edge of coverage irrespective of frequency. A similar antenna is taken for downlink transmit, with a TWTA of 20 W, giving an EIRP of 30 dBW.

Simplified link budgets are shown in Table A.4. Here we ignore link margins (for weather etc.) and consider only a single access which takes the full saturated downlink EIRP of the transponder (in practice, this may not be the case, and back off would additionally be applied). The resultant parameter of interest is the carrier-to-noise density, C/N_0, expressed in dB-Hz. This is often more useful than carrier-to-noise ratio, CNR, being independent of any modulation scheme. The term C/kT is sometimes used to mean the same thing. It may help to visualise C/N_0 as the CNR which would result if the signal were being detected within a 1 Hz bandwidth.

A.7.1 Large terminal power budget

It is shown (i) that the large station uplink can produce a C/N_0 of 102 dB-Hz at the satellite front end. In the satellite transponder bandwidth of 10 MHz (i.e. 70 dB-Hz), this yields 32 dB CNR, implying that the downlink power will be almost entirely wanted signal, with negligible noise contribution.

At the large-terminal receiver, the downlink C/N_0 is 92 dB-Hz (ii), implying 22 dB receive CNR over the full 10 MHz BW. This is adequate for most purposes, and indicates that noise is not a limitation in this system. It suggests that data rates up to at least 10 Mbit/s at negligible error rates should be achievable.

A.7.2 Small terminal (e.g. Manpack) budget

Consider the Manpack terminal uplink (iii), with an EIRP of only 31 dBW. The satellite received C/N_0 is 45 dB-Hz, which implies a transponder SNR (in 10 MHz) of -25 dB! Thus, with this single access, the satellite downlink power will be mostly thermal noise, and the wanted signal downlink EIRP is about $30 - 25 = 5$ dBW (neglecting here for the moment the small-signal suppression, which will reduce it a further 1 dB). This C/N_0 due to satellite front-end noise will appear at the downlink receiver together with the additional front-end noise of that receiver. The overall resultant C/N_0 is determined by the reciprocal of the sum of the reciprocals, although in practice one or other value may predominate. In the case of a large receive terminal, the downlink transmitted noise will still swamp local front-end noise, leaving C/N_0 of 45 dB-Hz.

We now consider the downlink to the Manpack terminal (iv). If the full satellite EIRP was devoted to the signal, the receiver C/N_0 would be 55 dB-Hz. This would apply with a large-station uplink, but if now our uplink is another Manpack, we have the reduction in signal EIRP of some 25 dB, which degrades the receiver C/N_0 from 55 to 30 dB-Hz. This poor figure predominates over the uplink C/N_0 of 45 dB-Hz to yield a resultant C/N_0 of approximately 30 dB-Hz.

With these figures must be included practical link margins. At SHF a realistic overall figure is 6 dB (for ground-satellite-ground), although an optimist might choose 4 dB, and a pessimist 10 dB. Here we take a loss of 3 dB per path, and also add in small signal suppression of 1 dB (where a weak uplink

Table A.4 *Specimen outline power budgets. Geostationary satellite at 8 GHz; weather margins not included here*

Large terminal

(i)   Uplink

$P = 1$ kW 10 m dish, $G = 58$ dB	EIRP	=	88 dBW
Path loss		=	202 dB
Satellite rx antenna gain		=	17 dB
Noise temp. T_N 1000 K ($= 30$ dBK) Boltzmann const. $- 229$ dBW/Hz/K	kT	=	-199 dBW/Hz
Resultant uplink C/N_0		=	102 dB-Hz

(ii)   Downlink

$P = 20$ W EC antenna, $G = 17$ dB	EIRP	=	30 dBW
Path loss		=	202 dB
Terminal rx antenna gain		=	58 dB
Noise temp. T_N 200 K ($= 23$ dBK) Boltzmann const. $- 229$ dBW/Hz/K	kT	=	-206 dBW/Hz
Resultant downlink C/N_0		=	92 dB-Hz

Small terminal (e.g. Manpack)

(iii)   Uplink

$P = 2$ W 45 cm dish, $G = 28$ dB	EIRP	=	31 dBW
Path loss		=	202 dB
Rx antenna gain		=	17 dB
Noise temp. T_N 1000 K ($= 30$ dBK) Boltzmann const. $- 229$ dBW/Hz/K	kT	=	-199 dBW/Hz
Resultant uplink C/N_0		=	45 dB-Hz

(iv)   Downlink

$P = 20$ kW EC antenna, $G = 17$ dB	EIRP	=	30 dBW
Path loss		=	202 dB
Rx antenna gain		=	28 dB
Noise temp. T_N 1000 K ($= 30$ dBK) Boltzmann const. $- 229$ dBW/Hz/K	kT	=	-199 dBW/Hz
Resultant downlink C/N_0		=	55 dB-Hz

signal is below broadband noise). The C/N_0 figures are combined reciprocally such that:

$$\frac{1}{(C/N_0)_{res}} = \frac{1}{(C/N_0)_{up}} + \frac{1}{(C/N_0)_{down}}$$

yielding the following results:

large–large	transponded C/N_0	99	dB-Hz
	downlink EIRP	30	dBW
	receiver C/N_0	89	dB-Hz
	resultant C/N_0	<u>88.6</u>	dB-Hz
large–small	transponded C/N_0	99	dB-Hz
	downlink EIRP	30	dBW
	receiver C/N_0	52	dB-Hz
	resultant C/N_0	<u>52</u>	dB-Hz
small–large	transponded C/N_0	42	dB-Hz
	downlink EIRP	1	dBW
	receiver C/N_0	60	dB-Hz
	resultant C/N_0	<u>42</u>	dB-Hz
small–small	transponded C/N_0	42	dB-Hz
	downlink EIRP	1	dBW
	receiver C/N_0	23	dB-Hz
	resultant C/N_0	<u>23</u>	dB-Hz

(It is seen that the path loss margin has greatest effect on the already poor small-to-small terminal links, and the uplink losses may have no effect when the transponder is saturated.)

These figures allow us to determine the capacity of the link. A practical modem may specify minimum operating C/N_0, otherwise we can estimate the maximum achievable data rate from a knowledge of the E_b/N_0 requirements of the particular modulation concerned. If the energy per data bit is E_b, and the baud rate is R bit/s, then the carrier power C is given by $C = E_b R$. Hence $C/N_0 = RE_b/N_0$. For most practical modulation schemes, a value of about 10 dB is required for E_b/N_0 in the absence of forward-error-correction coding, so the data rate is determined by subtracting 10 dB from the C/N_0 figures above. For example, if C/N_0 is 52 dB-Hz, R is $52 - 10 = 42$ dB-bits. $= 16$ kbit/s.

On the basis of the above figures, for the single access described, we arrive at the following approximate results for maximum data rate:

large–large	10 Mbit/s	*large–small*	16 kbit/s
	(i.e. transp bandwidth)		
small–large	1.6 kbit/s	*small–small*	20 bit/s

This assumes binary modulation and ideal Nyquist channel filters.

This shows that direct Manpack-Manpack communication is not feasible at a sensible data rate in this scenario, and it is necessary to route via a large anchor station, perhaps with baseband data regeneration. The above illustrations relate to only a single access; in practice, a number of terminals would require simultaneous power sharing of the satellite channel, and this further reduces the available capacity (it is most unlikely that a single Manpack would be able to demand exclusive use of an entire SHF 10 MHz transponder!).

As the number of users increases, the satellite downlink EIRP has to be shared among them (FDMA or CDMA may be assumed, although similar principles apply to TDMA also). This will further reduce the received C/N_0, and it can be seen that for downlinks to small terminals the already poor performance will further degrade. In order to increase the capacity, the downlink EIRP must be increased; the way to do this is through the use of spot beam antennas on the satellite. For example, if coverage were reduced from earth cover to the coverage of western Europe, additional gain of the order of 15 dB might be achieved. The penalty is of course the coverage restriction thus imposed.

A.8 VSAT link budget

This example is taken from an experimental VSAT system called CODE, operated at Ka-band via the ESA Olympus satellite. Calculation is performed for a small CODE VSAT terminal with a 0.9 m dish at the University of Surrey (UoS) at Guildford in the UK to a 2.5 m hub station located at DERA Defford, UK. The outbound link (hub-VSAT) is at 2.048 Mbit/s with QPSK modulation, and the inbound (VSAT-hub) is at 9.6 kbit/s information rate with half-rate Viterbi coding which produces a 19.2 k symbol/s rate for BPSK modulation. The hub–satellite link is planned for 99.9% availability, and the VSAT-satellite link for 99% availability. Note that at 20/30 GHz the rain fades are correspondingly larger than at Ku-band.

The link budgets printed in Tables A.5 and A.6 follow those presented previously but note that:

$$\frac{C}{N} = \frac{E_b}{N_0} \frac{R_b}{B_{if}}$$

and that:

$$\left(\frac{E_b}{N_0}\right)_{practical} = \left(\frac{E_b}{N_0}\right)_{Theory} - \text{coding gain} + \text{modem implementation margin} + \text{additional implementation margin}$$

Table A.5   Hub-VSAT link budget

Hub to VSAT data rate	2.048 Mbits/s
Modulation	QPSK
FEC	convolutional, rate = 1/2
Decoding	Viterbi, K = 7, rate = 1/2
HUB transmits on Channel 3	(28.6 GHz)
VSATs receive on Channel 1	(19.4 GHz)

TDS-6 hub station

RF power (300 W @ 5.6 dB backoff)	24.8 dBW
Antenna gain (2.5 m, 28.6 GHz)	55.8 dB
Antenna pointing loss	1.9 dB
Hub EIRP	73.0 dBW
Free-space loss (29 GHz)	213.0 dB
Uplink fade (99.9% year UK)	12.0 dB

Olympus satellite

Antenna gain	42.6 dB
Receiver noise temperature	32.4 dBK
Satellite G/T ($T_a = 300$ K)	10.2 dB/K
Transponder output backoff	3.6 dB
Transponder saturation EIRP (0.6° coverage zone)	52.1 dBW
Signal EIRP	48.5 dBW
C/T_u	−141.8 dBW/K
Free-space loss (19 GHz)	209.4 dB
Downlink fade (99.0% UK)	3.0 dB

UOS CODE terminal

Antenna gain (0.9 m, 19.5 GHz)	43.7 dB
Clear weather noise temperature	27.2 dBK
System noise temperature	28.2 dBK
System G/T	15.5 dB/K
Antenna pointing loss	0.4 dB
Received power	− 120.7 dBW
C/T_d	− 148.9 dBW/K
C/T_{total}	− 149.6 dBW/K
C/N_{0total}	79.0 dBHz
Data rate	63.1 dBHz
E_b/N_0 achieved	15.9 dB
E_b/N_0 theory ($BER = 10^{-8}$)	6.3 dB
Implementation loss	3.0 dB
E_b/N_0 required	9.3 dB
Link margin	6.6 dB

Table A.6   VSAT-Hub link budget

VSAT to hub data rate	9.6 kbit/s
Modulation	BPSK
FEC	convolutional, rate = 1/2
Decoding	Viterbi, K = 7, rate = 1/2
VSATs transmit on channel 1	(28.1 GHz)
hub receives on channel 3	(18.9 GHz)

UoS CODE terminal

RF power (50 mW @ 1 dB comp)	−13.0 dBW
Antenna gain (0.9 m, 28.1 GHz)	46.9 dB
Antenna pointing loss	0.9 dB
VSAT EIRP	33.0 dBW
Free-space loss (29 GHz)	213.0 dB
Uplink fade (99.9% year UK)	5.0 dB

Olympus satellite

Antenna gain	40.1 dB
Receiver noise temperature	32.4 dBK
Satellite G/T ($T_a = 300$ K)	7.7 dB/K
Received power (signal)	−144.9 dBW
Transponder output backoff	34.9 dB
Transponder saturation EIRP (spotbeam center)	53.7 dBW
Signal EIRP	18.8 dBW
C/T_u	−177.3 dBW/K
Free-space loss (19 GHz)	209.4 dB
Downlink fade (99.0% UK)	7.0 dB

TDS-6 hub station

Antenna gain (2.5 m 18.9 GHz)	52.6 dB
Clear weather noise temperature	27.1 dBK
System noise temperature	28.7 dBK
System G/T	23.9 dB/K
Antenna pointing loss	1.3 dB
Received power	−146.3 dBW
C/T_d	−175.0 dBW/K
C/T_{total}	−179.3 dBW/K
C/N_{total}	49.3 dBHz
Data rate	39.8 dBHz
E_b/N_0 achieved	9.5 dB

continued

Table A.6   continued

E_b/N_0 theory $(BER = 10^{-8})$	6.3 dB
Implementation loss	3.0 dB
E_b/N_0 required	9.3 dB
Link margin	**0.2 dB**

Appendix B

Domestic satellite systems design— a case study (INUKSAT system)

B. G. Evans

B.1 Introduction

Many countries have adopted INTELSAT lease transponder facilities to provide a domestic service. The types and extent of the services depend greatly on the needs of the country in question. In this case study of a domestic satellite system design, we have chosen a simple implementation but one that is representative in design terms of all domestic leased systems.

The study was performed for the Greenland Technical Organisation which wished to install a domestic satellite system (INUKSAT) to provide telephone and telegraph links between key communities in Greenland and to provide broadcast audio programme channels from the main station to all outstations. In addition, it was planned to link the system to Denmark by means of a Danish earth station operating within the system.

The existing telecommunications links in Greenland are shown in Figure B.1 and consist of:

(i)   Terrestrial microwave radio link (2 GHz) providing automatic telephone, telex and audio programme service for the main communities just north of Egedesmind (EGM) to the southernmost tip of Greenland.

(ii) An HF low-grade service to certain settlements on the north-western and eastern coasts.

(iii) The ICECAN cable which connects Iceland and Canada and accommodates six telephone channels for Greenland service.

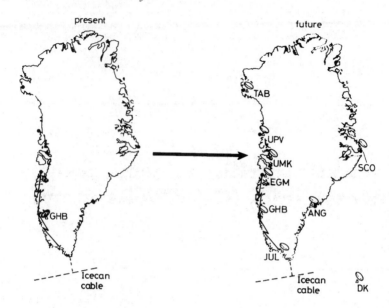

Figure B.1   Greenland telecommunications—present and future

(iv)   The DANSAT system which consists of an INTELSAT standard-B earth station located at Godthaab (GHB), providing ten telephone channels to the
joint Nordic standard-A earth station at Tanum in Sweden via the INTELSAT
AOR primary satellite.

In Figure B.1 we also show the long-term aims of the INUKSAT system
planned to consist of eight stations in Greenland and one earth station in
Denmark. A planned build-up is requested by establishing a main station and
four or five outstations operating on a leased INTELSAT transponder in the
first phase followed by adding the Danish station and the remaining earth
stations later.

B.2 Requirements

B.2.1 Traffic

The first stage in the planning of any domestic satellite system is to obtain the
traffic requirements (often a difficult process!). In this case there was an added
complication in the choice of the main station site. However, this will not be ela-
borated on here and EGM will be taken as the main station. The traffic and
channel summary is shown in Figure B.2 and Table B.1.

traffic configuration

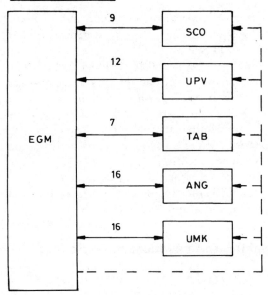

Figure B.2   INUKSAT phase 1 and 2 traffic and channel summary

Table B.1   Equipment requirements summary

Earth stations	FT		VAT		OVAT		Total two-way voice	Audio		Pilot		VFT		CT	SDc	ESc
	TX	RX	TX	RX	TX	RX		TX	RX	TX	RX	TX	RX			
EGM (inc. UMK)	–	12	38	38	22	10	60	1	–	1	–	1	–	1		1
EGM (excl. UMK)	–	9	28	28	16	7	44	1	–	1	–	1	–	1		1
SCO	2	–	6	6	1	3	9	–	1	–	1	–	1	1		1
UPV	3	–	6	6	3	6	12	–	1	1	1	–	1	1		1
TAB	1	–	6	6	–	1	7	–	1	–	1	–	1	1		1
ANG	3	–	10	10	3	6	16	–	1	–	1	–	1	1		1
UMK (optional)	3	–	10	10	3	6	16	–	1	–	1	–	1	1		1
TOTAL (inc. UMK)	12	12	76	76	32	32	120	1	5	2	5	1	5	6		6
TOTAL (excl. UMK)	9	9	56	56	23	23	88	1	4	2	4	1	4	5		5

B.2.2 Operating

The INUKSAT system will operate in a difficult environment to isolated earth stations. Access to most of the outstations, on a nonurgent basis, is planned as once per two-week period, and at SCO once in a three to four month period. Outstations and EGM will be staffed by technicians for a maximum of eight hours/day and on weekdays. Major outside work in Greenland is restricted to a six-month period each year starting in June and high winds and very low temperatures (-25 to $-50°C$) are experienced. Thus high antenna survival wind speeds ($\simeq 200$ mph) are required at some outstations. There is a requirement for monitoring the performance of outstations at the main station to ensure that emergency services can be activated.

The severe environmental problems can be totally overcome by the use of radomes to provide all-weather maintenance, protection from snow, ice and high winds. They will also guarantee continuity of service.

B.3 System performance characteristics

B.3.1 Performance objectives

(i) *SCPC telephony carriers*

Equivalent to that of 10 000 pWOp channel in any hour of the average day. Noise not to exceed 50 000 pWOp for more than 0.3% of worst month.

Allocations of 2000 pWOp for earth stations plus 1000 pWOp for interference into earth stations are common.

(ii) *Audio programme channel*

15 kHz programme channel whose weighted noise should not exceed -47 dBmOp.

(iii) *SCPC pilot channel*

In view of the importance of the pilot channel in maintaining correct control of system frequencies it is advisable to transmit the pilot at 3 dB higher than that of a normal SCPC channel. This gives more rapid recovery from deep fades while having insignificant impact on system capacity.

B.3.2 Satellite specifications

The full performance characteristics of INTELSAT leased transponders and the operational constraints placed on users are defined by INTELSAT document BG-28-7E. The important aspects are summarised in Table B.2. The transponder intermodulation spectrum is calculated by INTELSAT who will determine the final frequency plan prior to acceptance into the leased operation transponder.

INTELSAT publishes on a regular basis existing and planned placement of its satellites. The plan at the stage of study (1978) is shown in Table B.3. A check of visibility of all earth stations in the system must then be made in order to select the best satellite.

In Figure B.3 we have plotted the earth-station elevation angle as a function of satellite longitude for critical stations. Although the initial system would be feasible

Table B.2   INTELSAT leased transponders

	Quarter transponder	Half transponder	Unit
Available flux density	-84.0 ± 2.0 dB	-81.0 ± 2.0 dB	dBW/m^2
Available bandwidth	9.0	18.0	MHz
Available EIRP	11.5	14.5	dBW
Satellite receive sensitivity G/T	-18.6	-18.6	dB/K
Permitted transponder intermodulation			
EIRP density	-40.0	-40.0	dBW/4 kHz
Earth station off-beam	$42-25 \log_{10} \theta$ for $2.5° < \theta < 48°$		
EIRP density	0 for $> 48°$		dBW/40 kHz

Table B.3   INTELSAT AOR plan

Position	Function
325.5°E	Major path 1 (MP1)
329.0°E	Spare MP1
333.0°E	Spare primary
336.0°E	Primary
339.0°E	Spare
342.0°E	MP2
356.0°C	Residual life
359.0°E	Leases etc.

for satellites between 282°E–329°E (assuming a minimum elevation of 2°) when the Danish station is brought into service 308°E is the westerly limit (assuming a minimum elevation of 5°C for the main station). The use of the spare at 329°E looks the best option although there is some risk attached to this. A degree of security needs to be sought from INTELSAT that satellites will be available.

B.3.3 Earth-station parameters

(i) G/T

The specification called for either 30.7 or 31.7 dB/K at 5° elevation. It will be shown that only 31.7 dB/K will give the capacity required. The specification also puts an upper limit on the antenna diameter of 11 m (gain = 51.7 dB at 4 GHz). The total system receive noise is given by:

$$T_s = [\alpha T_a + (1 - \alpha) T_f] + T_{LNA} + \frac{T_{IPA}}{G_{LNA}}$$

for an 11 m dish at 4 GHz. $T_A = 16 + 84.5\gamma^{-0.6}$ where γ is the elevation angle. T_{LNA} for a cooled paramp is 45 K for a redundant pair. Thus for 5° elevation $T_s = 93$ K and $G/T = 32$ dB/K.

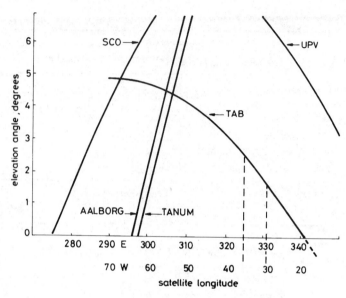

Figure B.3   Antenna elevation angle against satellite longitude

*Table B.4   Earth-station G/T at operating eleva-
tion angle*

Station	Elevation angle (deg)	G/T (dB/K)
EGM	11.16	32.3
GHB	15.65	32.5
DK	15.54	32.5
TAB	2.02	30.8
UPV	6.90	31.9
UMK	9.38	32.2
JWL	20.02	32.6
ANG	15.86	32.5
SCO	10.72	32.3

Full system earth station G/Ts are given in Table B.4

Note: Pointing errors must be checked, and this indicates the necessity to use step track in order to maintain the gain for the INTELSAT satellite stability ($\pm 0.5°$). Tracking error and wind effects will contribute approximately 0.25 dB to EIRP instability and the G/T by about 0.15 dB.

(ii) *EIRP and HPA ratings*

The EIRPs from each earth station are calculated in Section B.6. Assuming an antenna gain of 54.5 dB at 6 GHz for an 11 m antenna and a feeder loss of

Table B.5   *Power and total number of channels handled in expanded system*
           a   Voice activation factor not taken into account
           b   Includes pilot, VFT and data channels
           c   Includes timeshared data channels

Station	EIRP (dBW)	Number of channels (a)	Total EIRP (dBW)	Total HPA power (W)	HPA rating (W)
	57.7	17			
	63.4	14(b)			
EGM	59.0	18	80.9	550	3 kW
	58.6	22			
	57.7	22			
	58.4	12			
	74.7	programme			
JAB	61.8	11(c)	73.8	107	600
	60.9	6			
UPV	58.7	19(c)	72.0	71	600
	58.0	3			
UMK	58.5	23(c)	73.0	89	600
	57.8	6			
ANG	58.0	23(c)	71.6	65	600
	57.3	6			
SCO	58.4	13(c)	70.3	48	600
	57.7	3			

1 dB, the required HPA power/channel can then be calculated. In Table B.5 we give the power, together with the total number of channels to be handled by each station in the fully expanded system. Also given is the total HPA rating required. Back off of at least 6 dB is recommended.

B.3.4 System margin

The system margin to be added for each link separately includes the reduction of carrier owing to rain fading plus scintillation at the station elevation angle for the 0.3% of the time specified, together with the increase in system noise temperature produced by the rain. These data are usually difficult to obtain for most countries and must be extrapolated from rain data to attenuation using methods given in the CCIR. Here, the tropospheric scintillation is more important than rain, and data are extrapolated from measurements in the Canadian TELESAT system. Data are shown in Figure B.4 and the margins are given in Table B.7. It will be seen that short-term losses never exceed 2.4 dB, with the exception of TAB which is 4.7 dB. For an SCPC demodulation with threshold at 7.5 dB (C/N for 7000 pWOp) the operating point is thus set to 10 dB.

Table B.6   Quarter-transponder traffic matrix

From	To	Voice activated channels	Permanent channels	Link no.
	TAB	$6+1^1$	$4+1^2$	1
	UPV	$6+6$	0	2
EGM	UMK	$10+6$	0	3
	ANG	$10+6$	0	4
	SCO	$6+3$	0	5
TAB		$6+0$	1	6
UPV		$6+3$	3	7
UMK	EGM	$10+3$	3	8
ANG		$10+3$	3	9
SCO		$6+3$	2	10

[1] $x+y$ = voice activated + other voice activated
[2] $4+1$ = VFT + data + pilot at twice the EIRP (i.e. 4 total + programme)

B.3.5 Modulation parameters

For both voice and programme channels the modulation method chosen is SCPC companded FM. All channels, whether voice activated or not, have the same requirement for demodulated S/N. All voice-channel equipment is identical and thus deviations and channel bandwidths common. The programme channel, however, is single channel to all stations and thus must be designed for the worst link. The same is true for the VFT, time-shared data and pilot.

Specifications given are:

voice channels:	7000 pWOp($\equiv S/N = 51.6$ dB)
data channels:	BER of 1×10^{-5}
programme channel:	15 200 pWOp($\equiv S/N = 48.2$ dB)

Equipment available has the following specifications:

voice demodulator threshold	$= 7.5$ dB
programme channel threshold	$= 8.0$ dB
compander advantage	$= 17.5$ dB
CCIR Rec. 468-1 programme	$= -6.8$ dB
channel weighting advantage	$= 2.5$ dB

Speech channels: for voice circuits a C/N_0 of 54 dB-Hz has been shown to give adequate subjective performance. This leads to an r.m.s. deviation of 4.9 kHz and a filter noise bandwidth of 25 kHz to yield $C/N = 10$ dB. Available equipment has 30 kHz spacing.

Telegraph channels: A voice-type carrier can support a 24-channel, 50-baud telegraph circuit as per CCITT Rec. R35. A telegraph carrier can therefore be

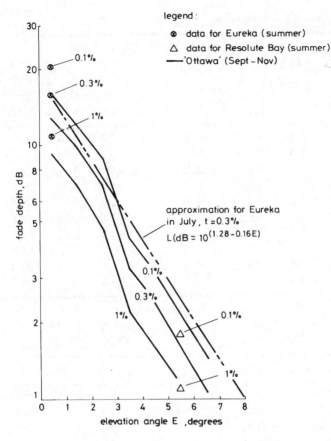

Figure B.4   Fade depth against elevation

regarded, so far as overall systems calculations are concerned, as equivalent to a full-time voice carrier.

Data channels: A voice-type carrier can support a 4.8 kbit/s data channel as per CCITT Rec. V27, with a BER of 1×10^{-5} for $C/N_0 = 54$ dBHz. A data carrier can therefore also be regarded as a voice carrier for systems calculations.

Pilot: Since loss of pilot means loss of system it is advisable to have C/N for pilot 3 dB higher than for the telephone carrier.

Programme channel: Parameters must be chosen such that the worst station margin (4.7 dB) is accommodated above the threshold (8 dB). Thus the operating $C/N = 12.7$ dB, where:

$$C/N_0 = 67.6 \text{ dB-Hz}$$
$$\Delta f_{rms} = 33.2 \text{ kHz}$$
$$\text{bandwidth} = 294 \text{ kHz (noise} = 310 \text{ kHz)}$$

Table B.7   Quarter-transponder link budget

Link	Temp. dB-K	Prop. dB	Sum dB	C/N dB	Voice channels EIRP/Ch.		EIRP/Ch.
					dBW	W	dBW
1	1.50	3.2	4.7	12.2	−4.6	0.34	63.2
2	1.00	1.3	2.3	10.0	−8.9	0.12	59.0
3	0.60	1.1	1.7	10.0	−9.3	0.11	58.6
4	0.60	0.9	1.5	10.0	−10.2	0.09	57.7
5	0.70	1.0	1.7	10.0	−0.5	0.11	58.44
6	0.70	3.9	4.6	12.1	−7.4	0.18	61.8
7	0.70	1.5	2.2	10.0	−9.5	0.10	58.7
8	0.70	1.0	1.7	10.0	−9.5	0.10	58.5
9	0.70	0.8	1.5	10.0	−9.5	0.10	58.0
10	0.70	1.0	1.7	10.0	−9.5	0.10	58.5

Voice W	Pilot W	VFT W	Data W	Programme W	Req. Sat W	Avail Sat W
6.98	0.68	0.34	0.34	4.71	13.06	14.12

Link	(C/N_0)U	C/I_0	(C/N_0)D	(C/N_0) total	(C/N_0) total
1	72.8	71.3	56.4	56.2	12.2
2	68.5	67.0	54.4	54.0	10.0
3	68.1	66.6	54.4	54.0	10.0
4	67.2	65.7	54.4	53.9	10.0
5	67.9	66.4	54.4	54.0	10.0
6	70.0	68.5	56.6	56.1	12.1
7	67.9	66.4	54.4	53.9	10.0
8	67.9	66.4	54.4	53.9	10.0
9	67.9	66.4	54.4	53.9	10.0
10	67.9	66.4	54.4	53.9	10.0

Voice threshold	Activity	Link margin (dB)
7.5	0.40	0.0

Programme channel to TAB

(C/N_0)U	C/I_0	(C/N_0)D	(C/N_0) total	(C/N_0) total	Link margin
81.2	82.7	67.9	67.6	12.7	0.0

B.4 System design

The satellite systems-design programme described in the Section B.6 can be run for a number of different parameters in order to optimise the design. The total satellite power necessary for voice transmission is evaluated knowing the number of full-time and voice-activated channels per link and using 0.4 as the voice-activation factor. The VFT and data channels are calculated on the basis of being equivalent in power to full-time voice channels on the worst link (EGM to TAB). The programme channel can be considered purely from the point of view of the EGM to TAB link. A final check compares the total required satellite power to that available.

B.4.1 Quarter transponder lease

The total number of channels is 123 in the first phase. On a 30 kHz channel separation this gives 3.7 MHz. The programme channel is less than 400 kHz and hence the total bandwidth is 4.1 MHz. The traffic matrix pertaining to this is shown in Table B.6 and the resulting link-budgets to which it refers in Table B.7.

It will be seen that all specified performance can be met with the following system margins:

For EGM links to and from

TAB: 0 dB
PUV: 0.2 dB
UMK: 0.8 dB
ANG: 1.0 dB
SCO: 0.8 dB

If one considers the additional problem of earth-station EIRP stability, this makes the system margins very minimal for secure operation.

Although not shown here, the performance for the extended phase of the system into half-transponder working also checks out with slightly better system margins.

B.4.2 Off-axis radiations

In leased systems INTELSAT specifies that earth-station off-axis EIRP density should not exceed 42-25 log θ dBW/40 kHz. The antenna specification was $G = 32 - 25 \log \theta$. Thus in the case of any carrier where spectrum is confined within 40 kHz this places a limit of 10 dBW (10 W) to the RF power per carrier. This is not exceeded for any of the telephony, data or telegraph carriers, but the programme channel has an EIRP of 74.5 dBW, e.g. RF power of 19 dBW. For much of the time the bandwidth occupied could be less than 40 kHz and therefore one must look to one of the following solutions:

- high-gain satellite transponder;
- increased antenna diameter;

- energy dispersal;
- alternative modulation methods;
- reduction in service availability

B.5 Coordination and earth-station siting

Before a new station is permitted to become operational, formal coordination according to articles of the ITU radio regulations must be effected by the administration responsible. These articles relate to the intersatellite interference and interference between terrestrial and satellite systems. The next step in the placing of the system would thus be to go through the coordination process as set out in the radio regulations for each of the stations. If results of this do indicate that interference might ensue, it becomes the responsibility of the administrations concerned to come to a uniformally acceptable solution. Even for part transponder leases, it is recommended that coordination be effected over the full satellite bands 3.7 to 4.2 and 5.925 to 6.425 GHz. This will permit operation in any INTELSAT transponder and to any position in the geostationary arc.

Every proposed earth-station site needs to be considered carefully with regard to potential interference, and site shielding used to a maximum. Ideally measurements should be carried out, but at least calculations must be performed.

B.6 Link-budget calculations

The basic design equations for the satellite links are as follows:

Uppath
$$\left(\frac{C}{N_0}\right)_u = (\text{EIRP}_E - B0_i) + \frac{G}{T_s} - L_u + K$$
$$= (S_{SAT} - B0_i) + G/T_s - 37 + K$$

where flux density $S_{SAT} = \text{EIRP} - 10 \log(4\pi d^2)$.

$$L_u = 20 \log\left(\frac{4\pi d}{\lambda}\right)$$

downpath
$$(C/N_0)_D = (\text{EIRP}_{SAT} - B0_i) - L_d + \frac{G}{T_{e/s}} + K$$

intermodulation
$$(C/N_0)_t \simeq 84.5 + 1.78 \times B0_0$$

specification
$$10 \log(N) = (C/N)_{\text{total}} - (C.N_0) \text{ reqd.}$$

$N =$ no. of telephony channels

$$(C/N_0)_{reqd} \simeq \left(\frac{C}{(N_u + N_0 + N_1)_0} + \text{margin}\right)$$

whence the above may be solved to calculate EIRP, G/T and number of telephony channels per carrier.

A computer programme has been produced to evaluate the above with the growing input data:

- satellite characteristics;
- earth-station latitude/longitude;
- propagation data;
- minimum C/N for each type of channel;
- channel bandwidth;
- demodulation threshold levels.

The output of the program then yields:

- EIRP of earth station;
- G/T of stations;
- number of channels;
- margin of power.

A/D	analogue to digital
AAF	antialiasing filter
ABM	apogee burst motor
ACC	antenna control console
ACE	attitude control electronics
AceS	Asian satellite cellular system
ACI	adjacent channel interference
ACSSB	amplitude companded single sideband
ACU	antenna control unit
ADCS	attitude determination and control system
ADM	adaptive delta modulation
ADPCM	adaptive differential PCM (CCITT 32 kbit/s)
ADS	automatic dependent surveillance
ADSL	asymmetric digital subscribers loop
AEF	apogee engine firing
AFC	automatic frequency control
AGC	automatic gain control
AIT	assembly integration and test
AJ	antijamming
AKM	apogee kick motor
AM	amplitude modulation
AM/PM	amplitude to phase conversion
AMBE	advanced multiband excited (codes)
AMPS	advanced mobile phone system
AMSAT	radio amateur satellite corporation
AMSC	American Mobile Satellite Corporation

ANC	active nutation control
ANIK	Canadian domestic system
ANSI	American National Standards Institute
AOR	Atlantic ocean region
AOTS	advanced orbital test satellite
AP	assembly of parties (INTELSAT)
APK	amplitude and phase keyed
APM	antenna positioning mechanism
APMT	Asian Pacific mobile telecommunication system
ARABSAT	Arab Space Communications Organisation
ARQ	automatic repeat request
ASAP	aria structure for auxiliary payloads
ASC	Asian satellite computer
ASIC	application-specific integrated circuit
ATM	asynchronous transfer mode
ATME	automatic transmission measuring equipment
ATS	application technology satellite
AUC	authentication centre
AUSSAT	Australian domestic satellite system
AVD	alternative voice/data
AWGN	additive white Gaussian noise
AZ	azimuth

BAe	British Aerospace
BAPTA	bearing and power transfer assembly
BCH	Bose-Chaudhuri-Hocquenghem (code)
BDF	baseband distribution frame
BER	bit error ratio
BG	board of governors (INTELSAT)
BG/PC	BG's advisory committee on planning (INTELSAT)
BG/T	BG's advisory committee on technical matters (INTELSAT)
BIP	bit interleaved parity
BO_i	input back off
BOL	beginning of life
BO_o	output back off
BPSK	binary phase-shift keying
BSkyB	British Sky Broadcasting
BSS	base station system
BSS	broadcasting satellite system
BTI	British Telecom International (UK signatory to INTELSAT)
BTR	bit time recovery
BW	bandwidth

c	velocity of light
C&W	Cable & Wireless Ltd
C/IM	carrier-to-intermodulation noise ratio
C/N	carrier-to-noise ratio
C/N_0	carrier power/noise density
C/N_{th}	carrier/thermal noise ratio
C/T	carrier-to-noise temperature ratio
CA	conditional access
CATV	community antenna television
CCD	charge coupled device
CCI	cochannel interference
CCIR	International Radio Consultative Committee (ITU)—now ITU-R
CCITT	International Telephone and Telegraph Consultative Committee (ITU)—now ITU-T
CCST	Consultative Committee on Satellite Telecommunications (BTI organised)
CDMA	code-division multiple access
CDV	cell-delay variation
CEPT	European Conference on Postal and Telecommunications Administrations
CES	coastal earth station
CFDM	companded frequency-division multiplex
CFM	compounded frequency modulation
CFRP	carbon-fibre-reinforced plastic
Ch	channel
CLR	cell-loss ratio
CME	circuit multiplication equipment
CME	circuit monitoring equipment
CMR	cell misinsertion ratio
CNES	Centre National d'Etudes Spatiales
CNET	Centre National d'Etudes des Telecommunications
Coax	coaxial cable
Codec	coder-decoder
COFDM	coded orthagonal frequency-division multiplex
Companding	use of dynamic range compression and expansion of a signal
COMSAT	US signatory to INTELSAT
COMSTAR	US domestic satellite system
COTS	commercial off-the-shelf
CPL	capillary-pumped lamps
CPSK	coherent phase-shift keying
CRC	cyclic redundancy check
CSC	common signalling channel
CSG	Centre Spatiale Guianais (Kourou)

CSM	communications system monitoring
CT	transit switching centre (international telephone exchange)
CTE	channel translating equipment
CVSD	continuously variable slope delta modulation
CW	continuous wave
D/A	digital to analogue
DA	demand assignment
DAB	digital audio broadcasting
DAMA	demand-assignment multiple access
DAMPS	digital AMPS
dB	decibel
DBFN	digital beamforming network
DBS	direct broadcasting satellite
dBW	decibels relative to one watt
DCC	data control channel
DCME	digital circuit multiplication equipment
DCP	data collection platform
DCPSK	differentially coherent phase-shift keying
DCRT	data collective receive terminal
DCT	discrete cosine transform
DEC	declination
Demod	demodulator
DF	direction finding
DG	director general
DGPS	differential GPS
DIO	delivery in orbit
DITEC	digital television communication system
DMA	direct memory access
DMAC	digital multiplexed analogue component
DNI	digital noninterpolated
DOD	depth of discharge
DOMSAT	domestic satellite system
DPSK	differential phase-shift keyed
DRSS	data-relay satellite system
DS	direct sequence (CDMA)
DSI	digital speech interpolation
DSR	differential slant range
DSR	digital satellite radio
DTH	direct-to-home
DVB	Digital Video Broadcasting (standards organisation)
DVB-S	DVB—satellite
e	number 2.71828
E_b/N_o	energy per bit/noise density

EBU	European Broadcasting Union
ECC	eccentricity
ECCM	electronic counter-countermeasures
ECOM	electronic computer-originated mail
ECS	European regional communications satellite system
ED	energy dispersal
EFC	earth fixed cells
EFS	error-free seconds, dS deciseconds
EFT	electronic fund transfer
EIR	equipment identity register
EIRP	equivalent isotropically-radiated power
EL	elevation
ELMSS	European land mobile satellite system
EMS	electronic message service
EO	executive organ (INTELSAT)
EOL	end of life
EPIRB	emergency position indicating radio beams
EQM	engineering qualification model
ER&S	explanatory research and study
ES	earth station
ESA	European Space Agency
ESC	engineering service circuit
ESD	electrostatic discharge
ESOC	European Space Operations Centre
ESRO	European Space Reserach Organisation
ESTEC	Euroepan Space Research Technology Centre
ETR	Eastern Test Range (NASA's Cape Canaveral)
ETSI	European Telecommunications Standards Institute
ETT	electrothermal thrusters
EUROSAT	a Swiss incorporated international company
EUROSPACE	European industrial space study group
EUTELSAT	European Telecommunications Satellite Organisation
FAC	Ford Aerospace Corporation
FANS	future air navigation system
FCC	Federal Communications Commission (US)
FDM/FM	frequency-division multiplexed/frequency modulated
FDMA	frequency-division multiple access
FDOA	frequency difference of arrival
FEC	forward error correction
FET	field-effect transistor
FFT	fast Fourier transform
FH	frequency hopping
FM	frequency modulation
FOV	field of view
FRC	frequency reuse character

FRR	flight reading review
FSK	frequency-shift keyed
FSS	fixed satellite service
G/T	gain-to-noise temperature ratio
GaAs	gallium arsenide
GBN-ARQ	goback-N-ARQ
GCE	Grad communications equipment
GCE	ground control equipment
GCS	ground control system
GEO	geostationary-earth orbit
GEO-IRS	geostationary-infrared sensor
GES	gateway earth-station
Globalstar	commercial LEO system
GMDSS	Global maritime distress and safety system
GMPCS	global mobile personal communication satellites
GNSS	global navigation satellite system
GOS	grade of service
GPS	global positioning system
GSM	global system for mobile
GTE	group translating equipment
GTO	geostationary tranfer orbit
HAC	Hughes Aircraft Company
HANE	high-altitude nuclear explosion
HDTV	high-definition television
HEMPT	high electron mobility transistor
HEO	highly-elliptic orbit
HLR	home location register
HPA	high-power amplifier
HPF	high-pass filter
HPM	hybrid phase modulation
Hz	hertz (cycle per second)
IADP	INTELSAT Asistance & Development Program
IATA	International Air Transport Association
IBS	INTELSAT business service
ICAO	International Civil Aviation Organisation
ICDSC	International Conference on Digital Satellite Communications
ICES	International Council for the Exploration of the Sea
ICN	idle channel noise
ICO	inclined circular orbit
IDD	international direct dialling
IDR	intermediate data rate
IDR	intermediate design review (INTELSAT)
IEE	Institution of Electrical Engineers

IETF	Internet engineering task force
IF	intermediate frequency
IFL	inter-facility link
IFRB	International Frequency Registration Board
IGO	International Government Organisation
IM	intermodulation
IMBE	improved multiband excitation
IMC	International Maintenance Centre
IMCO	Intergovernmental Maritime Consultative Organisation
IMO	International Maritime Organisation
IMP	intermodulation product
IMT 200 (FPLMTS)	integrated (future public land) mobile telecommunication system
IMUX	input multiplexer
INC	inclination
INMARSAT	International Maritime Satellite Organisation
INSAT	Indian domestic system
INTELSAT	International Telecommunications Satellite Organisation
INTERSPUTNIK	International Space Telecommunication Organisation (USSR based)
IOC	INTELSAT Operation Centre
IOR	Indian Ocean region
IOT	in-orbit test antenna
Iridium	commercial LEO system
IRR	internal rate of return
ISI	international symbol of interference
ISL	intersatellite link
ISO	International Standards Organisation
ISP	specific impulse
ISPC	International Sound Program Centre
ISRO	Indian Space Research Organisation
ITC	International Television Centre
ITMC	International Transmission Maintenance Centres
ITT	International Telephone & Telegraph Company
ITU	International Telecommunications Union
ITU-R	International Telecommunications Union—Radio
ITU-T	International Telecommunications Union—Technical
IUS	intertial upper stage
IWP	interim working party
JPEG	joint photographic expert group
k	Boltzmann constant (1.38×10^{-23} J/K)

K	Boltzmann constant logarithmic form (−228.6 dBW/Hz/K)
K	Kelvin (degree of absolute temperature)
k	kilo/thousand
KDD	Kokussai Denshin Denwa Co Ltd
kHz	kilohertz
kW	kilowatt
LANDSAT	Earth Resources Survey Satellite (US)
LCC	launch control centre
LD-CELP	low-delay code excited linear prediction
LEO	low-earth orbit
LEOP	launch and early operations
LHCP	left-hand circular polarisation
LMA	limited motion antenna
LMS	land mobile satellites
LNA	low-noise amplifier
LNB	low-noise block downconvertor
LNC	low-noise convertor
LNR	low-noise receiver
LO	local oscillator
LOP	line of position
LOS	line-of-sight
LPC	linear predictive coding
LPI	low probability of intercept
LRE	low-rate encoding (32 and 16 kbit/s speech etc)
LSI	large scale integration
M	mega/million
MAC	multiple analogue components. (C, D, D2 prefix denotes type)
MAP	mobile application part
MARECS	ESA, maritime satellite system
MARISAT	US, maritime satellite system
MASER	microwave amplification by stimulated emission of radiation
MAWG	Mutual Aid Working Group
MBA	multiple beam antenna
MBB	Messerschmitt-Bolkow-Blohm (Germany)
MCL	Mercury Communications Ltd
MCLF	multichannel loading factor
MCPC	multiple channel per carrier
MCS	maritime communications subsystem
MDA	mechanically despun antenna
MDF	main distribution frame

MEO	mid-earth orbit
MES	mobile earth station
MHz	megahertz
MLI	multilayer insulation
MLS	microwave landing system
MMDS	multichannel microwave distribution system
MMH	monomethyl-hydrazine
Mod	modulator
Modem	modulator-demodulator
Molnya	Russian HEO system
MOLNYA	USSR, domestic satellite
MPEG	Motion Picture Experts Group (standards organisation)
MS	meeting of signatories (INTELSAT)
ms	millisecond
MSDS	Marconi Space and Defence Systems—now MSS
MSK	minimum-shift keying
MSM	microwave switch matrix
MSS	mobile satellite service
MSSC	mobile satellite switching centre
MTBF	mean time before/between failures
MTTR	mean time to repair
Mux	multiplex
n	nano (10^{-9})
N	Newton
NASA	National Aeronautics and Space Administration (US)
NCS	network condition station
NEC	Nippon Electric Company (Japan)
NICAM	near instantaneous companding
NiH_2	nickel-hydrogen
NMC	network management centre
NMCC	network management and control centre
NMT	Nordic mobile telecom system
NPR	noise power ratio
NRZ	nonreturn-to-zero
ns	nanosecond
NTSC	system of colour television using 525 lines and 60 fields with a video bandwidth of 4.2 MHz (US, Canada, Japan)
O&M	operation and maintenance
OBC	onboard controller
OBDH	onboard data handler
OBN	out-of-band noise

OBP	onboard processing
OK-QPSK (OQPSK)	offset-keyed QPSK
Olympus	large experimental European satellite
OMC	operations & maintenance centre
OMT	orthomode transducer
OMUX	output multiplexer
ORTF	Office de Radiodiffusion et Television Francais
OSCAR	amateur radio communications satellite
OSI	open systems interconnect
OSPK	off-set keyed quadrature phase-shift keying
OSR	optical solar repeater
OTS	orbital test satellite (European)
p	pico (one trillionth)
PA	power amplifier
PAL	phase alternative line colour television system developed in Europe using 625 lines and 50 fields with video bandwidth of 5 MHz
PALAPA	Indonesian domestic satellite system
PAM	pulse amplitude modulation
PAPM	pulse amplitude phase modulation
Paramp	parametric low-noise amplifier
PC	printed circuit
PCM	pulse-code modulation
PCS	personal communication system (US cellular)
PDC	personal digital cordless
PDH	pleosynchronous digital hierarchy
PDM	pulse-duration modulation
PDR	preliminary design review
PERUMTEL	Perusahaan Ummum Telekomunikasi (Indonesia)
pF	picofarad
PFM	pulse-frequency modulation
PG	processing gain
PHILCOMSAT	Philippine Communications Satellite Corporation
PIM	passive intermodulation product
PIP	payload integration plan
PKM	perigee kick motor
PLMN	public land mobile network
PM	phase modulation
POH	part overhead
Polang	polarisation angle
POR	Pacific Ocean Region
P-P	peak to peak
PPL	phase-locked loop
PPM	pulse position modulation

PRN	pseudorandom noise
psi	pounds per square inch
PSK	phase-shift keying
PSTN	public-switched telephone network
PV	present value
pW	picowatt
pWp	picowatt psophometrically weighted
QA	quality assurance
QAM-PAM	quadrature amplitude modulation-pulse amplitude modulation
QFH	quadrafilter helix (antenna)
QM	qualification model
QOS	quality of service
QPSK	quadrature phase-shift keying
QSL	quasistatic load
RA	right ascension
Rad	radian
RAM	random access memory
RDF	repeater distribution frame
RELP	residual excited linear prediction
RET	reliable earth terminal
RF	radio frequency
RFP	request for proposal
RFQ	request for quotations
RHCP	right-hand circular polarisation
rms	root mean square
RN	reference noise
RN	Royal Navy
ROFA	region-oriented frequency
rpm	revolutions per minute
RS (RSC)	Reed–Solomon (code)
Rx	receive
RX	receiver
S/C	spacecraft
S/N	signal-to-noise ratio
S + DX	speech plus duplex (simultaneous transmission of speech and data)
SADA	solar array drive assembly
SADE	solar array drive electronics
SADM	solar array drive mechanism
SAN	satellite access node
SARPS	standards and recommended practices

SAW	surface acoustic wave
SBS	satellite business system (US domestic system)
SC	satellite control centre
SC	service channel
SCATHA	spacecraft charging at high altitudes scientific satellite (US)
SCC	Satellite control centre
SCPC	single-channel-per-carrier
SCR	silicon controlled rectifier
SDH	synchronous digital hierarchy
SEC	severely eroded cell
SECAM	sequential couleurs a memoire—A colour television system developed in France
SES	severely-eroded seconds
SES	ship earth station
SET	European Telecommunications Working Group of the CEPT
SEV	single event upset
SFC	satellite fixed cells
SG	supergroup
SIM	subscribers identity module
SIRIO	Italian Experimental Communications Satellite
SKYNET	UK Defence Communications Network
SMA	semi-major access
SMATV	satellite master antenna TV
SMS	shuttle mission simulator or satellite multiservice system
SNF	system noise figure
SNG	satellite newsgathering
SNIAS	Societe Nationale Industrielle Aerospatiale (France)
SNR	signal-to-noise ratio
SOFA	satellite-oriented frequency
SOH	section overhead
SOHO	small-office home office
SOLAS	safety of life at sea
SONET	synchronous optical network
SPADE	single-channel-per-carrier pulse-code modulation, multiple-access demand-assignment equipment
SPDT	single-pole double throw
SPEC	speech-predictive encoding system
SPM	signal-processing modem
SRARQ	selective repeat—automatic repeat request
SR-ARQ	selective reject-ARQ
SRC	UK, Science Research Council—now EPSRC
SRP	supergroup reference pilot

SS	satellite switching
SS/TDMA	satellite-switched time-division multiple access
SSB	single sideband
SSOG	satellite system operations guide
SSPA	solid-state power amplifier
SSRA	spread-spectrum random access
SSUS	spinning solid upper stage
STATSIONAR	USSR, communications satellite for geostationary orbit operation
STCC	Spacecraft Technical Control Centre
STE	supergroup translating equipment
STM	synchronous transport module
STS	space transportation system (space shuttle)
SWR	standing-wave ratio
SYMPHONIE	French–German experimental satellite system
T/V	thermal vacuum
TACS	total access conversation system
TASI	time assignment speech interpolation
TAT	transatlantic telephone cables
TC&R	telemetry command & ranging
TCP/IP	transport control protocol/internet protocol
TDA	tunnel diode amplifier
TDB	traffic database
TDD	time division duplex
TDM	time-division multiplex
TDMA	time-division multiple access
TDOA	time difference of arrival
TDRS	tracking data relay satellite
TE	threshold extension
TELECOM	French domestic satellite system
TELEGLOBE	Canadian signatory to INTELSAT
TELESAT	Canadian domestic system (ANIK)
TELESPAZIO	Societa per Azioni per le Comunicazioni Spaziali (Italy)
TELE-X	Sweden, experimental telecommunications satellite program
TES	transportable earth station
Tg	telegraph
TLS	transmitter location system
TM	thermal model
TMN	telecommunications network management
TNIP	terrestrial-network interface processor
TOCC	technical and operational control centre
Tp	telephone

TRMA	random-time multiple access
TRR	test reading review
TSF	through supergroup filter
TT&C	tracking, telemetry and command
TU	trisatory unit
TV	television (video and associated sound)
TVC	thrust vector control
TVRO	television receive-only
TWT	travelling wave tube
TWTA	travelling-wave-tube amplifier
TX	transmit
UET	unattended earth terminal
UMTS	universal mobile telecom system
UPS	uninterrupted power supply
UPT	universal personal telecommunications
USAT	ultra small aperture terminal
UW	unique word
VC	virtual container
VCO	voltage-controlled oscillator
VF	voice frequency
VFT	voice-frequency telegraph
VISTA	lowcost, thin route, telephony service (INTELSAT)
VLR	visitor location register
VLSI	very large scale integration
VOD	video on demand
VOLNA	USSR network
VSAT	very small aperture terminal
VSWR	voltage standing-wave ratio
WARC	World Administrative Radio Conference
WARC/ST	World Administrative Radio Conference on Space Telecommunication
WESTAR	US domestic satellite
WEU	Western European Union
WF	weighting function
Wg	waveguide
WTO	World Trade Organisation
XPD	cross-polar discrimination
XPI	crosspolar interference

Index